# Microbial Control of Vector-Borne Diseases

# Microbial Control of Vector-Borne Diseases

Edited by

Brij Kishore Tyagi

Dharumadurai Dhanasekaran

CRC Press
Taylor & Francis Group
Boca Raton London New York

CRC Press is an imprint of the
Taylor & Francis Group, an **informa** business

CRC Press
Taylor & Francis Group
6000 Broken Sound Parkway NW, Suite 300
Boca Raton, FL 33487-2742



Printed on acid-free paper

International Standard Book Number-13: 978-1-138-05581-0 (Hardback)



**Library of Congress Cataloging-in-Publication Data**

Names: Tyagi, B. K., editor. | Dhanasekaran, Dharumadurai, editor.
Title: Microbial control of vector-borne diseases / [edited by] Brij Kishore Tyagi and Dharumadurai Dhanasekaran.
Description: Boca Raton : Taylor & Francis, 2018. | Includes bibliographical references.
Identifiers: LCCN 2017039703| ISBN 9781138055810 (hardback : alk. paper) | ISBN 9781315165691 (ebook)
Subjects: | MESH: Insect Vectors | Insect Control | Communicable Disease Control | Environmental Microbiology
Classification: LCC SB975 | NLM QX 600 | DDC 363.7/8--dc23
LC record available at https://lccn.loc.gov/2017039703

**Visit the Taylor & Francis Web site at**
**http://www.taylorandfrancis.com**

**and the CRC Press Web site at**
**http://www.crcpress.com**

# Contents

## SECTION V Microbial Control of Flea Vector

# Foreword

Vector control is the primary intervention for most of vector-borne diseases, including malaria, dengue, and Zika, due to lack of effective drug and vaccine. For a long history, we had heavily relied on chemical insecticides to suppress the vector populations in hopes of reducing them below epidemiological thresholds that are required for disease transmission. Now, we learn that this has to be changed after seeing the rapid development of strong resistance to insecticides in vectors and the significant negative impact of chemical sprays on environment and nontarget species. We also realize the urgency to make such change because some old vector-borne diseases like dengue become more serious than before, new diseases like Zika emerge as global threats, and progress made in control of other diseases like malaria has now stalled and even reversed. There is a consensus that we need to better understand pathogen-vector interactions that determine the ability of vectors to transmit diseases and utilize those knowledge for developing novel tools and strategies with the potential to lead to sustainable disease control. One of the most promising areas is microbial control of vector-borne diseasesas addressed by this book.

Like vertebrate hosts, insect vectors have close contacts with microbes in nature. Some microbes form intimate relationships with vectors and play essential roles for vector survival, reproduction, or development; some just stay together with vectors as guests; and others infect vectors as pathogens, reduce insect fitness, and even kill them. All of the abovementioned three types of microbe-vector relationship can be utilized for vector control reduce insect fitness as introduced in different chapters of this book. The most straightforward approach to develop microbial control of vector-borne disease is to kill directly the vectors or their offspring to reduce the quantity of pests. Alternatively, efforts can be developed to reduce the quality of an insect to serve as a vector for human pathogens. It is worthy to note that those human pathogens transmitted by vectors are not harmful to vectors in most situations. After encountering vectors and becoming their guests, they complete replication, development, or both, inside the body of the insect, to prepare for their next journey in humans. In order for an insect to be able to transmit a specific human pathogen, its physiological environment and behavior have to match perfectly to the pathogen's requirements. This provides an opportunity for the other microbes to perturb or finely adjust to this environment, either naturally or artificially, such that insects are no long hospitable for pathogens or incapable for moving pathogens to human. To a certain degree, such modification of the insect's physiological microenvironment to break the linkage between vectors and pathogens is similar to the traditional environmental management for vector-borne disease control by preventing contact between vectors and human. Thus, microbial control of vector-borne disease can be accomplished by either reducing vector density or the ability of vectors in transmitting human pathogens.

While many novel microbe-based approaches have been demonstrated in the laboratory, exciting progresses have recently been made to provide proof of concept through field trial. One good example is the maternally transmitted endosymbiotic bacteria *Wolbachia*, which are estimated to be present in 65% of millions of insect species in nature. As selfish microbes, *Wolbachia* manipulate insect reproduction for their own benefit such that they can invade and spread into populations. Different insects may carry different *Wolbachia* strains, indicating that indicating that millions of *Wolbachia* strains may be present in nature. With the ability established to manually introduce a *Wolbachia* strain into an insect host to build a novel symbiosis in laboratory, we can make a naturally uninfected mosquito to carry *Wolbachia*, displace the existing *Wolbachia* with a novel strain, or add novel *Wolbachia* to create multiple strain combination. Importantly, some of those novel *Wolbachia* strains in mosquito can act like a vaccine to protect the mosquito from human pathogens. Recent field trials show that releasing *Wolbachia*-infected *Aedes aegypti* can result in invasion of *Wolbachia* in mosquito population and reduce its potential in transmitting dengue and Zika. This approach is attractive due to its low cost and sustainability in disease control because once a local mosquito population is modified to become pathogen resistant, disease transmission in this area will be reduced or blocked even given the migration of infected people from another endemic region into this control region. In a different trial, millions of *Wolbachia*-infected *Aedes albopictus* male mosquitoes are produced in a mosquito factory and released in the field every week to induce sterile matings with the wild type of mosquitoes, resulting in suppressing and even eliminating local populations. Due to these important progresses, the World Health Organization encourages endemic countries to continue developing *Wolbachia* as a practical tool for vector-borne disease control, resulting in ongoing field trials in ~20 countries or regions and the first success in registration of *Wolbachia* as a microbial pesticide in United States Environmental Protection Agency in 2017.

Due to its environment-friendly sustainability, and compatibility with the traditional approaches like vaccine and drug, microbial control of vector-borne disease is expected to play more important role in disease control and prevention in the near future. New microbes will be discovered in laboratories and gone through research and development and field trials, while specific microbes will be used to target each individual species of dominant disease vectors without negative impacts on nontarget species. Advances in biotechnology, artificial intelligence, automation, and real-time monitoring through web-based mapping service will facilitate the deployment of these approaches in field setting to accelerate the efforts for disease eradication.

**Dr. Zhiyong Xi, PhD**
*Associate Professor and Director*
*Microbiology and Molecular Genetics*
*SYSU-MSU Joint Centre of Vector Control for Tropical Diseases*
*College of Natural Science*
*Michigan State University*
*Michigan, USA*

# Preface

Vector-borne diseases such as malaria, dengue, chikungunya, schistosomiasis, human African trypanosomiasis, leishmaniasis, Chagas disease, yellow fever, onchocerciasis, Zika virus, and several different types of encephalitides including Japanese encephalitis are a major cause of human morbidity and mortality globally. More than 1 billion cases and over 1 million deaths are reported annually. These diseases, mostly rampant in tropical and subtropical regions of the world, account for over 17% of all infectious diseases. Distribution of these diseases is determined by a complex interrelationship among pathogen, vector, and human being, anchored by environmental and social factors, as well as global travel and trade, unplanned urbanization, and environmental challenges such as climate change and global warming. Some diseases, such as dengue, chikungunya, West Nile virus, and Zika virus, are emerging in countries where they were previously unknown. The recent spread of Zika virus, a mosquito-borne viral disease, across Americas, Europe, and parts of Asia, is a towering example of how rapidly some vector-borne diseases might disseminate over larger areas in a relatively short time period. Vector-borne diseases are spread mainly by the bite of insects and other arthropod vectors, such as mosquitoes, ticks, mites, triatomine bugs, tsetse flies, sandflies, and black flies, imposing heavy health and economic burdens, in addition to unmeasurable human misery and hardship, as many people who survive infection are left permanently debilitated, disfigured, maimed, or blind. Vectors of these diseases thrive under conditions where housing is poor, water is unsafe, and environments are contaminated with filth. Measures that control the vectors, the agents of diseases, provide an excellent but underutilized opportunity to help these people catch up.

For the past nearly 100 years, the vectors of these diseases were chemically controlled only to show resistance against the various chemicals that proved inhibitive in the long run and also because of their toxicity to both human and nontarget organisms as well as the environment. Subsequently, biological and environmental control methods were used in controlling these vectors but proved to no avail in emergencies of disease epidemics. Alternatively, following stringent research during the past four decades, microbial agents and tools have recently shown great promise, and the best example is *Bacillus thuringiensis* var. *israelensis* to control a wide range of vector and pest mosquitoes. Microbial control, defined simply as the use of microorganisms or their by-products by humans to suppress insect pest populations, implies that microorganisms like bacteria, actinobacteria, cyanobacteria, fungi, algae, and protozoa can bring about reduction in vector population by a variety of pathways without causing serious human health and environment concerns. Some of these are ready for field use, whereas others have already proven effective in reducing vector populations. In a similar way, many microorganisms interfere with the development of the disease causing pathogens in the vector and results in its depletion or reduction, bringing about a control of the disease before it breaks out.

This book presents a detailed overview of microbial biomolecules in meeting the challenges to control and prevent vector borne-diseases; autodissemination of current and future potential in the application of entomopathogens against mosquito-borne diseases; and bioprospecting of bacterial, actinobacterial, cyanobacterial, fungal metabolites, gut microbiota, and *Wolbachia* for mosquito control. Finally, genetically altered microbes and viruses are also used in the control of mosquito-borne diseases. Moreover, this book also provides a comprehensive account on microbial control of leishmaniasis, aquatic snail-borne diseases, blackfly-targeted onchocerciasis, and flea-borne Rickettsial diseases. This book will be eventually beneficial to future research programmers, planners, administrators, scientists, academicians, and researchers as well as the governments of various nations who are interested in fortifying and expanding their knowledge about microbial control of vectors in the fields of microbiology, biotechnology, entomology, biomedical science, public health, and environmental science.

The book is comprised of 21 chapters from multiple contributors around the world including the United States, Mexico, China, Turkey, Thailand, India, and the Kingdom of Saudi Arabia. We are grateful to all the contributors and leading experts for the submission of their stimulating and inclusive chapters in the preparation of this unique volume on microbial control of vector-borne diseases. The book content is divided into five sections, namely, microbial control of mosquitoes and mosquito-borne diseases, leishmanias, schistosomes, blackflies, and fleas.

We offer special thanks and appreciation to Renu Upadhyay, Shikha Garg, and Jennifer Blaise, editorial team members at CRC Press, for their encouragement and help in producing the book in a timely manner in its present form. We express our heartfelt gratitude to our respective universities for their concern, efforts, and support in publishing this volume.

**Professor Brij Kishore Tyagi**
**Dr. Dharumadurai Dhanasekaran**

# Editors

**Professor Dr. Brij Kishore Tyagi** (b 23rd July, 1951) is a renowned medical arthropodologist/entomologist of international repute, well known for his significant contributions on vector ecology, biology, and control as well as epidemiology and management of vector-borne diseases like malaria, filariasis, leishmaniasis, Japanese encephalitis, dengue and chikungunya, etc. He is best known for his innovative malariological research in the Indira Gandhi Nahar Project (IGNP) irrigated Thar Desert in Rajasthan, discovery of *Aedes albopictus* as a primary and only vector of dengue fever in Kerala State, and invention and development of various different vector mosquito traps/breeding preventers. He has discovered new species of mosquitoes, and published for the first time a complete catalogue of Indian mosquitoes (Culicidae). Recipient of many prestigious national and international awards/prizes, viz., World Health Organization-Special Programme for Research and Training in Tropical Diseases (WHO-TDR) First Prize in a worldwide competition (1995–1996), ICMR's Dr. M.O.T. Iyengar Memorial Award (2007) for outstanding biomedical research, he has published 28 books and more than 575 research titles, in addition to one Indian patent and two technology transfers to the State Government of Tamil Nadu. Dr. Tyagi has been a member of various expert committees in the India Council of Medical Research, Department of Biotechnology, etc. He had research collaborations with Brandeis University, USA, International Water Management Institute, Sri Lanka, World Health Organization, National Institute of Public Health, Hanoi (Vietnam), etc. Dr. Tyagi is also the originator of '*Conferences of Medical Arthropodology*'. He is editing about two dozen scientific journals, in addition to being a fellow/member of more than two dozen scientific bodies, both in India and abroad. He has visited several countries in all the continents. Prof. Tyagi retired recently in 2015 as the scientist 'G' (director) and director-in-charge of the ICMR's Centre for Research in Medical Entomology, Madurai, India, and is currently serving as a visiting professor at Bharathidasan University, Tiruchirapalli (TN) and Punjabi University, Patiala (Punjab), besides being the core faculty in the Department of Science and Technology-Science and Engineering Research Board (DST-SERB) School in Insect Biology at Hyderabad University, Hyderabad, India.

**Dr. Dharumadurai Dhanasekaran** is working as an assistant professor at the Department of Microbiology, School of Life Sciences, Bharathidasan University, Tiruchirappalli, India. He has experience in fields of actinobacteriology, mycology, and phycology. His current research focus is on actinobacteria, microalgae, fungi, and mushroom for animal and human health improvement. He has been awarded the University Grants Commission (UGC)-Raman Post-Doctoral Fellowship to visit the United States, and is currently working in the Department of Molecular, Cellular and Biomedical Sciences, University of New Hampshire, Durham, USA. He has qualified the Tamil Nadu State Eligibility Test for Lectureship in Life Science. He has deposited around 64 nucleotide sequences in GenBank and 5 bioactive compounds in PubChem, and published 92 research and review articles including one paper in the Nature Group journal *Scientific Report*. He has edited five books on antimicrobial compounds: *Synthetic and Natural Compounds*, CRC Press, Taylor & Francis Group, New York, and *Fungicides for Plant and Animal Diseases, Actinobacteria: Basics and Biotechnological Applications, Algae-Organisms for Imminent Biotechnology*, and *Microbial Biofilms—Importance and Applications* under In-tech open access publisher, Croatia, Eastern Europe. He guided 10 PhD candidates and organized several national level symposia, conference, and workshop programs. He received major research projects from Department of Biotechnology, University Grant Commission, Indian Council for Medical Research, and International Foundation for Science, Sweden. He is a member of the American Society for Microbiology, North American Mycology Association, Mycological Society of India, and National Academy of Biological Sciences, and a member of editorial boards in national and international journals, a doctoral committee member, a board of study member in microbiology, and a reviewer in the scientific Journals and research grants. As per the reports of *Indian Journal of Experimental Biology*, 51, 2013, Dr. Dharmadurai Dhanasekaran is rated in second position among the top five institutions in the field of actinobacteria research in India.

# Contributors

**Natarajan Amaresan**
C.G. Bhakta Institute of Biotechnology
Uka Tarsadia University
Bardoli, India

**Sankarappan Anbalagan**
Department of Zoology
Sethupathy Government Arts College
Ramanathapuram, Tamil Nadu, India

**Murugan Arunthirumeni**
Department of Biotechnology
School of Biosciences
Periyar University
Salem, Tamil Nadu, India

**Probal Basu**
Department of Zoology
Asutosh College
Kolkata, West Bengal, India

**Sameer S. Bhagyawant**
School of Studies in Biotechnology
Jiwaji University
Gwalior, India

**Sajal Bhattacharya**
Department of Zoology
Asutosh College
Kolkata, West Bengal, India

**Bhaskar Biswas**
Department of Chemistry
Raghunathpur College
Purulia, West Bengal, India

and

Department of Chemistry
Surendranath College
Kolkata, West Bengal, India

**Siddhardha Busi**
Department of Microbiology
School of Life Sciences
Pondicherry University
Puducherry, India

**Kshitij Chandel**
Defence Research and Development Establishment
Gwalior, India

**Arunachalam Chinnathambi**
Department of Botany and Microbiology
College of Science
King Saud University
Riyadh, Saudi Arabia

**Dhanasekaran Dharumadurai**
Department of Molecular
Cellular and Biomedical Sciences
University of New Hampshire
Durham, New Hampshire

**Madangchanok Imchen**
Department of Genomic Science
School of Biological Sciences
Central University of Kerala, Padannakkad
Kasaragod, India

**Hardik Naik Jinal**
C.G. Bhakta Institute of Biotechnology
Uka Tarsadia University
Bardoli, India

**Ganesan Kanimozhi**
Department of Botany and Microbiology
A.V.V.M. Sri Pushpam College (Autonomous)
Tamil Nadu, India

**Krishnan Kannabiran**
Department of Biomedical Sciences
School of Biosciences and Technology
VIT University
Tamil Nadu, India

**Loganathan Karthik**
Marine Biotechnology Laboratory
Shanghai Jiao Tong University
Shanghai, China

**Vitthalrao B. Khyade**
Malegaon Sheti Farm
Agricultural Development Trust
Baramati, Shardanagar
Maharashtra, India

**Arivarasan Vishnu Kirthi**
Unit of Nanotechnology and Bioactive
Natural Products
C. Abdul Hakeem College
Tamil Nadu, India

**Amrita Kumari**
Department of Zoology
Hindu Girls College
Sonepat, India

**Ranjith Kumavath**
Department of Genomic Science
School of Biological Sciences
Central University of Kerala,
Padannakkad
Kasaragod, India

**Ravikumar Mathrubutham**
Department of Botany
Government Arts College for Men
(Autonomous)
Chennai, India

**Saikat Mondal**
Department of Zoology
Raghunathpur College
Purulia, West Bengal, India

**Jamseel Moopantakath**
Department of Genomic Science
School of Biological Sciences
Central University of Kerala,
Padannakkad
Kasaragod, India

**Devarajan Natarajan**
Department of Biotechnology
Periyar University
Salem, Tamil Nadu, India

**Reehana Nazar**
P.G. and Research Department
of Microbiology
Jamal Mohamed College (Autonomous)
Tiruchirappalli, Tamil Nadu, India

**Thajuddin Nooruddin**
Department of Microbiology
School of Life Sciences
Bharathidasan University
Tiruchirappalli, Tamilnadu, India

**Didem Deliorman Orhan**
Department of Pharmacognosy
Gazi University
Ankara, Turkey

**Nilüfer Orhan**
Department of Pharmacognosy
Gazi University
Ankara, Turkey

**Annamalai Panneerselvam**
Department of Botany and Microbiology
A.V.V.M. Sri Pushpam College
(Autonomous)
Tamil Nadu, India

**Parasuraman Paramanantham**
Department of Microbiology
School of Life Sciences
Pondicherry University
Puducherry, India

**Subhaswaraj Pattnaik**
Department of Microbiology
School of Life Sciences
Pondicherry University
Puducherry, India

**Marimuthu Poornima**
Department of Botany and Microbiology
A.V.V.M. Sri Pushpam College
(Autonomous)
Tamil Nadu, India

**Chinnasamy Ragavendran**
Department of Biotechnology
Periyar University
Salem, Tamil Nadu, India

**Mohamudha Parveen Rahamathulla**
Department of Medical Lab Sciences
College of Applied Medical Sciences
Prince Sattam bin Abdulaziz University
Wadi Al Dawaser-Riyadh, Kingdom of
Saudi Arabia

**Eswara Rao**
Department of Genomic Science
School of Biological Sciences
Central University of Kerala, Padannakkad
Kasaragod, India

**Filiberto Reyes-Villanueva**
Instituto Politécnico Nacional
Centro de Biotecnología Genómica
Ciudad Reynosa, Tamaulipas, Mexico

**Mario A. Rodríguez-Pérez**
Instituto Politécnico Nacional
Centro de Biotecnología Genómica
Ciudad Reynosa, Tamaulipas, Mexico

**Alwarappan Sankaranarayanan**
C.G. Bhakta Institute of Biotechnology
Uka Tarsadia University
Bardoli, India

**Muthugounder Subramanian Shivakumar**
Department of Biotechnology
School of Biosciences
Periyar University
Salem, Tamil Nadu, India

**Brij Kishore Tyagi**
Department of Zoology
and Environment Science
Punjabi University
Patiala, Punjab, India

**Vijay Veer**
Defence Research Laboratory
Tezpur, Assam, India

**Govindraj Vengateshwari**
Department of Biotechnology
School of Biosciences
Periyar University
Salem, Tamil Nadu, India

**Perumal Vivekanandhan**
Department of Biotechnology
School of Biosciences
Periyar University
Salem, Tamil Nadu, India

**Viroj Wiwanitkit**
Department of Laboratory Medicine
Chulalongkorn University
Bangkok, Thailand

**Kamlesh Kumar Yadav**
Defence Research and Development
Establishment
Gwalior, India

**Mohamed Imran Mohamed Yousuff**
Department of Microbiology
School of Life Sciences
Bharathidasan University
Tiruchirappalli, Tamilnadu, India

# 1 Microbial Biomolecules
## *Challenges to Control and Prevent Vector-Borne Diseases*

*Madangchanok Imchen, Jamseel Moopantakath, Eswara Rao, and Ranjith Kumavath*

## CONTENTS



## ABSTRACT

Vector-borne diseases (VBDs) are caused majorly by arthropods which effects millions of people worldwide. Malaria alone is a dreadful disease in developing countries. However, the emergence of new VBDs and their resistance to standard drugs have posed a serious threat to the world. Hence, a new source of drugs to treat the dreadful VBDs is the need of the hour. It is surprising that one of the most ancestral innate immunities in every class of life is known as host defense peptides or antimicrobial peptides (AMPs) and have not been much in focus majorly due to lack of revenue generation. AMPs have shown to have promising anti-infective activates on a wide range of microbes including viral and cancer cells. In this chapter, we have compiled some of the research on microbial biomolecules targeting the vector and the pathogens along with the possible mode of action.

The later part of the chapter focuses on prevention of VBDs through mathematical models, regulatory measures, and community-level participation. Vectors are highly dependent on the environmental condition for its maturation and life cycles. Therefore, in order to make the best usage of various environmental data, models pertaining to the VBD outbreaks or the abundance of vectors can be built using environmental data and NASA's satellite-based Moderate Resolution Imaging Spectroradiometer in order to predict the most likely possibility of prevention using powerful algorithms. We have also discussed the importance and the problems faced in imparting knowledge through a public awareness program regarding various VBDs and the importance of community-level participation along with personal prophylaxis measures.

## 1.1 INTRODUCTION

### 1.1.1 Brief Introduction to Vector-Borne Diseases (VBDs)

Vector-borne diseases (VBDs) are those that are transmitted from an infected vector source to humans, plants, or animals. Biologically, vectors can be defined as organisms that carry disease-causing agents. Vectors are considered to be invertebrates, most commonly arthropods, since a majority of VBDs are caused by arthropods. However, vertebrates such as foxes, rats, certain bats, and a species of aquatic snail can also act as vectors. The disease-causing pathogen thrives within the vector, which is then transmitted to another biological body mostly through bites and stings or infestation of tissues. Since most vectors are arthropods ectothermic (cold blooded), they are highly influenced by the landscape and climatic conditions for their proliferation. Trade and commerce between countries has also led to the transmission of vectors to places previously unknown (World Health Organization [WHO] 2016).

### 1.1.2 Effects of VBDs on the Public

Vector-borne disease causes more than 1 million deaths annually, accounting for more than 17% of the overall infectious diseases. Emerging VBDs such as dengue have the potential of contracting to over 2.5 billion people. Malaria alone kills over 400,000 individuals mostly under the age of 5 (WHO 2016). Rural schools can be a breeding ground for malarial vector such as female *Aedes aegypti* and *Culex quinquefasciatus* (Olano et al. 2015). Besides the dreadful effects of VBDs in humans, the economic growth of a country could be hindered by vector-borne plant and animal diseases through reduced agricultural productivity and socioeconomic status. Tropical and subtropical areas experience the highest rate of VBD infections (Institute of Medicine 2008).

In order to control the emergence of VBDs, usage of insecticides to keep the vector under control can be followed. Malaria, dengue, and filariasis (WHO 2006) can be prevented by wise usage of insecticides where breeding of mosquitoes is most likely to occur. However, insects, as compared to microbes, can also gain resistance to insecticides (WHO 1998). This resistance could be due to changes in metabolic processes of the vector (Hemingway et al. 1998) through increased production of enzymes such as carboxylesterases, glutathione-S-transferases, and cytochrome P450-dependent monooxygenases, which are involved in sequestering, metabolism, and detoxification (Rivero et al. 2010). However, it should be noted that increased resistance of insects against insecticides does not always lead to increased transmission of VBDs. Insecticide-resistant *Culex quinquefasciatus* mosquitoes have been shown to exhibit reduced ability to transmit the filarial parasite *Wuchereria bancrofti* as compared to the wild type (Vontas et al. 2005).

## 1.2 MICROBIAL BIOMOLECULES AGAINST VECTOR-BORNE DISEASES

### 1.2.1 Introduction to Antimicrobial Peptides

The resistance of pathogens to various drugs is a serious threat. Pathogens gain resistance through different mechanisms such as plasmid encoding resistance genes or by overexpression of efflux pumps, which extrude drugs from the cells (Nikaido 2009). Antimicrobial peptides (AMPs), also known as host defense peptides (DHPs), are innate immune responses part of every class of life. AMPs are peptides that generally range from 15 to 50 amino acids and exhibit a broad range of action against pathogenic microbes. Some AMPs also act as anticancer peptides. Generally AMPs exhibit their activity due to the major difference between eukaryotic and prokaryotic cells. There are over 2400 AMPs, however, they do not show any correlation between amino acid residuals and their biological activity (Zhang et al. 2014). More than 90% of AMPs are positively charged. On the basis of amino acid residual composition, they are broadly classified into linear, cysteine rich, and specific amino acid rich AMPs. Classification based on secondary structure includes helical, sheet, mixed, and random coiled AMPs (Zhang et al. 2014). AMPs belonging to the magainin class cause osmotic lysis in various protozoa, leading to swelling of the cell until it bursts (Wu et al. 2015). This class of peptides is effective against several protozoa including *Trypanosoma cruzi*. The skin of amphibians is constantly exposed to environments that harbor an immense amount of microbes. Hence, they produce AMPs as a protective measure. One such AMP derived from *Rana temporaria*, a European frog, is temporin A and B peptides, which are composed of 13 amino acids. These peptides have anti-leishmania activity, leaving the healthy human erythrocytes intact (Mangoni 2006). AMPs also show inhibitory activity against various viruses through a range of mechanisms such as neutralization of virus by integrating with the host cell membrane or directly onto the viral envelope (Narayana and Chen 2015). They also inhibit a major viral protein, VP16, which is required by the virus for integration into the host nucleus.

### 1.2.2 Malaria

Malaria is caused by the protozoan *Plasmodia*. It starts with the female *Anopheles* mosquito infected with the infective form of *plasmodia* called sporozoites. When such an infected mosquito bites a mammal, the sporozoites are transmitted through its saliva into the mammal. The sporozoites then migrate to the liver cells called hepatocytes. In the liver cells, the sporozoites mature to the next phase of the life cycle and are called merozoites, followed by the rupture of hepatocytes, and finally release into the bloodstream (Vale et al. 2014). The asexual lifecycle of the plasmodia starts within the red blood cell (RBC). The merozoites develop into the ring stage followed by trophozoites that are metabolically active. The final stage is the development into schizonts that are responsible for infection of other healthy RBCs by realizing merozoites. The ring-form stage can also be developed into female and male gametocytes that can infect a healthy *Anopheles* mosquito during a blood meal. In the infected *Anopheles*, the gametocytes develop into ookinetes, oocysts, and finally sporozoites, which migrate to the salivary gland of the mosquito ready to infect a susceptible mammal in the next blood meal (Vale et al. 2014).

AMPs with broad-spectrum activity from various sources, including *Anopheles* mosquitoes, have been shown to exhibit antimalarial activates (Bell 2011). AMPs act against negatively charged prokaryotic cells. However, the antimalarial activity via inhibition of infected eukaryotic (mammalian) RBCs can sound contradicting. The selective antimalarial activity of AMPs could be linked to the changes brought about to the membrane of infected RBCs. Infection of RBCs by *Plasmodium falciparum* increases the contents of phosphatidylinositol and phosphatidic acid in the membrane and decreases sphingomyelin (Hsiao et al. 1991). Thus cationic AMPs have the potential and promising scope in the treatment of malaria as a new class of antimalarial drugs (Vale et al. 2014). Antimicrobial peptides can form channels (Krishna et al. 1990) through the formation of transbilayer bundles (Snook et al. 1998) or through dissipation of mitochondrial membrane potential or plasma membrane (Nagaraj et al. 2001). Fungal peptides efrapeptins, zervamicins, and antiamoebin inhibited the growth of *P. falciparum* in micromolar concentration. Efrapeptins inhibits mitochondrial $F_0F_1$ ATPase (Nagaraj et al. 2001).

Surfactants are compounds that weaken the surface tension of a given liquid. An example of a commonly used household surfactant would be soap and detergent. Surfactants have been found to be suitable candidature as an antimalarial. Rhamnolipids are produced by *Pseudomonas aeruginosa* that exhibit a low surface tension between 31.4 and 38.7 mN/m (millinewtons per meter) (Silva et al. 2015). It is biodegradable and shows low toxicity. The larvae of *Aedes aegypti* maintain balance on the water surface through air pockets in the trachea and hydrophobic region of the siphon (Christophers 1960). This hydrophobic balance is disturbed by rhamnolipids leading to difficulty of the larvae to stay on the water surface and expend more energy for active swimming to the surface (Silva et al. 2015).

Recent works by Li (2016) targeting fibrinogen-related protein 1 (*FREP1*) showed promising results as a antimicrobial. FREP1, which is produced in the midgut of mosquitoes after a blood meal, can attach to gametocytes and ookinetes. This can enable the parasite to penetrate the peritrophic matrix and epithelium. Thus, targeting the FREP1 using a nontoxin bioactive natural product *P*-orlandin from *Aspergillus niger* showed 92% inhibition of interaction between FREP1 and *Plasmodium falciparum*-infected cells. This disruption of interaction reduces the infection of mosquitoes by *Plasmodium*.

### 1.2.3 Chagas Disease (American Trypanosomiasis)

Chagas disease is named in honor of Brazilian physician Carlos Chagas. It is spread through the bite of triatomine bugs (kissing bugs) infected with protozoan *Trypanosoma cruzi*, and through blood or organ transfusion from an infected source. The disease is divided into three stages: The acute stage, which is characterized by malaise and fever lasting from 4 to 8 weeks. It is followed by the indeterminate phase, which can last up to 2 decades. Active replication of the protozoan takes places during this stage; however, the clinical symptoms are minimum. This can lead to the chronic stage of Chagas disease that leads to irreversible damage to the autonomous nervous and peripheral nervous system (Maguire 1987). This stage of Chagas disease is incurable. The hallmark of Chagas disease is progressive heart disease (Bestetti and Muccillo 1997; Hurwitz et al. 2011). *Trypanosoma cruzi* strains display a high rate of polymorphism (Martínez-Díaz et al. 2001) and thus it is difficult for a simple accurate detection method. Fungal entophytes thrive inside a plant host. Such entophytic fungus has been screened in large amounts by Higginbotham et al. (2013). Entophytic fungus (104 out of 2698 fungal entophytes, 3.9%) isolated from various plants in national parks throughout Panama showed high activity against *Trypanosoma cruzi*. The same research group isolated fungal cultures from sloth hair (a mammal generally found in the tropical forests of South and Central America). Interestingly, organic crude extracts from 8 out of 62 (12.9%) cultures were highly active against *Trypanosoma cruzi* (Higginbotham et al. 2014). Five anti-trypanosomal metabolites, actinoallolides A–E, were also isolated from *Actinoallomurus fulvus* MK10-036 (Inahashi et al. 2015).

Anti-*Trypanosoma cruzi* peptides have been tested in the past such as Dermaseptin 01 from the skin of the *Phyllomedusa hypochondrialis* frog (Brand et al. 2006) as well as fungal peptides such as efrapeptins (Cataldi de Flombaum and Stoppani 1981) and antiamoebin (Kumar et al. 1991). Efrapeptins and antiamoebin act by inhibition of ATPase of the protozoan. Extracts from *Aspergillus fumigatus* exhibited lysis of trypomastigote as high as 95% while leaving the healthy red blood cells intact (Furtado et al. 2005). Since the production of microbial bioactive molecules and compounds pose the limitation of limited quantity, artificial neural networks could contribute for optimization of optimum production of trypanocidal metabolites (Furtado et al. 2005).

### 1.2.4 Leishmaniasis

Leishmaniasis is transmitted by the bite of female phlebotomine sand flies through the transmission of intramacrophage protozoan of the genus *Leishmania*. Annually more than 200,000 new cases of visceral leishmaniasis are reported (WHO 2016).

However, most of these cases are concentrated in poor countries, such as Bangladesh, Nepal, India, and Brazil (Murray 2004). Hence, it is also known as the diseases of the poor that is most prevalent in Southeast Asia and Latin America (WHO 2014). There are three main types of leishmaniases: visceral leishmaniasis (VL), cutaneous leishmaniasis (CL), and muco-cutaneous leishmaniasis (ML) (Herwaldt 1999). The most severe among all the types of leishmaniases is the visceral leishmaniasis (VL), which is caused by *Leishmania donovani* (Davis et al. 2004). The pathogenesis of leishmaniasis is initiated by the breakdown of cell surface with the help of protease enzyme present on the surface of *Leishmania* species thus leading inside the host cell. Metalloprotease is present on the *Leishmania* promastigotes cell surface as a major surface protease (MSP), which helps in attachment of the protozoan to the sand fly gut (Sundar 2001). MSP also binds to the CR3 receptor on the macrophage, which aids in internalization of the promastigote. Treatments for leishmaniasis include Amphotericin-B and its lipid formulations, stibogluconate (pentostam) and meglumine antimoniate (glucantime). However, they are known to have severe side effects to the patients in addition to the high treatment costs, which is unaffordable to most in poor countries. Sodium antimony gluconate (SAG), which was an effective drug with antileishmanial effects, have been stopped in most countries due to the resistance developed by the pathogen (Sundar 2001).

Kojic acid (KA) is a water-soluble fungal metabolite produced by the *Aspergillus* species. Kojic acid has been shown to exhibit antiamastigote activity (Rodrigues et al. 2014). Macrophage infected with *L. amazonensis* is deprived of reactive oxygen species (ROS) and NO production (Olivier et al. 2005; Mukbel et al. 2007). However, treatment with Kojic acid reverses these inhibitory effects, which leads to production of $O_2^-$ leading to killing of the pathogen (Rodrigues et al. 2014).

### 1.2.5 Dengue and Japanese Encephalitis

Flavivirus is the causative agent of Japanese encephalitis and dengue. Flavivirus is an ssRNA virus carrying a genome of 10.6 to 11 kb that encodes for capsid, premembrane, and envelope protein, and other functions such as replication of RNA genome (Green and Rothman 2006). More than 70 variants of flavivirus strains cause various diseases. Some of the most dreadful viruses of the *Flavivirus* genus are dengue virus (DENV), yellow fever virus (YFV), West Nile virus (WNV), Japanese encephalitis virus (JEV), and tick-borne encephalitis virus (TBEV) (Rohde et al. 2008; Bollati et al. 2010). Morbidity and mortality rates of Japanese encephalitis and dengue are highest in southern and eastern Asia. Although both VBDs are caused by flavivirus, the vectors differ for dengue and Japanese encephalitis. The fresh water inhabitant *Aedes Aegypti* vectors the virus that causes dengue, whereas Japanese encephalitis is caused by *Culex* (*Culex tritaeniorhynchus, Culex vishnui*, and *Culex pseudovishnui*) often found in polluted water bodies. Symptoms are similar for both organisms, including severe headache, fever and vomiting, encephalitis (brain inflammation), meningitis, weakness, and movement disorders, which can develop over a number of days and may lead to coma and paralysis (El-Kafrawy et al. 2016; Kumar and Sharma 2016).

Various drugs to treat dengue have failed due to its adverse side effects to the patients. Drugs against dengue and Japanese encephalitis from microbial sources are in the infant stage. The various microbial products that exhibit dengue antiviral activity include bafilomycins, mycophenolic acid, and other fungal metabolites that work through a range of mechanisms such as inhibition of NS2B, ATPase, inosine 5′-monophosphate dehydrogenase (IMPDH), or through inhibition of endosome acidification to prevent the entry of the virus into the endosome. A number of *Streptomyces* sp. shows promising inhibition action against Japanese encephalitis causative virus (Ratnakomala et al. 2011). Its mode of action targets the ATPase enzyme, which inhibits the RNA helicase activity.

Chitins are an essential compound for mosquitoes and act as a protective layer in their body coverings and are required during the different growth stages of mosquitoes, especially during the transformation from larvae to pupae. Chitinase enzymes from *Streptomyces cacaoi* subsp. *cacaoi*-M20 targeting the chitin required for the larvae have shown to have insecticide activity against *Aedes* mosquitoes (Janaki et al. 2016). Metabolites of *Streptomyces* PO-02, PO-08, and PO-11 showed marked larvicidal efficacy via inhibitory activity on lipase. At concentration 500 μg/ml, inhibition of enzyme ranged between 12% and 58.50% (Prashith et al. 2012). Ethyl acetate extracts from various microbes such as *Py. sanguineus*, *Pe. virgulata*, *Streptomyces* sp. VITJS4 (Naine and Devi 2014), *Bacillus*, and *Pseudomonas* sp. (Nabar and Lokegaonkar 2015) have shown to exhibit *Aedes aegypti* larvicidal activity ranging from 98% to 100% at 550 ppm. Fungal mosquito pathogens such as *Lagenidium, Coelomomyces*, and *Culicinomyces* are also a promising tool to fight against the vector (Scholte et al. 2004). Besides the natural microbial products, synthesis of nanoparticles through the aid of microbes could pave a new dimension in the fight against VBDs. Nanoparticles are generally more effective than bulk compounds. Cerium oxide nanoparticle synthesis using *Aspergillus niger* showed activity against *Aedes aegypti* 0.250 mg/L (Gopinath et al. 2015).

### 1.2.6 West Nile Fever

West Nile fever is caused by ssRNA West Nile virus (WNV) belonging to the *Flavivirus* genus (Petersen and Marfin 2002). *Culex* mosquitoes mainly transmit it. The West Nile virus life cycle is maintained in a bird–mosquito–bird pattern, with birds being the main reservoir and arthropod vectors. WNV is mainly observed in high temperate regions. Chimeric protein from WNV is mainly responsible for humoral and cell-mediated immunity that can be used against the WNV itself. Fusion of *Salmonella typhimurium* fljB flagellin with EIII domain of the WNV envelope protein stimulates high immune response and activation of Toll-like receptor (TLR) (Huleatt et al. 2007). Such an approach could be used for the development of vaccine against WNV. Ethyl acetate extracts from Salinispora sp. SA6E, Salinispora sp. SA22E, and Rhodococcus sp. SA12E showed inhibition of West Nile protease NS3 inhibition 84%, 79%, and 93%, respectively (Abdelmohsen et al. 2014) (Table 1.1).

**TABLE 1.1**
**Various Microbial Products and Their Source along with Their Mode of Action**

| Source | Compound | Against | Action | Reference |
|---|---|---|---|---|
| *S. gougerotii* GT | 4S,10R-dihydroxy-11-methyl-dodec-2-en-1,4-olide | Dengue | Inhibits the expression of NS2B protease | Lin et al. 2016 |
| *M. variabilis* C-03 | Cyclo-(4-trans-6-dihydroxy-proline-L-leucine) | | | |
| *Streptomyces* sp. YIM56209 | Bafilomycins | | Inhibits endosome acidification | Bowman et al. 1988; Yu et al. 2011 |
| *Penicillium brevicompactum* | Mycophenolic acid | | Inhibits inosine 5′-monophosphate dehydrogenase (IMPDH), which affect DNA synthesis in virus | Bartman et al. 1981; Kang et al. 2014 |
| *Beauveria bassiana* | Fungal biomass | | Activates toll and JAK-STAT pathway-controlled effector genes and anti-dengue activity in *Aedes aegypti* | Dong et al. 2012 |
| *Streptomyces* sp., *Actinoplanes philippinensis, Kribbella flavida* | Secondary metabolites | Japanese encephalitis | ATPase inhibitor of RNA Helicase (40%–45%) | Ratnakomala et al. 2011 |
| *Streptomyces* sp. | | | 95%–100% inhibition of Virus NS3 at 0.05 mg and 0.1 mg/20 ml | Hatsu et al. 2002 |
| *Emericellopsis poonensis* | Antiamoebin I | *P. falciparum* | Efrapeptins inhibit mitochondrial $F_0F_1$ ATPase | Nagaraj et al. 2001 |
| *Tolypocladium niveum* | Efrapeptin C–G | | | |
| *Emericellopsis salmosynnemata* | Zervamicin IIA | | | |
| | Zervamicin IIB | | | |
| *Pseudomonas aeruginosa* | Rhamnolipids | Larvae of *A. aegypti* | Distribution of hydrophobic balance | Silva et al. 2015 |

(*Continued*)

**TABLE 1.1 (CONTINUED)**
**Various Microbial Products and Their Source along with Their Mode of Action**

| Source | Compound | Against | Action | Reference |
| --- | --- | --- | --- | --- |
| P-orlandin | *Aspergillus niger* | *P. falciparum* | Inhibits the interaction between fibrinogen-related protein 1 (FREP1) and *P. falciparum* infected cells | Li 2016 |
| Fungal extracts | Fungal entophytes | *T. cruzi* | | Higginbotham et al. 2013 |
| | Fungal cultures from sloth hair | | | Higginbotham et al. 2014 |
| Actinoallolides A–E | *Actinoallomurus fulvus* MK10-036 | *Trypanosoma cruzi* | | Inahashi et al. 2015 |
| Crude extract | *Aspergillus fumigatua* | *Trypanosoma cruzi* | Lysis of trypomastigote | Furtado et al. 2005 |
| Ethyl acetate extracts | Salinispora sp. SA6E, Salinispora sp. SA22E, Rhodococcus sp. SA12E | West Nile virus | Inhibits West Nile protease NS3 | Abdelmohsen et al. 2014 |
| Kojic acid (KA) | *Aspergillus species* | *L. amazonensis* | Reverses inhibitory effects that lead to production of $O_2^-$ and to killing of the pathogen | Rodrigues et al. 2014 |

## 1.3 VECTOR-BORNE DISEASE CONTROL AND PREVENTION

### 1.3.1 Development of Model for the Control and Prevention of VBDs

The survival and maturation of vectors require suitable environments. Thus, the spread of VBDs is directly related to environmental factors and the socioeconomic status of the society. For instance, malarial transmission is dependent on air temperature where the development and maturation cycle of the parasite decreases with an increase in air temperature (Alto and Juliano 2001). Similarly, correlations were seen between cutaneous leishmaniasis and air temperature (Chaves and Pascual 2006). The level of rainfall and the abundance of malarial vector are also well correlated (Yé et al. 2009). Such information based on rainfall and weather prediction can be used as an early warning sign (Chabot-Couture et al. 2014). Other environmental factors include humidity, water bodies, and latitude and longitude.

This mechanistic relationship between the environment and VBDs can be modeled using various algorithms and data sets. Studies on the outbreak of VBDs in correlation with weather and climate through data from NASA's satellite-based Moderate Resolution Imaging Spectroradiometer (MODIS) (Anyamba et al. 2014) have provided valuable information. Such models can be used to predict the probability of VBDs emergence in the near future. Thus, it can be used as an early warning sign to take proper preventative steps to tackle the problem beforehand. Spatial-risk models are used to estimate the vector abundance and VBD occurrences in a given geographical area (Eisen and Eisen 2011), whereas space–time risk models can be utilized for clustering of VBDs in order to identify patterns in outbreaks. Commonly used software for spatial-risk models and space–time risk models include ArcGIS and SaTScan. Technologies based on the Geographic Information System (GIS) and mapping and remote sensing (RS), along with powerful spatial-risk models and space–time risk models, provide tremendous aid to the knowledge of VBD spread patterns and generation of early warning signs.

### 1.3.2 Environmental Management and Community Participation

Various vector-borne disease outbreaks are due to poor sanitation and lack of implementation and maintenance of effective policies. Control of VBDs is hampered due to the emergence of mutant strains and alterations of different pathways including conformational change in the detoxification protein, mutation in the sodium channel, acetylcholinesterase, and GABA receptor genes in the mosquito, which are widely targeted for the inactivity. Various metabolic resistance mechanisms involving enzymes such as carboxylesterases, P450 monooxygenases, and glutathione S-transferases are involved in mosquitoes to inactivate insecticides (Hemingway et al. 2004). Thus, proper management along with community participation is required to check the spread of VBDs.

Strategically sound regulatory frameworks should be put forward into action to prevent the adverse effect of insecticides. Attempts have been made by various national and international bodies such as the Insecticides Act 1968 and Destructive Insects and Pests Act 1914 in India and the revised version of the International Code of Conduct on the Distribution and Use of Pesticides, from the Food and Agriculture Organization (FAO) of the United Nations, for effective use of insecticide (Ecobichon et al. 2001; Prasad 2001). However, the need for participation from every household in every community is a must in order to make any law effective. Hence, implementation of vector-borne disease programs for community awareness would be necessary. National and international health programs should be initiated and encouraged to prevent VBDs through proper education and hygiene. Some of the steps that could be taken include:

- Funding to underdeveloped and developing countries to establish various programs
- Optimal usage insecticides in order to prevent resistance towards the insecticides

- Development programs including proper planning of construction of dams and buildings to avoid mosquito growth
- Collaboration between various departments and communities to irradiate vector-borne disease, to include source reduction, sanitation, health education, and community surveys

### 1.3.3 Health Care Education

In developing countries like India, there have been several information, education, and communication (IEC) campaigns regarding vaccines and contraceptives. Such public education through mass media has substantially decreased the mortality rate in rural areas (Office of Registrar General, India 2011). However, education on VBDs has not been effective. Mass media like television, print, and Internet social media have a promising role in the spread of education on VBDs. However, it should be noted that measurement of awareness program efficiency through such mass media is challenging. Changes in the improvement before and after community-based education programs face more difficulties as compared to clinical trials since a control population is never kept, as it is not ethical to have one. Women from developing countries where VBDs are most common have the potential to play a major role in preventative measures, since most household work is taken care of by housewives. Imparting knowledge to such housewives and by giving the opportunity to participate in community-level policy making could bring about positive change (Kaur et al. 2015). However, it is in such rural areas that women are granted with fewer social privileges (Kishor and Gupta 2015).

In a study by Sreedevi et al. (2016), awareness programs in rural villages showed significant reduction in outdoor sleeping, increased cleaning of water containers, and covering of storage tanks. Such results show that the community participation relies on public as well as self-awareness (Sharma 2007). Schools can be a breeding ground for mosquitoes (Olano et al. 2015). Participatory school education on mosquito control and disease transmission showed interesting improvements among school students irrespective of the school, gender, or age (Deepthi et al. 2014). Such steps to impart the importance and knowledge about VBDs on schoolchildren in developing countries could help children who are most susceptible.

### 1.3.4 Personal Prophylactic Measures

Apart from the health care education and various programs organized by different bodies, it is necessary to keep oneself safe and hygienic to prevent being infected by VBDs. Following is a list of preventative measures to be followed where chances of VBDs are high:

- Mosquito repellents in various forms are available. Mosquito repellent creams, coils, or sprays should be used.
- In areas where insects are common, dressing should be with full sleeves along with socks and shoes.

- Mosquito nets should be used while sleeping at night. For babies, nets should be used during both day and night.
- Keep track of outbreaks and impart knowledge through various social medias.
- Avoid keeping stagnant water for too long and eliminate mosquito-breeding areas.
- Safely reheat meals and use proper hygienic storage for food.
- Drink boiled or filtered water from a reliable water source.
- Outdoor exposure during trekking or a prolonged exposure to dogs, cats, mice, or cattle should be followed by a bath, tick check, and washing and drying of clothes.
- Keep grass mowed and free of litter and woody piles.

## 1.4 CONCLUSION

One of the major hurdles faced while dealing with vector-borne diseases is the resistance gained by the vector and pathogenic organism. Though AMPs show promising results against VBDs, clinical trials on such AMPs are lacking. More emphasis on bioprospection of effective microbial biomolecules is of urgent need. Also, further works on promising biomolecules up to the clinical trials should be encouraged. In addition, victory against VBDs can be achieved only through proper measures between various regulatory bodies to bring about public awareness and community-level participation.

## REFERENCES

Abdelmohsen, U. R., Yang, C., Horn, H., Hajjar, D., Ravasi, T., and Hentschel, U. (2014). Actinomycetes from Red Sea sponges: Sources for chemical and phylogenetic diversity. *Marine Drugs*, 12(5), 2771–2789.

Alto, B. W., and Juliano, S. A. (2001). Temperature effects on the dynamics of *Aedes albopictus* (Diptera: Culicidae) populations in the laboratory. *Journal of Medical Entomology*, 38, 548–556.

Anyamba, A., Small, J. L., Britch, S. C., Tucker, C. J., Pak, E. W., Reynolds, C. A., Crutchfield, J., and Linthicum, K. J. (2014). Recent weather extremes and impacts on agricultural production and vector-borne disease outbreak patterns. *PLoS One*, 9(3), e92538.

Bartman, C. D., Doerfler, D. L., Bird, B. A., Remaley, A. T., Peace, J. N., and Campbell, I. M. (1981). Mycophenolic acid production by *Penicillium brevicompactum* on solid media. *Applied and Environmental Microbiology*, 41(3), 729–736.

Bell, A. (2011). Antimalarial peptides: The long and the short of it. *Current Pharmaceutical Design*, 17(25), 2719–2731.

Bestetti, R. B., and Muccillo, G. (1997). Clinical course of Chagas' heart disease: A comparison with dilated cardiomyopathy. *International Journal of Cardiology*, 60(2), 187–193.

Bollati, M., Alvarez, K., Assenberg, R., Baronti, C., Canard, B., Cook, S., Coutard, B., Decroly, E., de Lamballerie, X., Gould, E. A., and Grard, G. (2010). Structure and functionality in flavivirus NS-proteins: Perspectives for drug design. *Antiviral Research*, 87(2), 125–148.

Bowman, E. J., Siebers, A., and Altendorf, K. (1988). Bafilomycins: A class of inhibitors of membrane ATPases from microorganisms, animal cells, and plant cells. *Proceedings of the National Academy of Sciences*, 85, 7972–7976.

Brand, G. D., Leite, J. R. S., de Sá Mandel, S. M., Mesquita, D. A., Silva, L. P., Prates, M. V., and Kuckelhaus, S. A. (2006). Novel dermaseptins from *Phyllomedusa hypochondrialis* (Amphibia). *Biochemical and Biophysical Research Communications*, 347(3), 739–746.

Cataldi de Flombaum, M. A., and Stoppani, A. O. (1981). Influence of efrapeptin, aurovertin and citreoviridin on the mitochondrial adenosine triphosphatase from *Trypanosoma cruzi*. *Molecular and Biochemical Parasitology*, 3(3), 143–155.

Chabot-Couture, G., Nigmatulina, K., and Eckhoff, P. (2014). An environmental data set for vector-borne disease modeling and epidemiology. *PloS One*, 9(4), e94741.

Chaves, L. F., and Pascual, M. (2006). Climate cycles and forecasts of cutaneous leishmaniasis, a nonstationary vector-borne disease. *PLoS Med*, 3, e295

Christophers, S. (1960). *Aedes aegypti (L.), the yellow fever mosquito: Its life history, bionomics and structure*. Cambridge, UK: Cambridge University Press.

Davis, A. J., Murray, H. W., and Handman, E. (2004). Drugs against leishmaniasis: A synergy of technology and partnerships. *Trends in Parasitology*, 20(2), 73–76.

Deepthi, R., Naresh Kumar, S. J., Prasanna Kamath, B. T., and Rajeshwari, H. (2014). Participatory school health education on vector-borne diseases: Engaging children as change agents. *International Journal of Health Promotion and Education*, 52(2), 68–77.

Dong, Y., Morton, J. C., Ramirez, J. L., Souza-Neto, J. A., and Dimopoulos, G. (2012). The entomopathogenic fungus *Beauveria bassiana* activate toll and JAK-STAT pathway-controlled effector genes and anti-dengue activity in *Aedes aegypti*. *Insect Biochemistry and Molecular Biology*, 42(2), 126–132.

Ecobichon, D. J. (2001). Pesticide use in developing countries. *Toxicology*, 160(1), 27–33.

Eisen, L., and Eisen, R. J. (2011). Using geographic information systems and decision support systems for the prediction, prevention, and control of vector-borne diseases. *Annual Review of Entomology*, 56, 41–61.

El-Kafrawy, S. A., Sohrab, S. S., Ela, S. A., Abd-Alla, A. M., Alhabbab, R., Farraj, S. A., Othman, N. A., Hassan, A. M., Bergoin, M., Klitting, R., and Charrel, R. N. (2016). Multiple introductions of dengue 2 virus strains into Saudi Arabia from 1992 to 2014. *Vector-Borne and Zoonotic Diseases*, 16(6), 391–399.

Furtado, N. A. J. C., Duarte, M. C. T., de Albuquerque, S., Mello, C., and Bastos, J. K. (2005). Improvement of trypanocidal metabolites production by *Aspergillus fumigatus* using neural networks. *Microbiological Research*, 160(2), 141–148.

Gopinath, K., Karthika, V., Sundaravadivelan, C., Gowri, S., and Arumugam, A. (2015). Mycogenesis of cerium oxide nanoparticles using *Aspergillus niger*. *Journal of Nanostructure in Chemistry*, 5(3), 295–303.

Green, S., and Rothman, A. (2006). Immunopathological mechanisms in dengue and dengue hemorrhagic fever. *Current Opinion in Infectious Diseases*, 19(5), 429–436.

Hatsu, M., Tanaka, M., Utama, A., Shimizu, H., and Takamizawa, K. (2002). A Japanese encephalitis virus NS3 inhibitor produced by a *Streptomyces* sp. *Journal of Japanese Actinomycete Society*, 16(1), 6–8.

Hemingway, J., Hawkes, N., Prapanthadara, L. A., Jayawardenal, K. G. I., Ranson, H. (1998). The role of gene splicing, gene amplification and regulation in mosquito insecticide resistance. *Philosophical Transactions of the Royal Society of London, Series B, Biological Sciences*, 353, 1695–1699.

Hemingway, J., Hawkes, N. J., McCarroll, L., and Ranson, H. (2004). The molecular basis of insecticide resistance in mosquitoes. *Insect Biochemistry and Molecular Biology*, 34(7), 653–665.

Herwaldt, B. L. (1999). Leishmaniasis. *Lancet*, 354(9185), 1191–1199.

Higginbotham, S., Wong, W. R., Linington, R. G., Spadafora, C., Iturrado, L., and Arnold, A. E. (2014). Sloth hair as a novel source of fungi with potent anti-parasitic, anti-cancer and anti-bacterial bioactivity. *PloS One*, 9(1), e84549.

Higginbotham, S. J., Arnold, A. E., Ibañez, A., Spadafora, C., Coley, P. D., and Kursar, T. A. (2013). Bioactivity of fungal endophytes as a function of endophyte taxonomy and the taxonomy and distribution of their host plants. *PloS One*, 8(9), e73192.

Hsiao, L. L., Howard, R. J., Aikawa, M., and Taraschi, T. F. (1991). Modification of host cell membrane lipid composition by the intra-erythrocytic human malaria parasite *Plasmodium falciparum*. *Biochemical Journal*, 274, 121–132.

Huleatt, J. W., Foellmer, H. G., Hewitt, D., Tang, J., Desai, P., Price, A., Jacobs, A., Takahashi, V. N., Huang, Y., Nakaar, V., and Alexopoulou, L. (2007). A West Nile virus recombinant protein vaccine that coactivates innate and adaptive immunity. *Journal of Infectious Diseases*, 195(11), 1607–1617.

Hurwitz, I., Fieck, A., Read, A., Hillesland, H., Klein, N., Kang, A., and Durvasula, R. (2011). Paratransgenic control of vector borne diseases. *International Journal of Biological Sciences*, 7(9), 1334–1344.

Inahashi, Y., Iwatsuki, M., Ishiyama, A., Matsumoto, A., Hirose, T., Oshita, J., Sunazuka, T., Panbangred, W., Takahashi, Y., Kaiser, M., and Otoguro, K., (2015). Actinoallolides A–E, new anti-trypanosomal macrolides, produced by an endophytic actinomycete, *Actinoallomurus fulvus* MK10-036. *Organic Letters*, 17(4), 864–867.

Institute of Medicine. (2008). *Vector-borne diseases: Understanding the environmental, human health, and ecological connections: Workshop summary*. Washington, DC: National Academies Press.

Janaki, T., Nayak, B. K., and Ganesan, T. (2016) Larvicidal activity of *Streptomyces cacaoi* subsp. cacaoi-M20 against *Aedes aegypti*. *International Journal of Botany Studies*, 9(1), 47–49.

Kang, S., Shields, A. R., Jupatanakul, N., and Dimopoulos, G. (2014). Suppressing dengue-2 infection by chemical inhibition of *Aedes aegypti* host factors. *PLoS Neglected Tropical Diseases*, 8(8), e3084.

Kaur, M., Prinja, S., and Kumar, R. (2015). Evaluating the performance of health promotion interventions. *Indian Journal of Medical Research*, 142(2), 109.

Kishor, S., and Gupta, K. (2009). Gender equality and women's empowerment in India. National Family Health Survey (NFHS-3) India 2005-06.

Krishna, K., Sukumar, M., and Balaram, P. (1990). Structural chemistry and membrane modifying activity of the fungal polypeptides zervamicins, antiamoebins and efrapeptins. *Pure and Applied Chemistry*, 62(7), 1417–1420.

Kumar, A., Dhuley, J. N., and Naik, S. R. (1991). Evaluation of microbial metabolites for trypanocidal activity: Significance of biochemical and biological parameters in the mouse model of trypanosomiasis. *Japanese Journal of Medical Science and Biology*, 44(1), 7–16.

Kumar, C. S., and Sharma, S. K. (2016). Editorial: Dengue virus infection. *National Medical Journal of India*, 29(2), 61–63.

Li, J. (2016). Small fungal molecule metabolites targeting mosquito midgut FREP1 block malaria transmission. *FASEB Journal*, 30(1 Suppl.), 627–626.

Lin, C. K., Wang, Y. T., Hung, E. M., Yang, Y. L., Lee, J. C., Sheu, J. H., and Liaw, C. C. (2016). Butyrolactones and diketopiperazines from marine microbes: Inhibition effects on dengue virus type 2 replication. *Planta Medica*, 83(1-02), 158–163.

Maguire, J. H., Hoff, R., Sherlock, I., Guimarães, A. C., Sleigh, A. C., Ramos, N. B., Mott, K. E., and Weller, T. H. (1987). Cardiac morbidity and mortality due to Chagas' disease: Prospective electrocardiographic study of a Brazilian community. *Circulation*, 75(6), 1140–1145.

Mangoni, M. L. (2006). Temporins, anti-infective peptides with expanding properties. *Cellular and Molecular Life Sciences*, 63(9), 1060–1069.

Martínez-Díaz, R. A., Escario, J. A., Nogal-Ruiz, J. J., and Gómez-Barrio, A. (2001). Biological characterization of *Trypanosoma cruzi* strains. *Memórias do Instituto Oswaldo Cruz*, 96(1), 53–59.

Mukbel, R. M., Patten, C., Gibson, K., Ghosh, M., Petersen, C., and Jones, D. E. (2007). Macrophage killing of *Leishmania amazonensis* amastigotes requires both nitric oxide and superoxide. *American Journal of Tropical Medicine and Hygiene*, 76, 669–675.

Murray, H. W. (2004). Treatment of visceral leishmaniasis in 2004. *American Journal of Tropical Medicine and Hygiene*, 71(6), 787–794.

Nabar, B. M., and Lokegaonkar, S. (2015). Larvicidal activity of microbial metabolites extracted from extremophiles against vector mosquitoes. *International Journal of Mosquito Research*, 2(3), 161–165.

Nagaraj, G., Uma, M. V., Shivayogi, M. S., and Balaram, H. (2001). Antimalarial activities of peptide antibiotics isolated from fungi. *Antimicrobial Agents and Chemotherapy*, 45(1), 145–149.

Naine, S. J., and Devi, C. S. (2014). Larvicidal and repellent properties of *Streptomyces* sp. VITJS4 crude extract against *Anopheles stephensi*, *Aedes aegypti* and *Culex quinquefasciatus* (Diptera: Culicidae). *Polish Journal of Microbiology*, 63(3), 341–348.

Narayana, J. L., and Chen, J. Y. (2015). Antimicrobial peptides: Possible anti-infective agents. *Peptides*, 72, 88–94.

Nikaido, H. (2009). Multidrug resistance in bacteria. *Annual Review of Biochemistry*, 78, 119–146.

Office of Registrar General, India (ORGI). (2011). Maternal and child mortality and total fertility rate: Sample registration system. Accessed August 5, 2015. http://censusindia.gov.in/vital_statistics/SRS_Bulletins/MMR_release_070711.pdf.

Olano, V. A., Matiz, M. I., Lenhart, A., Cabezas, L., Vargas, S. L., Jaramillo, J. F., Sarmiento, D., Alexander, N., Stenström, T. A., and Overgaard, H. J. (2015). Schools as potential risk sites for vector-borne disease transmission: Mosquito vectors in rural schools in two municipalities in Colombia. *Journal of the American Mosquito Control Association*, 31(3), 212–222.

Olivier, M., Gregory, D. J., and Forget, G. (2005). Subversion mechanisms by which *Leishmania* parasites can escape the host immune response: A signaling point of view. *Clinical Microbiology Reviews*, 18, 293–305.

Petersen, L. R., and Marfin, A. A. (2002). West Nile virus: A primer for the clinician. *Annals of Internal Medicine*, 137(3), 173–179.

Prasad, S. S. (2001, November). *Country Report–India*. Prepared for the Meeting of the Programme Advisory Committee (PAC), Ayutthaya, Thailand.

Prashith Kekuda, T. R., Shobha, K. S., Onkarappa, R., Goutham, S. A., and Raghavendra, H. L. (2012). Screening biological activities of a *Streptomyces* species isolated from soil of Agumbe, Karnataka, India. *International Journal of Drug Development and Research*, 4(3), 104–114.

Ratnakomala, S., Ridwan, R., Lisdiyanti, P., and Andi, U. (2011). Screening of actinomycetes producing an ATPase inhibitor of Japanese encephalitis virus RNA helicase from soil and leaf litter samples. *Microbiology Indonesia*, 5(1), 3.

Rivero, A., Vezilier, J., Weill, M., Read, A. F., and Gandon, S. (2010). Insecticide control of vector-borne diseases: When is insecticide resistance a problem? *PLoS Pathogens*, 6(8), e1001000.

Rodrigues, A. P. D., Farias, L. H. S., Carvalho, A. S. C., Santos, A. S., do Nascimento, J. L. M., and Silva, E. O. (2014). A novel function for kojic acid, a secondary metabolite from *Aspergillus fungi*, as antileishmanial agent. *PloS One*, 9(3), e91259.

Rohde, J., Cousens, S., Chopra, M., Tangcharoensathien, V., Black, R., Bhutta, Z. A., and Lawn, J. E. (2008). 30 years after Alma-Ata: Has primary health care worked in countries? *The Lancet*, 372(9642), 950–961.

Scholte, E. J., Knols, B. G., Samson, R. A., and Takken, W. (2004). Entomopathogenic fungi for mosquito control: A review. *Journal of Insect Science*, 4(1), 19.

Sharma, A. K., Bhasin, S., and Chaturvedi, S. (2007). Predictors of knowledge about malaria in India. *Journal of Vector Borne Diseases*, 44(3), 189–197.

Silva, V. L., Lovaglio, R. B., Von Zuben, C. J., and Contiero, J. (2015). Rhamnolipids: Solution against *Aedes aegypti*? *Frontiers in Microbiology*, 6, 88.

Snook, C. F., Woolley, G. A., Oliva, G., Pattabhi, V., Wood, S. P., Blundell, T. L., and Wallace, B. A. (1998). The structure and function of antiamoebin I, a proline-rich membrane-active polypeptide. *Structure*, 6(6), 783–792.

Sreedevi, A., Burru, R. V., Rao, G. V., Yalamanchili, P., Subhaprada, C., Kumari, V., and Kala, S. (2016). Study on awareness about vector borne diseases and education about preventive measures in rural field practice areas of Kurnool medical college, Kurnool. *International Journal of Medical Science and Public Health*, 5(9), 1803–1807.

Sundar, S. (2001). Drug resistance in Indian visceral leishmaniasis. *Tropical Medicine and International Health*, 6(11), 849–854.

Vale, N., Aguiar, L., and Gomes, P. (2014). Antimicrobial peptides: A new class of antimalarial drugs? *Frontiers in Pharmacology*, 5, 275.

Vontas, J., Blass, C., Koutsos, A. C., David, J. P., Kafatos, F. C., Louis, C., Hemingway, J., Christophides, G. K., and Ranson, H. (2005). Gene expression in insecticide resistant and susceptible *Anopheles gambiae* strains constitutively or after insecticide exposure. *Insect Molecular Biology*, 14(5), 509–521.

World Health Organization (WHO). (1998). Test procedures for insecticide resistance monitoring in malaria vectors; bio-efficacy and persistence of insecticides on treated surfaces. WHO/CDS/CPC/MAL/9812.

World Health Organization (WHO). (2006). Pesticides and their application for the control of vectors and pests of public health importance. WHO/CDS/NTD/WHOPES/GCDPP /2006.1.

World Health Organization (WHO). (2014). Global Health Observatory: Report on neglected diseases: Leishmaniasis.

World Health Organization (WHO). (2016). *Vector-borne diseases*. http://www.who.int /mediacentre/factsheets/fs387/en/.

Wu, H., Ong, Z. Y., Liu, S., Li, Y., Wiradharma, N., Yang, Y. Y., and Ying, J. Y. (2015). Synthetic β-sheet forming peptide amphiphiles for treatment of fungal keratitis. *Biomaterials*, 43, 44–49.

Yé, Y., Hoshen, M., Kyobutungi, C., Louis, V. R., and Sauerborn, R. (2009). Local scale prediction of *Plasmodium falciparum* malaria transmission in an endemic region using temperature and rainfall. *Global Health Action*, 2(s2).

Yu, Z., Zhao, L. X., Jiang, C. L., Duan, Y., Wong, L., Carver, K. C., and Shen, B. (2011). Bafilomycins produced by an endophytic actinomycete *Streptomyces* sp. *YIM56209. Journal of Antibiotics*, 64(1), 159–162.

Zhang, M., Zhao, J., and Zheng, J. (2014). Molecular understanding of a potential functional link between antimicrobial and amyloid peptides. *Soft Matter*, 10(38), 7425–7451.

# *Section I*

## *Microbial Control of Mosquito Vector*

# 2 Autodissemination

## *Current and Future Potential in the Application of Entomopathogenic Fungi against Mosquito-Borne Diseases*

*Mario A. Rodríguez-Pérez and Filiberto Reyes-Villanueva*

## CONTENTS



## 2.1 INTRODUCTION

Chemical control is applied with the objective of killing as many insects as possible since the dichlorodiphenyltrichloroethane (DDT) was discovered in the last century. The benefits of chemical insecticides are undeniable in all respects, including those on public health, but insecticides are expensive and the actual effect upon target populations is unknown (Esu et al. 2010). For example, malathion reduces the *Aedes aegypti* densities after the first case of dengue has occurred; the product is assumed to have been "effective" if few or no cases of dengue are recorded afterward, without a mosquito survey program before and after treatment. This dogma, which has induced an intense selection of resistant genotypes (Fernandes et al. 2016), is still used to decide whether there was an effect of the chemical on dengue transmission.

Research on biological control of mosquito vectors against *Ae. aegypti* in particular has advanced with the use of ascomycetes Hypocreales (= deuteromycetes) fungi such as *Metarhizium anisopliae* and *Beauveria bassiana* that attack a wide variety of arthropods (Boucias and Pendland 1998). Both fungi are ubiquitous and infect hypogeous insects such as termites, ants, and collembolans, as well as multiple agricultural insect pests (Vega et al. 2009). The fungi also infect medically important dipterans such as the fly *Glossina morsitans*, and *Anopheles*, *Culex*, and *Aedes albopictus* mosquitoes, the latter being vectors of dengue, yellow fever, and other arboviruses that are highly prevalent in global public health (Patterson et al. 2016).

The two fungi were explored since 1970 as mortality factors on culicid larvae (Clark et al. 1968), and in the past decade have been investigated as potential agents of biological control by indirect contact on *Anopheles gambiae*, *Ae. aegypti*, *Ae. albopictus*, and *Culex quinquefasciatus* adults (Scholte et al. 2003, 2004a, 2005, 2007). There is also transmission of fungi through sexual behavior of insects (Kaaya and Okech 1990) and this horizontal transmission is common in nature (Bruck and Lewis 2002). However, when a sexual or other attractant is used to concentrate individuals of one sex and disseminate the fungi into the natural populations, the process is called *autodissemination* (Baverstock et al. 2010). This chapter describes the origin of the concept and how we examined it using these fungi on the house fly *M. domestica* and *Ae. aegypti* mosquitoes.

## 2.2 THE CONCEPT OF AUTODISSEMINATION

The transfer of a pathogen from one insect "donor" to another "recipient" is normal in nature. Collembolans ingest and transport conidia of *M. anisopliae* and *B. bassiana* attached to the cuticle without being infected. However, both external conidia and those ejected in fecal matter cause mortality on other insects such as *Tenebrio molitor* (Dromph 2001, 2003). The same occurs in other mycophagous insects such as the beetles Nitidulidae in cultivated corn. It is an interesting interaction because it is known that *Carpophilus freemani* carrying *B. bassiana* is concentrated five times more in corn ears infested by lepidopteran larvae, increasing the probability for the fungus to make contact with additional hosts (Rodríguez-del-Bosque et al. 1998). These examples show that autodissemination occurs naturally among different species where the donor is a true vector (Bruck and Lewis 2002). However, transmission is usual among conspecific individuals as well as in social insects where termites and ants infected or harboring conidia are detected by their congeners and cadavers buried or expelled from the nests (Rath 2010).

In the discipline of microbial control, experts pointed out that the transmission can be horizontal (in the same generation) and vertical (between generations) within a species and between species. It could be from a local (plant) to regional (landscape) scale. A textual quote is "horizontal transmission between individuals of the same species (auto dissemination) can occur through direct contact between contaminated and uncontaminated individuals" (Baverstock et al. 2010). The concept was proposed as horizontal transmission in applied studies and for the same species and generation. Transmission by copula or sexual behavior in this context is autodissemination. As far as we know, it was first documented in insects of medical relevance with the use of both *M. anisopliae* and *B. bassiana* against both genders of the tsetse fly *Glossina morsitans* by Kaaya and Okech (1990), then with the use of *Entomophtora muscae* against the house fly (Watson and Petersen 1993). It was not until 14 years later that it was again attempted in control of medically important vectors when Scholte and colleagues (2004b) used *M. anisopliae* against the malaria vector *An. gambiae*, with contamination from female to male and an infection rate that did not exceed 30%. Our research group has been investigating the concept by using these fungi against *Ae. aegypti* and *M. domestica* disseminated through copula (Garcia-Munguia et al. 2011, 2015; Reyes-Villanueva et al. 2011; Garza-Hernández et al. 2013, 2015).

## 2.3 LEVELS OF VIRULENCE AND SUBLETHAL EFFECTS

Multiple isolates of *M. anisopliae* and *B. bassiana* have been evaluated worldwide against mosquitoes by indirect exposure. Individuals were confined in a chamber constructed with petri dishes and using 10 micron pore filter paper impregnated with dried conidia (Garza-Hernández 2008). Mosquitoes have also been exposed in cages of different sizes: a black blanket impregnated with conidia in oil as a resting site for contamination (Paula et al. 2008). In the field, *M. anisopliae* has been explored against *An. gambiae* and *Cx. quinquefasciatus* in Africa by placing black cloth containing conidia on the roof of huts to contaminate the mosquitoes (Scholte et al. 2003) and there was another similar study conducted in Brazil against *Ae. aegypti* (Paula et al. 2011). Virulence with median lethal times ($LT_{50}$) of 3 to 6 days has been reported in these studies. When autodissemination was examined through copula, the estimator was in the range of 6 to 10 days (Garcia-Munguia et al. 2011, 2015; Reyes-Villanueva et al. 2011; Garza-Hernández 2013, 2015). Apart from mortality, the sublethal effects are a topic that is classical in pathology of insects, mainly in chronic infections (Boucias and Pendland 1998). When females of *Ae. aegypti* were contaminated by confining them for 48 hours with males exposed to 24 hours of $6 \times 10^8$ dry conidia/mL of a highly virulent *M. anisopliae* isolate, the fecundity was reduced up to 99% (Reyes-Villanueva et al. 2011). When *B. bassiana* was used against the same vector, the reduction was 96% (Garcia-Munguia et al. 2011). In addition, the ovarian cycle of *Ae. aegypti* was shortened from 5 to 3 days and from 7 to 2 days in *M. domestica* after females were contaminated with both fungi through copulation (Garcia-Munguia et al. 2015).

In terms of invested energy, the production of one ovule is expensive and infected females tend to save as many eggs as possible. The house fly accelerates its ovogenesis up to 2 days to save 30% of the remaining eggs due to mycosis (Figure 2.1). Recently, it was observed that *B. bassiana* infecting *Ae. aegypti* adults exerts its pathogenesis up to adults of F1, reducing survival by 70% and causing a 50% body miniaturization, with females having 1.41 mm of wing length in comparison to 2.97 mm of that of the control group (A. M. Garcia-Munguia, personal communication). It is known that other fungi such as *Aspergillus parasiticus* and *Aspergillus*

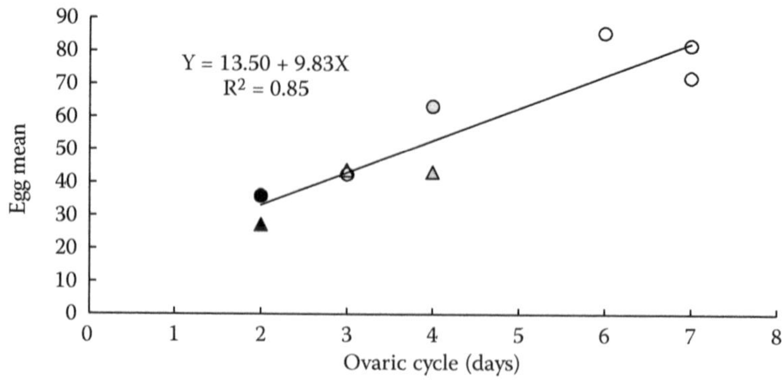


**FIGURE 2.1** Linear regression of the mean of eggs on the days of the ovarian cycle for 9 groups of *Musca domestica* (each of 30 flies); 3 groups were confined with 1 (black circles), 5 (in the line), and 10 male (gray) previously exposed to $3 \times 10^8$ conidia/mL of *Metarhizium anisopliae* (triangles); 3 other groups were confined with 1, 5, and 10 males exposed to *Beauveria bassiana* (circles). White circles are control groups (females confined with healthy males) (Garcia-Munguia et al. 2015).

*clavatus* in *Ae. aegypti* spread their infections to the adult where they affect fertility. In fact, by placing infected larvae of different instars along with healthy larvae, the resulting females showed reduced fecundity of 54% (Nnakumusana 1985; Seye et al. 2009). *M. anisopliae* is also autodisseminated from females to males with a median virulence of 7 days (Garza-Hernández 2008), suggesting that females contaminated with fungi can transfer conidia to other males during mating.

Although sexual aggressiveness is not reduced, 50% of *Ae. aegypti* males contaminated with each of the two fungi died in 3 to 4 days. When a male is exposed to *M. anisopliae* for 24 hours, its sexual behavior increases in the first 3 hours when exposed to virgin females. The males "capture" on average, seven females, whereas a healthy one captures only three females (Garza-Hernández et al. 2015). Furthermore, males contaminated with *M. anisiopliae* of low virulence inseminate six times more females than males contaminated with the fungus of high virulence (Reyes-Villanueva et al. 2011). However, in contaminated males, a 47% decrease in sperm production was observed 5 days postexposure to the fungus in comparison to healthy males in which production increased by 71%. Furthermore, from an average of 50,000 conidia attached to a male during the 24-hour confinement, about 10,000 and 5,000 conidia were transferred to the females with whom they copulate or intend to mate for the first and fifth time, respectively. This measurement is relevant because it was also shown that a topical application of 5000 conidia/mL on virgin females of the same age kills 50% of them in 3 days. Although females that have copulated for the fifth time only received 10% of the conidia transferred by the male, this amount is sufficient to kill all females exposed during 6 days, a time shorter than the period of viral incubation (Garza-Hernández et al. 2015). We have preliminarily observed that when males contaminated with *M. anisopliae* were released in the field, the density of host-seeking females in the houses is reduced by 40% to 60% (unpublished data).

## 2.4 EFFECT OF *METARHIZIUM ANISOPLIAE* ON TRANSMISSION

In theory, *Metarhizium anisopliae* applied by indirect exposure (black blankets) in African huts against *An. gambiae* reduces the transmission of malaria (Blanford et al. 2005). *M. anisopliae* contaminated *Ae. aegypti* fed with human blood containing dengue virus serotype 2 have reduced daily survival rate up to $P = 0.64$, which is less than 0.98 of healthy females and 0.96 of females with dengue virus and not contaminated with the fungus (Garza-Hernández et al. 2013). Of 50 female exposed to fungus and dengue virus, only 12% were positive to the virus in the heads at 7 days postinfection; and 7 days is the time reported in the classic literature for the extrinsic incubation period (*n*) for the dengue virus (Watts et al. 1987). Therefore, the fungus killed in 8 days the females copulating with the contaminated males, but of the 50% surviving females, only 6% were infective.

Using Ross-McDonald's vector capacity model (*C*), the daily survival rate (*P*) is the component that mostly affects *C* (Dye 1992). This parameter is important to calculate the "longevity factor" or infective life expectancy ($P^n/{-\ln P}$). Figure 2.2 depicts the infective life expectancy over time. It was 2 days for females contaminated with one or the other fungus. It was reduced to 1 day for females harboring fungus and virus, but lasted up to 7 days for healthy females that served as control.

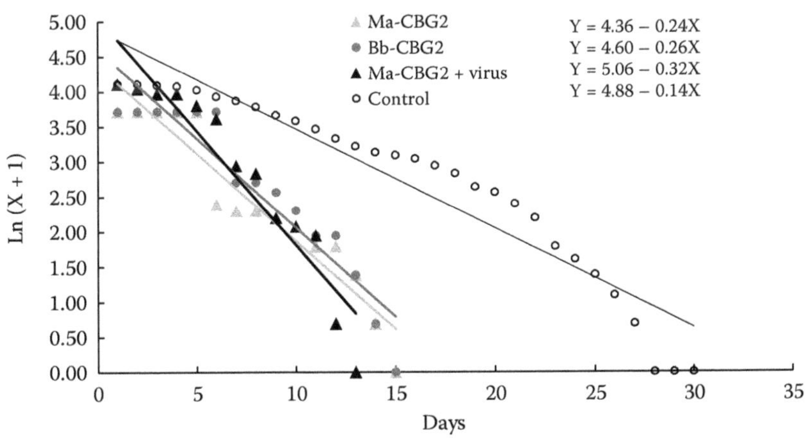


**FIGURE 2.2** Linear regressions of the daily number of surviving *Aedes aegypti* females confined with a virgin male previously exposed to $6 \times 10^8$ conidia/mL of *Metarhizium anisopliae*, *Beauveria bassiana*, and control (with healthy male) to calculate the daily survival rate (*P*). The analyses were from three previously published studies: *M. anisopliae* (Reyes-Villanueva et al. 2011), *B. bassiana* (Garcia-Munguia et al. 2011), and *M. anisopliae* + infection with dengue serotype 2 and control (Garza-Hernandez et al. 2013).

## 2.5 CONCLUDING REMARKS

*M. anisopliae* and *B. bassiana* autodisseminated by copulation of males to females in *Ae. aegypti* killed half of the females in 8 days and in the 50% of the surviving females, the sublethal effects on fecundity and body size suggest a drastic loss in their vector competence. This phenomenon is already being confirmed by release of males contaminated with *M. anisopliae* under field conditions.

The method to combat the dengue vectors based on release of Wolbachia-infected *Ae. aegypti* males can sterilize the wild females that have copulated with the released males, but it does not inhibit the biting behavior of the mated females, hence, there still remains the chance for dengue viruses to be transmitted to humans. The fungi autodisseminated can kill the wild female mosquitoes and interrupt the dengue transmission by the surviving females. This is an advantage over the Wolbachia-based method. The comparison of both technologies under field conditions would be an interesting research line to be explored.

## ACKNOWLEDGMENTS

We are indebted to Mr. Adebiyi A. Adeniran for reading and correcting an earlier version of the manuscript. We are also grateful to Dr. Alfonso J. Garza-Hernández (Universidad Autonoma Agraria Antonio Narro) for providing the original data sets to complete the present document.

This study was supported by Consejo Nacional de Ciencia y Tecnologia of Mexico (Fondo Sectorial de Investigacion en Salud y Seguridad Social SSA/IMSS/ISSSTE, grant S008-2013-200664). The funders had no role in study design, data collection and analysis, decision to publish, or preparation of the manuscript.

## REFERENCES

Baverstock J, Roy HE, Pell JK. 2010. Entomopathogenic fungi and insect behaviour: From unsuspecting hosts to targeted vectors. *Biocontrol* 55: 89–102.

Blanford S, Chan BHK, Jenkins N, Sim D, Turner RJ, Read AF, Thomas MB. 2005. Fungal pathogen reduces potential for malaria transmission. *Science* 308: 1638–1641.

Boucias DG, Pendland JC. 1998. *Principles of Insect Pathology*. New York: Kluwer Academic.

Bruck D, Lewis LC. 2002. *Carpophilus freemani* (Coleoptera: Nitidulidae) as a vector of *Beauveria bassiana*. *J Invertebr Pathol* 80: 188–190.

Clark TB, Kellen WR, Fukuda T, Lindegren JE. 1968. Field and laboratory studies on the pathogenicity of the fungus *Beauveria bassiana* to three genera of mosquitoes. *J Invertebr Pathol* 11: 1–7.

Dromph KM. 2001. Dispersal of entomopathogenic fungi by collembolans. *Soil Bio Biochem* 33: 2047–2051.

Dromph KM. 2003. Collembolans as vectors of entomopathogenic fungi. *Pedobiologia* 47: 245–256.

Dye C. 1992. The analysis of parasite transmission by bloodsucking insects. *Ann Rev Entomol* 37: 1–19.

Esu E, Lenhart A, Smith L, Horstick O. 2010. Systematic review: Effectiveness of peri-domestic space spraying with insecticide on dengue transmission. *Trop Med Int Health* 15: 619–631.

Fernandes DB, Viana-Medeiros PDF, Costa SA, Martins AJ, Pereira JB, and Valle D. 2016. Resistance status to the insecticides Temephos, Deltamethrin, and Diflubenzuron in Brazilian *Aedes aegypti* populations. *BioMed Res Int*, article ID 8603263. doi:10.1155/2016/8603263.

Garcia-Munguia AM, Garza-Hernandez JA, Rodriguez-Perez MA, Reyes-Villanueva F. 2011. Transmission of *Beauveria bassiana* from male to female *Aedes aegypti* mosquitoes. *Parasit Vectors* 4: 24.

Garcia-Munguia CA, Reyes-Villanueva F, Rodriguez-Perez MA, Cortez-Madrigal H, Acosta-Ramos M, Ibarra-Juarez LA, Velazquez-Machuca MA, Silva-Garcia JT, Rebollar-Plata M, Garcia-Munguia AM. 2015. Autodissemination of *Metarhizium anisopliae* and *Beauveria bassiana* in *Musca domestica* L. results in less oviposition and short gonotrophic cycle. *Southwest Entomol* 40: 519–552. http://www.bioone.org/doi/full/10.3958/059.040.0310.

Garza-Hernández JA. 2008. *Transmision de Metarhizium anisopliae en Aedes aegypti por comportamiento sexual*. Thesis, Universidad Autonoma de Nuevo Leon. Facultad de Ciencias Biologicas, Monterrey, Nuevo Leon Mexico.

Garza-Hernández JA, Reyes-Villanueva F, Russell Tanya L., Braks Marieta A. H., Garcia-Munguia Alberto M., Rodríguez-Pérez Mario A. 2015. Copulation activity, sperm production and conidia transfer in *Aedes aegypti* males contaminated by *Metarhizium anisopliae*: A biological control prospect. *PLoS Negl Trop Dis* 9(10): e0004144. doi:10.1371/journal.pntd.0004144.

Garza-Hernández JA, Rodríguez-Pérez MA, Salazar MI, Russell TL, Adeleke MA, de Luna-Santillana E. de J, Reyes-Villanueva F. 2013. Vectorial capacity of *Aedes aegypti* for dengue virus type 2 is reduced with co-infection of *Metarhizium anisopliae. PLoS Negl Trop Dis* 7(3): e2013. doi:10.1371/journal.pntd.0002013.

Kaaya GP, Okech MA. 1990. Horizontal transmission of mycotic infection in adult tsetse, *Glossina morsitans morsitans*. *Entomophaga* 35: 46–57.

Nnakumusana ES. 1985. Laboratory infection of mosquito larvae by entomopathogenic fungi with particular reference to *Aspergillus parasiticus* and its effects on fecundity and longevity of mosquitoes exposed to conidial infections in larval stages. *Curr Sci* 54: 1221–1228.

Patterson J, Sammon M, Garg M. 2016. Dengue, zika and chikungunya: Emerging arboviruses in the New World. *West J Emerg Med* 17: 671–679. doi:10.5811/westjem.2016.9.30904.

Paula AR, Brito ES, Pereira CR, Carrera MP, Samuels RI. 2008. Susceptibility of adult *Aedes aegypti* (Diptera: Culicidae) to infection by *Metarhizium anisopliae* and *Beauveria bassiana*: Prospects for dengue vector control. *Biocontr Sci Technol* 18: 1017–1025.

Paula AR, Carolino AT, Silva CP, Samuels RI. 2011. Susceptibility of adult female *Aedes aegypti* (Diptera: Culicidae) to the entomopathogenic fungus *Metarhizium anisopliae* is modified following blood feeding. *Parasit Vectors* 4: 91. doi:10.1186/1756-3305-4-91.

Rath AC. 2010. The use of entomopathogenic fungi for control of termites. *Biocontr Sci Technol* 10: 563–571.

Reyes-Villanueva F, Garcia-Munguia AM, Garza-Hernandez JA, Tamez-Guerra P, Howard AFP, Rodriguez-Perez MA. 2011. Dissemination of *Metarhizium anisopliae* of low and high virulence by mating behavior in *Aedes aegypti*. *Parasit Vectors* 4: 171.

Rodriguez-del-Bosque LA, Leos-Martinez J, Dowd PF. 1998. Effect of ear wounding and cultural practices on abundance of *Carpophilus freemani* (Coleptera: Nitidulidae) and other microcoleopterans in maize in northeastern Mexico. *J Econ Entomol* 91: 796–801.

Scholte E-J, Knols BGJ, Samson RA, Takken W. 2004a. Entomopathogenic fungi for mosquito control: A review. *J Insect Sci* 4: 19.

Scholte E-J, Knols BGJ, Takken W. 2004b. Autodissemination of the pathogenic fungus *Metarhizium anisopliae* amongst adults of the malaria vector *Anopheles gambiae* s.s. *Malaria J* 3: 45.

Scholte E-J, Ng'habi K, Kihonda J, Takken W, Paaijmans KP, Abdulla S, Killeen GF, Knols BGJ. 2005. An entomopathogenic fungus for control of adult African malaria mosquitoes. *Science* 308: 1641–1642.

Scholte E-J, Njiru BN, Smallegange RC, Takken W, Knols BGJ. 2003. Infection of malaria (*Anopheles gambiae* s.s.) and filariasis (*Culex quinquefasciatus*) vectors with the entomopathogenic fungus *Metarhizium anisopliae*. *Malaria J* 2: 29.

Scholte E-J, Takken W, Knols BGJ. 2007. Infection of adult *Aedes aegypti* and *Ae. albopictus* mosquitoes with the entomopathogenic fungus *Metarhizium anisopliae*. *Acta Tropica* 102: 151–158.

Seye F, Faye O, Ndiaye M, Njie E, Afoutou JM. 2009. Pathogenicity of the fungus, *Aspergillus clavatus*, isolated from the locust, *Oedaleus senegalensis*, against larvae of the mosquitoes *Aedes aegypti*, *Anopheles gambiae* and *Culex quinquefasciatus*. *J Insect Sci* 9: 53.

Vega FE, Goettel MS, Blackwell M, Chandler D, Jackson MA, Keller S, Koike M, Maniania NK, Monzon A, Ownley BH, Pell JK, Rangel DEN, Roy HE. 2009. Fungal entomopathogens: New insights on their ecology. *Fung Ecol* 2(4): 149–159. http://dx.doi.org/10.1016/j.funeco.2009.05.001.

Watson DW, Petersen JJ. 1993. Sexual activity of male *Musca domestica* (Diptera: Muscidae) infected with *Entomophtora muscae* (Entomophtorales: Entomophtoraceae). *Bio Contr* 3: 22–26.

Watts DM, Burke DS, Harrison BA, Whitmire RE, Nisalak A. 1987. Effect of temperature on the vector efficiency of *Aedes aegypti* for dengue 2 virus. *Am J Trop Med Hyg* 36: 143–152.

# 3 Inhibition of Mosquito Vectors of Malaria and Filariasis Using Marine Microorganisms and Their Associated Compounds

*Arivarasan Vishnu Kirthi and Loganathan Karthik*

**CONTENTS**



## 3.1 INTRODUCTION

Diseases that spread through vectors like mosquitoes are known as vector-borne diseases. They include malaria and filarial diseases. These vector-borne diseases are still known to occur in tropical and subtropical climatic regions where the sunlight is abundant. Also, the presence of no or poor drainage systems in developing countries encourage breeding and the spread of pathogenic parasites like *Plasmodium* and the Japanese encephalitis virus. Malaria and the other vector-based diseases contribute the major disease burdens in countries like India and China.

Mosquitoes, which belong to Diptera: Culicidae, act as the pathogens and parasites, which might result in the rising of epidemics or pandemics in the severity on human and animals (Benelli 2015, Mehlhorn et al. 2012).

The transmission of the major pathogens, which are responsible for the majority of the vector-borne diseases, are very much responsible for dog heartworm, West Nile virus, and Eastern equine encephalitis (World Health Organization [WHO] 2012).

Insecticides are targeted upon the larvae from the Culcidae group, in which the occurrence of organophosphate-like chemicals work against the transmission of mosquito-borne diseases (Benelli 2015, Lees et al. 2014). The constant usage of chemicals has resulted in the formation of ill-effects on the human population and the surrounding environment, as well as resistance in various vector species (Hemingway and Ranson 2000). Biological mass mosquitoes are the major contributor to the insect's biodiversity and the biological total mass, together constitute a total of 3500 different and variedly described species, out of which a few hundred are considered to be a serious threat to human medical and economical risk (Fang 2010).

The presence of mosquitoes have been shown to be found in almost all continents on which it has adapted to the corresponding land and habitats leading to a functional ecological role for mosquitoes. Mosquitoes have been found to have critically important roles in local water bodies especially the larval stages (Ricci et al. 2012). There are some symbiotic relationships among mosquitoes and some microbial strains, which might have critical effects on mosquito populations. The normal microbial flora of mosquitoes has been found to inhibit the growth and development of pathogens that are transmitted through the mosquito (Cirimotich et al. 2011, Ricci et al. 2012).

Therefore, there is an increasing number of vector species. The biodiversitification of the insecticidal world has led to the evolution of more than 3500 different species of insects, which has created severe medical and economical tension and risk to the human world (Feldhaar and Gross 2008). In terms of the spread of disease, malaria is an excellent example. The pathogen *Plasmodium* is spread through the female anapheline mosquitoes to the human population. Similar examples include yellow fever, dengue fever, West Nile virus, and chikungunya. The global human population is at increased risk to the spread and suffering from the infectious diseases spread by mosquitoes, and the occurrence of human infections are recorded every year by the World Health Organization (Fang 2010). Infections of the human population are largely dependent on the vectors of disease, which are interrelated to the vectors and their type. Infections that are spread by mosquitoes are highly infectious in terms of symptoms and clinical variations in patients.

There is the notion of controlling mosquito vectors using microorganisms, especially ones from terrestrial and aquatic regions. Here, the role of the microorganisms has been highly effective and have led to advantageous pathways over chemical control methodologies due to their increased host specificity and their ability of not affecting the nontarget organisms (both unicellular and the multicellular). Marine microorganisms have the potential to act as a source for commercially important biologically compounds (Deepika et al. 2012).

## 3.2 MOSQUITO BREEDING BURDEN: A WORLD SCENARIO

The main role of parasites is to cause continual disease by rapid multiplication of self-strains and rapid growth. Malaria is caused by transmission from the human host to the next by the Anopheles mosquitoes and there are an estimated 190 to 240 million cases every year in Sub-Saharan Africa (WHO 2009). The presence and rate of vector-based diseases like malaria and filariasis are increasing in developing and high-population countries like India and China. The Indian government is focused on certain states like

Orisa and Jharkhand and certain parts of northeastern states, including Arunachal Pradesh, Assam, Tripura, and Meghalaya, as reported by the World Health Organization and National Institute of Malaria Research (NIMR 2007, WHO 2008). Female anapheline mosquitoes, *Anopheles subpictus*, are the main vector in the coastal areas of southern India (Panicker et al. 1981). The study by Dhingra et al. (2010) noted 200,000 malarial deaths per year in the coastal areas of South India, many occurring before the age of 70 and in early childhood (Panicker et al. 1981).

The genera of microorganisms known as the actinomycetes have been known to produce a huge number of different and pharmaceutically important biologically active substances. The search for similar and more powerful bioactive substances has been very successful, which nowadays covers two-thirds of the naturally occurring antibiotics. Also many of the medically important strains of actinomycetes have been discovered (Okami 1952). The actinomycetes have been isolated using intensive screening programs for the past several years compared to the ones found in terrestrial regions than in the marine sediments; these strains of the actinomycetes showed varying degrees of salt tolerance and the production of the spores in huge quantities (Grein and Meyers 1958, Kurtböke et al. 2015).

## 3.3 THE ROLE OF ACTINOMYCETES IN VECTOR CONTROL

The presence of actinomycetes in the Earth ecosystem is widely distributed among the oceans and there are truly indigenously marine actinomycetes that have been described in the detail (Bull 2004, Bull et al. 2005). However, there are many related questions about the role of actinomycetes in ecological systems, their biological-geographical distribution, and evolution. However, it is clear that these marine actinomycetes are currently the major resource for biotechnological and pharmaceutical research and discovery of newer active metabolites (Bull 2004, Fenical and Jensen 2006, Fiedler et al. 2005, Jensen et al. 1991, Pathom-Aree et al. 2006).

The microorganisms that have been isolated from marine resources are gaining attention as the most potent source of the antibiotics and other biological, pharmaceutically important secondary metabolites (Newman et al. 2000). Microbial strains or the isolates belonging to the *Streptomyces* are found to be distributed evenly in both marine and terrestrial habitats and are commonly used for the production of the novel metabolites (Pathom-Aree et al. 2006). Studies on marine *Streptomyces* have focused on the production of the antibiotic and other biological products of pharmaceutical interest. There are certain reports on the biological presence and their effects on the local environment from the actinomycetes due to their increased quantitation (Imhoff and Stöhr 2003). Also, there are many similar reports that confirm the presence of actinomycetes in marine sediments even in the presence of human and aquatic pathogens (Patil et al. 2001, Peela et al. 2005).

Actinomycetaceae has been identified for its ability to produce secondary metabolites, which are known to be candidates for pharmaceutical substances using them and might reduce the pathogenicity of the microorganisms. The past few decades have seen marine and marine-based actinomycetes become one of the important sources of the pharmaceutically relative products. Actinomycetes got more attention and in turn provided different important bioactive compounds with a high commercial value and they continue to be routinely screened for newer bioactive compounds. The biological assays of these bioactive substances are isolated from *Streptomyces* and are therefore categorized as pharmacologically and agriculturally very active substances and are also autoregulators. Therefore, intensive research to advance this area is required with a focus on the biochemical and morphological characterization, and also the screening of the antagonistic effects on the larval stages of the mosquitoes using compounds from the actinomycetes.

The actinomycete strain are being regularly reported to have production sites for the compounds with medicinal usage. Spinosad from *Saccharopolyspora spinosa* (Snyder et al. 2007) and salinosporamide A from the marine *Salinispora tropica* have been reported to have strong antagosnostic activity against the erythrocytic phase during *Plasmodium* sp. growth (Prudhomme et al. 2008). Similarly, the larval instar stages of the *Culex tritaeniorhynchus* and *Culex gelidus* were the prime targets for the crude extracts of actinobacteria, with strong larvicidal, ovicidal, and repellent assay (Karthik et al. 2011).

Marine actinobacterial isolates are the special production factories for compounds that have quite interesting applications against the vectors of malaria and filariasis. The *Streptomyces* sp. VITDDK3 has been found to be effective against certain larval vector species due to the isolated compound known as (2S,5R,6R)-2-hydroxy-3,5, 6-trimethyloctan-4-one (Deepika et al. 2012). Similarly, 5-(2, 4-dimethylbenzyl) pyrrolidin-2-one, produced by marine *Streptomyces* VITSVK5 sp., showed activity against *R.* (*B.*) *microplus*, *A. stephensi*, and *C. tritaeniorhynchus* (Saurav et al. 2013).

Active biological metabolites obtained from the actinomycetes in huge quantities ranging from tetranectin, avermectins, faerifungin, macrotetrolides, and flavonoids were reported to be toxic and inhibitory in nature to the species genera *Micromonospora*, *Actinomadura*, *Actinoplanes*, *Micropolyspora*, *Nocardiopsis*, *Oerskonia*, *Thermomonospora*, *Sreptoverticillium*, and *Chainia*. Several pharmaceutical important compounds were isolated from actinomycetes such as tetranectin (Ando 1982), avermectins (Ando 1982), faerifungin (Copping and Menn 2000), macrotetrolides (Zizka et al. 1988), and flavonoids (Rao et al. 1990) produced by *Streptomyces aureus*, *Streptomyces avermitilis*, *Streptosporangium albidum*, *Streptomyces griseus*, and *Streptomyces sp.*, respectively. All of these compounds were reported to be toxic for mosquitoes.

Actinobacteria have a profound role in the marine aquatic environment; the degradation and turnover of various materials are a continual process which is further mediated by the action of the variety of microorganisms. There is the notion that the type of enzymes producing microorganisms might relate to the concentration of the natural substrate and the conditions of the environment (Das et al. 2006, Bertness et al. 2014). Also, actinobacteria are known for their contribution to the breakdown of the organic compounds and their recycling process (Weyland 1969).

The identification of pharmacologically important compounds for actinobacteria earmarked the epoch in the antibacterial research and their subsequent developments in the antibiotic chemotherapy. Since the identification of streptomycin, a large number of chemically derived antibacterial compounds have been isolated from the cultures of *Streptomyces* and *Streptoverticillium* (Miyadoh 1997). Also, some of the genera of actinomycetes have been focused upon, including *Actinomadura*,

*Actinoplanes*, *Amycolatopsis*, *Dactylosporangium*, *Kibdelosporangium*, *Microbispora*, *Micromonospora*, *Planobispora*, *Streptosporangium*, and *Planomonospora* (Lazzarini et al. 2000).

## 3.4 THE DREADED VECTOR-BORNE DISEASE: MALARIA

Vector control is an important requirement in an epidemic, as is the pressing need to increase the newer and improved mosquito management strategies within budget, powerful enough to remove the target organism, and safe for the nontargeted organisms and the environment. Mosquitoes transmit serious human illnesses, causing physiological resistance and adverse environmental outcomes in addition to excessive. Mosquitoes are a major public health threat as the vectors of serious human diseases including malaria, filariasis, Japanese encephalitis, dengue fever, chikungunya, and yellow fever, which cause substantial mortality and morbidity among people residing in tropical zones (Jang and Ahn 2002).

Malaria now is responsible for an estimated more than 300 million people falling ill, and 1 million deaths per year (WHO 2007). Malaria disease occurs mainly in the tropical and subtropical regions of the world and is the result of infections with *Plasmodium falciparum*, *P. vivax*, *P. ovale*, or *P. malariae* (Mueller et al. 2007), which are the main vectors for malaria transmitted to humans only by the *Anopheles* mosquito.

In India, there are nine vectors found to be responsible for malaria from the Anapheline species. There are about 2 to 3 million cases of malaria in the Indian subcontinent each year (Sharma et al. 2009). Control over the mosquito is being supported by many areas, but still there are certain challenges that involve the development of resistance to the insecticides and lack or absence of alternative, cost-effective, and safe insecticides. The utilization of insecticide has led to the resistance requiring the formation of strategies for the prolonged usage of highly effective vector control compounds. Many areas employ the combined usage multiple insecticides and phytochemicals for mosquito control (Bloland 2001).

Chemical insecticide sprays based on toxins from mosquitocidal bacteria are environment friendly and therefore there have been huge worldwide efforts to isolate novel mosquitocidal bacterial species with an increased efficacy (Federici et al. 2006, Park et al. 2007).

Efforts to improve efficacy for mosquito control using bacterial species, including *Bacillus sphaericus*, *Bacillus thuringiensis*, *Bravibacillus laterosporus*, and *Clostridium bifermentans*, are already reported to be lethal to vectors (Deepika et al. 2012, Favret and Yousten 1985, Orlova et al. 1998, Park et al. 2007, Patil et al. 2013, Singh and Prakash 2009, Thiéry et al. 1992).

## 3.5 SECOND IN VECTOR-BASED DISEASES: FILARIASIS

Mosquitoes (Diptera: Culicidae) are a huge threat to the human population, since they act as important pathogens and parasites (Mehlhorn et al. 2012). Mosquitoes are the main vectors of diseases and are the major mode of transmission of human pathogens more than any other group of insects, especially the arthropods. Mosquito-borne diseases are not only the major cause of the increased levels of morbidity and mortality, but also lead to great economic loss and the presence of the social chaos in both underdeveloped and developing countries where there are costs to health care and negative impacts on the tourism of the country. India carries almost 40% of filiariasis cases and their presence has led to the annual financial loss in USD 80 million (Hotez et al. 2004). According to the WHO, malaria accounts for 500 million infections and approximately 4 million deaths each year. Also, dengue fever has increased over the last 2 decades and more than 100 countries with a total population of 2 billion lives are at risk.

The filarial vector *Culex quinquefasciatus* is a widely distributed mosquito vector species in India and tropical regions of the world. It transmits lymphatic filariasis in many of the countries (Raghavendra et al. 2011). The remaining toxicity of the synthetic insecticides such as dichlorodiphenyltrichloroethane (DDT) and the benzene hexachloride are serious problems because these chemicals decompose slowly under natural influences (Kekuda et al. 2010). These chemicals remain in nature for an extended period of time.

The southern house mosquito (*Culex quinquefasciatus*) is a major threat that acts as the vector for lymphatic filariasis, a neglected tropical disease commonly known as elephantiasis. Also, *Filariodidea nematodesi*, otherwise known as *Wuchereria bancrofti*, is also responsible for lymphatic filariasis. At the world level, approximately 25 million men are suffering from genital disease and almost 15 million are also suffering due to lymphedema. The elimination of lymphatic filiariasis could prevent unnecessary suffering and could play an important key role in the elimination of the poverty (Michael and Bundy 1997, WHO 2013). Currently, the main ingredient that acts against the elimination of the larval stages of the mosquito are organophosphates and insect growth regulators. Plant-borne substances are a valuable arsenal against these so-called larvicides, which might be due to the reduced toxicity toward the vertebrates and increased biodegradability (Amer and Mehlhorn 2006a, 2006b; Benelli 2015).

The larval and the pupal stages of mosquito are most commonly predated upon. The list includes the copepods, odonate young instars, water bugs, tadpoles, and fishes. The usage of the biological methodology for the control over the larvivorous fishes is of particular relevance toward the malaria control to the urban areas of the developing countries (Gratz and Pal 1988, Hurst et al. 2004, Kim et al. 1994, Mulla 1961).

Mosquitoes are the only types of insects that have become a major problem for humans due to their ability to cause severe diseases like malaria, filiariasis, dengue, Japanese encephalitis, and other diseases leading to many deaths each year. The repeated usage of chemical insecticides and other nonenvironmental-friendly substances have led to the resurgence of resistant strains of the mosquito populations (Brown 1986), and environmental and human health concerns (Derelanko and Auletta 2014, Krieger 1991, Ray 1991).

## 3.6 CONCLUSION

Research on actinobacteria is at an early stage, but recent technological advancements in molecular biology and genomics have greatly improved and accelerated the capacity of ecology and evolution to inform bioprospecting programs. Further, research and research-aided programs will be focused at the levels of individuals, species, and then to huge population levels. Also, the natural product discovery and search in marine actinobacteria have shown exceptional promise. The research and discovery of marine actinobacterial species is different from the development and production of the end-product and is another subject. There have been many recent comments about the scarcity of new antibiotic entities, why their need has reached alarming proportions, and the reason for the withdrawal of many big pharmaceutical companies from this field.

## ACKNOWLEDGMENTS

The authors are thankful to management of Shanghai Jiao Tong University, Shanghai, China; and C. Abdul Hakeem College, Melvisharam, Vellore District, Tamil Nadu, India, for support and encouragement.

## REFERENCES

Amer A, Mehlhorn H. 2006a. Larvicidal effects of various essential oils against Aedes, Anopheles, and Culex larvae (Diptera, Culicidae). *Parasitology Research* 99:466–472.

Amer A, Mehlhorn H. 2006b. Repellency effect of forty-one essential oils against Aedes, Anopheles, and Culex mosquitoes. *Parasitology Research* 99:478.

Ando K. 1982. How to discover new antibiotics for insecticidal use. In *Natural Products: Proceedings of the 5th International Congress of Pesticide Chemistry*, Kyoto, Japan, 29 August–4 September, Elsevier.

Benelli G. 2015. Research in mosquito control: Current challenges for a brighter future. *Parasitology Research* 114:2801–2805.

Bertness MD, Bruno JF, Silliman BR, Stachowicz JJ. 2014. *Marine community ecology and conservation*. Sinauer Associates, Incorporated.

Bloland PB. 2001. *Drug resistance in malaria*. Geneva, Switzerland: World Health Organization, p. 22.

Brown A. 1986. Insecticide resistance in mosquitoes: A pragmatic review. *Journal of the American Mosquito Control Association* 2:123–140.

Bull AT. 2004. Bountiful oceans: Prospecting marine microbial diversity. *Screening Trends in Drug Discovery* 5:14–16.

Bull AT, Stach JE, Ward AC, Goodfellow M. 2005. Marine actinobacteria: Perspectives, challenges, future directions. *Antonie Van Leeuwenhoek* 87:65–79.

Cirimotich CM, Dong Y, Clayton AM, Sandiford SL, Souza-Neto JA, Mulenga M, Dimopoulos G. 2011. Natural microbe-mediated refractoriness to Plasmodium infection in Anopheles gambiae. *Science* 332:855–858.

Copping LG, Menn JJ. 2000. Biopesticides: A review of their action, applications and efficacy. *Pest Management Science* 56:651–676.

Das S, Lyla P, Khan SA. 2006. Marine microbial diversity and ecology: Importance and future perspectives. *Current Science* 1325–1335.

Deepika TL, Kannabiran K, Khanna VG, Rajakumar G, Jayaseelan C, Santhoshkumar T, Rahuman AA. 2012. Isolation and characterisation of acaricidal and larvicidal novel compound (2S,5R,6R)-2-hydroxy-3,5,6-trimethyloctan-4-one from Streptomyces sp. against blood-sucking parasites. *Parasitology Research* 111:1151–1163.

Derelanko MJ, Auletta CS. 2014. *Handbook of toxicology*. Boca Raton, FL: CRC Press.

Dhingra N, Jha P, Sharma VP, Cohen AA, Jotkar RM, Rodriguez PS, Bassani DG, Suraweera W, Laxminarayan R, Peto R. 2010. Adult and child malaria mortality in India: A nationally representative mortality survey. *The Lancet* 376:1768–1774.

Fang J. 2010. Ecology: A world without mosquitoes. *Nature* 466:432–435.

Favret ME, Yousten AA. 1985. Insecticidal activity of Bacillus laterosporus. *Journal of Invertebrate Pathology* 45:195–203.

Federici BA, Park H-W, Sakano Y. 2006. Insecticidal protein crystals of Bacillus thuringiensis. In: *Inclusions in prokaryotes*, 195–236. Berlin: Springer.

Feldhaar H, Gross R. 2008. Immune reactions of insects on bacterial pathogens and mutualists. *Microbes and Infection* 10:1082–1088.

Fenical W, Jensen PR. 2006. Developing a new resource for drug discovery: Marine actinomycete bacteria. *Nature Chemical Biology* 2:666–673.

Fiedler H-P, Bruntner C, Bull AT, Ward AC, Goodfellow M, Potterat O, Puder C, Mihm G. 2005. Marine actinomycetes as a source of novel secondary metabolites. *Antonie van Leeuwenhoek* 87:37–42.

Gratz N, Pal R. 1988. Malaria vector control: Larviciding. In *Malaria: Principles and practices of malariology*, 1213–1226. Edinburgh, UK: Churchill Livingstone.

Grein A, Meyers S. 1958. Growth characteristics and antibiotic production of actinomycetes isolated from littoral sediments and materials suspended in sea water. *Journal of Bacteriology* 76:457.

Hemingway J, Ranson H. 2000. Insecticide resistance in insect vectors of human disease. *Annual Review of Entomology* 45:371–391.

Hotez PJ, Remme JH, Buss P, George G, Morel C, Breman JG. 2004. Combating tropical infectious diseases: Report of the Disease Control Priorities in Developing Countries Project. *Clinical Infectious Diseases* 38:871–878.

Hurst TP, Brown MD, Kay BH. 2004. Laboratory evaluation of the predation efficacy of native Australian fish on Culex annulirostris (Diptera: Culicidae). *Journal of the American Mosquito Control Association* 20:286–291.

Imhoff JF, Stöhr R. 2003. Sponge-associated bacteria: General overview and special aspects of bacteria associated with Halichondria panicea. In *Sponges (Porifera)*, 35–57. Berlin: Springer.

Jang YS, Ahn YJ. 2002. Natural products, organic chemistry: Larvicidal activity of Brazilian plants against Aedes aegypti and Culex pipiens pallens (Diptera: Culicidae). *Journal of Applied Biological Chemistry* 45:131–134.

Jensen PR, Dwight R, Fenical W. 1991. Distribution of actinomycetes in near-shore tropical marine sediments. *Applied and Environmental Microbiology* 57:1102–1108.

Karthik L, Gaurav K, Rao KV, Rajakumar G, Rahuman AA. 2011. Larvicidal, repellent, and ovicidal activity of marine actinobacteria extracts against Culex tritaeniorhynchus and Culex gelidus. *Parasitology Research* 108:1447–1455. Epub 2010/12/15.

Kekuda TP, Shobha K, Onkarappa R. 2010. Potent insecticidal activity of two Streptomyces species isolated from the soils of the western ghats of Agumbe, Karnataka. *Journal of Natural Pharmaceuticals* 1:30–32.

Kim HC, Kim MS, Yu HS. 1994. Biological control of vector mosquitoes by the use of fish predators, Morocco oxycephalus and Misgurnus anguillicaudatus in the laboratory and semi-field rice paddy. *Korean Journal of Entomology* (Korea Republic).

Krieger R (ed.). 1991. *Hayes' Handbook of pesticide toxicology*. London: Academic Press.

Kurtböke Dİ, Grkovic T, Quinn RJ. 2015. Marine Actinomycetes in biodiscovery. In: *Springer handbook of marine biotechnology*, 663–676. New York: Springer.

Lazzarini A, Cavaletti L, Toppo G, Marinelli F. 2000. Rare genera of actinomycetes as potential producers of new antibiotics. *Antonie van Leeuwenhoek* 78:399–405.

Lees RS, Knols B, Bellini R, Benedict MQ, Bheecarry A, Bossin HC, Chadee DD, Charlwood J, Dabiré RK, Djogbenou L. 2014. Review: Improving our knowledge of male mosquito biology in relation to genetic control programmes. *Acta Tropica* 132:S2–S11.

Mehlhorn H, Al-Rasheid KA, Al-Quraishy S, Abdel-Ghaffar F. 2012. Research and increase of expertise in arachno-entomology are urgently needed. *Parasitology Research* 110: 259–265.

Michael E, Bundy D. 1997. Global mapping of lymphatic filariasis. *Parasitology Today* 13: 472–476.

Miyadoh S. 1997. *Atlas of actinomycetes*. Tokyo: Akasura Publishing Company.

Mueller I, Zimmerman PA, Reeder JC. 2007. Plasmodium malariae and Plasmodium ovale—The 'bashful' malaria parasites. *Trends in Parasitology* 23:278–283.

Mulla MS. 1961. Mosquito control investigations with emphasis on the integration of chemical and biological control in mosquito abatement. *Proceedings of 29th Annual Conference of the California Mosquito Control Association*, 1–4.

National Institute of Malaria Research (NIMR). 2007. In-depth review on malaria for National Vector Borne Disease Control Programme. New Delhi: National Institute of Malaria Research and National Vector Borne Disease Control Programme.

Newman DJ, Cragg GM, Snader KM. 2000. The influence of natural products upon drug discovery. *Natural Product Reports* 17:215–234.

Okami Y. 1952. Utilization of nitrogen compounds by Streptomycetaceae and its application to classification. *Japanese Journal of Medical Science and Biology* 5:265–275.

Orlova MV, Smirnova TA, Ganushkina LA, Yacubovich VY, Azizbekyan RR. 1998. Insecticidal activity of Bacillus laterosporus. *Applied and Environmental Microbiology* 64: 2723–2725.

Panicker K, Bai MG, Rao UB, Viswam K, Murthy US. 1981. Anopheles subpictus, vector of malaria in coastal villages of southeast India. *Current Science* 50:694–695.

Park H-W, Mangum CM, Zhong H, Hayes SR. 2007. Isolation of Bacillus sphaericus with improved efficacy against Culex quinquefasciatus. *Journal of the American Mosquito Control Association* 23:478–480.

Pathom-Aree W, Nogi Y, Sutcliffe IC, Ward AC, Horikoshi K, Bull AT, Goodfellow M. 2006. Dermacoccus abyssi sp. nov., a piezotolerant actinomycete isolated from the Mariana Trench. *International Journal of Systematic and Evolutionary Microbiology* 56:1233–1237.

Patil CD, Borase HP, Salunke BK, Patil SV. 2013. Alteration in Bacillus thuringiensis toxicity by curing gut flora: Novel approach for mosquito resistance management. *Parasitology Research* 112:3283–3288.

Patil R, Jeyasekaran G, Shanmugam S, Shakila RJ. 2001. Control of bacterial pathogens, associated with fish diseases, by antagonistic marine actinomycetes isolated from marine sediments. *IJMS* 30:264–267.

Peela S, Kurada VB, Terli R. 2005. Studies on antagonistic marine actinomycetes from the Bay of Bengal. *World Journal of Microbiology and Biotechnology* 21:583–585.

Prudhomme J, McDaniel E, Ponts N, Bertani S, Fenical W, Jensen P, Le Roch K. 2008. Marine actinomycetes: A new source of compounds against the human malaria parasite. *PloS One* 3:e2335. Epub 2008/06/05.

Raghavendra K, Barik T, Bhatt R, Srivastava H, Sreehari U, Dash A. 2011. Evaluation of the pyrrole insecticide chlorfenapyr for the control of Culex quinquefasciatus Say. *Acta Tropica* 118:50–55.

Rao KV, Chattopadhyay SK, Reddy GC. 1990. Flavonoids with mosquito larval toxicity. *Journal of Agricultural and Food Chemistry* 38:1427–1430.

Ray DE. 1991. Pesticides derived from plants and other organisms. *Handbook of Pesticide Toxicology* 2:585–636.

Ricci I, Damiani C, Capone A, DeFreece C, Rossi P, Favia G. 2012. Mosquito/microbiota interactions: From complex relationships to biotechnological perspectives. *Current Opinion in Microbiology* 15:278–284.

Saurav K, Rajakumar G, Kannabiran K, Rahuman AA, Velayutham K, Elango G, Kamaraj C, Zahir AA. 2013. Larvicidal activity of isolated compound 5-(2,4-dimethylbenzyl) pyrrolidin-2-one from marine Streptomyces VITSVK5 sp. against Rhipicephalus (Boophilus) microplus, Anopheles stephensi, and Culex tritaeniorhynchus. *Parasitology Research* 112:215–226. Epub 2011/10/20.

Sharma A, Parasher H, Singh O, Adak T. 2009. Species B of Anopheles culicifacies (Diptera: Culicidae) is reproductively less fit than species A and C of the complex. *Acta Tropica* 112:316–319.

Singh G, Prakash S. 2009. Efficacy of Bacillus sphaericus against larvae of malaria and filarial vectors: An analysis of early resistance detection. *Parasitology Research* 104:763.

Snyder DE, Meyer J, Zimmermann AG, Qiao M, Gissendanner SJ, Cruthers LR, Slone RL, Young DR. 2007. Preliminary studies on the effectiveness of the novel pulicide, spinosad, for the treatment and control of fleas on dogs. *Veterinary Parasitology* 150:345–351. Epub 2007/11/06.

Thiéry I, Hamon S, Gaven B, De Barjac H. 1992. Host range of Clostridium bifermentans serovar. malaysia, a mosquitocidal anaerobic bacterium. *Journal of the American Mosquito Control Association* 8:272–277.

Weyland H. 1969. Actinomycetes in North Sea and Atlantic ocean sediments. *Nature* 223:858.

World Health Organization (WHO). 2007. *Global plan to combat neglected tropical diseases 2008–2015*. Geneva, Switzerland: WHO.

World Health Organization (WHO). 2008. *World malaria report 2008*. Geneva, Switzerland: WHO.

World Health Organization (WHO). 2009. *World malaria report 2009*. Geneva, Switzerland: WHO.

World Health Organization (WHO). 2012. *Handbook for integrated vector management*. Geneva, Switzerland: WHO.

World Health Organization (WHO). 2013. Summary. In *Sustaining the drive to overcome the global impact of neglected tropical diseases: Second WHO report on neglected tropical diseases*, ix–xii. Geneva, Switzerland: WHO.

Zizka Z, Weiser J, Blumauerova M, Jizba J. 1988. Ultrastructural effects of macrotetrolides of Streptomyces griseus LKS-1 in tissues of Culex pipiens larvae. *Cytobios* 58:85–91.

# 4 Entomopathogenic Fungi-Mediated Biocontrol Mechanisms against Mosquito Vectors

## *Recent Trends and Future Perspectives*

*Subhaswaraj Pattnaik,*
*Parasuraman Paramanantham,*
*and Siddhardha Busi*

**CONTENTS**

## 4.1 INTRODUCTION

Advancement in technological expansion, massive industrialization, significant health care systems, and high throughput therapeutic approaches are available to treat the deadliest of diseases and disorders in a pragmatic way. Though there is a tremendous increase in facilities provided by health care systems, there still persist certain uphill tasks to take care of in order to complete eradication of disease-producing agents. Similar trends have also been observed in the case of vector-borne diseases, especially insect vector-mediated diseases. Among the insect vectors, mosquito vectors probably occupy the top position in spreading fatal diseases such as malaria, dengue, and many more (Mondzozo et al., 2011). Pathogenic microorganisms that possess a substantial impact on human health are mainly transmitted between hosts via vectors, particularly insects and ticks, and these vector-borne diseases cause a substantial public health burden by a complex interplay between environmental and socioeconomic factors (Messina et al., 2015).

### 4.1.1 Vector-Borne Diseases and Impact on Socioeconomic Status

Vector-borne diseases, especially malaria, are generally considered to be endemic in tropical and subtropical regions of the world, particularly in India, Sri Lanka, other parts of Southeast Asia, East African provinces, and Central Africa (Gunathilaka et al., 2016). The reason behind the epidemiology profile is the ecological parameters of vector-borne diseases such as malaria and dengue suggests they rely upon environmental factors such as temperature and rainfall (Naish et al., 2014). According to expert opinion, current climate changes, such as increase in temperature, have a greater role to play in expansion of the transmission range of vector-borne diseases including malaria at the community level associated with socioeconomic factors and lack of awareness toward these vector-borne diseases and their associated outcomes (Onyango et al., 2016). The transmission of mosquito-borne diseases such as dengue and malaria is triggered by socioeconomic factors, especially the economic status of the community. In addition to environmental and socioeconomic factors, lack of an integrative approach to tackle mosquito vectors and a lack of preventive measures are also responsible for transmission of vector-borne diseases in South Asian countries (such as India, Sri Lanka, Indonesia, and Bangladesh), African countries, and other tropical countries (Wijayanti et al., 2016).

### 4.1.2 Mosquito-Borne Diseases and Health Effects

In tropical countries, a highly anthropophilic mosquito vector is *Aedes aegypti* (L.), which is considered as a prolific epidemic vector of detrimental vector-borne diseases such as dengue and chikungunya in human beings. *A. aegypti* draws considerable attention worldwide because of its efficient invasiveness, high susceptibility, and capability to infect more than one individual within one gonotrophic cycle (Scholte et al., 2007). One of the close relatives of *A. aegypti* is *A. albopictus*, which is also considered to be important in transmitting several viral diseases such as Zika fever and different types of encephalitis (S. J. Lee et al., 2015). Malaria is one of the deadliest vector-borne diseases in tropical regions of Africa, Asia, and America. It is caused by the protozoan parasite *Plasmodium* sp., which causes infection through female mosquito vectors belonging to the genus *Anopheles* (Kamareddine, 2012).

### 4.1.3 Traditional Vector-Control Programs and Implications

From the last few decades, climate change and uncontrolled anthropogenic activities have allowed mosquito vectors to expand their geographical distribution all over the world with potential threat to human beings by transmitting dreadful diseases such as malaria, chikungunya, dengue, filariasis, encephalitis, and yellow fever (Bourtzis et al., 2014). As mosquito vector-borne diseases become challenging public health problems with substantial impact on socioeconomic status, under the influence of the World Health Organization (WHO) there is an increase in developing effective strategies to eradicate mosquito vectors. In this context, synthetic chemicals such as carbamates and organophosphates are the first to be used as insecticides against mosquito vectors (Govindarajan et al., 2013). However, the extensive and indiscriminate use of these synthetic insecticides nurtured several environmental and health concerns. Besides, prolonged use of these insecticides also contributes toward the development of resistance among the mosquito vectors against these synthetic insecticides and mosquito repellents (Mendki et al., 2015). In order to compensate for the problems associated with synthetic insecticides and to combat mosquito-borne diseases in an efficient manner, current research is focused on products and services from untapped natural reservoirs. The products obtained from plants and microbes are gaining much attention for their efficacy in mosquito vector control without any side effects or any environmental concerns (Benelli and Mehlhorn, 2016).

### 4.1.4 Biological Control of Mosquito Vectors

A wide variety of insecticides are traditionally used to control mosquito vectors with the control strategies involving killing of mosquitoes. The traditional knowledge on indigenous plants and plant products and their folkloric excellence in therapeutics including vector control strategies are well reported as potential repellant of disease-causing mosquito vectors (Tisgratog et al., 2016). Plants are a rich source of essential oils with tremendous therapeutic properties as antioxidants, antimicrobials, antiquorum sensing agents, and antibiofilm activities. In addition, the essential oils of plant origin are well versed with potential larvicidal activity against mosquito vectors for the deadliest of diseases such as malaria, filariasis, dengue, and chikungunya (Pavela et al., 2017). Plants and plant products serve as potent mosquitocides, mosquito repellents, and mosquito predators as well as oviposition deterrents irrespective of different varieties of mosquito vectors (Benelli et al., 2016). For the last few years, an interesting trend has been generated to utilize the world of nanotechnology toward synthesizing nanoparticles from natural plants and plant products for mosquito vector control in an efficient manner. In this regard, Muthukumaran et al. (2015) exploited the potency of silver nanoparticles synthesized from *Gmelina asiatica* leaf extract against *Anopheles stephensi*, *A. aegypti*, and *Culex quinquefasciatus* serving as vector for life-threatening diseases such as malaria, dengue, and filariasis, respectively (Muthukumaran et al., 2015).

The present scenario of biological control strategies for controlling vector-borne diseases are effective, safe, and ecofriendly in nature. However, the control measures provided by these strategies need to be effective for prolonged periods under different environmental conditions and to the uncertain behavioral conditions of mosquito vectors. In this regard, entomopathogenic fungi serve as promising alternatives to the existing biocontrol strategies and play an important role in future vector control programs.

### 4.1.5 Entomopathogenic Fungi as Alternative Tools in Control of Vector-Borne Diseases

The existing biological control measures for vector control programs have proven to be effective. However, certain limitations with respect to different physiological and environmental parameters exist. In order to implement a promising alternative strategy to combat mosquito vector transmission is a comprehensive and sustainable framework in the form of entomopathogenic fungi (Achee et al., 2015).

Insect pathogenic fungi or entomopathogenic fungi have the inherent property to infect and ultimately kill a wide range of agricultural pests, and the focus has now been shifted toward controlling mosquito vectors (Kanzok and Jacobs-Lorena, 2006). The targeted killing approach possessed by these entomopathogenic fungi could be exploited as promising alternatives to the existing biological control of vector-borne diseases. A wide variety of fungal entomopathogenic species of *Metarhizium*, *Beauveria*, *Lagenidium*, *Isaria*, *Tolypocladium*, *Aspergillus*, *Culicinomyces*, and *Coelomomyces* are frequently exploited as potent biocontrol agents against a wide variety of mosquito vectors (Santos et al., 2009; Bhan et al., 2013). Among these entomopathogens, *Beauveria bassiana* and *Metarhizium anisopliae* are the most prevalently used organisms as potential mosquitocides in vector control programs (Ortiz-Urquiza et al., 2010). The multidimensional field of nanotechnology has also been exploited using efficient entomopathogenic fungi to synthesize nanoparticles targeted toward effective mosquitocidal properties (Najitha Banu and Balasubramanian, 2014). A multifaceted and sustainable approach for vector control using the efficacy and targeted stature of entomopathogenic fungi could be exploited as an alternative to existing biocontrol strategies.

## 4.2 AN INSIGHT INTO THE WORLD OF ENTOMOPATHOGENIC FUNGI

### 4.2.1 Classification of Entomopathogenic Fungi

Entomogenous fungi are a group of fungi that keeps relationship with the insect host in either one or more interaction modes such as predation, competition, parasitism, mutualism, and commensalism. Entomopathogenic fungi are the subgroup of entomogenous fungi that undergo a predation, parasitism, and commensalism mode of association with the insect and cause diseases that lead to the death of the insect. These behavioral responses of an insect to entomopathogenic fungi directed numerous researchers to exploit these fungi as potential insect control agents to maintain pest balance in agricultural environments (Peng et al., 2008). So far, extensive studies have been conducted to bioprospect potential entomopathogenic fungi toward insect control. Still, morphological similarity-based classification are followed to classify entomopathogenic fungi. These fungi are most often found in the phylum Ascomycota, Zygomycota, Chytridiomycota, and Basidiomycota. Among the mentioned phylum, Ascomycota fungi are extensively studied and used as biological control agents. *Aschersonia*, *Beauveria*, *Culicinomyces*, *Hirsutella*, *Metarhizium*, *Nomuraea*, *Tolypocladium*, and *Verticillium* are the major genera investigated in the phylum of Ascomycota as biological insecticides. Especially, *Metarhizium* and *Beauveria* are two major genera considered as promising tools for insect control. These fungi are inhabited in soil and predominantly infect soil-dwelling insects. *Metarhizium* and *Beauveria* are the major two genera of entomopathogenic fungi that belong to the order Hypocreales of the phylum Ascomycota, and have a broad range of hosts, though considerable specificity occurs among individual isolates. *M. anisopliae* and *B. bassiana* species are important entomopathogenic fungi extensively evaluated as biological control agents due to their unique property of infecting and killing larval stages of insects (Erler and Ates, 2015). Both the fungi *M. anisopliae* and *B. bassiana* have been reported to occur naturally in more than 700 species of host. The documentation on the prevalence of the entomopathogenic fungi in the insect has usually been limited to species due to nontarget species such as some of the parasitoids. However, overall major reported insect taxa are found to be natural hosts for *M. anisopliae* and *B. bassiana* in temperate regions (Ownley et al., 2010).

### 4.2.2 Pathogenicity of Entomopathogenic Fungi

The infection of fungi to the host is considered as a part of the fungal life cycle where it undergoes multiplication to a significant amount of population by synthesizing large numbers of conidia or spores. The general mode of infection and transmission of entomopathogenic fungi is a series of processes that starts with the production of a large amount of spores or conidia, which are the major reason for the infection and wide distribution in the environment. The overspread of these conidia facilitates the chance to attack hosts that are present in the environment. When candida land on the cuticle of a suitable host, they initiate the germination of the conidia and penetrate into the cuticle using enzymatic and physical activities of the fungus. Invasion of fungi into the host body and the circulatory system will occur once the entomopathogenic fungi transfer through the cuticle of the external host skeleton. The entomopathogenic fungi resemble the entry of insect tissues, similar to plant pathogens, that includes formation of germ tubes, appresoria, and penetration pegs. Some of the entomopathogenic fungi are involved in the production of toxins to kill the insects. After the death of the host, the entomopathogenic fungi appear from the dead host and sporulation occurs on the outside of the cadaver. Sporulation even occurs internally when ambient humidity precludes external sporulation (Shahid et al., 2012). Moreover, the reproduction of entomopathogenic fungi on the cuticles of dead insects produces propagules, which disperse to infect other insects under certain environmental conditions. The primary steps involved in the infections of entomopathogenic fungi such as adhesion and germination are most often common in a wide range of host and fungal species, whereas toxin production varies considerably among host and fungal isolates. Among the toxins from entomopathogenic fungi, the toxin destruxins, produced as the secondary metabolite of entomopathogenic fungi, are acting as the best insecticidal compound produced within the hemocoel of infected hosts. Nevertheless, destruxins also possess multiple bioactive characteristics such as insecticidal, herbicidal, and antiviral activities (Fernandes et al., 2012).

#### 4.2.2.1 Entomopathogenicity and Molecular Mechanisms in *Beauveria bassiana* and *Metarhizium anisopliae*

Currently, the fungi *B. bassiana* and *M. anisopliae* are widely used as biological control agents for vectors like mosquitos. The ascomycete fungi *B. bassiana* and *M. anisopliae* possess the ability to exploit a range of environments, including soil plant, and insects. Interestingly, the fungi can live as saprophytes in the soil; in plants they behave as endophytes or as entomopathogens, which can affect a wide range of hosts (Boomsma et al., 2014). Understanding the life cycle of the biological agent and the mechanisms and mode of action in the insect vector is necessary to develop effective entomopathogenic formulations. Advancements in genomic technology facilitated quick and easier methods to study the life cycle, mechanisms, and mode of action of entomopathogenic fungi, which enable the establishment of the relation between host and pathogen. The infection cycle of *B. bassiana* and *M. anisopliae* occurs in seven steps (Figure 4.1):

1. Adhesion of candida to the host cuticle
2. Development of a germination tube and in certain cases an appressorium
3. The mechanical pressure and enzymatic degradation of cuticles
4. Stretching of hyphae to invade into the insect hemocoel from the cuticle, which can enable single-celled blastospores
5. The freely circulating cell in the hemocoel consuming the nutrient available in the hemolymph and secreting the toxins that can eventually kill the insect
6. It is necessary to overcome the immune response of the host by the fungi during the growth and especially during circulating around hemocoel
7. Breaching of the cuticle to allow inside conidia to escape from the insect body and to enable conidia to spread and initiate infection to other individuals (Valero-Jiménez et al., 2016)

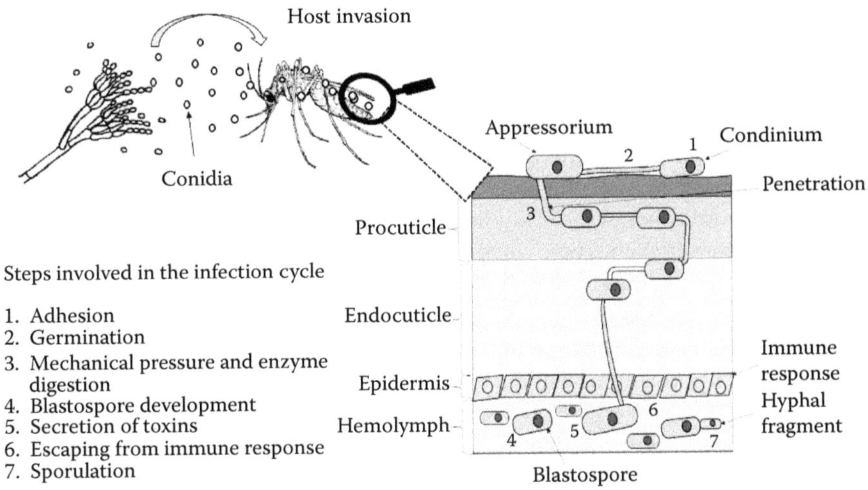


**FIGURE 4.1** Infection cycle of entomopathogenic fungi in mosquito vectors.

Pathogenicity of entomopathogenic fungi, which leads to overconsumption of the nutrient of the host, is the root cause for the death of the host. Nevertheless, they also produce secondary metabolites, derivatives from various intermediates in primary metabolites, which possess insecticidal activity. The fungal species producing these kinds of toxins has showed greater infection and ultimately resulting in the rapid killing of the host as compared to the fungal species that does not produce any secondary metabolites (Boomsma et al., 2014).

### 4.2.3 Interaction with the Host Immune System

In general, the immune system of the host is activated once the conidia attach to the cuticles of the host. Once the immune system of the host is activated, it participates in both the cellular and humoral defense to combat microbial infection. The host readily synthesizes the high phenolic compound and cuticular oxidases by the action of the immune system, which are involved in the defense mechanism by melanization around a penetration tube. When the fungus manages to escape from the host defense and successfully reaches the host hemocoel, it encounters hemocytes engaged in encapsulation, nodule formation, and phagocytic activities. The activation of the innate immunity of the host occurs via detection of the entomopathogenic fungus by changes in the properties of the cuticle basement membrane and biological substance associated with the fungal cell wall. Even though mosquitoes have a sophisticated defense system, entomogenous fungi manage to overcome the barriers of the host by continuing to grow even after having been phagocytized, and interestingly the propagating ability of granulocytes helps to overcome nodule formation. The entomopathogenic fungus also follows other special kinds of escape strategies from the host immune system by producing hyphal bodies in absence of a well-defined cell wall, and with its potential elicitors of detection, allows hyphal bodies to evade detection and phagocytosis even in immunocompetent hemocytes (Castrillo et al., 2005).

### 4.2.4 Genes Involved in the Pathogenicity of Entomopathogenic Fungi

In the infection cycle of the entomopathogenic fungi, several genes are involved in each step of the cycle (Table 4.1). Certain genes take part in a single-step virulence cycle process, whereas most of the genes express one or more steps of the pathogenic cycle of the entomopathogenic fungi. The genes that are involved in the infection cycle of entomopathogenic fungi *B. bassiana* and *M. anisopliae* are discussed. Successful adherence of conidia to the host is the primary step in the infection cycle of the entomopathogenic fungi. In earlier days, it was believed that the adherence process occurred through electrostatic interaction between conidia which possess net negative charge to the positively charged host surface. Studies have been conducted to evaluate the change in the adherence potential of conidia with the host by neutralizing the charge of conidia. The neutralization of the charge of the conidia does create much impact on the binding of the conidia to the insect cuticle (Boucias et al., 1988). After several developments in science and technology, it has been accepted that conidia adhere to an insect cuticle by hydrophobic surface interaction. Most often the entomopathogenic fungus *B. bassiana* employs this mode of interaction for the host adherence process. The two-hydrophobin genes such as *hyd1* and *hyd2* of *B. bassiana* express the encoding proteins that covalently aggregate on the surface of conidia (Zhang et al., 2011). In the entomopathogenic fungi *M. anisopliae*, two adhesion-like proteins were identified that mediate their adhesion through electrostatic interaction and hydrophobic surface interaction. Moreover, presence of these proteins on the surface of the conidial surface were found to be required for adhesion to the insect cuticle and involved in the processes of germination, blastospore formation, and virulence (Wang and St. Leger, 2007).

After successful adherence of conidia to the host, they germinate and enter into the host body when they encounter a favorable condition like temperature, exogenous carbon source, and humidity. The genes responsible for this biological process are still not well established, but in some strains such as *B. bassiana* Bb0062, the MAP kinase gene *Bbmpk1* and *Bbhog1* are involved in the formation of appressorium, which perforates the cuticle to invade the host (Zhang et al., 2010). In contrast, in the *Metarhizium* species the genes involved in the formation of appressoria are well studied. The genes *mpl*, *Mapls1*, *Mrskn7*, and *Magas1* are involved in the cuticle penetration process, which is second step of the infection cycle between the *Metarhizium* species and insect (Luo et al., 2013; Shang et al., 2015). Once the germ tube starts its development on the cuticles, they experience several environmental stress conditions such as high temperature and ultraviolet (UV) radiation as well as host immune responses. Hence, fungi are involved in the acceleration of penetration process in order to escape from these environmental factors and establish the successful infection of the host. The process of penetration occurs by two different modes of action: mechanical pressure and enzymatic degradation of the cuticles. Entomopathogenic fungi produce several enzymes, which facilitate pathogenicity in the host. The mosquito cuticles mainly consist of chitin alone with certain supporting proteins and lipids. As mentioned earlier, entomopathogenic fungi produce of several enzymes, particularly chitinases and proteases, to weaken the structural barrier and enable host invasion (Charnley and St. Leger, 1991). The cytochrome P450 gene

## TABLE 4.1
## Genes Involved in the Infection Cycle of Entomopathogenic Fungi *B. bassiana and M. anisopliae*

| Organism | Gene | Putative Function | Phenotype | Reference |
|---|---|---|---|---|
| *B. bassiana* | *hyd1* and *hyd2* | Hydrophobin | Reduced host adherence and virulence | Zhang et al., 2011 |
| | *Bbhog1* and *Bbmpk1* | MAP Kinase | No appressoria, reduced virulence and stress tolerance | Zhang et al., 2009, 2010 |
| | *Bbsnf1* | Protein kinase | Reduced virulence, acidification, and blastospore formation | Wang et al., 2014 |
| | *Chi1*, *Chi2*, *ChsA2* | Chitinase | Increased infection | Fang et al., 2005 |
| | *Wee 1* | Regulates Cdk1 activity | Altered cell morphology, reduced virulence and stress-tolerance | Qiu et al., 2015 |
| | *Bbbeas* | Beauvericin toxin | Reduced virulence and beauvericin production | Xu et al., 2008 |
| | *Ktr1*, *Ktr2*, *Ktr4* | α-1,2-Mannosyltransferase | Reduced conidiation and stress-tolerance | Wang et al., 2014 |
| | *ras1*, *ras2*, *ras3* | GTPase, signal transduction | Reduced virulence, germination, conidiation, stress tolerance | Xie et al., 2013; Guan et al., 2015 |
| | *Cdc14* | Cytokinesis | Reduced blastospore production, conidial yield, and stress tolerance | Wang et al., 2013a |
| | *catA-D*, *catP* | Catalase | Reduced virulence, stress-tolerance | Wang et al., 2013c |
| | *Bbpmr1* | ATPase, signaling | Reduced conidiation | Wang et al., 2013b |
| *M. anisopliae* | *Mest1* | Perilipin | Germination as well as appressorial differentiation | Wang et al., 2011 |
| | *dtxS1* | Destruxins | Reduced pathogenicity | Wang et al., 2012 |
| | *sidD* and *sidC* | Siderophores | Reduced virulence potential | Giuliano et al., 2015 |
| | *sg-B* and *sg-C* | Chitinases | Reduced virulent and aggressive functions | Baratto et al., 2006 |
| | *CBM1*, *CBM18*, *CBM50* | Chitin binding | Reduced host adaptation | Guillen et al., 2010 |
| | *MAT1-1* and *MAT1-2* | Homothallism | Less proliferation | Lee et al., 2010 |

*Bbcyp52x1* facilitates the process of enzymatic breakdown of the lipid layer present in the cuticles, which promotes the virulence nature of the entomopathogenic fungi (Zhang et al., 2012). As discussed earlier, proteins are the major component present in the cuticle of the host and break down of the protein serves as a major nutrient source for germination and hyphae development on the cuticle. Hence, extracellular synthesis of the protease enzymes by entomopathogenic fungi plays a vital role in the cuticle breakdown. In *B. bassiana*, subtilisin-like (*Pr1*), trypsin-like (*Pr2*), and *cdep1* are the genes that mediate the synthesis of cuticle-degrading proteases (Fang et al., 2008). Recent research has emphasized that the protein kinase gene *Bbsnf1* is also involved in the regulation of the cuticle penetration process; however, the exact mechanism of the gene *Bbsnf1* is currently not known (Wang et al., 2014).

Successful reaching of hyphae into the hemocoel from cuticles develops yeast-like growth. The entomopathogenic fungi forms single-cell hyphal bodies called blastopores. The formation of blastopores helps entomopathogenic fungi to escape from detection by the host immune system. They also proliferate by the budding process by employing hemolymph as a vehicle for the colonization of the host and also use a nutrient source (Lewis et al., 2009). The understating of genes responsible for the development of blastopores remains unknown, but certain genes show regulation of blastopore development. For example, the genes *wee1*, *cdc25*, *Fhk2*, and *Bbsnf1* mediate the development of the blastopores (Wang et al., 2015). Once the blastopore development process is complete, the entomopathogenic fungi switch over to the next step of the pathogenic cycle: colonization and killing of the host. Herein, the processes of colonization of the fungi and death to the host occur simultaneously. During the time of infection all nutrients for the development of the fungi like disaccharide trehalose, which possess abundant carbohydrates, present in the insect hemolymph will be utilized. The entomopathogenic fungi produce trehalase to hydrolyze the trehalose present in the hemolymph of the host that result in the increase availability of glucose for the consumption for entomopathogenic fungi. The α-glucoside transporter gene *Bbagt1*, the mannitol dehydrogenase genes *Bbmpd* and *Bbmtd*, and the *BbCrwA* regulator gene are involved the colonization and the nutrient-depletion process in the host (Luo et al., 2014).

In the infection process, entomopathogenic fungi experience different pH conditions depending on the environment inside the host. It has been well understood that the synthesizing enzymes have certain optimal pH at which they can exhibit higher activity. Hence, the expression of genes by the entomopathogenic fungi needs to adapt to the pH inside of host body. The pH-dependent regulation of expression is mediated by the response element in the genome. The transcription factors like *BbpacC* enable its expression according to ambient pH. Whereas at a higher pH, it undergoes upregulation of alkaline-expressed genes and repression of acid-expressed genes (Zhou et al., 2014). Nevertheless, entomopathogenic fungi also secrete toxic metabolites to speed the death of the host. The expression of the gene *Bbbeas* leads to the synthesis of beauvericin synthase, which is found to be an important virulence factor in entomopathogenic fungi infection cycle. Certain genes like *Ras1*, *Ras2*, and *Ras3* mediate virulence by affecting multistress tolerance, germination, and growth (Xie et al., 2013).

### 4.2.5 Factors That Affect the Virulence of Entomopathogenic Fungi

The effect of entomopathogenic fungi on vector control activity has directly influenced several factors, such as availability of water, ions, fatty acids, nutrients on the cuticle surface, the physiological state of the host, and environmental factors (e.g., temperature, humidity, UV exposure, and pH). These factors are considered major hurdles that affect the successful infection of entomopathogenic fungi to mosquitos. Most of these environmental factors affect the viability of the fungus, as well as persistence and virulence (Jaronski, 2010). Nevertheless, the virulence of entomopathogenic fungi is directly influenced by the genetic makeup of the specific isolates. In the case of humidity, optimum humidity is required for the infection by mycopathogens, whereas at a higher humidity the environment is made critical for sporulation, germination, and invasion of the host. In the case of temperature, the optimum temperature for the growth of the fungus is not necessarily the same for development of the disease. Generally, optimum temperature for the effective infection to the host is between 20°C and 30°C. Change in the temperature may distinctly affect the development of the fungal disease but not necessarily be involved in the mortality of the fungi. Moreover, if the temperature affects germination of the conidia, this directly influences the concentration of infective units. For example, *B. bassiana* isolates showed lower infection when acridids actively increased their body temperature by habitat selection (Inglis et al., 1997). Recently, genetic manipulation has been implemented to enhance the ability to kill insects and tolerate adverse environmental conditions. These genetically modified entomopathogenic fungi are effectively employed as potential vector control agents. Genetic modification is an effective method to enhance the killing efficacy of the isolate and tolerance efficiency from the environmental conditions. In this regard, several studies has been conducted to improve spore delivery to mosquitos, including UV protectants and humidity stabilizers, and using different growth substrates that increase sporulation, virulence, and stress resistance. For example, the enhancement of the tolerance level of the temperature has been achieved in *Metarhizium anisopliae* by directed evolution using continuous culture (de Crecy et al., 2009).

## 4.3 RECENT TRENDS IN CONTROL OF MOSQUITO VECTORS BY ENTOMOPATHOGENIC FUNGI

Mosquito vector-borne diseases such as malaria, dengue, and chikungunya are the most widespread and dreadful diseases, with an enormous socioeconomic impact and severe mortality rate (Ramirez et al., 2009). To circumvent the problems associated with synthetic mosquito control agents, more effective biocontrol agents with high specificity toward the control of particular mosquito species is needed. In this context, exploitation of species-specific entomopathogenic fungi in mosquito control programs provides a promising and effective alternative. The development and exploitation of entomopathogenic fungi as classical and augmentative alternatives toward vector control has a rich history of several decades with success and limitations.

In this regard, high throughput and integrated management strategies using the entomopathogenic fungi as weapons to control vector-borne diseases is of paramount importance (Lacey et al., 2015). Entomopathogenic fungi are widely described as natural biocontrol agents with immense agricultural, horticultural, and, especially, medical importance.

In vector transmission control programs, the application of microbes, especially the exploitation of entomopathogenic fungi as biocontrol agents, draws considerable attention. As a result of which, highly intensive and focused research is being carried out using a variety of entomopathogenic fungi toward controlling transmission of mosquito vectors and particularly toward longevity and fecundity of both larvae and adult mosquitoes (Abdul-Ghani et al., 2011). Among the variety of entomopathogenic fungal species used in mosquito vector control, *M. anisopliae* and *B. bassiana* are the most frequently exploited as the most prevalent and potential vector control agents.

### 4.3.1 Control of Mosquito Vectors by *Metarhizium anisopliae*

Among the various mosquito species transmitting different vector-borne diseases, *Aedes albopictus*, commonly known as the Asian tiger mosquito, holds a prominent position owing to its potential to transmit several viral diseases including encephalitis like that of its other relative *Aedes* sp. Being an epidemiologically important vector, its transmission control is of considerable importance. In 2007, Scholte et al. evaluated the infection efficacy of the entomopathogenic fungus *M. anisopliae* against *A. aegypti* and *A. albopictus* and suggested its application as a promising alternative to traditional biocontrol agents in vector control programs. In this context, an attempt was initiated by Lee et al. (2015) to control the Asian tiger mosquitoes by 12 soil-borne entomopathogenic fungi belonging to 6 different genera of fungi. Among the fungal isolates, *M. anisopliae* JEF-003 exhibited the most significant larvicidal activity with more than 90% of mortality, suggesting its application in the management of controlling the dreadful disease caused by Asian tiger mosquitoes (Lee et al., 2015). In 2012, Leles et al. suggested the effectiveness of *M. anisopliae* IP46 against the adults, larvae, and even the eggs of *A. aegypti*, which eventually opens a new window for the application of these entomopathogenic fungi directly on the breeding sites of *A. aegypti* and thus its successful exploitation as an effective biocontrol agent. The prospective pathogenicity profile of *M. anisopliae* against aquatic larvae of *A. aegypti* can be attributed to the upregulation of proteinases *Pr1* and *Pr2*, adhesions *Mad1* and *Mad2*, and chitinase *chit1* in the presence of the mosquito larvae (Khan et al., 2012; Butt et al., 2013). In integrated mosquito vector management programs, the potential of the entomopathogenic fungus *M. anisopliae* against the dengue vector *A. aegypti* under field conditions was established by Carolino et al. (2014).

African countries are one of the critical centers for the transmission of mosquito-borne diseases and the subsequent ailments. Among the vector-borne diseases, malaria is the most widespread and vulnerable disease causing death to thousands of people annually. Different species of *Anopheles* mosquitoes are the agents of malarial infection of which *Anopheles gambiae* is the most prevalent African malaria vector.

Factors like longevity and the blood-feeding capacity of *A. gambiae* enhance its ability to transmit *Plasmodium* parasites. Though an array of control strategies are being applied to control transmission of *A. gambiae*, the exploitation of the entomopathogenic fungus *M. anisopliae* is the most effective strategy of biocontrol owing to its impact on longevity, feeding, and fecundity of mosquito vectors in a species-specific manner (Scholte et al., 2006). In Tanzania, the most predominant malaria vector causing malaria is *A. arabiensis* with similar mortality rates like that of its related *Anopheles* sp. Lwetoijera et al. (2010) conducted a field study in Tanzania with an aim to control the transmission of adult *A. arabiensis* mosquito vectors using the highly efficient entomopathogenic fungus *M. anisopliae* IP46. The results depicted the effective mortality rate of wild type *A. arabiensis* by the application of *M. anisopliae* IP46 (Lwetoijera et al., 2010). In tropical and subtropical countries, malaria is the most dreadful and life-threatening vector-borne disease with the cause of highest mortality rate annually to its name. In India and certain West Asian countries, *Anopheles stephensi* L. is the prime vector responsible for causing malaria and subsequent lethal effects. To circumvent the problems associated with *A. stephensi* L., an improved and integrative approach is required, and in this context Murugan et al. (2012) made an attempt. They established the significant larvicidal and pupicidal properties of *M. anisopliae* against the malaria fever mosquito *A. stephensi*, and it is thus an ideal ecofriendly approach in vector control programs (Murugan et al., 2012).

*Culex pipiens* (the common house mosquito) is one of the important members of the blood-feeding mosquito family Culicidae and is the vector of several life-threatening diseases such as Japanese encephalitis and meningitis. In 2014, Benserradj and Mihoubi evaluated the prospective larvicidal activity of *M. anisopliae* against *C. pipiens* larvae. The results suggested the efficacy of *M. anisopliae* as a potential biocontrol agent against *Culex* mosquito vector in laboratory conditions (Benserradj and Mihoubi, 2014).

### 4.3.2 Control of Mosquito Vectors by *Beauveria bassiana*

Large-scale resistance to available synthetic insecticides and the diversity shown by mosquito vectors propelled the scientific community to search for promising strategies for biological control of the adult and larval stages of mosquito vectors. In this regard, the use of entomopathogenic fungi as biocontrol agents has drawn considerable attention because of the complex mode of action of the fungi as a result of which the resistance shown by the vectors become lowered (Vogels et al., 2014). *B. bassiana* is the first species to be exploited as a control strategy of both larval and adult stages of different mosquito species.

The high rate of morbidity and mortality in humans is generally caused by severe epidemiological diseases such as malaria, chikungunya, dengue, yellow fever, and arboviral encephalitis. The main credit for transmitting such life-threatening diseases goes to different species of mosquito vectors. *Culex quinquefasciatus* is a wild-type mosquito vector causing severe health consequences by transmitting parasitic viruses to the human host and thereby promotes the transmission of malaria and related vector-borne diseases. Howard et al. (2010) showed the promising attributes of the entomopathogenic fungus *B. bassiana* in controlling the transmission of *C. quinquefasciatus* by increased mortality and depleting the blood-feeding capability of the vector.

Owing to the resistance shown by these mosquito vectors to traditional mosquitocides, use of microbial mosquitocides is an alternative strategy. In this regard, Ragavendran et al. (2017) evaluated the larvicidal and pupicidal effect of the entomopathogenic fungus *B. bassiana* against *A. stephensi*, *C. quinquefasciatus*, and *A. aegypti*. The results suggested the exploitation of *B. bassiana* as a target-specific biocontrol agent against a range of host mosquito vectors (Ragavendran et al., 2017). During the last few years, dengue has become a global health concern worldwide, with special reference to India, with a high rate of infection and mortality. In addition to that, the resistance shown by the dengue vector *A. aegypti* against several biocides is of critical interest. In this regard, the eminent surveillance of *B. bassiana* can be exploited as a promising alternative to control *A. aegypti* and ultimately control the transmission of dengue (Darbro et al., 2011). The efficacy of *B. bassiana* as a potent biocontrol agent against the dengue vector *A. aegypti* can be attributed to its implication in mortality, deterioration of blood-feeding behavior, and controlling the fecundity of *A. aegypti* under laboratory as well as semifield conditions (Darbro et al., 2012). Transient reverse genetic analyses of entomopathogenic fungal infection to the dengue vector *A. aegypti* showed a marked increase in the expression of a variety of antimicrobial chemicals and dengue virus restriction factor genes, which is responsible for controlling the transmission of *A. aegypti*. Dong et al. (2012) suggested the mechanism of vector control by *B. bassiana* infection is mediated by the amelioration of the mosquito's signaling defense systems such as Toll, immune deficiency (IMD), and Janus kinase-signal transducer and activator of transcription (JAK-STAT) signaling pathways (Dong et al., 2012).

*Anopheles gambiae* is the most prevalent malaria-causing mosquito vector in African countries. Trypsin-modulating oostatic factor (TMOF), a decapeptide hormone synthesized by the ovaries and the neuroendocrine system, regulates the digestion process in adult and larval mosquitoes. TMOF is currently being developed and exploited as an alternative mosquitocide. Entomopathogenic fungi offer a means for targeting mosquitoes with TMOF. Kamareddine et al. (2013) made an attempt to express TMOF in the entomopathogenic fungus *B. bassiana* and observed its enhanced efficacy against *A. gambiae* owing to its increased virulence properties and reducing the fecundity. Farenhorst et al. (2010) investigated the enhanced efficacy of traditional mosquitocides, such as permethrin, in combination with the entomopathogenic fungi *B. bassiana* and *M. anisopliae*. The results suggested the pragmatic approach of infecting the mosquitoes with entomopathogenic fungi yielding in additional promising tool to control permethrin-resistant *A. gambiae* in West African countries (Farenhorst et al., 2010).

In Western Hemisphere countries, especially Central America, Mexico, Venezuela, and Colombia, the main vector for transmitting malaria is *Anopheles albimanus*. Though novel and advanced strategies have been developed to control mosquito vectors, breeding suppression of *A. albimanus* is still underdefined. In this regard, a successful attempt was made to exploit the environment-friendly and target-specific entomopathogenic fungi *B. bassiana* and *M. anisopliae* toward the effective control of larval as well as adult form of *A. albimanus* (Vazquez-Martinez et al., 2013). The larvicidal properties of *B. bassiana* was successfully employed against the larval form of *A. stephensi* in the laboratory conditions, and from the results it has been suggested that development of suitable formulation of *B. bassiana* could be used as a promising alternative to mosquito vector control programs (Veys-Behbahani et al., 2014).

### 4.3.3 Control of Mosquito Vectors by Other Entomopathogenic Fungi

Apart from *M. anisopliae* and *B. bassiana*, several other species of entomopathogenic fungi are employed for the control of mosquito vectors (Table 4.2).

## 4.4 TECHNOLOGICAL EXPANSION AND GENETIC ENGINEERING STRATEGIES TO ENHANCE MOSQUITO VECTOR CONTROL PROGRAMS

### 4.4.1 Novel and Integrative Formulation in Entomopathogenic Fungi: A New Pragmatic Approach

Nanotechnology provides a multidimensional avenue to revolutionize the world of science, be it in the agricultural sector, the food industry, or therapeutics. The tool of nanotechnology is now targeted toward integrated management of vector-borne diseases. In an attempt to enhance efficacy of entomopathogenic fungi in mosquito vector control, a green synthesis approach for synthesis of silver nanoparticles (AgNPs) using *M. anisopliae* was followed to target the rural malaria vector *Anopheles culicifacies*. The mycosynthesized AgNPs exhibited an ecofriendly and significant alternative to control *A. culicifacies* by virtue of its larvicidal and pupicidal properties (Amerasan et al., 2016). During the last few decades, vector control programs have built a dimension in their arsenal that aims to develop some novel formulations to facilitate the spreading and infection efficacy of entomopathogenic fungi against the larvae of the malaria-causing mosquito vector *Anopheles* sp. In this context, Bukhari et al. (2011) developed certain formulations such as aqueous and oil formulations or dry formulations. The results suggested the synthetic oil formulation was highly effective against the mosquito larvae and it also significantly increases the persistence of fungal spores during the infection process (Bukhari et al., 2011). Neem oil exhibited severe toxicity to mosquito vectors such as *A. stephensi*, *C. quinquefasciatus*, and *A. aegypti* larvae. In addition to this, the entomopathogenic fungus *M. anisopliae* is well known for its vector control efficacy. To exploit the toxicity profile of both neem oil and entomopathogenic fungus, a novel formulation was recently developed and tested against the mosquito vector. The results suggested the presence of neem oil enhances the efficacy of *M. anisopliae* in controlling the larvae of the dengue vector *A. aegypti* (Gomes et al., 2015). The increased efficiency shown by different formulations in entomopathogenic fungi is basically due to the increased mortality rate and decreased fecundity of the majority of mosquito vectors such as species of *Aedes*, *Anopheles*, and *Culex* under laboratory conditions (Greenfield et al., 2015).

## TABLE 4.2
## Control of Mosquito Vectors by Other Entomopathogenic Fungi

| Entomopathogenic Organism | Phylum | Active against Adult/ Larvae of Vector | Proposed Action/Effect on Host Mosquito Vectors | Target Vector Species | Reference |
|---|---|---|---|---|---|
| *Leptolegnia chapmanii* | Oomycota | Larvae | Fecundity, number of gonotrophic cycles, fertility | *Aedes aegypti* | Pelizza et al., 2012 |
| *Leptolegnia chapmanii* | Oomycota | Larvae | Larvicidal | *Culex quinquefasciatus* | Choi and Jung, 2015 |
| *Leptolegnia chapmanii* | Oomycota | Larvae | Larvicidal | *Aedes aegypti* | Gutierrez et al., 2017 |
| *Lagenidium giganteum* | Oomycota | Larvae | Larvicidal | *Aedes aegypti* | Maldonado-Blanco et al., 2011 |
| *Tolypocladium cylindrosporum* | Ascomycota | Eggs, larvae, adults | Ovicidal, larvicidal, mortality | *Aedes aegypti* | Rocha et al., 2015 |
| *Aspergillus clavatus* | Ascomycota | Larvae | Mortality | *Culex quinquefasciatus* | Bawin et al., 2016 |
| *Chrysosporium lobatum* | Ascomycota | Larvae | Mortality | *Anopheles stephensi*, *Culex quinquefasciatus* | Mohanty and Prakash, 2009 |
| *Lecanicillium* sp. | Ascomycota | Adults | Mortality | *Aedes* spp. | Ishii et al., 2015 |
| *Culicinomyces clavosporus* | Deuteromyces | Larvae | Mortality | *Anopheles quadrimaculatus*, *A. stephensi*, *Culex quinquefasciatus* | Scholte et al., 2004; Choi and Jung, 2015 |
| *Culicinomyces clavosporus* | Deuteromyces | Adults | Adulticide | *Culex quinquefasciatus*, *Anopheles stephensi*, *Aedes aegypti* | Singh and Prakash, 2012 |
| *Culicinomyces* spp. | Deuteromyces | Larvae | Larvicidal | *Aedes aegypti* | Rodrigues et al., 2017 |

Imidacloprid is a well-known insecticide targeted towards controlling mosquito vectors. However, the synthetic nature and possible side effects shown by the insecticide makes it improper in vector control programs. In the quest for a novel formulation and high efficacy of imidacloprid, a successful attempt was made to combine the potent entomopathogenic fungus *M. anisopliae* with the insecticide imidacloprid to enhance the efficacy of vector control against the dengue vector *A. aegypti* (Paula et al., 2011).

### 4.4.2 Genetic Engineering Strategies

The infection cycle of any entomopathogenic fungi starts with the adherence followed by penetration through the insect cuticle, which is an interlinked network of proteins and chitin. The invasion mechanism of entomopathogenic fungi through the host cuticle is based upon the secretion of a diverse range of hydrolases such as proteases and chitinases, which eventually enhance the penetration efficacy of the entomopathogen (Fan et al., 2010). In the era of technological advancement and engineered prospective to enhance the efficacy of target-specific invasion and penetration rate, an attempt was initiated to formulate a hybrid protease engineered by the fusion of the chitin-binding domain of the host insect and subtilisin-like protease from *B. bassiana*. As per the hypothesis, the hybrid protease bearing *B. bassiana* showed an enhancement in target specificity in binding chitin followed by release of a greater amount of peptides. The hybrid protease also increased the insecticidal property of *B. bassiana* as compared to the wild-type *B. bassiana*. In order to combat malaria-causing *Plasmodium* sp. in mosquito vectors by entomopathogenic fungi, a transgenic strategy was followed to develop recombinant strains of the entomopathogenic fungus *M. anisopliae* specifically expressing molecules targeting the sporozoites. The results suggested the developed strategy might act as a powerful weapon to combat malaria (Fang et al., 2011).

The development of an engineered transgenic fungus *M. anisopliae* focuses upon the gene manipulation of a specific gene expression profile, which eventually controls the transmission of *Plasmodium falciparum* in *A. gambiae*. The prime advantage shown by the genetically engineered entomopathogen is the slow and sustained killing of target mosquito vectors (Chong et al., 2013).

## 4.5 FUTURE AVENUES AND CHALLENGES OF ENTOMOPATHOGENIC FUNGI-BASED VECTOR CONTROL MEASURES

The exploration of efficient entomopathogenic fungi for controlling mosquito vectors has been investigated for several decades. Recently, biological controlling of mosquito-borne diseases has been employed to reduce the prolonged use of chemical insecticides, which are currently used as mosquito-controlling agents. Biological strategies to eradicate mosquito vectors controlling have gained significant attention compared to that of their counterpart chemical-based strategies due to several advantages such as eco-friendly nature, safety, cost effectivity, sustainability, and can target a range of different mosquito vectors. Nevertheless, a potential entomopathogenic fungi should also possess series of properties, including (a) the ability to kill both larval and adult stages of the mosquito vectors, (b) the ability to reduce the number of applications per season, (c) active dispersion of adult female to predominantly unoccupied breeding sites, (d) enhanced activity and persistence in the mosquito population, (e) selectively kills the mosquitoes and no other organisms, (f) significant activity in environmental factors such as salinity, temperature, and relative humidity, (g) fewer formulation steps and cost-effective mass production, (h) prolonged shelf-life period, and (i) should not cause any harmful effect to human and nontargeted organisms. To date, several potential entomopathogenic fungi were reported but very few of them are fulfilling all the mentioned properties to be an effective mosquito vector controlling agents. Hence, to facilitate entomopathogenic fungi, genetic modification has been employed that helps entomopathogenic fungi to fulfill the above mentioned properties.

Entomopathogenic fungi are a vast group of fungi present in the natural environment among which less were discovered and limited methods were developed to control mosquito vectors. Propagation of entomopathogenic fungi in the environment is a bottleneck in the development of an effective formulation. Genetically engineered entomopathogenic fungi are the potential alternative to overcome the limitations in vector control programs. Recent advancement in science and technology in the field of molecular biology can achieve the desired target in the improvement of vector control programs using entomopathogenic fungi.

## REFERENCES

Abdul-Ghani, R., Al-Mekhlafi, A. M., and Alabsi, M. S. 2011. Microbial control of malaria: Biological warfare against the parasite and its vectors. *Acta Trop.*, 121, 71–84.

Achee, N. L., Gould, F., Perkins, T. A., Reiner Jr., R. C., Morrison, A. C., Ritchie, S. A., Gubler, D. J., Teyssou, R., and Scott, T. W. 2015. A critical assessment of vector control for dengue prevention. *PLoS Negl. Trop. Dis.*, 9(5), e0003655.

Amerasan, D., Nataraj, T., Murugan, K., Panneerselvam, C., Madhiyazhagan, P., Nicoletti, M., and Benelli, G. 2016. Myco-synthesis of silver nanoparticles using *Metarhizium anisopliae* against the rural malaria vector *Anopheles culicifacies* Giles (Diptera: Culicidae). *J. Pest. Sci.*, 89, 249–256.

Baratto, C. M., Dutra, V., Boldo, J. T., Leiria, L. B., Vainstein, M. H., and Schrank, A. 2006. Isolation, characterization, and transcriptional analysis of the chitinase *chi2* Gene (DQ011663) from the biocontrol fungus *Metarhizium anisopliae var. anisopliae*. *Curr. Microbiol.*, 53(3), 217–221.

Bawin, T., Seye, F., Boukraa, S., Zimmer, J., Raharimalala, F. N., Ndiaye, M., Compere, P., Delvigne, F., and Francis, F. 2016. Histopathological effects of *Aspergillus clavatus* (Ascomycota: Trichocomaceae) on larvae of the southern house mosquito, *Culex quinquefasciatus* (Diptera: Culicidae). *Fungal Biol.*, 120, 489–499.

Benelli, G., and Mehlhorn, H. 2016. Declining malaria, rising of dengue and zika virus: Insights for mosquito vector control. *Parasitol. Res.*, 115, 1747–1754.

Benelli, G., Jeffries, C. L., and Walker, T. 2016. Biological control of mosquito vectors: Past, present, and future. *Insects*, 7, 52.

Benserradj, O., and Mihoubi, I. 2014. Larvicidal activity of entomopathogenic fungi *Metarhizium anisopliae* against mosquito larvae in Algeria. *Int. J. Curr. Microbiol. Appl. Sci.*, 3(1), 54–62.

Bhan, S., Shrankhla., Mohan, L., and Srivastava, C. N. 2013. Larvicidal toxicity of Temephos and entomopathogenic fungus, *Aspergillus flavus* and their synergistic activity against malaria vector, *Anopheles stephensi*. *J. Entomol. Zool. Stud.*, 1(6), 55–60.

Boomsma, J. J., Jensen, A. B., Meyling, N. V., and Eilenberg, J. 2014. evolutionary interaction networks of insect pathogenic fungi. *Annu. Rev. Entomol.*, 59, 467–485.

Boucias, D. G., Pendland, J. C., and Latge, J. P. 1988. Nonspecific factors involved in attachment of entomopathogenic deuteromycetes to host insect cuticle. *Appl. Environ. Microbiol.* 54(7), 1795–805.

Bourtzis, K., Dobson, S. L., Xi, Z., Rasgon, J. L., Calvitti, M., Moreira, L. A., Bossin, H. C., Moretti, R., Baton, L. A., Hughes, G. L., Mavingui, P., and Gilles, J. R. L. 2014. Harnessing mosquito–Wolbachia symbiosis for vector and disease control. *Acta Trop.*, 132S, S150–S163.

Bukhari, T., Takken, W., and Koenraadt, C. J. M. 2011. Development of *Metarhizium anisopliae* and *Beauveria bassiana* formulations for control of malaria mosquito larvae. *Parasite. Vector.*, 4, 23.

Butt, T. M., Greenfield, B. P. J., Greig, C., Maffeis, T. G. G., Taylor, J. W. D., Piasecka, J., Dudley, Ed., Abdulla, A., Dubovskiy, I. M., Garrido-Jurado, I., Quesada-Moraga, E., Penny, M. W., and Eastwood, D. C. 2013. *Metarhizium anisopliae* pathogenesis of mosquito larvae: A verdict of accidental death. *PLoS One*, 8(12), e81686.

Carolino, A. T., Paula, A. R., Silva, C. P., Butt, T. M., and Samuels, R. I. 2014. Monitoring persistence of the entomopathogenic fungus *Metarhizium anisopliae* under simulated field conditions with the aim of controlling adult *Aedes aegypti* (Diptera: Culicidae). *Parasite. Vector.*, 7, 198.

Castrillo, L. A., Roberts, D. W., and Vandenberg, J. D. 2005. The fungal past, present, and future: Germination, ramification, and reproduction. *J. Invertebr. Pathol.*, 89(1), 46–56.

Charnley, A. K., and St. Leger, R. J. 1991. The role of cuticle-degrading enzymes in fungal pathogenesis in insects. In: Cole, G. T., and Hoch, H. C. (Eds.) *The Fungal Spore and Disease Initiation in Plants and Animals*. Boston: Springer US, pp. 267–286.

Choi, K. S., and Jung, H. Y. 2015. Mosquito control using entomopathogenic fungi. *Korean J. Mycol.*, 43(2), 77–87.

Chong, W. C., Basir, R., and Fei, Y. M. 2013. Eradication of malaria through genetic engineering: The current situation. *Asian Pac. J. Trop. Med.*, 2013, 85–94.

Darbro, J. M, Graham, R. I., Kay, B. H., Ryan, P. A., and Thomas, M. B. 2011. Evaluation of entomopathogenic fungi as potential biological control agents of the dengue mosquito, *Aedes aegypti* (Diptera: Culicidae). *Biocontrol Sci. Technol.*, 21(9), 1027–1047.

Darbro, J. M., Johnson, P. H., Thomas, M. B., Ritchie, S. A., Kay, B. H., and Ryan, P. A. 2012. Effects of *Beauveria bassiana* on Survival, Blood-Feeding Success, and Fecundity of *Aedes aegypti* in Laboratory and Semi-Field Conditions. *Am. J. Trop. Med. Hyg.*, 86(4), 656–664.

de Crecy, E., Jaronski, S., Lyons, B., Lyons, T. J., and Keyhani, N. O. 2009. Directed evolution of a filamentous fungus for thermotolerance. *BMC Biotechnol.*, 9(1), 74.

Dong, Y., Morton Jr., J. C., Ramirez, J. L., Souza-Neto, J. A., and Dimopoulos, G. 2012. The entomopathogenic fungus *Beauveria bassiana* activate toll and JAK-STAT pathway-controlled effector genes and anti-dengue activity in *Aedes aegypti*. *Insect Biochem. Mol. Biol.*, 42, 126–132.

Erler, F., and Ates, A. O. 2015. Potential of two entomopathogenic fungi, *Beauveria bassiana* and *Metarhizium anisopliae* (Coleoptera : Scarabaeidae), as biological control agents against the June beetle. *J. Insect. Sci.*, 15(1), 1–6.

Fan, Y., Pei, X., Guo, S., Zhang, Y., Luo, Z., Liao, X., and Pei, Y. 2010. Increased virulence using engineered protease-chitin binding domain hybrid expressed in the entomopathogenic fungus *Beauveria bassiana*. *Microb. Pathog.*, 49, 376–380.

Fang, W., Leng, B., Xiao, Y., Jin, K., Ma, J., Fan, Y., Feng, J., Yang, X., Zhang, Y., and Pei, Y. 2005. Cloning *of Beauveria bassiana* chitinase gene *Bbchit1* and its application to improve fungal strain virulence. *Appl. Environ. Microbiol.*, 71, 363–370.

Fang, W., Scully, L. R., Zhang, L., Pei, Y., and Bidochka, M. J. 2008. Implication of a regulator of G protein signalling (BbRGS1) in conidiation and conidial thermotolerance of the insect pathogenic fungus Beauveria bassiana. *FEMS Microbiol. Lett.*, 279(2), 146–156.

Fang, W., Vega-Rodriguez, J., Ghosh, A. K., Jacobs-Lorena, M., Kang, A., and St. Leger, R. J. 2011. Development of transgenic fungi that kill human malaria parasites in mosquitoes. *Science*, 331(6020), 1074–1077.

Farenhorst, M., Knols, B. G. J., Thomas, M. B., Howard, A. F. V., Takken, W., Rowland, M., and N'Guessan, R. 2010. Synergy in efficacy of fungal entomopathogens and permethrin against West African insecticide-resistant *Anopheles gambiae* mosquitoes. *PLoS One*, 5(8), e12081.

Fernandes, É. K. K., Bittencourt, V. R. E. P., and Roberts, D. W. 2012. Perspectives on the potential of entomopathogenic fungi in biological control of ticks. *Exp. Parasitol.*, 130, 300–305.

Giuliano, G. D. B., Gibson, D. M., and Krasnoff, S. B. 2015. Intracellular siderophore but not extracellular siderophore is required for full virulence in Metarhizium robertsii. *Fungal. Genet. Biol.*, 82, 56–68.

Gomes, S. A., Paula, A. R., Ribeiro, A., Moraes, C. O. P., Santos, J. W. A. B., Silva, C. P., and Samuels, R. I. 2015. Neem oil increases the efficiency of the entomopathogenic fungus *Metarhizium anisopliae* for the control of *Aedes aegypti* (Diptera: Culicidae) larvae. *Parasite. Vector.*, 8, 669.

Govindarajan, M., Sivakumar, R., Rajeswary, M., and Yogalakshmi, K. 2013. Chemical composition and larvicidal activity of essential oil from *Ocimum basilicium* (L.) against *Culex tritaeniorhynchus*, *Aedes albopictus* and *Anopheles subpictus* (Diptera: Culicidae). *Exp. Parasitol.*, 134, 7–11.

Greenfield, B. P. J., Peace, A., Evans, H., Dudley, Ed., Ansari, M. A., and Butt, T. M. 2015. Identification of *Metarhizium* strains highly efficacious against *Aedes*, *Anopheles* and *Culex* larvae. *Biocontrol Sci. Technol.*, 25(5), 487–502.

Guan, Y., Wang, D. Y., Ying, S. H., and Feng, M. G. 2015. A novel RasGTPase (Ras3) regulates conidiation, multi-stress tolerance and virulence by acting upstream of Hog1 signaling pathway in *Beauveria bassiana. Fungal Genet. Biol.*, 82, 85–94.

Guillen, D., Sanchez, S., and Rodriguez-Sanoja, R. 2010. Carbohydrate-binding domains: Multiplicity of biological roles. *Appl. Microbiol. Biotechnol.*, 85(5), 1241–1249.

Gunathilaka, N., Abeyewickreme, W., Hapugoda, M., and Wickremasinghe, R. 2016. Determination of demographic, epidemiological, and socio–economic determinants and their potential impact on malaria transmission in Mannar and Trincomalee districts of Sri Lanka. *Malaria J.*, 15, 330.

Gutierrez, A. C., Rueda Paramo, M. E., Falvo, M. L., Lopez Lastra, C. C., and Garcia, J. J. 2017. *Leptolegnia chapmanii* (Straminipila: Peronosporomycetes) as a future biorational tool for the control of *Aedes aegypti* (L.). *Acta Trop.*, 169, 112–118.

Howard, A. F. V., N'Guessan, R., Koenraadt, C. J. M., Asidi, A., Farenhorst, M., Akogbeto, M., Thomas, M. B., Knols, B. G. J., and Takken, W. 2010. The entomopathogenic fungus *Beauveria bassiana* reduces instantaneous blood feeding in wild multi-insecticide-resistant *Culex quinquefasciatus* mosquitoes in Benin, West Africa. *Parasite. Vector.*, 3, 87.

Inglis, G. D., Johnson, D. L., and Goettel, M. S. 1997. Effects of temperature and sunlight on mycosis (Beauveria bassiana) (Hyphomycetes: Sympodulosporae) of grasshoppers under field conditions. *Environ. Entomol.*, 26(2), 400–409.

Ishii, M., Takeshita, J., Ishiyama, M., Tani, M., Koike, M., and Aiuchi, D. 2015. Evaluation of the pathogenicity and infectivity of entomopathogenic hypocrealean fungi, isolated from

wild mosquitoes in Japan and Burkina Faso, against female adult *Anopheles stephensi* mosquitoes. *Fungal Ecol.*, 15, 39–50.
Jaronski, S. T. 2010. Ecological factors in the inundative use of fungal entomopathogens. In: Helen, Roy (Eds.) *The Ecology of Fungal Entomopathogens*, Springer Netherlands, pp. 159–185.
Kamareddine, L. 2012. The biological control of the malaria vector. *Toxins*, 4, 748–767.
Kamareddine, L., Fan, Y., Osta, M. A., and Keyhani, N. O. 2013. Expression of trypsin modulating oostatic factor (TMOF) in an entomopathogenic fungus increases its virulence towards *Anopheles gambiae* and reduces fecundity in the target mosquito. *Parasite Vector.*, 6, 22.
Kanzok, S. M., and Jacobs-Lorena, M. 2006. Entomopathogenic fungi as biological insecticides to control malaria. *Trends Parasitol.*, 22(2), 49–51.
Khan, S., Guo, L., Maimaiti, Y., Mijit, M., and Qiu, D. 2012. Entomopathogenic fungi as microbial biocontrol agent. *Mol. Plant. Breed.*, 3(7), 63–79.
Lacey, L. A., Grzywacz, D., Shapiro-Ilan, D. I., Frutos, R., Brownbridge, M., and Goettel, M. S. 2015. Insect pathogens as biological control agents: Back to the future. *J. Inverteb. Pathol.*, 132, 1–41.
Lee, S. C., Ni, M., Li, W., Shertz, C., and Heitman, J. 2010. The evolution of sex: A perspective from the fungal kingdom. *Microbiol. Mol. Biol. Rev.*, 74(2), 298–340.
Lee, S. J., Kim, S., Yu, J. S., Kim, J. C., Nai, Y. S., and Kim, J. S. 2015. Biological control of Asian tiger mosquito, Aedes albopictus (Diptera: Culicidae) using *Metarhizium anisopliae* JEF-003 millet grain. *J. Asia-Pac. Entomol.*, 18, 217–221.
Leles, R. N., D'Alessandro, W. B., and Luz, C. 2012. Effect of *Metarhizium anisopliae* conidia mixed with soil against the eggs of *Aedes aegypti. Parasitol. Res.*, 110, 1579–1582.
Lewis, M. W., Robalino, I. V., and Keyhani, N. O. 2009. Uptake of the fluorescent probe FM4-64 by hyphae and haemolymph-derived in vivo hyphal bodies of the entomopathogenic fungus Beauveria bassiana. *Microbiology*, 155(9), 3110–3120.
Luo, S., He, M., Cao, Y., and Xia, Y. 2013. The tetraspanin gene MaPls1 contributes to virulence by affecting germination, appressorial function and enzymes for cuticle degradation in the entomopathogenic fungus, Metarhizium acridum. *Environ. Microbiol.*, 15 (11), 2966–2979.
Luo, S., He, M., Cao, Y., and Xia, Y. 2013. The tetraspanin gene MaPls1 contributes to virulence by affecting germination, appressorial function and enzymes for cuticle degradation in the entomopathogenic fungus, Metarhizium acridum. *Environ. Microbiol.*, 15(11), 2966–2979.
Luo, Z., Qin, Y., Pei, Y., and Keyhani, N. O. 2014. Ablation of the creA regulator results in amino acid toxicity, temperature sensitivity, pleiotropic effects on cellular development and loss of virulence in the filamentous fungus Beauveria bassiana. *Environ. Microbiol.*, 16(4),1122–1136.
Lwetoijera, D. W., Sumaye, R. D., Madumla, E. P., Kavishe, D. R., Mnyone, L. L., Russell, T. L., and Okumu, F. O. 2010. An extra-domiciliary method of delivering entomopathogenic fungus, *Metharizium anisopliae* IP 46 for controlling adult populations of the malaria vector, *Anopheles arabiensis. Parasite Vector.*, 3, 18.
Maldonado-Blanco, M. G., Leal-Lopez, E. Y., Ochoa-Salazar, O. A., Elias-Santos, M., Galan-Wong, L. J., and Quiroz-Martinez, H. 2011. Effects of culture medium and formulation on the larvicidal activity of the mosquito pathogen *Lagenidium giganteum* (Oomycetes: Lagenidiales) against *Aedes aegypti. Acta. Trop.*, 117, 114–118.
Mendki, M. J., Singh, A. P., Tikar, S. N., Parashar, B. D., Veer, V., Shukla, S. V., and Prakash, S. 2015. Repellent activity of *N*, *N*-diethylphenylacetamide (DEPA) with essential oils against *Aedes aegypti*, vector of dengue and chikungunya. *Int. J. Mosquito Res.*, 2(3), 17–20.

Messina, J. P., Brady, O. J., Pigott, D. M., Golding, N., Kraemer, M. U. G., Scott, T. W., William Wint, G. R., Smith, D. L., and Hay, S. I. 2015. The many projected futures of dengue. *Nat. Rev. Microbiol.*, 13, 230–239.

Mohanty, S. S., and Prakash, S. 2009. Effects of culture media on larvicidal property of secondary metabolites of mosquito pathogenic fungus *Chrysosporium lobatum* (Moniliales: Moniliaceae). *Acta. Trop.*, 109, 50–54.

Mondzozo, A. E., Musumba, M., McCarl, B. A., and Wu, X. 2011. Climate change and vector-borne diseases: An economic impact analysis of malaria in Africa. *Int. J. Environ. Res. Pub. Health*, 8, 913–930.

Murugan, K., Kovendan, K., Vincent, S., and Barnard, D. R. 2012. Biolarvicidal and pupicidal activity of *Acalypha alnifolia* Klein ex Willd. (Family: Euphorbiaceae) leaf extract and Microbial insecticide, *Metarhizium anisopliae* (Metsch.) against malaria fever mosquito, *Anopheles stephensi* Liston. (Diptera: Culicidae). *Parasitol. Res.*, 110, 2263–2270.

Muthukumaran, U., Govindrajan, M., Rajeswary, M., Hoti, S. L. 2015. Synthesis and characterization of silver nanoparticles using *Gmelina asiatica* leaf extract against filariasis, dengue and malaria vector mosquitoes. *Parasitol. Res.*, 114, 1817–1827.

Naish, S., Dale, P., Mackenzie, J. S., McBride, J., Mengersen, K., and Tong, S. 2014. Climate change and dengue: A critical and systematic review of quantitative modeling approaches. *BMC Infect. Dis.*, 14, 167.

Najitha Banu, A., and Balasubramanian, C. 2014. Myco-synthesis of silver nanoparticles using *Beauveria bassiana* against dengue vector, *Aedes aegypti* (Diptera: Culicidae). *Parasitol. Res.*, 113, 2869–2877.

Onyango, E. A., Sahin, O., Awiti, A., Chu, C., and Mackey, B. 2016. An integrated and vulnerability assessment framework for climate change and malaria transmission in East Africa. *Malaria J.*, 15, 551.

Ortiz-Urquiza, A., Riveiro-Miranda, L., Santiago-Alvarez, C., and Quesada-Moraga, E. 2010. Insect-toxic secreted proteins and virulence of the entomopathogenic fungus *Beauveria bassiana*. *J. Inverteb. Pathol.*, 105, 270–278.

Ownley, B. H., Gwinn, K. D., and Vega, F. E. 2010. Endophytic fungal entomopathogens with activity against plant pathogens: Ecology and evolution. *Biol. Cont.*, 55(1), 113–128.

Paula, A. R., Carolino, A. T., Paula, C. O., and Samuels, R. I. 2011. The combination of the entomopathogenic fungus *Metarhizium anisopliae* with the insecticide Imidacloprid increases virulence against the dengue vector *Aedes aegypti* (Diptera: Culicidae). *Parasit. Vectors.*, 4, 8.

Pavela, R., Maggi, F., Ngahang Kamte, S. L., Rakotosaona, R., Rasoanaivo, P., Nicoletti, M., Canale, A. and Benelli, G. 2017. Chemical composition of *Cinnamosma madagascariensis* (Cannelaceae) essential oil and its larvicidal potential against the filariasis vector *Culex quinquefasciatus* Say. *S. Afr. J. Bot.*, 108, 359–363.

Pelizza, S. A., Scorsetti, A. C. and Tranchida, M. C. 2012. The sublethal effects of the entomopathic fungus *Leptolegnia chapmanii* on some biological parameters of the dengue vector *Aedes aegypti*. *J. Insect Sci.*, 13(22), 1–8.

Peng, G., Wang, Z., Yin, Y., Zeng, D. and Xia, Y. 2008. Field trials of *Metarhizium anisopliae* var. acridum (Ascomycota : Hypocreales) against oriental migratory locusts, *Locusta migratoria manilensis* (Meyen) in Northern China. *Crop Prot.*, 27(877), 1244–1250.

Qiu, L., Wang, J. J., Ying, S. H. and Feng, M. G. 2015. Weel and Cdc25 control morphogenesis, virulence and multistress tolerance of *Beauveria bassiana* by balancing cell cycle-required cyclin-dependent kinase 1 activity. *Environ. Microbiol.*, 17, 1119–1133.

Ragavendran, C., Dubey, N. K., and Natarajan, D. 2017. *Beauveria bassiana* (Clavicipitaceae): A potent fungal agent for controlling mosquito vectors of *Anopheles stephensi*, *Culex quinquefasciatus* and *Aedes aegypti* (Diptera: Culicidae). *RSC Adv.*, 7, 3838–3851.

Ramirez, J. L., Garver, L. S., and Dimopoulos, G. 2009. Challenges and approaches for mosquito targeted malaria control. *Curr. Mol. Med.*, 9(2), 116–130.

Rocha, L. F. N., Sousa, N. A., Rodrigues, J., Catao, A. M. L., Marques, C. S., Fernandes, E. K. K., and Luz, C. 2015. Efficacy of *Tolypocladium cylindrosporum* against *Aedes aegypti* eggs, larvae and adults. *J. Appl. Microbiol.*, 119, 1412–1419.

Rodrigues, J., Luz, C., and Humber, R. A. 2017. New insights into the *in vitro* development and virulence of *Culicinomyces* spp. as fungal pathogens of *Aedes aegypti*. *J. Inverteb. Pathol.*, 146, 7–13.

Santos, A. H., Hua Tai, M. H., Rocha, L. F. N., Silva, H. H. G., and Luz, C. 2009. Dependence of *Metarhizium anisopliae* on high humidity for ovicidal activity on *Aedes aegypti*. *Biol. Cont.*, 50, 37–42.

Scholte, E. J., Knols, B. G. J., Samson, R. A., and Takken, W. 2004. Entomopathogenic fungi for mosquito control: A review. *J. Insect Sci.*, 4, 19–42.

Scholte, E. J., Knols, B. G. J., and Takken, W. 2006. Infection of the malaria mosquito *Anopheles gambiae* with the entomopathogenic fungus *Metarhizium anisopliae* reduces blood feeding and fecundity. *J. Invertebr. Pathol.*, 91, 43–49.

Scholte, E. J., Takken, W., and Knols, B. G. J. 2007. Infection of adult *Aedes aegypti* and *Ae. albopictus* mosquitoes with the entomopathogenic fungus *Metarhizium anisopliae*. *Acta Trop.*, 102, 151–158.

Shahid, A. A., Rao, A. Q., Bakhsh, and Husnain, T. 2012. Entomopathogenic fungi as biological controllers: New insights into their virulence and pathogenicity. *Arch. Biol. Sci. Belgrade.*, 64(1), 21–42.

Shang, Y., Chen, P., Chen, Y., Lu, Y., and Wang, C. 2015. MrSkn7 controls sporulation, cell wall integrity, autolysis, and virulence in Metarhizium robertsii. *Eukaryot. Cell*, 14(4), 396–405.

Singh, G. and Prakash, S. 2012. Evaluation of culture filtrates of *Culicinomyces clavisporus*: Mycoadulticide for *Culex quinquefasciatus*, *Aedes aegypti* and *Anopheles stephensi*. *Parasitol. Res.*, 110, 267–272.

Tisgratog, R., Sanguanpong, U., Grieco, J. P., Ngoen-Kluan, R., and Chareonviriyaphap, T. 2016. Plants traditionally used as mosquito repellants and the implication for their use in vector control. *Acta. Trop.*, 157, 136–144.

Valero-Jiménez, C. A., Wiegers, H., Zwaan, B. J., Koenraadt, C. J. M., and Van Kan, J. A. L. 2016. Genes involved in virulence of the entomopathogenic fungus *Beauveria bassiana*. *J. Invertebr. Pathol.*, 133, 41–49.

Vazquez-Martinez, M. G., Rodriguez-Meneses, A., Rodriguez, A. D., and Rodriguez, M. H. 2013. Lethal effects of *Gliocladium virens*, *Beauveria bassiana* and *Metarhizium anisopliae* on the malaria vector *Anopheles albimanus* (Diptera: Culicidae). *Ann. Bioeth.*, 23(9), 1098–1109.

Veys-Behbahani, R., Sharififard, M., Dinparast-Djadid, N., Shamsi, J., and Fakoorziba, M. R. 2014. Laboratory evolution of the entomopathogenic fungus Beauveria bassiana against Anopheles stephensi larvae (Diptera: Culicidae). *Asian Pac. J. Trop. Dis.*, 4(2), S799–S802.

Vogels, C. B. F., Bukhari, T., and Koenraadt, C. J. M. 2014. Fitness consequences of larval exposure to Beauveria bassiana on adults of the malaria vector Anopheles stephensi. *J. Invertebr. Pathol.*, 119, 19–24.

Wang, B., Kang, Q., Lu, Y., Bai, L., and Wang, C. 2012. Unveiling the biosynthetic puzzle of destruxins in Metarhizium species. *Proc. Natl. Acad. Sci. USA*, 109(4), 1287–1292.

Wang, C., and St. Leger R. J. 2007. The MAD1 adhesin of Metarhizium anisopliae links adhesion with blastospore production and virulence to insects, and the MAD2 adhesin enables attachment to plants. *Eukaryot. Cell*, 6(5), 808–816.

Wang, J., Liu, J., Hu, Y., Ying, S. H., and Feng, M. G. 2013a. Cytokinesis-required Cdc14 is a signaling hub of asexual development and multi-stress tolerance in *Beauveriabassiana*. *Sci. Rep.*, 3, 3086.

Wang, J. J., Qiu, L., Cai, Q., Ying, S. H., and Feng, M. G. 2015. Transcriptional control of fungal cell cycle and cellular events by Fkh2, a forkhead transcription factor in an insect pathogen. *Sci. Rep.*, 5(1), 10108.

Wang, J., Zhou, G., Ying, S. H. and Feng, M. G. 2013b. P-type calcium ATPase functions as a core regulator of *Beauveria bassiana* growth, conidiation and responses to multiple stressful stimuli through cross-talk with signalling networks. *Environ. Microbiol.*, 15, 967–979.

Wang, S., Fang, W., Wang, C. and St. Leger, R. J. 2011. Insertion of an Esterase gene into a specific locust pathogen (*Metarhizium acridum*) enables it to infect caterpillars. *PLoS Pathog.*, 7(6), e1002097.

Wang, X. X., He, P. H., Feng, M. G., and Ying, S. H. 2014. BbSNF1 contributes to cell differentiation, extracellular acidification, and virulence in *Beauveria bassiana*, a filamentous entomopathogenic fungus. *Appl. Environ. Microbiol.*, 98, 8657– 8673.

Wang, Z. L., Zhang, L. B., Ying, S. H., and Feng, M. G. 2013. Catalases play differentiated roles in the adaptation of a fungal entomopathogen to environmental stresses. *Environ. Microbiol.*, 15, 409–418.

Wijayanti, S. P. M., Porphyre, T., Chase-Topping, M., Rainey, S. M., McFarlane, M., Schnettler, E., Biek, R., and Kohl, A. 2016. The importance of socio-economic versus environmental risk factors for reported dengue cases in Java, Indonesia. *PLoS Neglect. Trop. Dis.*, 10(9), e0004964.

Xie, X. Q., Guan, Y., Ying, S. H., and Feng, M. G. 2013. Differentiated functions of Ras1 and Ras2 proteins in regulating the germination, growth, conidiation, multistress tolerance and virulence of *Beauveria bassiana. Environ. Microbiol.*, 15, 447–462.

Xu, Y., Orozco, R., Wijeratne, E. M. K., Gunatilaka, A. A. L., Stock, S. P., and Molnár, I. 2008. Biosynthesis of the cyclooligomerdepsipeptide beauvericin, a virulence factor of the entomopathogenic fungus *Beauveria bassiana. Chem. Bio.*, 15, 898–907.

Zhang, S., Widemann, E., Bernard, G., Lesot, A., Pinot, F., Pedrini, N., and Keyhani, N. O. 2012. CYP52X1, representing new cytochrome P450 subfamily, displays fatty acid hydroxylase activity and contributes to virulence and growth on insect cuticular substrates in entomopathogenic fungus *Beauveria bassiana. J. Biol. Chem.*, 287(16), 13477–13486.

Zhang, S., Xia, Y. X., Kim, B., and Keyhani, N. O. 2011. Two hydrophobins are involved in fungal spore coat rodlet layer assembly and each play distinct roles in surface interactions, development and pathogenesis in the entomopathogenic fungus, *Beauveria bassiana. Mol. Microbiol.*, 80, 811–826.

Zhang, Y., Zhang, J., Jiang, X., Wang, G., Luo, Z., Fan, Y., Wu, Z., and Pei, Y. 2010. Requirement of a mitogen-activated protein kinase for appressorium formation and penetration of insect cuticle by the entomopathogenic fungus *Beauveria bassiana. Appl. Environ. Microb.*, 76, 2262–2270.

Zhang, Y., Zhao, J., Fang, W., Zhang, J., Luo, Z., Zhang, M., Fan, Y., and Pei, Y. 2009. Mitogen-activated protein kinase hog1 in the entomopathogenic fungus *Beauveria bassiana* regulates environmental stress responses and virulence to insects. *Appl. Environ. Microb.*, 75, 3787–3795.

Zhou, Y. H., Hou, L., Zhang, Y. J., Fan, Y. H., Luo, Z. B., Jin, B., Zhou, Q. S., Li, Y. J., Wang, Y., and Pei, Y. 2014. Expression and promoter characterization of BbPacC, a pH response transcription factor gene of the entomopathogenic fungus *Beauveria bassiana. Microbiology*, 160(2), 353–361.

# 5 Bioprospecting of Novel Fungal Secondary Metabolites for Mosquito Control

*Perumal Vivekanandhan, Murugan Arunthirumeni, Govindaraj Vengateswari, and Muthugounder Subramanian Shivakumar*

## CONTENTS

## 5.1 INTRODUCTION

### 5.1.1 Vector-Borne Diseases

Vector-borne diseases cause illnesses that are transmitted by blood-sucking arthropod pests like mosquitoes, lice, fleas, bed bugs, and ticks. These vectors carry pathogens that are responsible for diseases like malaria, dengue, schistosomiasis, human African trypanosomiasis, leishmaniasis, Chagas disease, yellow fever, Japanese encephalitis, onchocerciasis, chikungunya, West Nile virus, lymphatic filariasis, and Zika (World Health Organization [WHO], 2016). Among the arthropods, mosquitoes are the principal vectors of several diseases. *Aedes* mosquito vectors are responsible for transmitting dengue fever, Rift Valley fever, and Zika virus. *Anopheles* mosquito vectors spread malaria, while *Culex* mosquitoes are responsible for Japanese encephalitis, lymphatic filariasis, West Nile fever, and chikungunya. Sand flies are responsible for visceral leishmaniasis. Ticks can spread Crimean-Congo hemorrhagic fever, Lyme disease relapsing fever (borreliosis), rickettsial diseases (spotted fever and q fever), tick-borne encephalitis, and tularaemia. Triatomine bugs are responsible for Chagas disease (also called American trypanosomiasis). Tsetse flies spread sleeping sickness (African trypanosomiasis). Fleas spread plague (transmitted by fleas from rats to humans) and rickettsiosis. Blackflies are responsible for onchocerciasis (river blindness) disease (WHO, 2017).

### 5.1.2 Populations Affected by Vector-Borne Disease

Vector-borne diseases account for more than 17% of all infectious diseases, causing more than 1 million deaths annually (WHO, 2016). More than 2.5 billion people in over 100 countries are at risk of contracting dengue alone, and malaria causes more than 400,000 deaths every year globally, most of them children under 5 years of age (WHO, 2016). Other diseases such as Chagas disease, leishmaniasis, and schistosomiasis affect hundreds of millions of people worldwide. Worldwide, 500 million people are suffering from the aforementioned diseases and 1 million people die every year (Wariko et al., 2012). Vector-borne diseases produce economic loss and public health problems. They are severe problems in India and other developing countries; in India 2500 death are reported every year (National Vector Borne Disease Control Program [NVBDCP], 2013; Innovative Vector Control Consortium, 2015).

### 5.1.3 Mosquitoes

Mosquitoes are insects belonging to the order Diptera. Currently, there are about 3500 mosquito species that have been identified. Among these mosquitoes, those belonging to three genera, namely, *Anopheles*, *Aedes*, and *Culex* are economically important as vectors of diseases.

### 5.1.4 *Anopheles* Mosquito and Malaria

Malaria is one of the most widespread diseases spanning several continents worldwide. Africa and Southeast Asian countries have been especially affected by this disease. Several species of *Anopheles* mosquitoes transmit malaria. *Anopheles gambiae* and *Anopheles stephensi* are two species that are primary vectors of this disease in Africa and Southeast Asia, respectively. Female *Anopheles* mosquito transmits plasmodium from one individual to the next through biting. According to WHO reports, people in 91 countries are affected by malaria (WHO, 2015). Sub-Saharan African countries are particularly severely affected by malaria (WHO, 2015). Malaria is caused by four parasites, namely, *Plasmodium falciparum*, *P. vivax*, *P. malariae*, and *P. ovale*; among these *P. falcifparum* is mainly responsible for mortality in humans.

### 5.1.5 *Aedes* Mosquito and Dengue

*Aedes aegypti* is a day biting mosquito that is responsible for the transmission of diseases such as dengue fever, chikungunya, Rift Valley fever, yellow fever, and Zika virus (WHO, 2016). The disease spreads in tropical and subtropical regions in the world. However, the world's fastest growing vector-borne disease is dengue, with a 30-fold increase in disease incidence over the last 50 years. Dengue fever (or dengue hemorrhagic fever) is caused by dengue virus. This virus belongs to the genus *Flavivirus* in the family *Flaviviridae*, and includes serotypes 1-4 (Den-1, Den-2, Den-3, and Den-4) (WHO, 2010). Recent WHO estimates suggest that 50 million cases of dengue fever occur annually, with more than 2.5 billion people living in risk areas worldwide (WHO, 2016). Dengue viruses are overwhelmingly transmitted to man through the bite of dengue-bearing female *Aedes* mosquitoes. The main dengue vector worldwide is the *Ae. aegypti* mosquito, which is well adapted to urbanized areas and breeds often in man-made containers (Braks et al., 2004; Scott et al., 1993). The National Vector Borne Disease Control Program estimated there are 74,454 of people affected by dengue with 2000 deaths every year in India (NVBDCP, 2013). This disease has dramatically increased in recent decades because of poor hygiene, development of insecticide resistance, and due to the lack of effective drugs for curing dengue.

### 5.1.6 *Culex* Mosquito

There are two species of Culex mosquitoes that are vectors of diseases: *Culex pipens*, present in the temperate regions, and *Culex quinquefasciatus*, which is also called the southern house mosquito. It is a night biting mosquito and female *Culex* mosquitoes are responsible for the transmission of Japanese encephalitis, lymphatic filariasis, and West Nile fever (WHO, 2016). These diseases are endemic in 73 countries and it is anticipated that 1.39 billion people are at risk (Weekly Epidemiological Record, 2012). Globally about 120 million people have been infected by filariasis disease worldwide.

### 5.1.7 Insecticides for Mosquito Control

The classes of insecticides most commonly used for vector control programs are organochlorines, organophosphorus, carbamates, and pyrethroids (N'guessan et al., 2007). The first synthetic organochlorine insecticide to be commercialized, DDT (dichlorodiphenyltrichloroethane), was central by the WHO-led global malaria eradication campaign (1955–1969) (Najera, 1999). Except in Sub-Saharan Africa, this resulted in the elimination of the disease in North America, Europe, and parts of Asia (Gordon and Gannon, 2000). Although agricultural use of DDT has now ceased due to environmental persistence and reduced efficacy against resistant insects, extensive use for malaria control continues as a cost-effective and safe insecticide for indoor residual spraying (IRS) (WHO, 2016).

More recently, pyrethroids have been widely used for mosquito control. They are the only class of insecticides recommended by the WHO for impregnation of long-lasting insecticidal bed nets (LLINs) (N'guessan et al., 2007) and are also available for IRS. Pyrethroids, such as permethrin and deltamethrin, and etofenprox, are neurotoxic compounds that target the sodium channels in neuronal membrane of insects. Carbamates and organophosphates act by binding to the acetylcholinesterase at the nerve junction (WHO, 1957). Some of the carbamates that are commonly used in mosquito control programs are recommended by WHO (WHO, 2005, 2010, 2016). Though several insecticides have been produced for use in agriculture, only a few of these have been permitted by WHO for mosquito control. The primary reason being most of these insecticides are highly toxic to mammals and do not degrade quickly leading to biomagnification in the food chain. Insecticide used for controlling mosquito larvae is still scarce, as larvae are aquatic. The commonly used mosquito larvicides are temephos belonging to organophosphate class and pyriproxifen, which is an insect growth regulator (Yoshiaki Kono et al., 1997). There is a huge lacuna in this area that needs to be filled in molecules of newer chemistries.

In addition, there are several personal protection kits that are used as mosquito repellents, in which the primary ingredient is DEET, a chemical that is very effective as a repellent. There have been a few reports of skin allergies due to long term use of this product. There have been several studies for alternative molecules. Essential oils from plants have been studied, though they are effective, but they are costly and have to be used in increasing frequency, which is the main bottleneck in commercializing these molecules.

Vector control, personal protection, and community participation are the pillars of WHO strategies for mosquito control. Integrated vector management (IVM) has been advocated as a recommended approach for combating vector-borne diseases (WHO, 2008b). IVM is defined as “a rational decision-making process for optimal use of resources for vector control” (WHO, 2008). The objective of vector control is to interrupt transmission of disease by preventing human contact with pathogen-bearing mosquitoes, eliminating breeding sites, killing the mosquito larvae, or reducing the longevity of adult mosquitoes (Van den Berg, 2009). IVM uses a combination of chemical insecticides, insect growth regulators, repellents, and insecticide-impregnated nets (LLIN) along with indoor residual spraying to reduce the incidence of mosquito–human contact. In addition, legislative laws are also enacted for mosquito control in the IVM approach.

Presently, a combination of IRS and insecticide treated nets (ITNs) affords enhanced protection to populations severely affected by malaria (Kleinschmidt et al., 2009). Moreover, the growing resistance of malaria vectors to available insecticides is a major cause for concern and an increasing threat to mosquito control programs (N’guessan et al., 2007; Pages et al., 2007; Sharp et al., 2007). In order to increase the effectiveness of chemical insecticides for mosquito control, larval source management using biocontrol agents have been advocated (Keiser et al., 2005).

Larvivorous fishes and bacterial pathogens such as *Bacillus thuringiensis* israelensis and *Bacillus sphaericus* are examples of biological agents that are used to kill larvae (Walker and Lynch, 2007). In addition temephos is also being used in the management of mosquito larvae. Effective and sustained malaria vector control requires clear commitment from national authorities, including long-term support from funding partners (Komatsu et al., 2010). There has to be a coordinated approach toward periodic surveys to detect the baseline susceptibility of mosquitoes. This will help in managing insecticide resistance in mosquitoes (Coleman and Hemingway, 2007).

### 5.1.8 Insecticide Resistance in Mosquitoes

Insecticide resistance is defined as “the development of an ability in a strain of some organism to tolerate doses of a toxicant that would prove lethal to a majority of individuals in a normal population of the same species” (WHO, 1957; Coleman and Hemingway, 2007). Alternatively, a resistant phenotype has been defined as “an insect that survives a dose of insecticide that would normally have killed it” (Zlotkin, 1999; Hemingway et al., 2002). This ability to survive is a result of repeated exposure to one particular insecticide for several generations (Ramkumar and Shivakumar, 2015). Resistance is a multidimensional biological phenomenon that depends on the interaction of multiple influences for its development (WHO, 1957). The *WHO Global Technical Strategy for Malaria 2016–2030* highlights insecticide resistance as a major obstacle to achieving malaria control targets (WHO, 2015). Insecticide resistance is genetically inherited and is passed on to the progeny (Bruce-Chwatt, 1985). This inheritance is the result of mutation or gene duplication resulting in altered physiology and behavior of individual, thus promoting survival value. The resistance phenotypes quickly increase in the population, which ultimately results in development of insecticide resistance to a particular chemical insecticide. The level of resistance in insect populations is dependent on the amount and frequency of insecticides used and the inherent characteristics of the insect species selected. Mosquitoes, for instance, are endowed with all attributes suited for rapid resistance development including high reproductive potentials and short life cycles producing several generations per season with abundant progeny (WHO, 1957; Nauen, 2006). Mostly resistance in a particular species is considered to occur throughout the control area, but in reality, insecticide resistance can be local in nature and is very heterogeneous even over very small distances. It often develops within a small part of the population of one species of *Anopheles* and assumes different patterns depending on the type of selection pressure (WHO, 1957). In Guatemala, sampling sites for *Anopheles albimanus* only a few kilometers apart varied not only by presence or absence of resistance, but also by level of resistance and the mix of mechanisms responsible for resistance (Brogdon and McAllister, 1998). Currently, major emphasis in research into vector resistance is double pronged. The first approach strives toward understanding the molecular mechanisms underlying resistance with the view of developing novel vector control methods that avoid or minimize resistance problems. The second approach to research involves rational resistance management, which is developing and implementing control methods that minimize the likelihood that vectors will evolve strong resistance to important insecticides (Hemingway and Ranson, 2000).

## 5.2 INSECTICIDE RESISTANCE MECHANISMS IN INSECTS

There are different ways for insect pests to develop resistance to synthetic insecticides. Continuous use of synthetic chemical insecticide for control of an insect finally results in enhanced enzymatic metabolism, altered target-site insensitivity, behavioral modification, reduced penetration, or absorption of the toxicant (Kranthi, 2005). Insects may develop more than one of the aforesaid mechanisms at the same time or the mechanism acts (e.g., biochemical metabolism) on different groups of insecticides and results in a cross-resistance.

### 5.2.1 Evidence of Insecticide Resistance in Mosquitoes

Chemical control has been the main effective measure to reduce the population of these disease vectors since the 1950s (Hemingway et al., 2006). Four classes of chemical insecticides are the mainstay of vector control programs, namely, organochlorines, organophosphates, carbamates, and pyrethroids. Pyrethroids account for approximately 25% of world insecticide market and are used extensively because they kill insects rapidly and have low mammalian toxicity. As per WHO, insecticide resistance is defined as "development of an ability in a strain of an organism to tolerate doses of toxicants, which would prove lethal to a majority of individuals in a normal (susceptible) population of the same species." Elucidation of resistance mechanisms becomes crucial to guide the use of permethrin and the development of its substitutes, and should be considered one of the most challenging issues in modern applied entomology (Hansen et al., 2012; Muthusamy et al., 2013; Nazni et al., 2000).

However, a major control problem is that these species have developed resistance to all major insecticides, including pyrethroids (Mebrahtu et al., 1997), which has led to failures in vector-borne disease control, especially in dengue control (Strode et al., 2008). Resistance mechanisms in mosquitoes have been extensively studied and are known to be predominantly classified into two classes: metabolic resistance (degradation of the active ingredient by detoxification enzymes) and target-site resistance (mutations in the target proteins). A number of genes associated with insecticide resistance, including cytochrome P450, esterases, elevated glutathione S-transferase (GST), and sodium channel gene were identified (Hu et al., 2011; Nardini et al., 2012). In earlier studies, many researchers identified multiple resistance mechanisms, such as P450- and GST-mediated detoxification and target-site insensitivity, which were involved in pyrethroid resistance in this *Cx. quinquefasciatus* strain (Xu et al., 2008). Many studies have been conducted throughout the world to understand the mechanisms of pyrethroid resistance in insects, especially in mosquitoes. Biochemical and molecular methods have been used to detect resistance mechanisms in these vectors. Using biochemical methods, the increase of enzyme activities of P450-mediated monooxygenases and mixed function oxidases (MFOs) have been reported to play a role in the metabolism of pyrethroids (Muthusamy and Shivakumar, 2015). Permethrin is a broad-spectrum pyrethroid insecticide and has been widely used in the vector control program throughout the world. As a consequence, it is believed that permethrin resistance has developed in mosquitoes, especially among filariasis vectors throughout world. In the present study, an effort was taken to understand the possible resistance mechanisms to permethrin in the filariasis vectors *Cx. quinquefasciatus* larvae and adult mosquitoes.

### 5.2.2 Insecticide Resistance in Mosquitoes

A small percentage of the overall amount of insecticides is used in public health. However, there is widespread resistance among many vector species of public health importance. The first report on insecticide resistance published about 100 years ago (Melander, 1914) noticed that certain populations of scale insects were becoming less susceptible to lime sulfur than they had been in the past. Even when researchers increased the amount of the active ingredient by 10 times, 74% of them still survived (Melander, 1914). Resistance to insecticides is defined as the development of the ability to survive doses of insecticides that previously were lethal to the majority of individuals in a population (IRAC, 2011). The level of resistance in insect vector populations is dependent both on the volume and frequency of applications of insecticides used against them and the inherent characteristics of the insect species involved (Hemingway and Ranson, 2000). Resistance became a major obstacle to the global malaria eradication program of the 1950s and 1960s and rendered some insecticides, particularly dieldrin, useless (Busvine, 1969). Presently, insecticide resistance is widespread and it is reported in nearly two-thirds of malaria endemic countries. It affects all major vector species and all classes of insecticides (WHO, 2012). Insecticide resistance is expected to directly affect the reemergence of vector-borne diseases and threaten disease control in areas where vector-borne diseases already exist (Brogdon and McAllister, 1998). Insecticide resistance was called "the biggest single obstacle in the struggle against vector-borne diseases" (WHO, 1976). There have been reported cases of failure of some malaria control strategies due to pyrethroid resistance (Hargreaves et al., 2000; Guessanet al., 2007). Therefore, it is very important to preserve useful insecticides by slowing and preventing the development of resistance in mosquitoes. To achieve this goal, it is necessary to understand and monitor the development of insecticide resistance and to find ways of preventing resistance development. In response to the spread of insecticide resistance in mosquito species, WHO has proposed various guidelines to encourage countries to plan and implement insecticide resistance management strategies (WHO, 2012).

### 5.2.3 Modes of Insecticide Resistance

According to Hemingway and Ranson (2000), resistance mechanisms are either metabolic change in detoxification enzyme activity that leads to detoxification of the insecticide or target-site dependent (i.e., changes in sensitivity of the target site, for example, mutations in the sodium channel genes). Furthermore, these mechanisms can occur together in an organism or population. Increased enzymatic detoxification is considered to be the most common resistance mechanism (Oppenoorth, 1984; Price, 1991; Scott, 1999), although both kinds of resistance have contributed significantly to the rise in insecticide resistance (Hemingway and Ranson, 2000). Another resistance mechanism is based on changes in behavior by the target insect behavioral resistance in the presence of a particular insecticide. For example, DDT causes behavioral changes in mosquitoes by decreasing the rate of mosquito entry into houses, increasing the rate of early exit from houses, and by causing a shift in biting times (Pates and Curtis, 2005). Reduced penetration of insecticides due to the formation of thicker cuticles is also considered to be an evolved mode of resistance (Hemingway and Karunaratne, 1998).

### 5.2.4 Insecticide Resistance Mechanisms

#### 5.2.4.1 Metabolic Resistance Mechanisms

Metabolic resistance is the most common resistance mechanism that occurs in insects. It occurs due to changes in a mosquito's enzyme system that results in a more rapid detoxification of the insecticide than normal, preventing the insecticide from reaching the intended site of action. In mosquitoes, three enzyme systems are believed to be important: the esterases, the monooxygenases, and the glutathione S-transferases. Glutathione S-transferase is often involved in DDT resistance. Several reports have shown esterases to be involved in organophosphate, carbamate, and, to a lesser extent, pyrethroid resistance. Similarly, monooxygenases have been associated with pyrethroid resistance, the activation and/or detoxication of organophosphates and, to a lesser extent, carbamate resistance. Most insects possess these enzyme systems to help them detoxify naturally occurring foreign materials. However, they are often enhanced in resistant insect strains enabling them to metabolize or degrade insecticides before they are able to exert a toxic effect. Metabolic resistance mechanisms have been identified in vector populations for all major classes of insecticides currently used for vector control, including organophosphates, carbamates, pyrethroids, and DDT (Hemmingway and Ranson, 2000; IRAC, 2011).

#### 5.2.4.2 Target-Site Resistance

The second most common resistance mechanism encountered in insects is target-site resistance. Target-site resistance occurs due to the insensitivity of an active site to which an insecticide would normally bind and act. This insensitivity is due to any change in the protein structure of the target site leading to a lesser affinity or lesser action upon the molecule. This results in the insects being unaffected or less affected by the insecticide, compared to susceptible insects. For the main insecticides used in public health, this type of resistance may occur in either the voltage sensitive sodium channel, causing DDT and pyrethroid resistance, or the acetyl cholinesterase (AChE), causing organophosphate and carbamate resistance.

#### 5.2.4.3 Behavioral Resistance

Behavioral resistance describes any modification in insect behavior that helps to avoid the lethal effects of insecticides. Insecticide resistance in mosquitoes is not always based on biochemical mechanisms such as metabolic detoxification or target-site mutations, but may also be conferred by behavioral changes in response to prolonged spraying programs. Behavioral resistance does not have the same importance as physiological resistance but might be considered to be a contributing factor, leading to the avoidance of lethal doses of an insecticide. A behavioral response is either dependent or independent on a stimulus. If mosquitoes avoid a treated place due to sensing the insecticide, it is considered to be a behavioral change dependent on a stimulus, whereas the selective and sustained occupation of an untreated area can be considered as stimulus independent response.

#### 5.2.4.4 Cross-Resistance

Cross-resistance occurs when a resistance mechanism, that allows insects to resist one insecticide, also confers resistance to compounds within the same class, and may occur between chemical classes. The phenomenon of cross-resistance is a relatively frequent one in vector populations. For example, DDT and pyrethroid insecticides are chemically unrelated but both act on the same target site (sodium channel). Past use of DDT has resulted in several insect species developing resistance to DDT by the *kdr* mutation at the target site (Hu et al., 2011; Nardini et al., 2012). Where these mutations have been retained in the population, the insects have some resistance to all pyrethroids in addition to DDT. Cross-resistance can also occur between organophosphate and carbamate insecticides when resistance results from altered AChE.

## 5.3 BIOLOGICAL CONTROL

### 5.3.1 Plant-Derived Secondary Metabolites

Many herbal insecticides of plants are used for products against vectors. Several laboratory and field tests have already been carried out in this area and some potentially bioactive compounds were isolated and tested against larvicides of plant origin like falcarinol, azadirachtin, geranial pipernonaline, and octacosane. Bioactive compounds were extracted from plant parts and also essential oils from seeds. These kinds of plant-derived materials are safer and cheaper than synthetic chemical pesticides (see Table 5.1).

Mosquitoes mostly breed in water in the larval stage and are attractive targets for pesticides because it is easy to deal with them. Many researchers have reported on the effectiveness of plant extract against mosquito larvae. Medicinal plants like *Jatropha curcas* have played major roles in the treatment of various diseases, including bacterial and fungal infections. All parts of Jatropha (seeds, leaves, bark) have been used in traditional medicine and for veterinary purposes for a long time (Dalziel, 1955; Duke, 1988). Larvicidal activity of stem/bark and leaf extracts of *Jatropha curcas* have been used against the dengue-carrying mosquito *Aedes aegypti* (Gutierrez et al., 2014). Extract from many medicinal plants such as *Artemisia annua* leaf parts possess a potent larvicidal activity against mosquito larvae and biological activity was obtained in $LC_{50}$ value 16.85 ppm after 24 h and 11.45 ppm after 48 h of exposure (Sharma et al., 2006).

Panneerselvam and Murugan (2013) reported the repellent and ovicidal potential of the crude hexane, ethyl acetate, benzene, and aqueous and methanol solvent extracts from the medicinal plants *Andrographis paniculata*, *Cassia occidentalis*, and *Euphorbia hirta* against the medically important mosquito vector. The adulticidal and smoke toxicity activity of acetone, ethyl acetate, benzene, chloroform, and methanol leaf extracts of *Cipadessa baccifera* against *Cx. quinquefasciatus*, *Ae. aegypti*, and *An. stephensi* were also determined (Ramkumar et al., 2014). Traditional medicinal plant has been used for pest control (Kuppusamy and Murugan, 2006). *Clausena dentata* (Willd) is a small tree plan, belonging to the family of Rutaceae, and found in India, Sri Lanka, and China. It is popularly known as Anai chedi in Tamil. *C. dentata* is used by the local people of Yercaud and Boda Hills for its medicinal and nutritive value. The phytochemical studies of the plant have revealed the presence of volatile oils of four furanoid terpenic compounds: α-clausenan, rosefuran (γ-clausenan), and diclausenans A and B (Rao and Subramanian, 1934). Manjari et al. (2014) previously

**TABLE 5.1**
**Plant Secondary Metabolites Showing Insecticidal and Repellent Activity against Insects**

| Plant Metabolite | Activity | Reference |
|---|---|---|
| Monoterpene alcohol | Repellent activity | (Junker et al., 2011) |
| Diterpine | Repellent and antifeedant activity | (Schmelz et al., 2011) |
| Azadirachtin | Insecticidal activity | (Alouani et al., 2009) |
| Nicotine | Insecticidal activity | (Regnault-Roger et al., 2005) |
| Carbaryl | Insecticidal activity | (Regnault-Roger et al., 2005) |
| Rotenone | Insecticidal activity | (Regnault-Roger et al., 2005) |
| Pyrethrum | Insecticidal activity | (Regnault-Roger et al., 2005) |
| Linalool | Insecticidal activity | (Alouani et al., 2009) |
| Ryania | Insecticidal activity | (Uysal and Agar, 2005; Jaya et al., 2014) |

reported that *C. dentata* leaf acetone extracts had a good larvicidal activity against *Cx. quinquefasciatus*, *Ae. aegypti*, and *An. stephensi*. Medicinal plants have been recognized as important natural insecticides (Knio et al., 2008).

Seed and leaf extracts from *Eucalyptus globules* show 100% and 80% mortality to the tested larva of *Cx. pipiens* (Sheeren et al., 2006). Larvicidal activity of solvent crude extract from *Cassia fistula* shows toxicity against the larvae of *Cx. quinquefasciatus* (Trevors et al., 1987). *Solanum xanthocarpum* plant root show activity against *Cx. pipiens* (Mohan et al., 2006). Leaf extract from *Ocimum basilicum* showed plant toxicity against two mosquitoes *An. Stephensi* and *Cx. quinquefasciatus*. These factors are ecofriendly and biodegradable insecticides. Most plant-derived metabolites are harmless to nontargeted organisms and the environment compared to chemical insecticides. The family of Papaveraceae possesses various types of activity against many species of mosquitoes. Plant species belonging to Rutaceae, Asteraceae, Umbelliferae, Myrtaceae, and Lamiaceae families have insecticidal molecules (Roger, 1997). Hence, there is constant need for developing bioactive compounds from plant material to give good larvicidal activity, which are also expected to reduce biological risk and minimize the accumulation in the environment.

### 5.3.2 Microbial Secondary Metabolites

The genetically altered microbial pesticides from entomopathogens like bacteria virus and fungi. These microbial pesticides are alternatives to chemical insecticides and increased specificity and ecological safety so it used as a vector control. Many studies have demonstrated that microbes produced insecticidal toxins important in pest control. The conventional practice of using chemical pesticide is troubling due to their toxic effects on human beings and livestock, and pest outbreaks on beneficial non-target organisms like fish and invertebrate. At present it is absolutely necessary to limit the use of chemical pesticides against the pest control and environmental risk. Among the safe and alternatives considered, entomopathogens are playing a major role in the regulation of pest population. Microbes are a rich source of bioactive chemicals and are alternatives for mosquito control agents. Dhanasekaren et al. (2010) reported secondary metabolites from actinobacteria were screened for larvicidal activity against *Anopheles* mosquitoes. Approximately 942 microbes and their derived secondary metabolites were active compound against the insects. Extracellular secondary metabolites from 350 fungi and 94 actinobacteria were screened for larvicidal activity against *Cx. quinquefasciatus*, *Anapheles stephensi*, and *Ae. aegypti* (Vijayan and Balaraman, 1991). The 133 fungal metabolites and 35 from bacterial species were active against the mosquitoes.

#### 5.3.2.1 Bacterial Metabolites

Bacteria-derived metabolites are the most common and cheaper pesticide toxic to several species like butterflies, beetles, flies, and mosquitoes. Secondary metabolites from bacterial isolates showed mosquito larvicidal activity against *Ae. aegypti* and *Cx. quinquefasciatus* larvae. Bacterial pathogens are used as vector controls because of the spore-forming, rod-shaped bacteria in the genus of *Bacillus*. Many types of bacteria are isolated from soils and they have many insecticidal properties (Table 5.2).

**TABLE 5.2**
**Bacterial Secondary Metabolites for Insecticidal Activity**

| Bacterial Metabolites | Activity | Reference |
|---|---|---|
| Beta-exotoxin | Insecticidal activity | (Kanaoka et al., 1978) |
| Bassianolide | Insecticidal activity | (Kanaoka et al., 1978) |
| Cyclodepsipeptide | Insecticidal activity | (Vyas et al., 2006) |

*Bacillus thuringiensis* (Bt) is a soil-borne bacteria that has been used for natural insect control. It consists of a spore, which gives it persistence, and a protein crystal within the spore, which is toxic to several species of caterpillars, including gypsy moth, cabbage looper, and tomato hornworm, and *Spodoptera litura. B. thuringiensis* is harmless to other nontargeted organisms. When the insects are consumed by certain bacteria, the toxic crystal is released in the insects gut and blocking the system that protects the insect midgut from its own digestive enzymes. The stomach is penetrated, and the insect dies by poisoning from the stomach contents and the spores themselves. Bactur, a bioworm caterpillar killer from *B. thuringiensis var israelensis*, has larvicidal activity against *Aedes* mosquitoes. Secondary metabolites from *Pseudomonas* spp. were effective against *Ae. aegypti* larvae and also toxic against the lepidopteran pest (Murty et al., 1994).

Many antibiotics are derived from bacteria and fungi that possess insecticidal properties. Streptomyces produce secondary metabolites that have antibiotic properties. Abamectin and avermectin insecticidal toxins from *Steptomyces avermectinus* have mosquitocidal activity (Goodfellow and Donell, 1989). Steptomycis sp. VITDDK3, (2S,5R,6R)-2-Hydroxy-3-5-6-trimethyloctan-4-one compounds have larvicidal activities against the mosquito vector (Moncheva et al., 2002). Streptomyces NRRL 30582, and the munumbicins E-4 and E-2 are most biologically active against *P. falciporum* protozoan (Berdi, 1989). Actimycin A, patulin, and piericidins are some antibiotics from bacterial-derived molecules. Actinobacteria have the many insecticidal properties to control the insects, including the house fly *Musca domestica* (Gadelhak et al., 2005). Mosquito survey and larvicidal activity of actinobacterial isolates against Culex larvae (Rajesh et al., 2015). Wolbachia-produced metabolites have insecticidal activity against the *Aedes* mosquito, which spreads dengue virus (Frentiu et al., 2010).

### 5.3.2.2 Viral Metabolites

Viruses can be used as pesticides to control moths, flies, mosquitoes, beetles, ants, and nematodes. Baculoviridae are a family of viruses known to infect only insects and some related pests like arthropods. Baculoviruses are used as microbial biopesticides are so specific in their action in that they infect and kill only one or a few species of Lepidoptera larvae, making them good candidates for management of crop pests with minimal effects on nontarget. Nuclear polyhedrosis viruses (NPVs), a subgroup of baculoviruses, have insecticidal properties against fourth instar larvae of *Spodoptera litura* and *Helicoverpa armigera.* Mainly NPV metabolites are genetically engineered viral pesticides used to control most crop-damaging insect caterpillars. *Lymanteria dispar* and *Oryctes rhinoceros* are insects that can be infected with NPV virus. The granulovirus of the codling moth *Cydia pomonella*, or CpGV, is a good example of a commercially successful viral insecticide. Advantages in using viral pesticide are that they are safe for humans and other nontarget organisms, reduce pesticide residues, and its unique mode of action can minimize risk of resistance to chemical insecticides. Viral enzymes are highly toxic and change the feeding behavior in medical and agricultural pests so there is a need to promote awareness for viral mediated pest control method to public.

## 5.4 CLASSIFICATION OF ENTOMOPATHOGENIC FUNGI

Entomopathogenic fungi (EPF) are widely used as biocontrols against insects like lepidopteran and mosquito vector. It is very important that the fungal infection occurs in the insects or other species, and that the cultivated crops are not affected cultivated crops. *Gibellula* is a species of fungus that can parasitize a spider, and some fungal species, including *Cordyceps* and *Erynia*, can be used against ants. EPF species are found in the divisions Ascomycota, Zycomycota, Deuteromycota, Chytridiomycota, and Oomycota. Many of the genera of EPF belong to the order of Entomophthorales (division: Zygomycota) or Hyphomycetes (also called) Deuteromycota.

### 5.4.1 Ascomycota and Deuteromycota

Ascomycota is a large fungal division said to have 64,000 species (Kirk et al., 2008). Its members are commonly known as the sac fungi or ascomycetes. This group is mainly identified based on characters such as fungal mycelia, ascospores and their fruiting body. These kinds of fungal species are known to have mutualistic relationships with environments. Insects are infected with more than 300 fungal species that are present in *Cordyceps*, which is the best known ascomycete. Many ascomycetes are of commercial importance, as they are used in yeasts in the baking process, brewing, wine fermentation, and truffles and morels. The majority of entomopathogenic species are *Ascomycota*, which develop a parasitic phase that infects the insect body. After killing the insect, the spores will develop surrounding the insect body; these are able to maintain the hyphal growth, even after the death of the insect (Evans, 1988). Ascosphaera fungal species causes chalbrood disease in bees. Spores of Ascosphaera are infected with bee larvae that germinate causing infection (Klich, 2007). Many insect-pathogenic fungal strains were already reported, namely, *Metarhizium*, *Hirsutella*, *Beauria*, *Paecilomyces*, *Tolypocladiun Aschersonia*, *Culicinomyces*, *Lecanicillium*, *Sorosporella* in addition fungal orders namely *Pleosporales*, *Myriangiales* and *Ascosphaerales* also show pathogenicity to insect pests (Samson et al., 1988).

### 5.4.2 Zygomycota

Zygomycota are organized as a single phylum and class Zygomycetes. Fungi from this group are mostly entomopathogens. These fungi reproduce mainly by asexual reproduction producing spores (Alexopoulos et al., 1996). Approximately 1050 species are known to be part of the division of Zycomycota. They mostly live in soil, plant, and animal material, and some are parasites of insects, plants, and animals. Some other species form symbiotic relationships with plants, functioning as substrates such as fruit, soil, and dung (Mucorales). Cooper and Sweeney (1986) reported *Smittium morbosum* fungal species work against mosquito vectors. The phylum of Trichomycetes associates of arthropods especially flies with aquatic larvae (Diptera) (Lichtwardt et al., 2001). Subphylum mucotina, from Trichomycetes, order Harpellales, are fungi that exclusively inhabit the gut of arthropods (Horn and Lichwardt, 1981).

### 5.4.3 Chytridiomycota

Chytridiomycota, often called chytrids, are unique among all fungi in that they are motile at least once in their life cycle. The majority of chytrids are found as saprophytic organisms, especially in freshwaters and wet soils, but there are also some marine species (Gleason et al., 2011). Single flagellum zoospores and gametes are settled and a thallus is grown, which then converts into a coenocytic hyphae or a resting spore. This phyla of fungus is especially important for species pathogenic on aquatic organisms (Sparrow, 1960). Zoospores also have the ability to adapt and respond to environmental changes like water loss and collapse of the cell (Gleason and Lilje, 2009). Most of the usual insects infecting Chytridiomycetes are contained in Blastocladiales genus Coelomomyces. Myriophagus (Chytridiales) are insect pathogens found on pupae of Diptera and Coelomycidium found on mosquitos, and blackflies contain some other species (Samson et al., 1988).

### 5.4.4 Oomycota

Oomycota (or oomycetes) are known as water molds and involve devastating plant pathogens. The cell wall of oomycota lack chitin, as in fungi, but they are made up of a mix of cellulosic compounds and glycan. Oomycota reproduce by thicker-walled spore but the conformed structure is not found in fungi. Alexopoulos et al. (1996) reported oomycota having the cell walls containing cellulose. This contrasts with the fungi chitin as a cell wall component. Many oomycota have parasitic and saprophytic feeding with dead insect tissue and organic particles (Secombes and van West, 2008). Oomycota infect a broad range of organisms like plant, fungi, insect arthropods, protists, and vertebrate including humans (Kamoun, 2003).

### 5.4.5 Ascomycota

Ascomycota and Basidiomycota together form the subkingdom of Dikarya, which exhibits a dikaryotic phase (Hibbett et al., 2007). Basidiomycota consist of some of the most familiar fungi, mushrooms, brackets, puffballs, shelf fungi, plant rusts, and smuts. There are approximately 15,000 known species, distinguished by the presence of a club-shaped reproductive organ called the basidium. Basidiomycota can undergo both asexual and sexual reproduction. Sexual reproduction in Basidiomycota takes place in the fruitbody. Basidiomycota are common and important agents of wood decay, decomposers of leaf litter, animal dung, and mycorrhizal on forest trees. Basidiomycetes have been noticed in a few studies to act as insect pathogenics. Some researchers reported the *Septobasidium* genera of *Septobasidiales* and *Uredinella* have infection-causing agents on insects (Samson et al., 1988).

## 5.5 FUNGAL METABOLITES

Fungal genera belong to the genera *Beauveria*, *Lagenidium*, *Coelomomyces*, *Culicinomyces*, and *Metarhizium* have been considered to control the vector borne disease (Kamareddine, 2012). Most entomopathogenic fungi parasitize insect hosts including *Metarhizium*, *Beauveria*, *Verticillium*, and *Nomuraea* (Singkaravanit et al., 2010). The fungi include 70 species in 7 strains that have insecticidal properties (Dhanasekaren and Thangaraj, 2014). *Streptomycetes*, *Micromonospora*, *Actinomadura*, and *Nocardiopsis* have the highest insecticidal properties (Boucias and Pendland, 1998; Roberts and St. Leger, 2004) and these fungal entomopathogens are used as microbial control agents. Many fungal metabolites are more toxic to *Cx. quinquefacciatus* than to *An. stephensi* and *Ae. aegypti. Tricophyton*, *L. giganteum*, *C. lobatum*, and *Chrysosporum tripicum* fungi show the highest mortality against *Cx. quinquefaciatus* and *An. stephensi* (Mohanty and Prakash, 2009). Seye (2009) reported *Aspergillus clavatus* is pathogenic to *Ae. aegypti*, *An. gambiae*, and *Cx. quinquefasciatus*. The main route of entrance of the entomopathogens is through integument and it may also infect the insect by ingestion method or through wounds (Holder et al., 2005). Roberts and Humber (1981) reported that entomopathogenic fungi infected a wide range of insects including lepidopteran larvae, aphids, and thrips, which are of great concern in agriculture. *Metarhizium anisopliae* Sorokin and other entomopathogenic fungi are being examined as potential biological insect control agents (Lacey et al., 2001).

Singh and Prakash reported (2012a, 2012b) isolated metabolites from *Trichophytom ajelloi* produced adulticidal activity in the mosquitoes *An. stephensi*, *Ae. aegypti*, and *Cx. quinquefacciatus. Culicinomyces clavisporus* had fungal-derived metabolites activity against *An. stephensi*, *Ae. aegypti*, and *Cx. quinquefacciatus. Fusarium oxyporum*-derived metabolites conjugated with nanoparticles showed the highest activity against three mosquito species: *An. stephensi*, *Ae. aegypti*, and *Cx. quinquefacciatus* (Soni and Prakash, 2010, 2012). *Metarhizium anisopliae*, *conidia*, and metabolites test against the *An. gambiae s.s.*, *An. funestus*, *An. stephensi*, and *Cx. quinquefacciatus* (Farenhorest et al., 2009). *Metarhizium anisopliae* and *Beauveria bassiana* have been studied and applied in controlling of rice brown planthoppers, rice bugs, coconut beetles, grasshoppers, and termites (Chinh et al., 2001). Vey et al. (2001) reported fungi-producing secondary metabolites, derivatives from various intermediates in primary metabolism, some of which have insecticidal activities. Moraes et al. (2001) reported *Aspergillus flavus*, an entomopathogenic fungus, was pathogenic against mosquito species *Ae. fluviatilis* and *Cx. quinquefasciatus*. Secondary metabolites like bassianolide, a cyclodepsipeptide produced by the fungus *Beauveria bassiaria*, elicits atonic symptoms in silkworm larvae (Kanaoka et al., 1978). The pathogenicity of *M. anisopliae* isolate obtained from dead cadavers of mole formula cricket (*Gryllotalpa orientalis*) was reported to be pathogenic to cockroaches (*Periplaneta americana*) (Wakil et al., 2012). Murali Mohan et al. (1999) reported *P. americana* insects affected with *Beauveria bassiana* fungi. EPF are a major component of integrated pest management techniques as biological control agents against insect pests and other arthropods and forestry and agriculture field (Inglis et al., 2000). An example of a forest pest infected with entomopathogenic fungi, *Erynia Canadensis* was *Cinara pinea* present in pine forests. *Beaveria bassiana* thwart *Dendrolimus punctatus* in Masson pine. Fighting vector-borne disease by using chemical/synthetic pesticides is becoming increasingly difficult because of increasing resistance of mosquitoes. Therefore, the alternative use of microbial pesticide is a natural product that is safe and effective against the vector. In addition many advantages in using microbial control agents, such as human safety, other nontarget organisms, the aquatic system, and pesticide residues are minimized in food, and biodiversity would be increased in managed ecosystems (Shahid et al., 2012).

## 5.6 INSECTICIDAL SECONDARY METABOLITES

Entomopathogenic fungi produce several secondary metabolites that cause insecticidal activity (Table 5.3). Many potential bioactive compounds have been isolated from entomopathogenic fungi. Vyas et al. (2006) reported the entomopathogenic fungi *Paecilocyces fumosoroseus*, *Fusarium moniliforme*, and *Beauveria bassiana* produce compounds like beauvericin, including cyclode peptide and enniatin against mosquito larval activity. Beauvericin compound was isolated from *Beauveria bassiana* but it is also produced in several other fungi like Fusarium. Beauvericin is a cyclic hexade peptide with alternating N-methyl-phenylalanyl and D-hydroxy-iso-valeryl residues. Mainly it is active against gram-positive bacteria and mycobacterium, and it is also involved in programmed cell death in vertebrates. Secondary metabolites of destruxins, derived from *Metarhizium anisopliae*, are bioactive agents

**TABLE 5.3**
**Fungal Secondary Metabolites from Entomopathogenic Fungi for Insecticidal Activity**

| Fungal Metabolites | Activity | Reference |
|---|---|---|
| Beauvericin | Insecticidal activity | (Hamill et al., 1969; Vyas et al., 2006) |
| Bassianolide | Insecticidal activity | (Kanaoka et al., 1978) |
| Cyclodepsipeptide | Insecticidal activity | (Vyas et al., 2006) |
| Enniatin | Insecticidal activity | (Vyas et al., 2006) |
| Destruxins | Insecticidal activity | (Sowjanya et al., 2008) |
| Trichothecenes | Insecticidal and antimicrobial activity | (Jarvis et al., 1988; Douglas et al., 1989) |
| Zearalenone | Insecticidal activity | (Chattopadhyay et al., 2012) |
| Fumonisins | Insecticidal and antimicrobial activity | (Blandino and Reyneri, 2009) |
| Aflatoxin | Insecticidal activity | (Uysal and Agar, 2005; Jaya et al., 2014) |

and have insecticidal activity against *Spodoptera litura* larvae. Mainly they are cyclode peptides comprised of hydroxyl acid and five amino acids.

## 5.7 ADVANTAGES OF BIOLOGICAL PESTICIDE

Microbial pesticides are less harmful, pose less environmental risk, are target specific, do not cause pollution, are relatively cheaper, and the pest never builds resistance when compared to the chemical pesticide. Biopesticides are biodegradable and easy to recycle, because they are made from natural material like plant and microbes. When the pesticide reaches the soil after treatment, other microorganisms can easily recycle it. Some pests build resistance against chemical pesticides. Resistance is the ability of the insect's offspring to tolerate the pesticide that kills its ancestor. But it will not happen if using a biopesticide, because a biopesticide contains several active toxins that work at the same time. This active toxin is produced in the secondary metabolites of the plant and microbes and is used as the material to make biopesticide. Biopesticide are different from chemical pesticides because chemical pesticides contain only one active toxin. It is difficult for pests to tolerate at the same time the several active toxins that are contained in biopesticides. Therefore, microbial insecticides offer effective alternatives for the control of many insect pests. Biological control is the use of natural enemies to manage mosquito populations and other parasites, and it is ecofriendly for the control of insect vectors.

## 5.8 ISOLATION AND FORMULATION OF FUNGAL SECONDARY METABOLITES

### 5.8.1 Extraction of Secondary Metabolites

With improvement in screening, separation, and isolation techniques, the number of natural compounds discovered from natural sources exceeds 1 million. Among them, 50% to 60% are produced by plants in the form of alkaloids, flavonoids, terpenoids, steroids, and carbohydrates, and 5% of these products have a microbial origin. In terms of the biological activities shown by the isolated metabolites, of all the natural products isolated from plants only 20% to 25% show biological activity, whereas approximately 10% of microbial metabolites shows this activity. Microorganisms produce many compounds with biological activity. From the 22,500 biologically active compounds so far obtained from microbes, about 40% are produced by fungi (Demain, 2014). Isolation of secondary metabolites from fungi mainly requires between 5- to 15-day-old culture. Researchers have used extraction methods using various solvents for the isolation of secondary metabolites from fungi. In their methods, ethyl acetate is used for removal of the organic layer and then the crude extracts are analyzed by thin-layer chromatography (TLC) and nuclear magnetic resonance (NMR) for separation and structural determination of metabolites. Sometimes culture is extracted with two solvents (chloroform and ethyl acetate). After that the solvent is dried over anhydrous sodium sulfate, filtered, then distilled to give a semisolid extract that is used for further analysis (TLC) (Hamed, 2015). Fungal metabolites from *Aspergillus* spp. is usually extracted using ethyl acetate and then separated using separating funnels. The metabolites that are now dissolved in ethyl acetate solvent are separated by using separating funnels. In the separating funnel, ethyl acetate is added. To that add some ethyl acetate, shake well, and allow it to settle for a few minutes. Later, two layers were observed in the separating funnel. The bottom layer is the broth layer, which is discarded, and the upper layers of ethyl acetate with metabolites are collected, which is called the organic layer. The organic layer was separated and concentrated by using a rotary evaporator (under reduced pressure 23–27 mm Hg at 48°C with the resultant residue stored at room temperature for further experiments. The washes are repeated three times to extract the complete metabolites. This extract was separated by a rotary evaporator for the collection of crude extract. The organic layer was allowed to dry and the crude metabolites are used for further analysis (Swathi et al., 2013). Structural identifications of the secondary metabolites were made by spectroscopic analysis.

Nonribosomal peptides, alkaloids, terpenes, and polyketides are the main classes of fungal secondary metabolites whose expression and secretion appear to be controlled by various genetic and cellular regulatory mechanisms (Hoffmeister and Keller, 2007; Chanda et al., 2009). These molecules are a network of regulatory molecules, which is required by the fungi to meet various ecological challenges encountered during the lifetime.

All chemical reagents were commercially available and treated with standard methods before use. $^{1}H$ NMR and $^{13}C$ NMR spectra were recorded in CDC13 on a Bruker AVANCE 600 MHz, and chemical shifts are given in parts per million (ppm) relative to tetramethylsilane (TMS). ESI-MS spectra were obtained on the Mariner System 5304 mass spectrometer (LU Yi-hui et al., 2016).

### 5.8.2 Secondary Metabolite Formation as a Resistance Trait in Fungi

Studies from various disciplines like molecular genetics, biochemistry, and ecology have shown that secondary metabolites from fungi provide putative resistance mechanisms against other species and pathogens. However, concrete, supporting experimental evidence on their protective role is lacking. Nevertheless earlier reviews have interpreted the principle function of secondary metabolites to be resistance characters or to even have evolved defenses against their natural enemies (Vining, 1990; Domain and Fang, 2000).

### 5.8.3 Interactions with Insect Competitors

Some secondary metabolites are involved in fending off competing microbes, secretion of insecticidal mycotoxins such as aflatoxin B1 (Chinnici and Bettinger, 1984; Rohlfs and Obmann, 2009), and producing allelopathy among plants. they grow on insects which also exploit the same resource patch (Hodge and Mitchell, 1997; Rohlfs et al., 2005; Trienens et al., 2010). Insects also have evolved several mechanisms to fend off fungal infection with the help of mutualistic microbes (Rohlfs and Kürschner, 2010). Insects also have a very robust detoxification system comprised of a battery of enzymes, prominent among them are cytochrome P450 monooxygenases (Niu et al., 2008) by which they detoxify mycotoxins.

*Aspergillus flavus*, *A. fumigatus*, and *A. nidulans* bring about infection in *Drosophila melanogaster* (Trienens et al., 2010). Survival of *Drosophila* larvae in the presence of the Aspergillus nuclear protein LaeA in *A. nidulans* mutant was almost as high as under mold-free conditions, which suggests that LaeA controls the expression of almost all insecticidal properties of *A. nidulans* in this system. This shows LaeA does not control all genomic regions where secondary metabolite gene clusters are located (Perrin et al., 2007; Kosalková et al., 2009), which may be one reason for the species-specific effects of the various LaeA *Aspergillus* mutants on the survival of *Drosophila* larvae. The reason for this effect is hypothesized as a result of higher diversity of secondary metabolites in produced *A. flavus* and *A. fumigatus*, possibly reflecting greater diversity in secondary metabolite clusters, which may lead to a higher proportion of metabolic pathways that are not controlled by LaeA and, hence, still continue to produce insecticidal compound. Fungal secondary metabolites are not applicable for direct use in the environment or testing and development of metabolic products. So the secondary metabolic products or compounds formulated are generally preferred. Formulations are used very frequently in the pharmaceutical industry.

## 5.9 FUTURES OF MYCOINSECTICIDES

### 5.9.1 Mycoinsecticides

Mycoinsecticides are usually formulated products with live conidia as the active ingredient. The conidia germinate on contact with the cuticle of the insect, produce a penetrating germ tube and establish a systemic infection that kills the host in 7 to 21 days depending on conditions, especially temperature and dose. At death, the host insect is a hard cadaver full of mycelium that grows out through the cuticle to form a new generation of conidia on the outside of the cadaver. This process and the initial germination and invasion are dependent on very high humidity and warm conditions. In epigeal insects such as locusts, it is necessary to formulate the conidia in oil to overcome the humidity barrier at infection. The live conidia are produced by solid substrate fermentation on sterile moist rice, and then used directly as natural granules, or removed by sieving and formulated as a powder, granule, or oil concentrate. Ideally 1 to 3 kg of substrate will produce enough conidia after 2 to 3 weeks incubation to treat a hectare. Several basal filamentous fungal lineages are parasites of invertebrates, implying that fungal–arthropod interactions constitute ancient paradigms for examining pathogenesis.

*Beauveria bassiana* (Bals.) *Vuill.* and *Metarhizium robertsii* (formerly *Metarhizium anisopliae* Metsch.) species are the most widely studied fungal pathogens of insects. Both have a broad arthropod host range and are capable of infecting an astonishingly wide range of targets spanning most insect orders (e.g., Lepidoptera, Coleoptera, Hymenoptera, Diptera, and others). Even extending beyond insects they are efficient in infecting other arthropods such as ticks and mites (subclass Acari) (Kirkland et al., 2004a, 2004b, 2005). The systematic and phylogeny of *Beauveria* spp. and *Metarhizium* complex species have been well characterized (Bischoff et al., 2009; Rehner et al., 2011), and from an evolutionary perspective, it is clear that these fungi and their arthropod hosts have been engaged in a long-term struggle in relation to pathogen tactics designed to overcome ever-changing host defenses. Both fungi are also saprophytes and form intimate associations with plants. They are essentially facultative pathogens, that is, they do not require insect hosts to complete their life cycle, with the exception of sexual stages that have only sparsely been reported on (Li et al., 2001; Sung et al., 2006), and can efficiently sporulate on a wide range of artificial substrates.

## 5.10 CONCLUSION

Mosquito-borne diseases are a major health issue in the developing world. Chemical insecticides remain the most important vector control method. Due to use the same or different synthetic chemical insecticide causes a lot of side effects on the natural environment and nontarget insects. Many insect vectors evolved insecticide resistance to the key classes of insecticides, particularly pyrethroids. To preserve effectiveness of insecticides, resistance management should be followed. In this respect, appropriate resistance strategies involving the use of microbial insecticides, especially entomopathogenic fungi and their secondary metabolites, can be useful in the future. Microbial consortium and their metabolites might became more effective than chemical insecticides. There is active research on identification of novel fungal metabolites, and bioprospecting their ability to control mosquitoes. In the field as well as testing, the stability of the compound can be useful in designing ecofriendly insecticides for future.

## REFERENCES

Alexopoulos, C. J., Mims, C. W., Blackwell, M. 1996. *Introductory Mycology*. New York: John Wiley & Sons.

Alouani, A., Rehimi, N., Soltani N. 2009. Larvicidal activity of a neem tree extract (Azadirachtin) against mosquito larvae in the Republic of Algeria. *Jordan J. Bio. Sci.*, 2(1).

Beier, J. C., Keating, J., Githure, J. I., Macdonald, M. B., Impoinvil, D. E., Novak, R. J. 2008. Integrated vector management for malaria control. *Malaria J.*, 7(1), S4.

Berdi, J. 1989. The discovery of new bioactive metabolites, screening and identification. *Prog. Ind. Microbiol.*, 27, 3–25.

Bischoff, J. F., Rehner, S. A., Humber, R. A. 2009. A multilocus phylogeny of the *Metarhizium anisopliae*. *Mycologia*, 101(4), 512–530.

Blandino, M., Reyneri, A. 2009. Effect of fungicide and foliar fertilizer application to winter wheat at anthesis on flag leaf senescence, grain yield, flour bread-making quality and DON contamination. *European J. Ag.*, 30, 275–282.

Bocquene, G., Franco, A. 2005. Pesticide contamination of the coastline of Martinique. *Marine Poll. Bull.*, 51, 612–619.

Boucias, D. G., Pendland, J. C., Latge, J. P. 1988. Nonspecific factors involved in the attachment of entomopathogenic Deuteromycetes to host insect cuticle. *Appl. Environ. Microbiol.* 54(7), 1795–1805.

Braks, M. A. H., Honorio, N. A., Lounibos, L. P., Lourenc, R., Juliano, A. S. 2004. Interspecific competition between two invasive species of container mosquitoes, *Aedes aegypti* and *Aedes albopictus* (Diptera: Culicidae) in Brazil. *An. Entomol. Soc. Am.*, 97, 130–139.

Brengues, C., Hawkes, N. J., Chandre, F., McCarroll, L., Duchon, S. 2003. Pyrethroid and DDT cross-resistance in *Aedes aegypti* is correlated with novel mutations in the voltage-gated sodium channel gene. *Med. Vet. Entomol.*, 17, 87–94.
Brogdon, W. G., McAllister, J. C. 1998. Insecticide resistance and vector control. *Emerg. Infec. Dis.*, 4, 605–613.
Brown, A. W. A. 1986. Insecticide resistance mosquitoes: A pragmatic review. *J. Am. Mosq. Con. Ass.*, 2, 123–140.
Bruce-Chwatt, L. J. 1985. *Essential Malariology*. 2nd ed. John Willey & Sons, New York, 166–179.
Busvine, J. R. 1969. The impact on malaria of insecticide resistance in Anopheles mosquitoes. *Trans. Roy. Soc. Trop. Med. Hyg.*, Suppl. No., 4, 19–30.
Chanda, A., Roze, L. V., Kang, S., Artymovich, K. A., Hicks, G. R. 2009. A key role for vesicles in fungal secondary metabolism. *Proc. Nat. Acad. Sci.*, 106, 19533–19538.
Chanda, E., Masaninga, F., Coleman, M., Sikaala, C., Katebe, C., MacDonald, M., Baboo, K.S., Govere, J., Manga, L. 2008. Integrated vector management: The Zambian experience. *Malaria J.*, 7, 164.
Chattopadhyay, P., Chatterjee, S., Gorthi, S., Sen, S. K. 2012. Exploring agricultural potentially of serratia entomophila AB2: Dual properties of biopesticide and fertilizers. *Brit. Biotechnol. J.*, 2(1), 1–12.
Chinh, T. K., Quyen, H. T., Tu, H. T. M. 2001. Selecting medium for rearing and multiplicating *Metarhizium anisopliae* fungus for controlling termite, Coptotermes formosanus. *Proceeding, International Workshop On Biology Hanoi*, Vol. 2, 77–81.
Chinnici, J. P., Bettinger, D. A. 1984. Effects of aflatoxin B1 and caffeine on viability in natural strains of *Drosophila melanogaster*. *J. Invert. Pathol.*, 44, 263–266.
Coleman, M., Hemingway, J. 2007. Insecticide resistance monitoring and evaluation in disease transmitting mosquitoes. *J. Pesticide. Sci.*, 32, 69–76.
Cooper, R., Sweeney, A. W. 1986. Laboratory studies on the recycling potential of the mosquito pathogenic fungus Culicinomyces. *J. Inver. Pathol.*, 48, 152–158.
Corriveau, R., Philippon, B., Yebakima, A. 2003. *La Dengue dans les Departements Francais d'Amerique. Comment Optimiser la Lutte Contre cette Maladie*. IRD Edition. Expertise Collegiale, Paris.
Dalziel, J. M. 1955. *The Useful Plants of West-Tropical Africa*. Crown Agents for Oversea Governments and Administration, London, 147.
David, J. P., Coissac, E., Melodelima, C., Poupardin, R., Riaz, M. A. 2010. Transcriptome response to pollutants and insecticides in the dengue vector *Aedes aegypti* using next-generation sequencing technology. *B.M.C. Genomic.*, 11, 216.
Demain, A. L. 2014. Importance of microbial natural products and the need to revitalize their discovery. *J. Ind. Microbiol. Biotechnol.*, 41, 185–201.
Demain, A. L., Fang, A. 2000. The natural function of secondary metabolites. In: Sheper, T. (Ed.), *Advances in Biochemical Engineering/Biotechnology*. Springer, Heidelberg, 1–39.
Denholm, I., Rowland, M. W. 1992. Tactics for managing pesticide resistance in arthropods: Theory and practice. *An. Rev. Entomol.*, 37, 91–112.
Dhanasekaran, D., Sakthi, V., Thajuddin, N., Panneerselvam, A. 2010. Preliminary evaluation of *Anopheles* mosquito larvicidal efficacy of mangrove actinobacteria. *Int. J. App. Biol. Pharm. Technol.*, 1, 374–381.
Dhanasekaran, D., Thangaraj, R. 2014. Microbial secondary metabolites are an alternative approaches against insect vector to prevent zoonotic diseases. *Asian Pac. J. Trop. Dis.*, 4(4), 253–261.
Djouaka, R. F., Bakare, A. A., Coulibaly, O. N., Akogbeto, M. C., Ranson, H. 2008. Expression of the cytochrome P450, CYP6P3 and CYP6M2 are significantly elevated in multiple pyrethroid resistant populations of *Anopheles gambiae* s.s. from Southern Benin and Nigeria. *B.M.C. Genomic.*, 9, 538.

Duke, J. A. 1988. *CRC Handbook of Medicinal Herbs*. CRC Press, Boca Raton, FL, 253–254.

Eberbach, P. L., Douglas, L. A. Herbicide effects on the growth and nodulation potential of Rhizobium trifolii with *Trifolium subterraneum*. *Plant Soil*, 119, 15–23.

Evans, H. C. 1988. Coevolution of entomogenous fungi and their insect hosts. In K. A. Pirozynski, D. L. Hawksworth (Eds.). *Coevolution of Fungi with Plants and Animals*.

Farenhorst, M., Mouatcho, J. C., Kikankie, C. K., Brooke, B. D., Hunt, R. H., Thomas, M. B., Loekemoer, L. L., Knols, B. G. J., Coetzee, M. 2009. Fungal infection counters insecticide resistance in African malaria mosquitoes. *Proc. Nat. Acad. Sci.*, 106, 17443–17447.

Frentiu, F. D., Robinson, J., Young, P. R., McGraw, E. A., O Neill, S. L. 2010. Wolbachia-mediated resistance to dengue virus infection and death at the cellular level. *PLoS ONE*. 5:e13398 10.1371/journal.pone.0013398.

Gadelhak, G., Khaled, A., El-Tarabily, K. A., Fatma, K., Al-Kaabi, F. K., 2005. Insect control using chitinolytic soil actinomycetes as biocontrol agents. *Int. J. Agri. Biol.*, 7(4), 627–633.

Georghiou, G. P., Taylor, C. E. 1977. Genetic and biological influences in the evolution of insecticide resistance. *J. Eco. Entomol.*, 70(3), 319–323.

Gleason, F. H., Kupper, F. C., Amon, J. P., Picard, K., Gachon, C. M. M., Marano, A. V. Lilje, O. 2011. Zoosporic true fungi in marine ecosystems: A review. *Marine. Freshwater Res.*, 62(4), 383–393.

Gleason, F. H., Lilje, O. 2009. Structure and function of fungal zoospores: Ecological implications. *Fungal Ecol.*, 2(2), 53–59.

Gleiser, R. M., Zygadlo, J. A. 2009. Essential oils as potential bioactive compounds against mosquitoes. *Rec. Ad. Phytochem.*, 37, 53–76.

Goodfellow, M., O'Donnell, A. G. 1989. Search and discovery of industrially significant actinomycetes. In *Microbial Products: New Approaches* ed. Baumberg, S., Hunter, I.S., Rhodes P.M., pp. 343–383. Cambridge, UK: Cambridge University Press.

Gordon, D. F., Gannon, J. C. 2000. The global infectious disease threat and its implications for the United States. National Intelligence Council http://www.dni.gov/nic/special _globalinfectious.html.

Gutierrez, P. M. Jr., Antepuesto, A. N., Eugenio, B. A. L., Santos, M. F. L. 2014. Larvicidal activity of selected plant extracts against the dengue vector *Aedes aegypti* Mosquito. *Int. Res. J. Biol. Sci.*, 3(4), 23–32.

Hamed, A., Maria, N., Karolina, N., Nico, M., van, S., Wilfred, F. M. R. 2015. Functional and compositional responses in soil microbial communities along two metal pollution gradients: Does the level of historical pollution affect resistance against secondary stress? *Biol. Fertil. Soils*, 51, 879–890.

Hamill, R. L., Higgens, C. E., Boaz, H. E., Corman, M. 1969. The structure of beauvericin, a new depsipeptide antibiotic toxic to *Artemia salina*. *Tetrahedron Lett.*, 45, 4255–4258.

Hansen, I. A., Marcombe, S., Mathieu, R. B., Pocquet, N., Riaz, M. A., Poupardin, R., Sélior, S., Darriet, F., Reynaud, S., Yébakima, A., Corbel, V., David, J.-P., Chandre, F. 2012. Insecticide resistance in dengue vector *Aedes aegypti* from Martinique: Distribution, mechanisms and relations with environmental factors. *PLoS One*, 7(2), e30989.

Hemingway, J., Beaty, B. J., Rowland, M., Scott, T. W., Sharp, B. L. 2006. The Innovative Vector Control Consortium: Improved control of mosquito-borne diseases. *Trends Parasitol.*, 22(7), 308–312.

Hemingway, J., Field, L., Vontas, J. 2002. An overview of insecticide resistance. *Science*, 298 (5591), 96–97.

Hemingway, J., Karunaratne, S. 1998. Mosquito carboxylesterases: A review of the molecular biology and biochemistry of a major insecticide resistance mechanism. *Med. Vet. Entomol.*, 12: 1–12.

Hemingway, J., Ranson, H. 2000. Insecticide resistance in insect vectors of human disease. *Ann. Rev. Entomol.*, 45, 371–391.

Hibbett, D. S., Binder, M., Bischoff, J. F., Blackwell, M., Cannon, P. F., Eriksson, O. E. Zhang, N. 2007. A higher-level phylogenetic classification of the Fungi. *Mycol. Res.*, 111(5), 509–547.

Hodge, S., Mitchell, P. 1997. Inhibition of *Drosophila melanogaster* and D. hydei by *Aspergillus niger. Dros. Info. Serv.*, 80, 6–7.

Hoffmeister, D., Keller, N. P. 2007. Natural products of filamentous fungi: Enzymes, genes, and their regulation. *Nat. Prod. Rep.*, 24, 393–416.

Holder, D. J., Keyhani, N. O. 2005. Adhesion of the entomopathogenic fungus *Beauveria bassiana* to substrata. *App. Env. Microbiol.*, 71(9), 5260–5266.

Horn, B. W., Lichtwardt, R. W. 1981. Studies on the nutritional relationship of larval *Aedes aegypti* (Diptera: Culicidae) with *Smittium culisetae* (Trichomycetes). *Mycologia.*, 724–740.

Hu, Z., Du, Y., Nomura, Y., Dong, K. A. 2011. Sodium channel mutation identified in *Aedes aegypti* selectively reduces cockroach sodium channel sensitivity to type I, but not type II pyrethroids. *Insect Biochem. Mol. Biol.*, 41(1), 9–13.

Inglis, G. D., Ivie, T. J., Duke, G. M., Goettel, M. S. 2000. Influence of rain and conidial formulation on persistence of *Beauveria bassiana* on potato leaves and Colorado potato beetle larvae. *Bio. Control.*, 18, 55–64.

Innovative Vector Control Consortium. 2015. Annual Report.,120–125.

Insecticide Resistance Action Committee (IRAC). 2011. *Prevention and management of insecticide resistance in vectors of public health importance.* 2nd ed.

Jarvis, M. J., Russell, M. A., Benowitz, N. L., Feyerabend, C. 1988. Elimination of nicotin from body fluids: Implication for noninvasive measurement of Tobacco smoke exposure. *Am. J. Public. Health.*, 78(6), 696–698.

Junker, R. R., Gershenzon, J., Unsicker, S. B. 2011. Floral odor bouquet loses its ant repellent properties after inhibition of terpene biosynthesis. *J. Chem. Ecol.*, 37(12), 1323–1331.

Kamareddine, L. 2012. The biological control of the malaria vector. *Toxins*, 4, 748–767.

Kamoun, S. 2003. Molecular genetics of pathogenic oomycetes. *Eukaryotic Cell.*, 2(2), 191–199.

Kanaoka, M., Isogai, A., Murakoshi, S., Suzuki, A., Tamura, S. 1978. Bassianolide, a new insecticide cyclodepsipeptide from *Beauveria bassiana* and *Verticillium lecanii. Agric. Bio. Chern.*, 42, 629–635.

Keiser, J., Singer, B. H., Utzinger, J. 2005. Reducing the burden of malaria in different eco-epidemiological settings with environmental management: A systematic review. *Lancet Inf. Dis.*, 5(11), 695–708.

Kirk, P. M., Cannon, P. F., Minter, D. W., Stalpers, J. A. 2008. *Dictionary of the Fungi* (10th ed.). Wallingford, UK: CAB International.

Kirkland, B. H., Cho, E. M., Keyhani, N. O. 2004a. Differential susceptibility of *Amblyomma maculatum* and *Amblyomma americanum* (Acari: Ixodidea) to the entomopathogenic fungi *Beauveria bassiana* and *Metarhizium anisopliae. Biol. Control*, 31, 414–421.

Kirkland, B. H., Eisa, A., Keyhani, N. O. 2005. Oxalic acid as a fungal acaracidal virulence factor. *J. Med. Entomol.*, 42, 346–351.

Kirkland, B. H., Westwood, G. S., Keyhani, N. O. 2004b. Pathogenicity of entomopathogenic fungi *Beauveria bassiana* and *Metarhizium anisopliae* to Ixodidae tick species *Dermacentor variabilis*, *Rhipicephalus sanguineus*, and *Ixodes scapularis. J. Med. Entomol.*, 41, 705–711.

Kleinschmidt, I., Schwabe, C., Shiva, M., Segura, J. L., Sima, V., Mabunda, S. J., Coleman, M. 2009. Combining indoor residual spraying and insecticide-treated net interventions. *Am. J. Trop. Med. Hyg.*, 81(3), 519–524.

Kleinschmidt, I., Sharp, B., Benavente, L. E., Schwabe, C., Torrez, M., Kuklinski, J., Morris, N., Raman, J., Carter, J. 2006. Reduction in infection with *Plasmodium falciparum* one year after the introduction of malaria control interventions on Bioko Island, Equatorial Guinea. *Am. J. Trop. Med. Hyg.*, 74(6), 972–978.

Klich, M. A. 2007. *Aspergillus flavus*: The major producer of aflatoxin. *Mol. Plant Pathol.*, 8(6), 713–722.

Knio, K. M., Usta, J., Dagher, S., Zournajian, H., Kreydiyyeh, S. 2008. Larvicidal activity of essential oils extracted from commonly used herbs in Lebanon against the seaside mosquito, *Ochlerotatus caspius*. *Bioresour. Technol.*, 99, 763–768.

Komatsu, R., Korenromp, E. L., Low-Beer, D., Watt, C., Dye, C., Steketee, R. W., Nahlen, B. L., Lyerla, R., Garcia-Calleja, J. M., Cutler, J., Schwartlander, B. 2010. Lives saved by Global Fundsupported HIV/AIDS, tuberculosis and malaria programs: Estimation approach and results between 2003 and end-2007. *B.M.C. Inf. Dis.*, 10, 109.

Kosalková, K., García-Estrada, C., Ullán, R. V., Godio, R. P., Feltrer, R. 2009. The global regulator LaeA controls penicillin biosynthesis, pigmentation and sporulation, but not roquefortine C synthesis in *Penicillium chrysogenum*. *Biochimie.*, 91, 214–225.

Kranthi, K. R. 2005. Insecticide resistance monitoring, mechanisms and management manual. *Central Institute for Cotton Research, India*, 78–82.

Kuppusamy, C., Murugan, K. 2006. Mosquitocidal effect of ethanolic extracts of *Andrographis paniculata* Nees on filarial vector *Culex quinquefasciatus* Say (Diptera: Culicidae). In: International Conference on Diversity of Insects: Challenging Issues in Management and Conservation, 30 January–3 February 2006, Tamil Nadu, India, 194.

Lacey, L. A., Frutos, R., Kaya, H. K., Vail, P. 2001. Insect pathogens as biological control agents: Do they have a future? *Bio. Control.*, 21(3), 230–248.

Lengeler, C. 2004. Insecticide-treated bed nets and curtains for preventing malaria. *Cochrane Database Syst. Rev.*, (2), CD000363.

Lengeler, C., Sharp, B. 2003. Indoor residual spraying and insecticide-treated nets. In: Global Health Council, Reducing Malaria's Burden: Evidence of Effectiveness for Decision Makers, White River Junction, VT, 17–24.

Li, Z. Z., Li, C. R., Huang, B., Fan, M. Z. 2001. Discovery-and demonstration of the teleomorph of *Beauveria bassiana* (Bals.), an important entomogenous fungus. *Chin. Sci. Bull.*, 46, 751–753.

Lichtwardt, R. W., Cafaro, M. J., White, M. M. 2001. The Trichomycetes: Fungal associates of arthropods. Revised edition. Published on the Internet: http://www.nhm.ku.edu/~fungi.

Lindsay, S. W., Snow, R. W., Broomfield, G. L., Janneh M. S., Wirtz, R. A., Greenwood, B. M. 1989. Impact of permethrin-treated bednets on malaria transmission by the *Anopheles gambiae* complex in The Gambia. *Med. Vet. Entomol.*, 3(3), 263–271.

Lu, Y.-H., Li, S., Shao, M.-w., Xiao, X.-H., Kong, L.-C., Jiang, D.-H., Zhang, Y.-L. 2016. Isolation, identification, derivatization and phytotoxic activity of secondary metabolites produced by *Cladosporium oxysporum* DH14, a locust-associated fungus. *J. Integ. Agri.* 15(4), 832–839.

Manjari, M. S., Karthi, S., Ramkumar, G., Muthusamy, R., Natarajan, D., Shivakumar, M. S. 2014. Chemical composition and larvicidal activity of plant extracts from *Clausena dentata* (Willd) (Rutaceae) against dengue, malaria and filariasis vectors. *Parasitol. Res.*, 113, 2475–2481.

Mebrahtu, Y. B., Norem, J., Taylor, M. 1997. Inheritance of larval resistance to permethrin in Aedes aegypti and association with sex ratio distortion and life history variation. *Am. J. Trop. Med. Hyg.*, 56, 456–465.

Mohan, L., Sharma, P., Shrivastava, C. N. 2006. Evaluation of Solanum xanthocarpum extract as a synergist for cypermethrin against larvae of filarial vector Culex quinquefasciatus (Say). *Entomol Res.*, 36, 220–225.

Mohanty, S. S., Prakash, S. 2000. Laboratory evaluation of *Trichophyton ajelloi*, a fungal pathogen of *Anopheles stephensi* and *Culex quinquefasciatus*. *J. Am. Mos. Con. Ass.*, 16 (3), 254–257.

Mohanty, S. S., Prakash, S. 2002. Efficacy of *Chrysosporium lobatum* against larvae of malaria vector *Anopheles stephensi* in the laboratory. *Curr. Sci.*, 83(12), 1585–1588.
Mohanty, S. S., Prakash, S. 2008. Laboratory and field evaluation of the fungus *Chrysosporium lobatum* against the larvae of the mosquito *Culex quinquefasciatus. Parasitol. Res.*, 102 (5), 881–886. http://dx.doi.org/10.1007/s00436-007-0843.
Mohanty, S. S., Prakash, S. 2009. Effects of culture media on larvicidal property of secondary metabolites of mosquito pathogenic fungus *Chrysosporium lobatum* (*Moniliales: Moniliaceae*). *Acta Trop.*, 109(1), 50–54.
Mohanty, S. S., Prakash, S. 2010. Comparative efficacy and pathogenicity of keratinophilic soil fungi against *Culex quinquefasciatus* larvae. *J. Microbiol.*, 50(3), 299–302.
Moncheva, P., Tishkov, S., Dimitrova, N., Chipeva, V., Nikolova, S. A., Bogatzevska, N. 2002. Characteristics of soil actinobacteria from Antarctica. *J. Cul. Coll.*, 3, 3–14.
Moraes, A. M. L. D., Costa, G. L. D., Barcellos, M. Z. C., Oliveira, R. L., Oliveira, P. C. 2001. The entomopathogenic potential of *Aspergillus* spp. in mosquitoes vectors of tropical diseases. *J. Basic. Microbiol.*, 41(1), 45–49.
Murali Mohan, C. H., Aruna Lakshmi, K., Uma Devi, K. 1999. Laboratory evaluation of the pathogenicity of three isolates of the entomopathogenic fungus *Beauveria bassiana* (Bals.) *Vuillemin* on the American cockroach (Periplaneta americana). *Bio. Sci. Technol.*, 9, 29–33.
Murty, M. G., Srinivas, G., Seker, V. 1994. Production of mosquitocidal exotoxin by a *Pseudomonas fluorescens. J. Invert. Pathol.*, 64, 68–70.
Muthusamy, R., Shivakumar, M. S. 2015. Resistance selection and molecular mechanisms of cypermethrin resistance in red hairy caterpillar (*Amsacta albistriga Walker*). *Pestic. Biochem. Physiol.*, 117, 54–61.
Muthusamy, R., Suganya, R., Gowri, M., Shivakumar, M. 2013. Biochemical mechanisms of organophosphate and pyrethroid resistance in red hairy caterpillar *Amsacta albistriga* (Lepidoptera: Arctiidae). *J. Saudi Soc. Agric. Sci.*, 12, 47–52.
N'guessan, R., Corbel, V., Akogbeto, M., Rowland, M. 2007. Reduced efficacy of insecticide treated nets and indoor residual spraying for malaria control in pyrethroid resistance area, Benin. *Emer. Inf. Dis.*, 13, 199–206.
Najera, J. A. 1999. Malaria control: Achievements, problems and strategies. World Health Organization, Geneva.
Nardini, L., Christian, R. N., Coetzer, N., Ranson, H., Coetzee, M., Koekemoer, L. L. 2012. Detoxification enzymes associated with insecticide resistance in laboratory strains of *Anopheles arabiensis* of different geographic origin. *Parasit. Vect.*, 5, 111–113.
National Vector Borne Disease Control Programme. 2013. Dengue Cases and Deaths in the Country.
Nauen, R. 2006. Insecticide resistance in public health pests: A challenge for effective vector control. *Pub. Heal. J.*, 18, 8–15.
Nazni, W. A., Kamaludin, M. Y., Lee, H. L., Rogayah, T., Sa'diyah, I. 2000. Oxidase activity in relation to insecticide resistance in vectors of public health importance. *Trop. Biomed.*, 17, 69–79.
Niu, G., Wen, Z., Rupasinghe, S. G., Zeng, R. S., Berenbaum, M. R., Schuler, M. A. 2008. Aflatoxin B1 detoxification by CYP321A1 in *Helicoverpa zea. Arch. Insect Biochem. Physiol.*, 69, 32–45.
Oppenoorth, F. J. 1984. Biochemistry and genetics of insecticide resistance. In: Kerkut, G. A., Gilbert, L. I. (Eds). *Com. Insect. Phy. Biochem. Pharmacol.* vol 12. Pergamon Press, Oxford., 731–773.
Pages, F., Orlandi-Pradines, E., Corbel, V. 2007. Vectors of malaria: Biology, diversity, prevention and individual protection. *Med. Mal. Inf.*, 37(3), 153–161.
Panneerselvam, C., Murugan, K. 2013. Adulticidal, repellent, and ovicidal properties of indigenous plant extracts against the malarial vector, Anopheles stephensi (Diptera: Culicidae). *Parasitol. Res.*, 112, 679–692.

Pates, H., Curtis, C. 2005. Mosquito behavior and vector control. *Annu. Rev. Entomol.* 50, 53–70. 10.1146/annurev.ento.50.071803.130439.

Pavela, R., Kaffková, K., Kumšta, M. 2014. Chemical composition and larvicidal activity of essential oils from different *Mentha* L and *Pulegium* species against *Culex quinquefasciatus* Say (Diptera: Culicidae). *Plant Prot. Sci.*, 50, 36–42.

Perrin, R. M., Fedorova, N. D., Bok, J. W., Cramer, R. A. Jr., Wortman, J. R. 2007. Transcriptional regulation of chemical diversity in *Aspergillus fumigatus* by LaeA. *PLoS Path.*, 3, 50.

Phillips, A. J., Anderson, V. L., Robertson, E. J., Secombes, C. J., Van West. 2008. New insight into animal pathogenic oomycetes. *Trends. Microbiol.*, 16, 13–19.

Pluess, B., Tanser, F. C., Lengeler, C., Sharp, B. L. 2010. Indoor residual spraying for preventing malaria. *Cochrane Database System. Rev.*, 4, CD006657.

Poupardin, R., Reynaud, S., Strode, C., Ranson, H., Vontas, J. 2008. Crossinduction of detoxification genes by environmental xenobiotics and insecticides in the mosquito *Aedes aegypti*: Impact on larval tolerance to chemical insecticides. *Insect Biochem. Mol. Biol.*, 38, 540–551.

Price, N. R. 1991. Insect resistance to insecticides: Mechanism and diagnosis. *Comp. Biochem. Physiol.*, 100, 319–326.

Protopopoff, N., Van, B. W., Marcotty, T., Van, H. M., Maes, P., Baza, D. D., Alessandro, U., Coosemans, M. 2008. Spatial targeted vector control is able to reduce malaria prevalence in the highlands of Burundi. *Am. J. Trop. Med. Hyg.*, 79(1): 12–18.

Raghavendra, K., Barik, T. K., Sharma, P., Bhatt, R. M., Srivastava, H. C., Sreehari, U., Dash, A. P. 2011. Chlorfenapyr: A new insecticide with novel mode of action can control pyrethroid resistant malaria vectors. *Malaria J.*, 10, 16.

Rajesh, K., Dhanasekaran, D., Tyagi, B. K. 2015. Mosquito survey and larvicidal activity of actinobacterial isolates against *Culex* larvae (*Diptera: Culicidae*). *J. Saudi Soc. Agri. Sci.*, 14, 116–122.

Ramkumar, G., Karthi, S., Muthusamy, R., Natarajan, D., Shivakumar, M. S. 2014. Adulticidal and smoke toxicity of *Cipadessa baccifera* (Roth) plant extracts against *Anopheles stephensi*, *Aedes aegypti*, and *Culex quinquefasciatus*. *Parasitol. Res.*, doi:10.1007/s00436-014-41735.

Ramkumar, G., Shivakumar, M. 2015. Laboratory development of permethrin resistance and cross-resistance pattern of *Culex quinquefasciatus* to other insecticides. *Parasitol. Res.*, 114, 2553–2560.

Rao, B. S. S., Subramanian, K. S. 1934. The occurrence of furan derivatives in volatile oils. *Proc. Indian Acad. Sci.*, 1, 189–200.

Regnault-Roger, C., Philogene, B. J. R., Vincent, C. 2005. *Biopesticides of Plant Origin.* Intercept–Lavoisier, Andover, UK, 1–15.

Rehner, S. A., Minnis, A. M., Sung, G. H., Luangsa-ard, J. J., Devotto, L., Humber, R. A. 2011. Phylogeny and systematics of the anamorphic, entomopathogenic genus *Beauveria*. *Mycologia*, 103, 1055–1073.

Riaz, M. A., Poupardin, R., Reynaud, S., Strode, C., Ranson, H. 2009. Impact of glyphosate and benzo [a] pyrene on the tolerance of mosquito larvae to chemical insecticides. Role of detoxification genes in response to xenobiotics. *Aqua. Toxicol.*, 93, 61–69.

Roberts, D. W., Humber, R. A. 1981. Entomogenous fungi. *Biology of Conidial Fungi.*, 2, 201–236.

Roberts, D. W., St. Leger, R. J. 2004. *Metarhizium* spp., cosmopolitan insect-pathogenic fungi: Mycological aspects. *Adv. Appl. Microbiol.*, 54, 1–70.

Roger, R. C. 1997. The potential of botanical essential oils for insect pest control. *Integrated Pest. Mgt. Rev.*, 2, 25–34.

Rohlfs, M. 2005. Clash of kingdoms or why Drosophila larvae positively respond to fungal competitors. *Front. Zool.*, 2, 2.

Rohlfs, M., Kürschner, L. 2010. Saprophagous insect larvae, *Drosophila melanogaster*, profit from increased species richness in beneficial microbes. *J. Appl. Entomol.*, 134, 667–671.
Rohlfs, M., Obmann, B., Petersen, R. 2005. Competition with filamentous fungi and its implications for a gregarious life-style in insects living on ephemeral resources. *Ecol. Entomol.*, 30, 556–563.
Rohlfs, M., Obmann, B. 2009. Species-specific responses of dew fly larvae to mycotoxins. *Mycotox. Res.*, 28, 103–112.
Saavedra-Rodriguez, K., Urdaneta-Marquez, L., Rajatileka, S., Moulton, M., Flores, A. E. 2007. A mutation in the voltage-gated sodium channel gene associated with pyrethroid resistance in Latin American *Aedes aegypti*. *Insect Mol. Biol.*, 16, 785–798.
Samson, R. A., Evans, H. C., Latge, J. P. 1988. *Atlas of Entomopathogenic Fungi*. Springer-Verlag GmbH and Co. KG.
Schmelz, E. A., Kaplan, F., Huffaker, A., Dafoe, N. J., Vaughan, M. M., Ni, X. Z., Rocca, J. R., Alborn, H. T., Teal, P. E. 2011. Identity, regulation, and activity of inducible diterpenoid phytoalexins in maize. *Proc. Natl. Acad. Sci.*, 29(13), 5455–5460.
Scott, J. G. 1999. Cytochromes P450 and insecticide resistance. *Insect. Biochem. Mol. Biol.* 29, 757–777.
Scott, T. W., Chow, E., Strickman, D., Kittayapong, P., Wirtz, R. A. 1993. Blood feeding patterns of *Aedes aegypti* (Diptera: Culicidae) collected in a rural Thai village. *J. Med. Entomol.*, 30, 922–927.
Seye, F., Faye, O., Ndiaye, M., Njie, E., Marie Afoutou, J. 2009. Pathogenicity of the fungus, *Aspergillus clavatus*, isolated from the locust, Oedaleus senegalensis, against larvae of the mosquitoes *Aedes aegypti*, *Anopheles gambiae* and *Culex quinquefasciatus*. *J. Insect. Sci.*, Vol. 9 | Article 53.
Shahid, M. R., Farooq, J., Mahmood, A., Ilahi, F., Riaz, M., Shakeel, A., Valentin Petrescu-Mag, I., Farooq, A. 2012. Seasonal occurrence of sucking insect pest in cotton ecosystem of Punjab, Pakistan. Advances in Agriculture & Botanics: *Int. J. Bioflux Soc.*, Vol. 4 | Issue 1.
Sharma, P., Mohan, L., Srivastava, C.N. 2006. Phytoextract-induced developmental deformities in malaria vector. *Bioresour. Technol.*, 97, 1599–1604.
Sharp, B. L., Ridl, F. C., Govender, D., Kuklinski, J., Kleinschmidt, I. 2007. Malaria vector control by indoor residual insecticide spraying on the tropical island of Bioko, Equatorial Guinea. *Malaria J.*, 6, 52.
Sheeren, M. E. 2006. Larvicidal effects of *Eucalyptus* extracts on the larvae of *Culex pipiens* mosquito. *Int. J. Agri. Biol.*, 8, 896–7.
Singh, G., Prakash, S. 2012a. Efficacy of the *Trichophyton ajelloi* and *Lagenidium giganteum* metabolites against mosquitoes after flash chromatography. *Parasitol. Res.*, 110(5), 2053–2060.
Singh, G., Prakash, S. 2012b. Evaluation of culture filtrates of *Culicinomyces clavisporus*: Mycoadulticide for *Culex quinquefasciatus*, *Aedes aegypti* and *Anopheles stephensi*. *Parasitol. Res.*, 110(1), 267–272.
Singh, J. P., Prakesh, B., Dubey, N. K. 2014. Insecticidal activity of Ageratum conyzoides coleus aromaticus Benth and hyptis suaverlens (L) poit essentiai oil as fumicants against strored grain insect Tribolium castaneum herbst. *J Food. Sci. Technol.*, DOI:10.1007/s13197-012-0698-8.
Singkaravanit, S., Kinoshita, H., Ihara, F., Nihira, T. 2010. Cloning and functional analysis of the second geranylgeranyl diphosphate synthase gene influencing helvolic acid biosynthesis in *Metarhizium anisopliae*. *App. Microbiol. Biotechnol.*, 87(3), 1077–1088.
Soni, N., and Prakash, S. 2010. Effect of *Chrysosporium keratinophilum* metabolites against *Culex quinquefasciatus* after chromatographic purification. *Parasitol. Res.*, 107(6), 1329–1336.
Soni, N., and Prakash, S. 2012. Efficacy of fungus mediated silver and gold nanoparticles against *Aedes aegypti* larvae. *Parasitol. Res.*, 110(1): 175–184.

Sowjanya, K. S., Padmaja, V., Murthy, Y. L. N. 2008. Insecticidal activity of Destruxin a mycotoxins from M. anisoplae against Spodoptera litura. *Manag. Sci.*, 64, 119–125.
Sparrow, F. K. 1960. *Aquatic Phycomycetes*. USA: Arbor.
Sree, S., Padmaja, V., and Yellajyohula. 2008. Insecticidal activity of destruxin, a mycotoxin from Metarhizium anisopliae (Hypocreales), against Spodoptera litura (Lepidoptera: Noctuidae) larval stages. *Pest Manag. Sci.*, 64(2), 119–125.
Strode, C., Wondji, C. S., David, J. P., Hawkes, N. J., Lumjuan, N., Parakrama, N. D. R., Karunaratne, S. H. P., Hemingway, J., Black, W. C. IV, Ranson, H. 2008. Genomic analysis of detoxification genes in the mosquito *Aedes aegypti. Insect Biochem. Mol. Biol.*, 38, 113–123.
Swathi, J., Sowjanya, K., Narendra, K., Krishna, A. 2013. Bioactivity assay of an isolated marine *Fusarium* spp. *Int. J. Biosci. Biotechnol.*, 5, 179–186.
Townson, H., Nathan, M. B., Zaim, M., Guillet, P., Manga, L., Bos, R., Kindhauser, M. 2005. Exploiting the potential of vector control for disease prevention. *Bull. WHO*, 83(12), 942–947.
Trevors, J. T., Barkey, T., Bourquin, A. W. 1987. Gene transfer among bacteria in soil and aquatic environments: A review. *Can. J. Microbiol.*, 33, 191–198.
Trienens, M., Keller, N. P., Rohlfs, M. 2010. Fruit, flies and filamentous fungi experimental analysis of animal–microbe competition using *Drosophila melanogaster* and *Aspergillus* as a model system. *Oikos*. doi:10.1111/j.1600-0706.2010.18088.x.
Uysal, H., Agar, G. 2005. Selenium protective activity against aflatoxin B1 adverse effects on *Drosophila melanogaster. Braz. Arch. Biol. Technol.*, 48(2), 227–233.
Van den Berg, H. 2009. Global status of DDT and its alternatives for use in vector control to prevent disease. *Env. Health Persp.*, 117, 1656–1663.
Vey, A., Hoagland, R., Butt, T. M. 2001. Toxic metabolites of fungal biocontrol agents. In: *Fungi as biocontrol agents: Progress, problems and Potential*. CAB. *Int Walling ford.*, 311–346.
Vijayan, V. K., Balaraman, K. 1991. Metabolites of fungi and actinobacteria active against mosquito larvae. *In. J. Med. Res.*, 93, 115.
Vining, L. C. 1990. Function of secondary metabolites. *Ann. Rev. Microbiol.*, 44, 395–427.
Vyas, N., Dua, K. K., Prakash, S. 2006. Laboratory efficacy of metabolites of *Lagenidium giganteum* (Couch) on *Anopheles stephensi* (Liston) after filterations by column chromatography. *J. Commun. Dis.*, 38(2), 176–180.
Wakil, W., Yasin, M., Qayyum, M. A., Asim, M. 2012. Combined toxicity of *Metarhizium anisopliae* with sublethal doses of *chlorpyrifos, fipronil* and *chlorantraniliprole* against *Periplaneta americana* (Dictyoptera: Blattidae). *Pakistan Entomol.*, 34(1), 59–63.
Walker, K., Lynch, M. 2007. Contributions of Anopheles larval control to malaria suppression in tropical Africa: Review of achievements and potential. *Med. Vet. Entomol.*, 21(1), 2–21.
Wariko, R., Ray, A., Sandhu, J. K., Samal, R., Wahab, N., Kumar, S. 2012. Larvicidal and irritant activities of hexane leaf extracts of *Citrus sinensis* against the dengue vector *Aedes aegypti* L. *Asian. Pac. J. Trop. Biomed.*, 152–155.
Weekly epidemiological record Relevé épidémiologique hebdomadaire, 2012. 6 APRIL 2012, 87th YEAR / 6 AVRIL 2012, 87e ANNÉE No. 14, 2012, 87, 129–144.
World Health Organization (WHO). 1957. *Insecticides. 7th report of the expert committee on insecticides. WHO Technical Report Series 125*. World Health Organization, Geneva.
World Health Organization (WHO). 2005. Report of the Eighth WHOPES Working Group Meeting. Geneva. Guidelines for laboratory and field testing of mosquito larvicides (WHO/CDS/ WHOPES/GCDPP/2005.13). Geneva, Switzerland.
World Health Organization (WHO). 2008a. *Instruction for determining the susceptibility or resistance of adult mosquitoes to organochlorine, organophosphate and carbamate insecticides*. WHO/VBC/81.806.

World Health Organization (WHO). 2008b. *Position statement on integrated vector management.* World Health Organization, Geneva. WHO_HTM_NTD_VEM_2008.2_eng .pdf, accessed April 5, 2015.

World Health Organization (WHO). 2010. *Vector control for malaria and other mosquito-borne diseases. WHO Technical Report Series 857.* World Health Organization, Geneva.

World Health Organization (WHO). 2012. Global information system on alcohol and health. Geneva (www.who.int/globalatlas/DataQuery/default.asp).

World Health Organization (WHO). 2015. *Instruction for determining the susceptibility or resistance of adult mosquitoes to organochlorine, organophosphate and carbamate insecticides.* WHO/VBC/81.806.

World Health Organization (WHO). 2016. *Instructions for determining the susceptibility or resistance of adult mosquitoes to organochlorine, organophosphate and carbamate insecticides.* WHO/VBC/81.805.

World Health Organization (WHO). 2017. *Report of the WHO informal consultation on the evaluation and testing of insecticides.* CTD/WHO PES/IC/96.1.

Xu, Y., Yang, M., Sun, J., Qian, J., Zhang, D., Sun, Y., Ma, L., Zhu, C. 2008. Glycogen branching enzyme: A novel deltamethrin resistance associated gene from *Culex pipiens* pallens. *Parasitol. Res.*, 103(2), 449–458.

Yebakima, A. 1991. *Recherche sur Aedes aegypti et Culex pipiens en Martinique. Ecologie Larvaire, Resistance aux Insecticides, Application a la Lutte. The se de Doctorat d'Etat es Sciences*, Universite. Montpellier II, 210.

Yoshiaki, K., Omata-Iwabuchi, K., Takahashi, M. 1997. Change in susceptibility to pyriproxifen, a JH mimic during late larval and early pupal stage of *Culex pipienes* molestus. *Med. Entomol. Zoo.*, 48(2), 85–89.

Zlotkin, E. 1999. The insect voltage-gated sodium channel as target of insecticides. *Ann. Rev. Entomol.*, 44, 429–455.

# 6 Mosquito Larvicidal Properties of Mycogenic Silver Nanoparticles Derived from *Aspergillus niger* against *Culex quinquefasciatus*

*Marimuthu Poornima, Ganesan Kanimozhi, and Annamalai Panneerselvam*

## CONTENTS

## 6.1 INTRODUCTION

India is endemic to mosquito-borne diseases due to favorable ecological conditions. Mosquitoes have the ability of carrying and transmitting human and animal diseases across countries causing hundreds of millions of deaths annually (Dhanasekaran et al., 2010). Among several species of mosquitoes, *Culex quinquefasciatus* is the main periodic vector of filarial parasite accredited with human lymphatic filariasis transmission (Oduola et al., 2006).

It has proven to be very difficult to control mosquitoes due to their remarkable ability to adapt to various environments, their close contact with humans, and their reproductive biology. The rich organic content, stagnant water, low illumination, and small orifices of coconut shells on rubber plantations favor intense breeding. Several insecticides, namely, DDT, dieldrin, organophosphrus, fenithothion, and propoxur, were widely used in India to mitigate this dangerous problem (Zhang et al., 2012). Persistent application of the synthetic chemical products mostly available in local markets causes undesirable consequences, including production of resistant strains of mosquitoes, ecological imbalance, and elimination of nontarget organisms in the environment (Mathivanan et al., 2010). Therefore, a demand stems out for the synthesis of bio origin mosquito repellent (Niraimathi et al., 2010).

In this juncture, the field of nanotechnology is one of the most active areas of research in modern material sciences (Shameli et al., 2012). Nanoparticles have vast application in pharmaceutical, industrial, and biotechnological fields (Suman et al., 2013). Silver nanoparticles (AgNPs) are emerging as one of the fastest growing materials due to their unique physical, chemical, and biological properties; small size; and high specific surface area. AgNPs are reported to possess antiviral, antibacterial, and antifungal properties (Rogers et al., 2008).

Currently, fungi are being utilized in nanotechnology for the production of nanoparticles. Synthesis using fungi has shown that this environmental friendly and renewable source can be used as an effective reducing agent for synthesis of AgNPs. This biological reduction of metal could be utilized for a clean, nontoxic, and environmentally acceptable green approach to producing metal nanoparticles. Fungi-based biosynthesis of nanoparticles is one of the best biological methods to prepare silver nanoparticles (Dhanasekaran and Thangaraj, 2013).

The use of environmentally benign materials such as silver nanoparticles offer numerous benefits of ecofriendliness and compatibility for larvicidal application. In these circumstances, an improvised method using biologically synthesized silver nanoparticles were evaluated for the destruction of mosquito larvae. The present study describes the synthesis of fungal *Aspergillus niger* nanoparticles for larvicidal and mode of action. Hence, this study is a laboratory evaluation of mycosynthesized AgNPs tested against larvae of the filarial vector *Cx. quinquefasciatus*.

## 6.2 MATERIALS AND METHODS

### 6.2.1 Mycosynthesis of Silver Nanoparticles

The *Aspergillus niger* obtained were grown aerobically in liquid broth containing malt extract powder, glucose, yeast extract, and peptone. The culture flasks were incubated on room temperature at 27°C. The biomass was harvested after 120 hours of growth through a plastic sieve followed by extensive washing with sterile double-distilled water to remove any medium components from the biomass.

Typically, 15 g of biomass (wet weight) were brought into contact with 100 mL of sterile double-distilled water for 48 hours at 27°C ± 1°C in an Erlenmeyer flask and agitated at 150 rpm. After incubation, the cell filtrate was filtered by Whatman filter paper No. 1 into 100 mL of cell filtrate, a carefully weighed quantity of silver nitrate was added, and the reaction was carried out in darkness. A control with only the cell filtrate and without silver ion was run along with the test sample.

### 6.2.2 Characterization of Silver Nanoparticles

#### 6.2.2.1 Ultraviolet-Visible (UV-Vis) Spectroscopy

The bioreduction of pure $AgNO_3$ are monitored using ultraviolet-visible (UV-Vis) spectroscopy at regular intervals. During the reduction, 0.1 ml of samples were taken and diluted several times with Millipore water. After dilution, it was centrifuged at 800 rpm for 5 minutes. The supernatant was scanned in the wavelength ranging from 190 to 1100 nm by using arithmetic instrument model spectrophotometer.

#### 6.2.2.2 Fourier Transform Infrared Spectroscopy (FTIR)

A known weight of sample (1 mg) was taken in a mortar and pestle and ground with 2.5 mg of dry potassium bromide (KBr). The powder so obtained was filled in a 2 mm internal diameter micro-cup and loaded onto a Fourier transform infrared spectroscopy (FTIR) set at 26°C ± 1°C. The samples were scanned using infrared in the range of 4000 to 400 $Cm^{-1}$ using a Fourier transform infrared spectrometer (Thermo Nicolet Model-6700). The spectral data obtained were compared with the reference chart to identify the functional groups present in the sample.

#### 6.2.2.3 X-Ray Diffraction (XRD)

The air-dried nanoparticles were coated onto an x-ray diffraction (XRD) grid and analyzed for the formation of silver nanoparticles by a Philips X-Ray Diffractometer with Philips PW 1830 X-Ray Generator operated at a voltage of 40 kv and current of 30 mA with cu kαl radiation. The diffracted intensities were recorded from 10° to 80° of 2θ angles.

The average particle size of silver nanoparticles synthesized by the present green method was calculated using the Debye-Scherrer equation (Cullity, 1956):

$$D = K\lambda/\beta \cos \theta$$

where

D = crystallite size of AgNPs particles
λ = wavelength of x-ray source (0.1541 nm) used in XRD
β = full width at half maximum of the diffraction peak
$K$ = Scherrer constant with value from 0.9 to 1
θ = the Bragg angle

#### 6.2.2.4 Scanning Electron Microscopy (SEM)

This study was undertaken to know the size and shape of the silver nanoparticles biosynthesized. About 25 μl of the sample was pipette out and loaded on a stub provided for scanning electron microscopy (SEM) analysis. After loading the sample, the stub was fixed to a holder. The details regarding applied voltage, magnification used, and the size of the content of the images were implanted on the photographs itself.

### 6.2.3 Mosquito Larvae Rearing of *Culex quinquefasciatus*

*Culex quinquefasciatus* larvae were collected from stagnant water in the Sirumalai Hills (Tamil Nadu, India). The larvae thus collected were identified by the district malaria control department of Pudukkottai. They were maintained and reared in the laboratory inside the insect rearing cage. Water in the rearing container was refreshed every day by removing a little quantity of water from the rearing buckets and replacing with fresh water.

### 6.2.4 Mosquito Larvicidal Bioassay

Bioassays were conducted separately at four different test concentrations (2, 4, 6, and 8 ppm) of AgNPs. Early first to fourth instar larvae of *Cx. quinquefasciatus* were transferred in 100 ml of sterilized tape water to a 250 mL conical flask. Nanoparticles were added to the conical flasks at different concentrations (2, 4, 6, and 8 ppm). For each test, 25 instar larvae were released in 100 ml. Similarly, the control (without AgNPs) was run to test the natural mortality. Thereafter, larval mortality was recorded after different hours of treatment (1, 8, 16, and 24 hours), the experiment time.

### 6.2.5 Dose–Response Bioassay

The larvicidal bioassays were performed at 28°C to 34°C. Based on the screening results, synthesized AgNPs were subjected to dose–response bioassays for larvicidal activity against the larvae of *Cx. quinquefasciatus.* The effectiveness of silver nanoparticles as mosquito larvicides was determined from all the 25 first to fourth instar larvae with exposure to time periods. No food was offered to the larvae during the experiment.

The larvae were considered dead if they showed no sign of swimming movement even after gentle touching with a glass rod, as described in the World Health Organization's technical report series. The dead larvae in three replicates were combined and expressed as a percentage mortality of each concentration. Percentage mortality was calculated as follows:

$$\text{Percentage mortality} = \frac{\text{Number of dead larvae}}{\text{Number of larvae introduced}} \times 100$$

### 6.2.6 Statistical Analysis

The dose–response data were subjected to probit regression analysis (Finney, 1971). The average larval mortality data were subjected to probit analysis for calculating $LC_{50}$, $LC_{90}$, and other statistics at 95% confidence intervals of upper confidence limit (UCL) and lower confidence limit (LCL), and chi-square values were calculated using a statistical software package (Statplus ver. 5.8.0).

### 6.2.7 Morphological Variation

After treatment with a lethal dosage ($LC_{50}$) of each fungal nanoparticles synthesis, the dead larvae were studied for morphological alterations under confocal laser scanning microscopy. The morphological variation in the silver nanoparticle treated larvae compared to the control larvae was identified using CSLM.

#### 6.2.7.1 Confocal Laser Scanning Microscopy (CLSM)

Confocal laser scanning microscopy (CLSM) is a valuable tool for detailed study of small, complex structures at high resolution close to the diffraction limit. CLSM depends on the degree of induced fluorescence of the examined samples (Lee et al., 2009). Fluorescence is aided either by staining with appropriate dyes or can be obtained using the autofluorescence properties of the examined structures.

## 6.3 RESULTS

### 6.3.1 Formation of Silver Nanoparticles

The fungal filtrate changed in color from pale yellow to dark brown by the addition of silver nitrate. The change in color was noted by visual observation in the cell filtrate when it was incubated with $AgNO_3$ solution (Figure 6.1). The color changed to light brown within an hour and then later changed to dark brown after two hours of incubation. The brown color could be due to the excitation of surface plasmon vibrations, typical of the silver nanoparticles (Ahmad et al., 2003; Krishnaraj et al., 2010).

### 6.3.2 Characterization

#### 6.3.2.1 UV-VIS Spectroscopy

A strong resonance centered at 426 nm was clearly observed (Figure 6.2) and increased in intensity with time. It might arise from the excitation of longitudinal plasmon vibrations in silver nanoparticles in the solution. Reduction of silver ions

**FIGURE 6.1** Cell filtrates with control (left) and $AgNO_3$ (right).

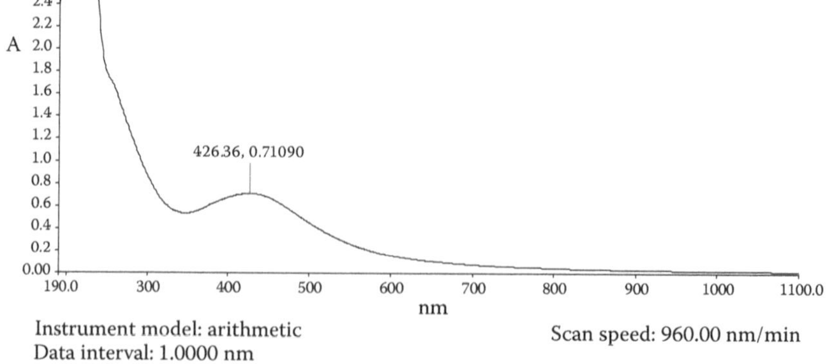


**FIGURE 6.2** UV analysis of AgNP synthesized using *Aspergillus niger.*

present in the cell filtrate observed by the UV-VIS spectroscopy revealed the presence of silver nanoparticles may be correlated with the UV-VIS spectra. These changes were attributed to the excitation of surface plasmon resonance (SPR) in the metal nanoparticles (Natarajan et al., 2010).

#### 6.3.2.2 FTIR

FTIR absorbance spectrum was observed at 3463.83 $cm^{-1}$, 2352.23 $cm^{-1}$, 2076.30 $cm^{-1}$, 1638.10 $cm^{-1}$, 1356.28 $cm^{-1}$, and 675.99 $cm^{-1}$, which indicates the functional group (Table 6.1) of the fungus component is involved in the reduction and acts as a capping agent (Figure 6.3).

**TABLE 6.1**
**FTIR Analysis of Biomolecules Involved in Reduction $AgNO_3$ into Ag Ions Using *Aspergillus niger***

| Wave Number ($cm^{-1}$) | Bond | Functional Group |
|---|---|---|
| 3463.83 | O-H stretch, H-bonded | Alcohols, phenols |
| 2352.23 | C-H stretch | Alkanes |
| 2076.30 | -NCS | Isothiocyanate |
| 1638.10 | N-H bend | Primary amines |
| 1356.28 | N-O symmetric stretch | Nitro compounds |
| 675.99 | C-Br stretch | Alkyl halides |

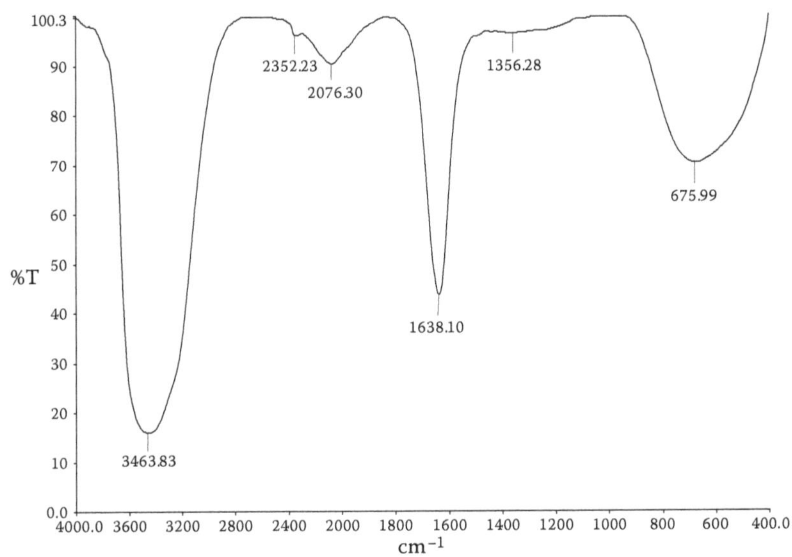


**FIGURE 6.3** FTIR analysis of AgNP synthesized using *Aspergillus niger*.

### 6.3.2.3 XRD

A comparison of our XRD spectrum with the standard confirmed that the silver particles formed in our experiments were in the form of nanocrystals. The assigned peaks at 2θ values were 32.49, 46.42, 57.75 and 76.83 (Figure 6.4). The XRD results clearly show that the silver nanoparticles formed by the reduction of Ag+ ions by the fungi *Aspergillus niger* are crystalline in nature.

It was found that the average size from XRD data and using the Debye-Scherrer equation was 10 ± 1 nm. The presence of structural peaks in XRD patterns and average crystalline size around 10 ± 1 nm clearly illustrates that AgNPs synthesized were nanocrystalline in nature.

### 6.3.2.4 SEM Analysis

The SEM micrographs of nanoparticles obtained in the filtrate showed that silver nanoparticles are spherical to roughly spherical shaped in the size range 5 to 10 nm (Figure 6.5). The rough morphology of the nanoparticles provides excellent larvicidal activity for the synthesized nanoparticles.

## 6.3.3 Larvicidal Activity

Considerable mortality was evident after the treatment of silver nanoparticle for all larvae. Mortality was increased as concentration increased (Figure 6.6). The percentage mortality of first instar was 100% at 6 μl/100 ml (6 ppm) concentration (crude extract) at 24 hours and mortality was further increased in second instar. A similar trend has been noted for all the instars (third and fourth).

The data obtained from the present study clearly indicate that silver nanoparticles could provide excellent larval control of *Cx. quinquefasciatus*. The statistical values of biolarvicidal efficacies of AgNPs synthesized by using the fungus *A. niger* against all the four instars larvae of *Cx. quinquefasciatus* are enumerated in Tables 6.2 to 6.5.

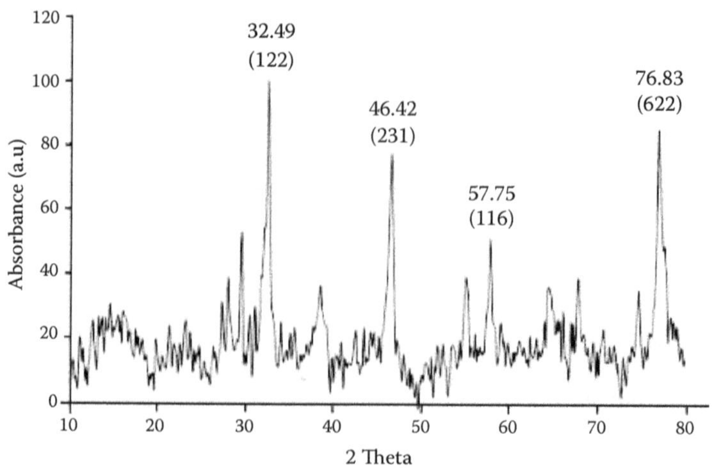


**FIGURE 6.4** XRD analysis of AgNP synthesized using *Aspergillus niger*.

**FIGURE 6.5** SEM image of AgNP synthesized using *Aspergillus niger*.

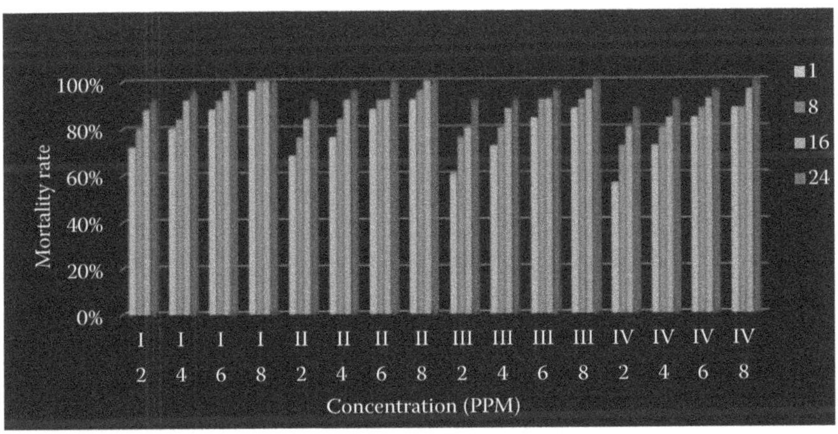


**FIGURE 6.6** The mortality rate of mosquito larvae treated against AgNP synthesized using *Aspergillus niger*.

The comparative values of $LC_{50}$ and $LC_{90}$, and LCL and UCL values for all the four instars of larvae, which were treated against 2, 4, 6, and 8 ppm concentration of test solution during different exposure times (1, 8, 16, and 24 hours), are enumerated in Table 6.6 and also represented in probit graphs (Figures 6.7 to 6.10).

A mortality or 100% was observed at 8 ppm concentration against all instars treated with *Cx. quinquefasciatus*. At 6 ppm concentration, mortality against first and

**TABLE 6.2**
**Biolarvicidal Efficacies of First Instar Larvae**

| Exposure Time | Concentrations (ppm) | Actual Percent | Log10 Concentration | Total Larvae | Total Dead Larvae | Probit Value | Chi-Square Value ($\chi^2$) |
|---|---|---|---|---|---|---|---|
| 1 hour | 2 | 0.72 | 0.301 | 25 | 18 | 0.6943 | 0.0237 |
| | 4 | 0.8 | 0.6021 | 25 | 20 | 0.8388 | 0.0448 |
| | 6 | 0.88 | 0.7782 | 25 | 22 | 0.8981 | 0.0091 |
| | 8 | 0.96 | 0.9031 | 25 | 24 | 0.9293 | 0.0252 |
| 8 hours | 2 | 0.8 | 0.301 | 25 | 20 | 0.7643 | 0.0415 |
| | 4 | 0.84 | 0.6021 | 25 | 21 | 0.8939 | 0.0813 |
| | 6 | 0.92 | 0.7782 | 25 | 23 | 0.9401 | 0.0108 |
| | 8 | 0.99 | 0.9031 | 25 | 25 | 0.9620 | 0.0203 |
| 16 hours | 2 | 0.88 | 0.301 | 25 | 22 | 0.8631 | 0.0082 |
| | 4 | 0.92 | 0.6021 | 25 | 23 | 0.9445 | 0.0159 |
| | 6 | 0.96 | 0.7782 | 25 | 24 | 0.9703 | 0.0027 |
| | 8 | 0.99 | 0.9031 | 25 | 25 | 0.9818 | 0.0016 |
| 24 hours | 2 | 0.92 | 0.301 | 25 | 23 | 0.9097 | 0.0028 |
| | 4 | 0.96 | 0.6021 | 25 | 24 | 0.9797 | 0.0099 |
| | 6 | 0.99 | 0.7782 | 25 | 25 | 0.9931 | 0.0002 |
| | 8 | 0.99 | 0.9031 | 25 | 25 | 0.9970 | 0.0012 |

**TABLE 6.3**
**Biolarvicidal Efficacies of Second Instar Larvae**

| Exposure Time | Concentrations (ppm) | Actual Percent | Log10 Concentration | Total Larvae | Total Dead Larvae | Probit Value | Chi-Square Value ($\chi^2$) |
|---|---|---|---|---|---|---|---|
| 1 hour | 2 | 0.68 | 0.301 | 25 | 17 | 0.6575 | 0.0192 |
| | 4 | 0.76 | 0.6021 | 25 | 19 | 0.8058 | 0.0651 |
| | 6 | 0.88 | 0.7782 | 25 | 22 | 0.8707 | 0.0024 |
| | 8 | 0.92 | 0.9031 | 25 | 23 | 0.9065 | 0.0050 |
| 8 hours | 2 | 0.76 | 0.301 | 25 | 19 | 0.7427 | 0.0099 |
| | 4 | 0.84 | 0.6021 | 25 | 21 | 0.8707 | 0.0271 |
| | 6 | 0.92 | 0.7782 | 25 | 23 | 0.9206 | 0.0000 |
| | 8 | 0.96 | 0.9031 | 25 | 24 | 0.9460 | 0.0051 |
| 16 hours | 2 | 0.84 | 0.301 | 25 | 21 | 0.8288 | 0.0037 |
| | 4 | 0.92 | 0.6021 | 25 | 23 | 0.9233 | 0.0002 |
| | 6 | 0.92 | 0.7782 | 25 | 23 | 0.9561 | 0.0341 |
| | 8 | 0.99 | 0.9031 | 25 | 25 | 0.9716 | 0.0086 |
| 24 hours | 2 | 0.92 | 0.301 | 25 | 23 | 0.9097 | 0.0028 |
| | 4 | 0.96 | 0.6021 | 25 | 24 | 0.9797 | 0.0099 |
| | 6 | 0.99 | 0.7782 | 25 | 25 | 0.9931 | 0.0002 |
| | 8 | 0.99 | 0.9031 | 25 | 25 | 0.9970 | 0.0012 |

**TABLE 6.4**
**Biolarvicidal Efficacies of Third Instar Larvae**

| Exposure Time | Concentrations (ppm) | Actual Percent | Log10 Concentration | Total Larvae | Total Dead Larvae | Probit Value | Chi-Square Value ($\chi^2$) |
|---|---|---|---|---|---|---|---|
| 1 hour | 2 | 0.6 | 0.301 | 25 | 15 | 0.5856 | 0.0087 |
| | 4 | 0.72 | 0.6021 | 25 | 18 | 0.7518 | 0.0336 |
| | 6 | 0.84 | 0.7782 | 25 | 21 | 0.8293 | 0.0034 |
| | 8 | 0.88 | 0.9031 | 25 | 22 | 0.8736 | 0.0011 |
| 8 hours | 2 | 0.76 | 0.301 | 25 | 19 | 0.7410 | 0.0121 |
| | 4 | 0.8 | 0.6021 | 25 | 20 | 0.8468 | 0.0648 |
| | 6 | 0.92 | 0.7782 | 25 | 23 | 0.8931 | 0.0202 |
| | 8 | 0.92 | 0.9031 | 25 | 23 | 0.9191 | 0.0000 |
| 16 hours | 2 | 0.8 | 0.301 | 25 | 20 | 0.7924 | 0.0017 |
| | 4 | 0.88 | 0.6021 | 25 | 22 | 0.8898 | 0.0027 |
| | 6 | 0.92 | 0.7782 | 25 | 23 | 0.9286 | 0.0020 |
| | 8 | 0.96 | 0.9031 | 25 | 24 | 0.9491 | 0.0031 |
| 24 hours | 2 | 0.92 | 0.301 | 25 | 23 | 0.9001 | 0.0109 |
| | 4 | 0.92 | 0.6021 | 25 | 23 | 0.9512 | 0.0255 |
| | 6 | 0.96 | 0.7782 | 25 | 24 | 0.9696 | 0.0023 |
| | 8 | 0.99 | 0.9031 | 25 | 25 | 0.9788 | 0.0031 |

**TABLE 6.5**
**Biolarvicidal Efficacies of Fourth Instar Larvae**

| Exposure Time | Concentrations (ppm) | Actual Percent | Log10 Concentration | Total Larvae | Total Dead Larvae | Probit Value | Chi-Square Value ($\chi^2$) |
|---|---|---|---|---|---|---|---|
| 1 hour | 2 | 0.56 | 0.301 | 25 | 14 | 0.5511 | 0.0035 |
| | 4 | 0.72 | 0.6021 | 25 | 18 | 0.7414 | 0.0155 |
| | 6 | 0.84 | 0.7782 | 25 | 21 | 0.8293 | 0.0034 |
| | 8 | 0.88 | 0.9031 | 25 | 22 | 0.8784 | 7.0865 |
| 8 hours | 2 | 0.72 | 0.301 | 25 | 18 | 0.7167 | 0.0003 |
| | 4 | 0.8 | 0.6021 | 25 | 20 | 0.8145 | 0.0065 |
| | 6 | 0.88 | 0.7782 | 25 | 22 | 0.8606 | 0.0109 |
| | 8 | 0.88 | 0.9031 | 25 | 22 | 0.8881 | 0.0018 |
| 16 hours | 2 | 0.8 | 0.301 | 25 | 20 | 0.7784 | 0.0149 |
| | 4 | 0.84 | 0.6021 | 25 | 21 | 0.8791 | 0.0436 |
| | 6 | 0.92 | 0.7782 | 25 | 23 | 0.9202 | 0 |
| | 8 | 0.96 | 0.9031 | 25 | 24 | 0.9423 | 0.0082 |
| 24 hours | 2 | 0.88 | 0.301 | 25 | 22 | 0.8631 | 0.0082 |
| | 4 | 0.92 | 0.6021 | 25 | 23 | 0.9445 | 0.0159 |
| | 6 | 0.96 | 0.7782 | 25 | 24 | 0.9703 | 0.0027 |
| | 8 | 0.99 | 0.9031 | 25 | 25 | 0.9818 | 0.0016 |

**TABLE 6.6**
**Comparative Values of $LC_{50}$ and $LC_{90}$, and LCL and UCL**

| Instar | Time (hours) | Percentage of $LC_{50}$ | Percentage of $LC_{90}$ | LCL | UCL |
|---|---|---|---|---|---|
| I | 1 | 0.9623 | 6.1757 | 0.0012 | 1.9566 |
| | 8 | 0.7758 | 4.7171 | 0.0001 | 1.6965 |
| | 16 | 0.4376 | 2.8904 | 0.044 | 4.3553 |
| | 24 | 0.5396 | 1.1512 | 0.0762 | 3.82 |
| II | 1 | 1.0809 | 7.0984 | 0.0016 | 2.1275 |
| | 8 | 0.7771 | 5.4009 | 0 | 1.7568 |
| | 16 | 0.5049 | 3.7835 | 0.0642 | 3.9722 |
| | 24 | 0.5396 | 1.1512 | 0.0762 | 3.82 |
| III | 1 | 1.4474 | 8.2879 | 0.0334 | 2.5194 |
| | 8 | 0.6082 | 6.5309 | 0.0779 | 4.7485 |
| | 16 | 0.5055 | 4.9256 | 0.0556 | 4.5939 |
| | 24 | 0.1858 | 1.998 | 0.0022 | 15.9833 |
| IV | 1 | 1.6846 | 8.1125 | 0.2028 | 2.6809 |
| | 8 | 0.5817 | 8.1677 | 0.0541 | 6.256 |
| | 16 | 0.5364 | 5.2792 | 0.0632 | 4.5496 |
| | 24 | 0.4376 | 2.8904 | 0.044 | 4.3553 |

*Note:* LC, lethal concentration; LCL, lower confidential limit; UCL, upper confidential limit.

second instars of *Cx. quinquefasciatus* were 100%, whereas third and fourth instar was 96%. At 4 ppm concentration, mortality against first and second instars of *Cx. quinquefasciatus* were 96%, and third and fourth instar was 92%. At 2 ppm concentration, mortality against first and second instars of *Cx. quinquefasciatus* were 92%, and third and fourth instar was 88%.

### 6.3.4 Morphological Variation

After treatment with a lethal dosage of each fungal nanoparticle synthesis, the dead larvae were studied for morphological alterations under confocal microscopy. Morphological changes in body segments (including the head, thorax, and abdomen) and other organs (such as the eyes, antennae, mouth brushes, setae, saddle, and anal gills) were observed (Figures 6.11 to 6.14). After 5 minutes of exposure, abnormal evidence of excitation, restlessness, and sluggishness was initially observed.

The larvae showed major changes in the head, epithelial layer, and cuticle membrane, and complete destruction in the abdominal region as observed. However, no variation was found in the control. These histology observations clearly indicate that the fungal toxic nanoparticles specifically affect the abdominal region of *Cx. quinquefasciatus*. The mechanism that causes the death of the larvae could be the ability of the nanoparticles to penetrate through the oral cavity as well as the body membrane.

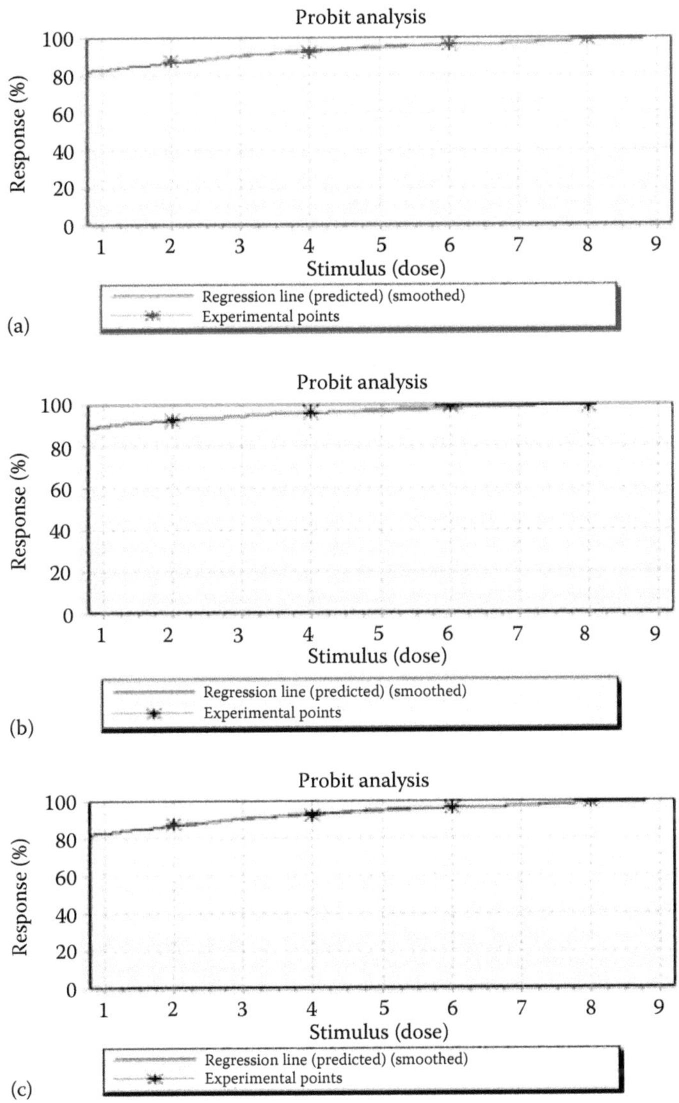


**FIGURE 6.7** Probit graph of first instar larvae. Exposure time: (a) 1 hour, (b) 8 hours, (c) 16 hours. *(Continued)*

The silver nanoparticles in the intracellular space can bind to sulfur-containing proteins or to phosphorus-containing compounds like DNA, leading to the denaturation of some organelles and enzymes. Subsequently, the decrease in membrane permeability and disturbance in proton motive force causes loss of cellular function and finally cell death.

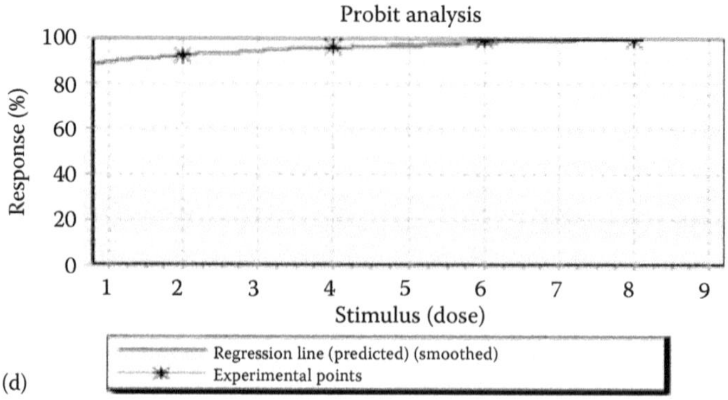


(d)

**FIGURE 6.7 (CONTINUED)** Probit graph of first instar larvae. Exposure time: (d) 24 hours.

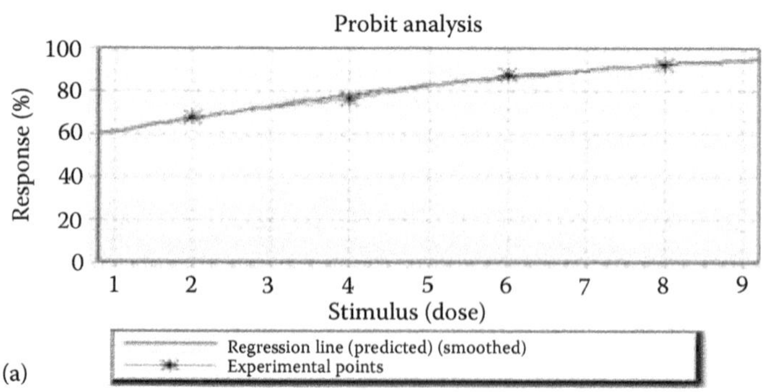


(a)

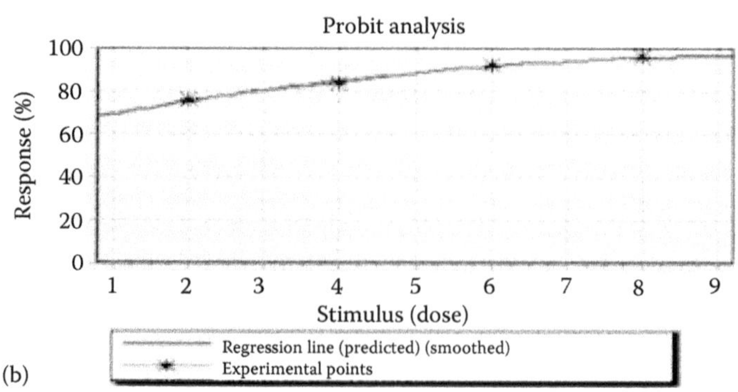


(b)

**FIGURE 6.8** Probit graph of second instar larvae. Exposure time: (a) 1 hour, (b) 8 hours.

*(Continued)*

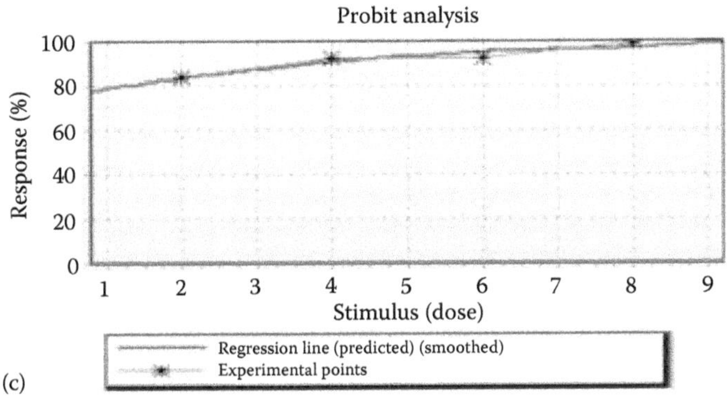


(c)

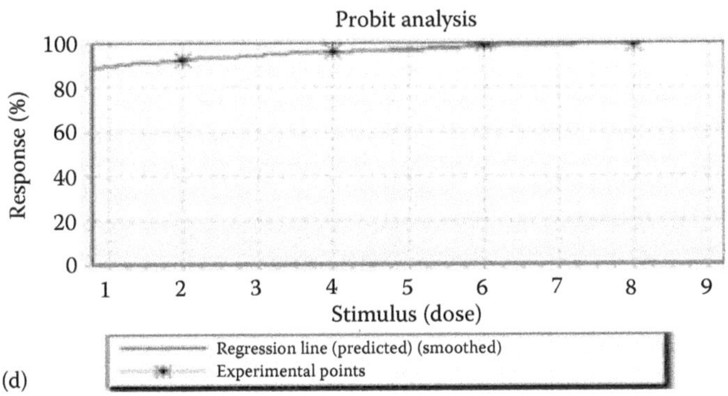


(d)

**FIGURE 6.8 (CONTINUED)** Probit graph of second instar larvae. Exposure time: (c) 16 hours, (d) 24 hours.

## 6.4 DISCUSSION

Over the last 5 decades the indiscriminate use of synthetic insecticides in agriculture and public health programs for the control of pest species has created multifarious problems, namely, insecticide resistance, environmental pollution, and toxic hazards to humans and other nontarget organisms. In an attempt to overcome these problems, great emphasis has been recently placed on the research and development of forms of pest control of microbial origin. Studies on natural products as larvicides have indicated that they could provide possible alternative to synthetic insecticides.

Another drawback with use of chemical insecticides is the failure of many vector control campaigns resulting in the vector resurgence in epidemic zones. The use of commercially available conventional synthetic insecticides has raised serious ecological, economical, and environmental problems. They also contribute toward the development of resistance in the target insect species. All these factors led to the search for safer and more compatible alternatives among which natural products are of first importance.

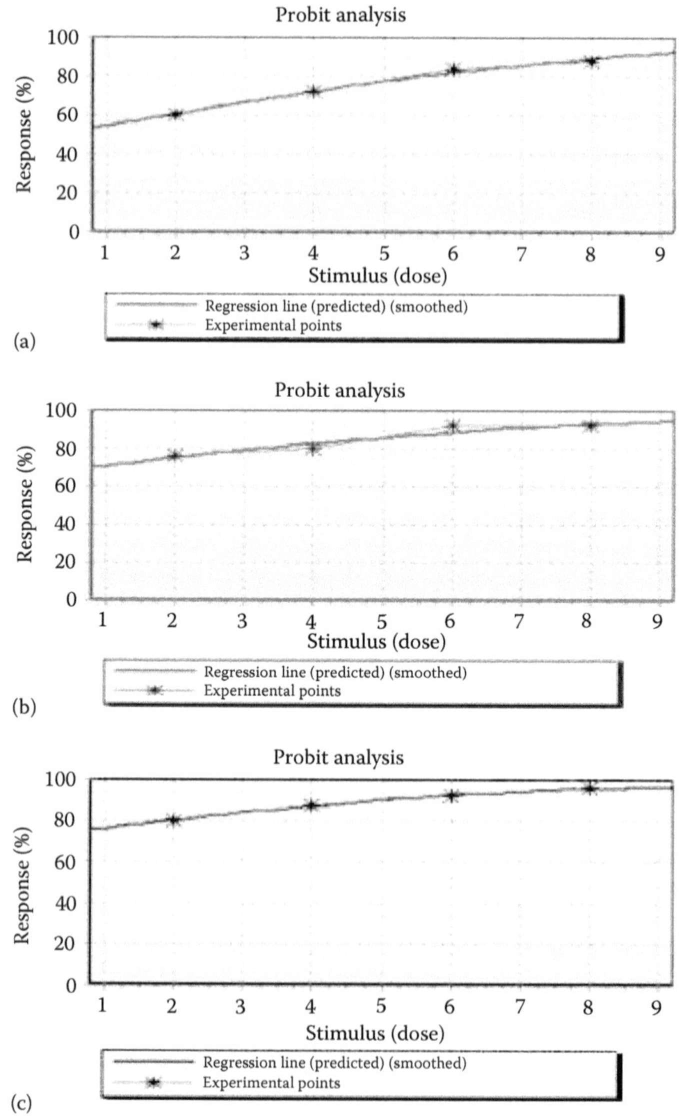


**FIGURE 6.9** Probit graph of third instar larvae. Exposure time: (a) 1 hour, (b) 8 hours, (c) 16 hours. *(Continued)*

The synthesized AgNPs were characterized by using UV-VIS, FTIR, XRD, and SEM analysis. The organisms have shown potential for extracellular synthesis of spherical silver nanoparticles in the range of 5 to 10 nm. Since silver nanoparticles are potential agents for various biological applications including antimicrobial, their application as a mosquito larvicidal agent was investigated. In the present study, younger larvae are more susceptible than older ones.

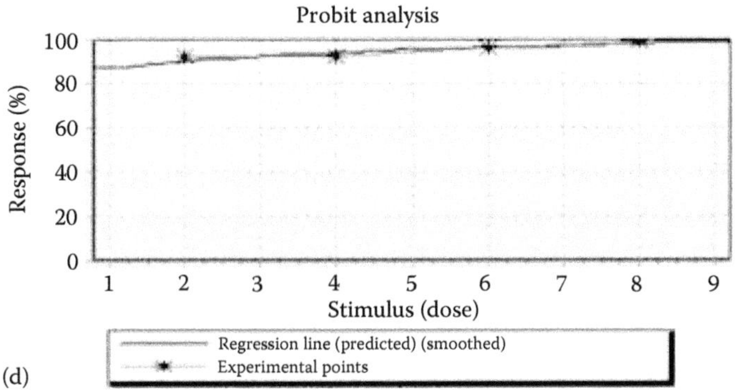


(d)

**FIGURE 6.9 (CONTINUED)** Probit graph of third instar larvae. Exposure time: (d) 24 hours.

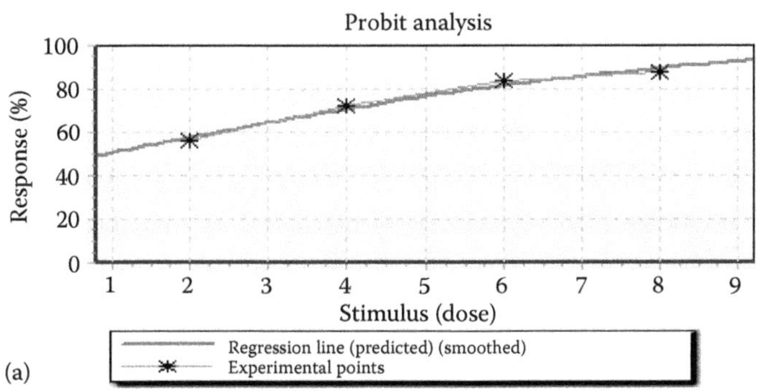


(a)

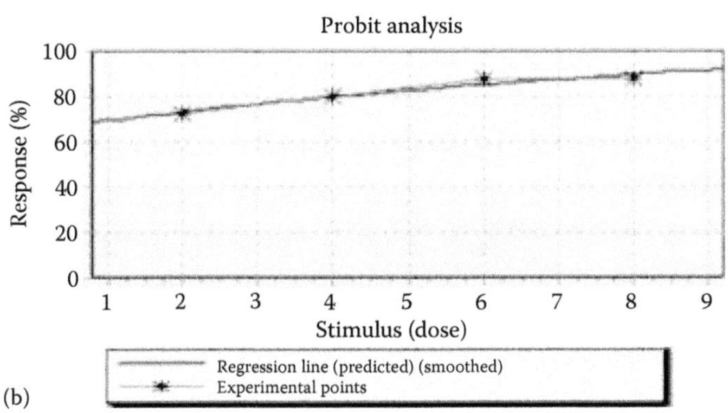


(b)

**FIGURE 6.10** Probit graph of fourth instar larvae. Exposure time: (a) 1 hour, (b) 8 hours.

*(Continued)*

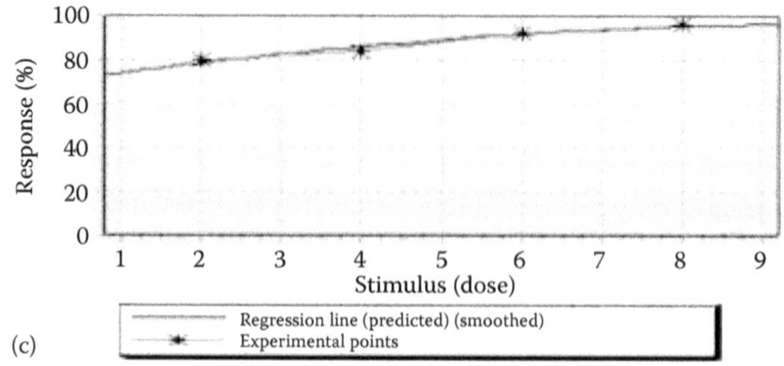


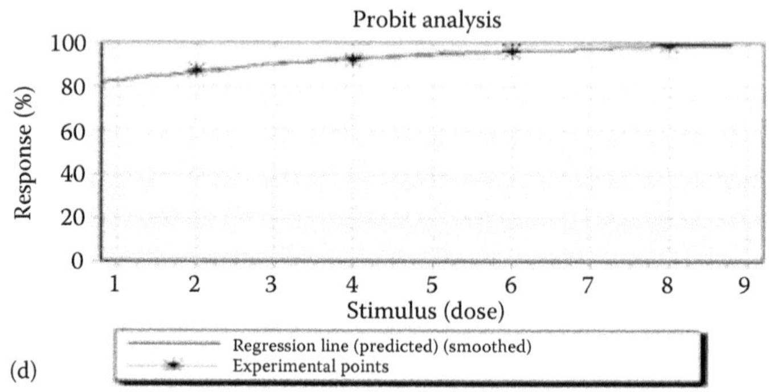


**FIGURE 6.10 (CONTINUED)** Probit graph of third instar larvae. Exposure time: (c) 16 hours, (d) 24 hours.

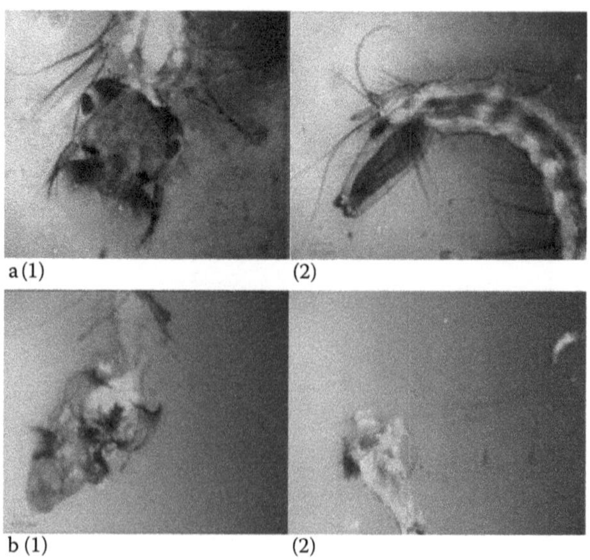


**FIGURE 6.11** CLSM view of first instar. (a) Untreated mosquito larvae: (1) Head and (2) anterior region. (b) Treated mosquito larvae.

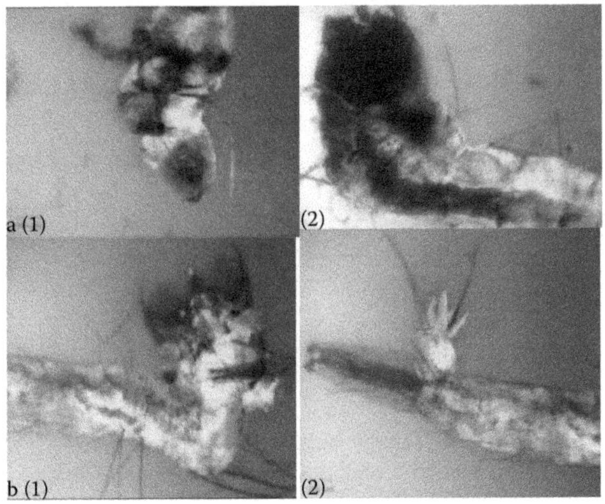


**FIGURE 6.12** CLSM view of second instar. (a) Untreated mosquito larvae: (1) head and (2) anterior region. (b) Treated mosquito larvae.

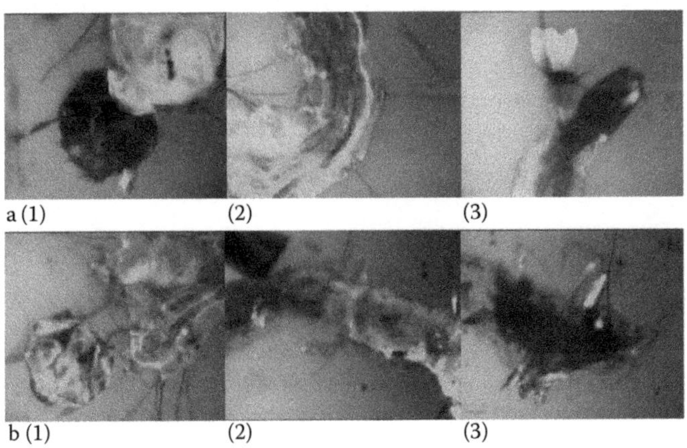


**FIGURE 6.13** CLSM view of third instar. (a) Untreated mosquito larvae: (1) head, (2) abdominal region, and (3) anterior region. (b) Treated mosquito larvae.

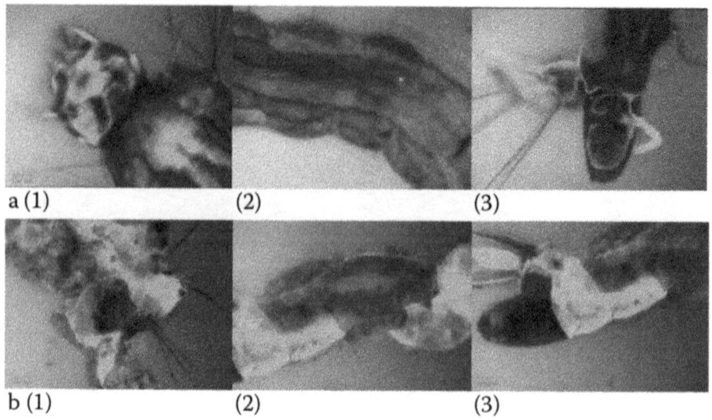


**FIGURE 6.14** CLSM view of fourth instar. (a) Untreated mosquito larvae: (1) head, (2) abdominal region, and (3) anterior region. (b) Treated mosquito larvae.

Late fourth instars that have ceased feeding or feed little before pupation are much less susceptible because of lack of ingestion of a lethal dose in a short period of time. The data obtained from the present study clearly indicate that silver nanoparticles could provide excellent larval control of *Cx. quinquefasciatus*.

## ACKNOWLEDGMENTS

The authors sincerely acknowledge the Secretary and Correspondent Thiru. K. Thulasiah Vandayar and the Principal Dr. S. Udayakumar, A.V.V.M. Sri Pushpam College (Autonomous) Poondi, Thanjavur, for their support. They also thank Dr. D. Dhanasekaran, assistant professor, Department of Microbiology, Bharathidasan University, Tiruchirappalli, India.

## REFERENCES

Ahmad A, Mukherjee P, Senapati S, Mandal D, Khan MI, Kumar R, Sastry M, 2003. Extracellular biosynthesis of silver nanoparticles using the fungus *Fusarium oxysporum*. *Colloid Surf B Biointerfaces*, 28: 313–318.

Cullity, BD, 1956. *Elements of x-ray diffraction*. Addison-Wesley Company, USA.

Dhanasekaran D, Sakthi V, Thajuddin N, Panneerselvam A, 2010. Preliminary evaluation of *Anopheles* mosquito larvicidal efficacy of mangrove actinobacteria. *Int J Appl Biol Pharm Technol*, 1(2): 374–381.

Dhanasekaran D, Thangaraj R, 2013. Evaluation of larvicidal activity of biogenic nanoparticles against filariasis causing *Culex* mosquito vector. *Asian Pac J Trop Dis*, 3(3): 174–179.

Finney DJ, 1971. *Probit analysis*. Cambridge University, London, 68–78.

Gaddeyya G, Shiny Niharika P, Bharathi P, Ratna Kumar PK, 2012. Isolation and identification of soil mycoflora. *Pelagia Res Lib*, 3(4): 2020–2026.

Krishnaraj C, Jagan EG, Rajasekar S, Selvakumar P, Kalaichelvan PT, Mohan N, 2010. Synthesis of silver nanoparticles using *Acalypha indica* leaf extracts and its antibacterial activity against water borne pathogens. *Colloid Surf B Biointerfaces*, 76: 50–56.

Lee SE, Kim JE, Lee HS, 2009. Insecticide resistance in increasing interest. *Agric Chem Biotechnol*, 44: 105–112.

Mathivanan T, Govindarajan M, Elumalai K, Krishnappa K, Annandan A, 2010. Mosquito larvicidal and phytochemical properties of *Ervatamia coronaria* Staf. *(Family: Apocynaceae). J Vector Borne Dis*, 47: 178–180.

Natarajan K, Selvaraj S, Murty VR, 2010. Microbial production of silver nanoparticle. *Digest J Nanomat Biostruct*, 5: 135–140.

Niraimathi S, Balaji N, Venkataramanan N, Govindarajan M, 2010. Larvicidal activity of alkaloid from Sida acuta against *Anopheles subpictus* and *Culex tritaeniorhynchus. Int J Curr Res*, 11: 34–38.

Oduola AO, Awe OO, 2006. Behavioural biting preference of *Culex quinquefasciatus* in human host in Lagos metropolis Nigeria. *J Vector Borne Dis*, 43: 16–20.

Rogers JV, Parkinson CV, Choi YW, Speshock JL, Hussain SM, 2008. A preliminary assessment of silver nanoparticle inhibition of monkeypox virus plaque formation. *Nanoscale Res Lett*, 3: 129–133.

Shameli K, Ahmad MB, Zamanian A, Sangpour P, Shabanzadeh P, Abdollahi Y, 2012. Green biosynthesis of silver nanoparticles using Curcuma longa tuber powder. *Int J Nanomed*, 7: 5603–5610.

Suman TY, Elumali D, Kaleena PK, Rajasree SRR, 2013. GCMS analysis of bioactive components and synthesis of silver nanoparticle using *Ammannia baccifera* aerial extract and its larvicidal activity against malaria and filariasis vectors. *Ind Crop Prod*, 47: 239–245.

Zhang H, Cui H, Yao S, Zhang K, Tao H, Meng H, 2012. Ionic liquid stabilized non-spherical gold nanofluids synthesized using a one-step method. *Nanoscale Res Lett*, 7(1): 583.

# 7 Larval Toxicity of Entomopathogenic Bacteria *Bacillus cereus* against *Aedes aegypti*, *Anopheles stephensi*, and *Culex quinquefasciatus* Mosquitoes

*Chinnasamy Ragavendran and Devarajan Natarajan*

**CONTENTS**

## 7.1 BACKGROUND

Vector control is a key component of various strategies aiming at the interruption of transmission of mosquito-borne diseases (e.g., malaria, dengue, chikungunya, filariasis, Japanese encephalitis, and leishmaniasis), which cause thousands of deaths per year in India (Benelli and Mehlhorn, 2016; Benelli et al., 2016). India reports 1.48 million malarial cases and about 1173 deaths; 1.4 million suspected and 1985 confirmed chikungunya cases; 5000 Japanese encephalitis cases and approximately 1000 deaths; and 383 dengue cases and 6 deaths during 2006 and 2007 (World Health Organization [WHO], 2007; Gopalan and Das, 2009; Dhiman et al., 2010). Mosquitoes belonging to the order Diptera and family *Culicidae*, include the genera of medical importance. *Aedes aegypti* is known to carry dengue, yellow fever, and chikungunya; and malaria is carried by *Anopheles stephensi* (WHO, 2012).

Classical larvicides, such as organophosphorus, carbamate, and pyrethroid insecticides, are less persistent, as they break down quickly in the environment (Patil et al., 2011). The most commonly used larvicides are the organophosphorus compounds such as temephos, fenthion, and chlorpyrifos, which are highly active against mosquito larvae and other aquatic insects (Amer and Mehlhorn, 2006a). Temephos is the larvicide of choice for *Ae. aegypti* and *Anopheles* control, whereas fenthion and chlorpyrifos are used against *Culex sp*. In situations where mosquitoes have developed resistance to all conventional larvicides, research and development of natural larvicidal products mostly focus on the use of larvicidal oils (Amer and Mehlhorn, 2006b), botanical extracts (Rahuman et al., 2009), and microbial larvicides (secondary metabolites), or more expensive insect growth regulators as alternatives (Priyanka et al., 2001; Das and Mukherjee, 2006).

A promising alternative is biological control, which is being employed against both agricultural pests as well as mosquitoes (Porter et al., 1993). Well-known bacterial agents have been used successfully for mosquito control, namely, *Bacillus thuringiensis* var. *israelensis* and *B. sphaericus* (Balaraman et al., 1983, 1987; Armengol et al., 2006; Medeiros et al., 2005). Few other bacterial strains such as *Bacillus alvei* and *B. brevis* (Balaraman et al., 1979; Khyami-Horani et al., 1999), *B. circulans* (Darriet and Hougard, 2002), *B. subtilis* (Das and Mukherjee, 2006), and *B. laterosporus* (Shida et al., 1996) have been reported to produce mosquitocidal toxins.

### 7.1.1 Vector-Borne Disease

Mosquito vectors are responsible for transmitting many diseases such as malaria, dengue, chikungunya, Japanese encephalitis, and lymphatic filariasis, which are mostly transmitted by three genera: *Culex*, *Aedes*, and *Anopheles*.

#### 7.1.1.1 *Aedes* Species

*Aedes aegypti* female mosquitoes are responsible for spreading dengue and chikungunya. Dengue is prevalent throughout the tropics and subtropics. The World Health Organization estimates that around 2.5 billion people are at risk of dengue. Infections have dramatically increased in recent decades due to increased urbanization, trade, and travel. No effective drug or vaccine is available so far. The only solution is to prevent the disease-carrying mosquito from breeding and biting humans. Dengue is a mosquito-borne viral disease transmitted mainly by *Ae. aegypti* and, to a lesser extent, *Ae. albopictus* (Sujitha et al., 2015). Recently, dengue transmission has strongly increased in urban and semiurban tropical areas worldwide, becoming a major international public health concern. Its medical importance is mainly due to its aggressive daytime human-biting behavior and to its ability to transmit many diseases. It works as a vector for many viruses, including dengue, yellow fever, West Nile, Japanese encephalitis, St. Louis encephalitis (Flaviridae, genus *Flavivirus*), chikungunya, Eastern equine encephalitis, Venezuelan equine encephalitis, Western equine encephalitis, Ross River virus, Sindbis virus, Mayaro virus, Getah virus (Togaviridae, genus *Alphavirus*), Potosi virus, San Angelo virus, La Crosse encephalitis virus, Jamestown Canyon virus (Bunyaviridae, genus *Bunyavirus*), Rift Valley fever (Bunyaviridae, genus *Phlebovirus*), and Orungo virus (Reoviridae, genus *Orbivirus*) (Paupy et al., 2009). Over 2.5 billion people are now at risk from dengue. The World Health Organization estimates that there may be 50 million to 100 million of dengue infections worldwide every year. There are four distinct, but closely related, serotypes of the virus that cause dengue: DEN-1, DEN-2, DEN-3, and DEN-4 (WHO, 2015).

#### 7.1.1.2 *Anopheles* Species

Malaria is caused by *Plasmodium* sp. parasites, which are transmitted to humans through the bites of infected *Anopheles* mosquitoes, biting mainly between dusk and dawn (WHO, 2014).

Malaria spread by *Anopheles* sp. is a huge threat for public health and socioeconomic development in resource-limited settings of the world (Karunamoorthi, 2012). According to the latest estimates, there were about 198 million cases of malaria in 2013 and an estimated 584,000 deaths. Malaria mortality rates have fallen globally by 47% since 2000 and by 54% in African regions. Most deaths occur among children living in Africa, where a child dies every minute from malaria (WHO, 2014).

#### 7.1.1.3 *Culex* Species

*Culex quinquefasciatus* is commonly known as the southern house mosquito vector of lymphatic filariasis, a neglected tropical disease commonly known as elephantiasis. Lymphatic filariasis is caused by Filarioidea nematodes, namely, *Wuchereria bancrofti*, which is responsible for 90% of cases, *Brugia malayi*, and *Brugia timori*. Nowadays, more than 1.4 billion people in 73 countries are living in areas where lymphatic filariasis is transmitted and at risk of being infected. Globally, an estimated 25 million men suffer with genital disease and over 15 million people are afflicted with lymphoedema. Eliminating lymphatic filariasis can prevent unnecessary suffering and contribute to the reduction of poverty (WHO, 2014). In recent years, important efforts have been conducted to propose plant- and microbial-borne bioactive compounds to act as valuable alternatives to synthetic larvicides, due to their reduced toxicity toward vertebrates and high biodegradability (Benelli et al., 2015).

### 7.1.2 Mosquito Control Measures

Mosquito control measures involve physical changes to the mosquito larval breeding habitat, but mosquito suppression can also be achieved through directly treating the breeding sites with physical, chemical, and biological agents (Table 7.1). Chemical larviciding and biological control, particularly using larvivorous fish, were important

**TABLE 7.1**
**Components of Integrated Vector Control Program**

| Type | Intervention | Targets | Products |
|---|---|---|---|
| Community education | Behavioral change, application of all other interventions | All vectors | |
| Environmental management and sanitation | Natural environment changes | Mosquitoes, snails, blackflies | |
| | Physical barriers to breeding sites | Vectors of filariasis | Polystyrene in standing water bodies |
| Biological control | Larvivorous fishes predators and competitors larviciding | Mosquitoes | Microbial larvicides, organophosphates, neem extracts, and other herbal insecticides |
| Chemical control | Space spraying | Urban mosquitoes | Pyrethroids, organophosphates |
| | Indoor residual spraying | Vectors of malaria, leishmaniasis, lymphatic filariasis | Carbamates and DDT |
| | Insecticide-treated materials | | Aerosols, coils, mats, repellents, and natural products |
| | Household products | Mosquitoes and flies | |

*Source:* World Health Organization (WHO), 2003, Guidelines for integrated vector management, WHO Regional Office for Africa, Harare, Zimbabwe.

to vector-borne disease control programs in the early part of the 20th century (Gratz and Pal, 1988).

#### 7.1.2.1 Physical Method

Physical control is the second method of reducing mosquito populations. Physical control means the modification and management of the environment of the mosquitoes. Vector control remains the primary defense against diseases like malaria, dengue fever, chikungunya, and Zika viral infections. The success of vector control relies on the assumption that vector density is related to disease transmission.

- Removal of stagnant water/treatment with biocide oils to prevent egg—The best method of mosquito control is preventing the development of the eggs into adult mosquitoes by reducing the source of breeding.
- Insecticide-treated bed nets (ITNs)—Insecticides are used for treating bed nets to kill mosquitoes, as well as other insects. Only pyrethriod insecticides are approved for use on ITNs. These insecticides have been shown to pose very low health risk to humans and other mammals.
- Personal protection creams/sprays—These types of insecticides are used as repellent in some human personal protection creams for preventing biting of mosquitoes.

#### 7.1.2.2 Chemical Method

Chemical insecticide may be synthetic or made of natural products. Natural products are derived from plants, usually called botanical insecticides or phytomolecules, and include alkaloids, nicotine, rotenone and other rotenoids, pyrethroids, and neem extracts. Twelve insecticides from four classes—organochlorines (e.g., aldrin, DDT, dieldrin, and endosulfan), organophosphates (e.g., chlorpyrifos, dichlorvos, malathion, dimethoate, and phorate), carbamates (e.g., carbaryl, aldicarb, methiocarb, methomyl), and pyrethroids—have been recommended for indoor residual spraying (IRS), but only pyrethroids have been approved for treating bed nets (Trevors et al., 1987). Chemical insecticides are leads to contaminate water and food sources, poisoning nontarget fauna and flora. Increased public concern of the possible adverse environmental effects associated with the heavy use of chemical insecticides has prompted the examination of alternative methods for insect or pest control. Since the mid-1950s, there have been numerous reports of reduced *Anopheles* sp. susceptibility to malathion, fenitrothion, propoxur, and bendiocarb, and resistance to all four classes of insecticides has been found in *Anopheles* sp. in different regions of world (Dhanasekaran and Thangaraj, 2014).

#### 7.1.2.3 Biological Methods

Biological methods consist of the utilization of natural enemies of targeted mosquitoes and of biological toxins to achieve effective vector management. They are typically most feasible with easily identifiable breeding places. The major biological control agents are

- Plants
- Larvivorous fish
- Invertebrate predators
- Nematodes
- Protozoa and fungi
- Bacteria

#### *7.1.2.3.1 Plants Act as Mosquitocidal Activity*

Mosquito control has become increasingly difficult with the indiscriminate use of synthetic chemical insecticides, which has had an adverse impact on the environment and disturb the ecological balance. The majority of chemical pesticides are harmful to man and animals, and some are not easily degradable and spread toxic effects. The search for herbal preparations that do not produce any adverse effects on the nontarget organisms and are easily biodegradable remain a research priority for scientists associated with vector control alternatives (Govindarajan, 2010). Phytomolecules derived from plant sources can act as larvicides, insect growth regulators, repellents, and oviposition attractants, and can play an important role in the interruption of the transmission of mosquito-borne diseases at the individual as well as community level (Kamaraj et al., 2008). Several groups of phytochemicals such as alkaloids, steroids, terpenoids, essential oils, and phenolics from different plants have been reported for insecticidal activities (Shaalan et al., 2005). Larvicidal activity of acetone extracts of *Murraya koenigii*, *Coriandrum sativum*, *Ferula asafetida*, and *Trigonella foenum graceum* reported maximum activity ranging from 25 to 900 ppm against *Ae. aegypti* (Harve and Kamath 2004). Sosan et al. (2001) reported larvicidal activities of essential oils of *Ocimum gratissimum*, *Cymbopogon citrus*, and *Ageratum conyzoides* against *Ae. aegypti* and achieved 100% mortality at 120, 200, and 300 ppm concentrations, respectively.

#### *7.1.2.3.2 Animals (Larvivorous fish)*

Biological control of mosquitoes using vertebrates has mainly focused on the role of larvivorous fish that consume the aquatic larvae of mosquitoes (Griffin and Knight, 2012). Larvivorous fish have been demonstrated to be very effective at reducing mosquito larval populations in many parts of the world and in a variety of habitats (Van Dam and Walton, 2007). Especially, larvivorous fish (belonging to the genus *Gambusia* and *Poecilia* (Poeciliidae)) have been introduced in more than 60 countries for mosquito control purposes (Das and Prasad, 1991; Kweka et al., 2011; Chobu et al., 2015). However, introducing larvivorous fish are often considered a threat to native aquatic fauna, including amphibians, are highlighting the need to consider the ecological cost of introducing predatory species to contribute to mosquito control (Rupp, 1996).

In recent years, biopesticides having attributes of activity and efficacy, specificity, and selectivity, low mammalian toxicity, environmental acceptability, safe for nontargets and beneficial biota, and economic viability have been sought (Baruah et al., 2008). The most important component of such an approach is the biological suppression of the insect pest by employing pathogens like bacteria, virus, and fungus. The microbial metabolites have are highly selective and established as an alternative to eco-destabilizing chemical insecticides, especially against lepidopteran insects (Hussain et al., 2002).

#### *7.1.2.3.3 Invertebrate Predators*

Biocontrol potential of dragonfly nymph of *Brachythemis contaminata* against the larvae of *Cx. quinquefasciatus*, *Ae. aegypti*, and *An. stephensi* was conducted by Singh et al. (2003). *Toxorhyncites splendens* is a nonblood sucking predatory mosquito larvae that was found to be effective killing the larvae of *Anopheline* and *Culicine* (Amalraj and Das, 1998).

#### *7.1.2.3.4 Entomopathogens*

The extensive use of chemical insecticide to control insect pests has been found to have harmful effects on people, and there is an urgent need to discover and develop new entomopathogens to these insects. Therefore, entomopathogenic bacteria have a potency against mosquitoes that can be a good alternate in vector control (Prasad et al., 2014). In the current scenario, many researchers reported that microbes like *Aspergillus niger*, *A. niger* 2587, *Cladosporium tropicum*, *Agaricus bisporus*, *Beauveria bassiana*, *Pencillium* sp. and *Vibrio* sp., *Bacillus thuringiensis*, and *Escherichia coli* synthesized metabolites dynamically to control the mosquito larvae (Soni and Prakash, 2012; Dhanasekaran and Thangaraj, 2013; Najitha Banu and Balasubramanian, 2014; Najitha Banu et al., 2014).

*7.1.2.3.4.1 Entomopathogenic Bacteria* Entomopathogenic bacteria are capable of causing infection in arthropod hosts. The entomopathogenic *Bacillus thuringiensis* subsp. *israelensis* (Bti) has been proven as an effective biocontrol agent against the larvae of many mosquito species worldwide (Ben-Dov, 2014). Microbial insecticides are essentially nontoxic and nonpathogenic to nontarget organisms. The safety offered by microbial insecticides is their greatest advantage. Bacteria and fungi have been shown to kill mosquitoes to varying degrees (Su et al., 2001). Bti and *B. sphaericus* (Bs) are being used in worldwide field tests designed to control mosquitoes' population (Philip, 2001).

Bacterial larvicides have been used for the control of nuisance and vector mosquitoes for more than two decades. The discovery of bacterium like *Bacillus* sp., which is highly toxic to dipteran larvae, has been opened for the use of potential biolarvicides in mosquito eradication programs worldwide (Kalfon et al., 1984). The mosquitocidal activity of the highly active genus of *Bacillus* sp. resulted in their development as commercial larvicides. They are now used in many countries in different regions of the world to control vector and nuisance mosquito species (Wirth et al., 2001). The major advantages of biolarvicides are reduced application costs, and safety to the environment, human beings, animals, and other nontarget organisms. Bti and Bs are entomopathogenic bacteria producing upon sporulation a parasporal crystal toxic to some invertebrates, mostly insects and nematodes

(Feitelson et al., 1992). Bti and Bs strains produce toxic proteins such as Cyt1A, Cry4A, Cry4B, Cry11A, and binary toxin (Bin) during sporulation (Federici et al., 2003). These toxins ingested by mosquito larvae enter the midgut cells (receptor mediated) and lead to cytolysis. A few other bacterial strains, including *Brevibacillus alvei*, *Brevibacillus brevis*, *Bacillus laterosporus*, *Bacillus circulans*, and *Bacillus subtilis*, have proven as potent biolarvicides against mosquitoes.

*7.1.2.3.4.2 Entomopathogenic Nematode* Entomopathogenic nematodes typically dwell in soil and are found throughout the world. They effectively control a variety of economically important insect pests and have excellent potential for control of tropical insect pests (Grewal et al., 2001). Nematodes from the family Mermithidae (Enoplea: Mermithida) parasitize terrestrial and aquatic invertebrates, including many insect disease vector species (Poinar, 1979). Many insect disease vectors are parasitized by mermithid nematodes during their aquatic larval stage, including blackflies and mosquitoes (Petersen, 1975). Seven genera of Mermithidae are parasites of mosquitoes: *Culicimermis*, *Strelkovimermis Empidomermis*, *Hydromermis*, *Octomyomermis*, *Perutilimermis*, and *Romanomermis* (Poinar et al., 2001) and over 25 species have been described (Blackmore, 1994). *Octomyomermis muspratti* was isolated from various *Aedes* and *Culex* sp. larvae from tree holes in Livingstone, Zambia (Muspratt, 1945). Subsequent colonies were established and *O. muspratti* was found to be capable of parasitizing *Ae. aegypti*, *Ae. polynesiensis*, *Cx. pipiens*, and *An. stephensi*. The nematode *Strelkovimermis spiculatus* was found to be a promising biological control agent against *Cx. quinquefasciatus* in Cuba (Rodriguez et al., 2003). *Romanomermis iyengari* and *Strelkovimermis spiculatus* are the important mermithid nematodes that naturally parasitize larval forms of various mosquito species (Sanad et al., 2013).

*7.1.2.3.4.3 Protozoa and Entomopathogenic Fungi* Microsporidians such as *Nosema*, *Thelohania*, *Parathelohania*, and *Amblyospora* have been studied for mosquito control efficacy. Selective infection of *Anopheles* larvae with some ciliates belonging to the genus Lamborella was first reported from forest areas of Assam. Natural infection was found in the immature larvae of *An. barbirostris* and *An. philippinensis*. However, none of these agents are yet ready for field application. *Anopheline* larvae are parasitized by *Thelohania* sp. (Kudo, 1929; Sen, 1941). *Nosema algerae* were infective to *Cx. quinquefasciatus*, *Ae. aegypti*, and *An. stephensi* (Gajanana et al., 1979).

Fungi belonging to Oomycota (e.g., *Lagenidium* sp., *Leptolegnia* sp., *Pythium* sp., and *Crypticola* sp.), Chytridiomycota (e.g., *Coelomomyces* sp.), Zygomycota (e.g., *Entomophthora* sp. and *Conidiobolus* sp.) and Deuteromycota (e.g. *Culicinomyces* sp, *Beauveria* sp., and *Metarhizium* sp.) phyla have reported to parasitize mosquito larvae and cause epizotics in mosquito populations (Scholte et al., 2004). *L. giganteum* induce infection with high mortality in most medically important mosquitoes like *Anopheles*, *Culex*, and *Ades* species (Guzman and Axtell, 1986). *Culicinomyces* was found to infect *Cx. quinquefasciatus* and *An. aegypti* (Kali, 2015).

The different types of fungi like *Coelomomyces*, *Lagenidium*, *Metarrhizium*, *Culinomyces*, *Tolypocladium* (Roberts and Strand, 1977), *Aspergillus terreus*

(Ragavendran and Natarajan, 2015), and *Beauveria bassiana* (Ragavendran et al., 2017) have been isolated and revealed better results against *Aedes*, *Culex*, and *Anopheles* mosquitoes. The entomopathogenic fungus *Metarhizium anisopliae* was found to be effective against *An. gambiae* (malaria vector) and *Cx. quinquefasciatus* (filariasis vector). Insect pathogenic fungi of the *Hypocrella/Aschersonia* group might be useful as agents for pest control. *Culicinomyces* known to disrupt mosquito populations were reported by Chandrahas and Rajagopalan (1979). *Lagenidium giganteum* was highly pathogenic against different larval stages of *Cx. quinquefasciatus*, *Cx. tritaeniorhynchus*, *An. culicifacies*, *An. stephensi*, and *An. subpictus*. Fungal metabolites of *Chrysosporium tropicum* was evaluated against *Cx. quinquefasciatus* larvae and found to have promising effects (Priyanka and Prakash, 2003). *Trichophyton ajelloi*, a fungus isolated from soil, caused high larval mortality in *An. stephensi* and *Cx. quinquefasciatus* (Mohanty and Prakash, 2003). Residual sprays of fungal biopesticides might replace or supplement chemical insecticides for malaria control, mainly in areas of high insecticide resistance (Blanford et al., 2005).

With this background, the present chapter deals with the isolation and identification of native entomopathogenic bacteria *Bacillus cereus* and tests its larvicidal potential against target mosquitoes.

## 7.2 MATERIALS AND METHODS

### 7.2.1 Soil Sample Collection

Soil samples (2 to 3 cm below the top soil) of 20 g of were collected from agricultural land Salem District, Tamil Nadu (using a sterile spatula) and stored in sterile screw-capped vials (capacity, 2 mL) and brought to the laboratory for storage (–20°C) (Geetha et al., 2007).

### 7.2.2 Bacterial Isolation

One gram of soil was weighed, transferred to a sterile vial containing 10 ml of sterile water and kept on a rotary shaker at 100 rpm for 30 min to dislodge the bacterial cells from the soil particles. The supernatant was diluted 10-fold and 0.1 ml was taken for serial dilution on a plate containing the nutrient agar medium (Sambrook et al., 1989). The nutrient agar (NA) contained 5 g glucose (bacteriological), 5 g peptone, 5 g NaCl, 3 g beef extract (HiMedia, India) per liter of distilled water. The plates were incubated at 30°C for 48 h, and bacterial colonies that appeared were purified on the nutrient agar. Each of the purified colonies was subcultured on nutrient agar slants, allowed to grow for 72 h on a bacteriological incubator and stored at 4°C. These bacterial isolates were screened for mosquito larvicidal activity.

### 7.2.3 Test Insects

Third and fourth instar larvae of *Ae. aegypti*, *Cx. quinquefasciatus*, and *An. stephensi* were collected from the field, in and around Salem (latitude 21°2′54″ N, longitude 76° 32′3″ E, elevation 209 m), Tamil Nadu, India. The larvae were kept in chemically sterilized plastic enamel trays containing dechlorinated tap water and maintained as per the modified method of Patil et al. (2010). All the experiments were carried out at 28 ± 2°C and 75%–85% relative humidity under 14:10 light and dark cycles (Mitchell and Smith, 1985). Larvae were fed with a diet of finely ground brewer's yeast and dog biscuits (3:1) (Dinesh et al., 2015).

### 7.2.4 Preliminary Larvicidal Activity

A loopful of bacteria from the nutrient slant was inoculated into 10 ml of nutrient broth and incubated (30°C) on a rotary shaker (200 rpm) for 72 h. After incubation, 1 ml of each sample from the whole culture was used to screen for mosquito larvicidal activity (Geetha et al., 2011). Bioassay was conducted in beakers containing 25 third and fourth instar larvae of *Cx. quinquefasciatus*, *An. stephensi*, and *Ae. aegypti* in 30 ml chlorine-free tap water. Appropriate controls were maintained without the addition of the bacterial culture, but did contain 1 ml of uninoculated nutrient broth. After 24 h of exposure, mortality was recorded by counting the number of dead larvae present in the respective container. A bacterial isolate was considered as potent if it caused 100% mortality of the test larvae. The potential bacterial cultures were further screened to find out whether the bacterial cells or their metabolites exhibited larvicidal activity. The cells and culture supernatant (CS) were separated by centrifuging the active cultures at 8000 rpm for 20 min followed by performing the bioassay tests.

### 7.2.5 Extraction of Crude Extract

The broth culture of *B. cereus* were extracted by soaking in an orbital shaker for 3 days with the n-hexane:ethyl acetate (1:1) solvents. The residue of fermented broth was centrifuged at 10,000 rpm for 20 min to remove the cells. The supernatants were adjusted to a pH of 2.0 with 12 N HCl. The precipitates were collected by centrifugation at 12,000 g (Remi, India) for 20 min. After each extraction, the solutions were filtered through Whatman No. 1 filter paper, and the solvents were evaporated in a rotary evaporator at reduced pressure (Lin et al., 2011). The solvent extract was removed, dried properly, dissolved in dimethyl sulfoxide (DMSO) and was used for larvicidal bioassay.

### 7.2.6 Larvicidal Activity of Solvent Extract

Larvicidal assay of secondary metabolites from *B. cereus* were performed against third and fourth instar of *Ae. aegypti*, *An. stephensi*, and *Cx. quinquefasciatus* using 1 mg/l concentration. DMSO (1000 µl) served as a control. Twenty-five larvae were used for all the experiments and tested in triplicates (total, $n = 75$). The lethal concentration ($LC_{50}$ and $LC_{90}$) was calculated using probit analysis (1971). For mortality studies, 25 larvae each of third and fourth instar were placed in 100 ml containers containing 1 mg/l concentration of the metabolites. Percentage mortality in the treatments and controls was corrected using Abbott's (1925) formula:

$$\text{Percentage of mortality} = \frac{\text{Number of dead larvae/pupae}}{\text{Number of larvae/pupae introduced}} \times 100$$

### 7.2.7 Identification of the Potential Bacterial Strain

Morphological, biochemical, and physiological characteristics of the potential mosquitocidal isolate were identified according to *Bergey's Manual of Systematic Bacteriology* (Sneath, 1986). Morphological characters were determined by visual examination of the plate culture and also by light microscopy (Labomed Microscope, Germany). Gram staining was performed according to the procedure described by Claus (1992). Endospore staining was carried out based on the modified method of Prescott et al. (1996). Catalase activity was determined by bubble production in a 3% (v/v) hydrogen peroxide solution. The tests were performed in the following manner: oxidase, citrate utilization, Voges Proskauer test, methyl red, and indole tests. The isolated bacteria were analyzed for the growth parameters at varied temperatures (30°C and 50°C) and various concentrations of NaCl (2%, 4%, 6%, and 8%).

### 7.2.8 Molecular Identification of Bacteria

Genomic DNA was extracted and purified as per the modified protocol of Grimont and Grimont (1995) and Clarridge (2004). Luria broth (LB) was inoculated with a loopful of cultures and incubated overnight. Next, 1.5 mL of culture from LB was transferred in a microcentrifuge tube and centrifuged at 10,000 rpm for 15 min and the supernatant was discarded. The pellet was resuspended with 450 μL of TE buffer. To this, 45 μL of 10% SDS was added with 5 μL of lysozyme, gently mixed, then incubated for 1 hr (at 37°C). To this, 500 μL of phenol:chloroform was added, mixed well by inverting the tubes until the two phases were completely mixed and centrifuged at 10,000 rpm for 10 min. The aqueous phase was transferred into a new micro centrifuge tube, to this equal volume of phenol: chloroform was added and centrifuged again at 10,000 rpm for 10 min. The aqueous phase was transferred into a new microcentrifuge tube; to this, 50 μL of sodium acetate was added and mixed thoroughly. Further, 300 μL of isopropanol was added and gently mixed to precipitate the DNA. The mixture was centrifuged at 10,000 rpm for 10 min. After discarding the supernatant, the pellet was washed with 70% ethanol for 30 sec and centrifuged at 8000 rpm for 1 to 2 min. The obtained pellet was resuspended in 100 μL of TE buffer and stored at 4°C for further analysis. The sample (5 μL) was allowed to run in 1% agarose gel to confirm the presence of genomic DNA, and isolated DNA samples were stored at –20 °C until use.

### 7.2.9 Identification of Bacterial Isolate by 16S rRNA Analysis

The 16S rRNA gene was amplified using the primers, 27 Forward (5′AGAGTT TGATCCTGGCTCAG 3′), and 1525 Reverse (5′-AAGGAGGTGWTCCARCC) (Weisberg et al., 1991). Genomic DNA was extracted from bacterial strains using thermal shock (Sambrook et al., 1989). The polymerase chain reaction (PCR) mixture contained 4 μl of Milli-Q water, 2 μl of each primer (Priority Life Science, Coimbatore, India), and 10 μl of Taq polymerase (Merck, India). To this mixture, 2 μl of the DNA template was added. In the control tube, 2 μl of ultrapure water was added instead of DNA. The total reaction volume was 20 μl. The reaction was amplified in a Thermal Cycler (Techne, United Kingdom). The PCR condition were as follows: denaturation at 94°C for 5 min, 35 cycles of denaturation at 94°C for 60 sec, annealing at 55°C for 30 sec, extension at 72°C for 90 sec, and final extension step at 72°C for 5 min was also included. The size of the PCR products was determined by agarose gel electrophoresis using appropriate (1 Kb Merck India) size markers.

### 7.2.10 DNA Sequence Analysis

DNA sequencing of amplified 16S rRNA from bacterial culture was performed at Xcelris Lab, Ahmedabad, India, using the primers of 16S rRNA under Applied Biosystems' DNA sequencing instrument, model 3730xl/AB13730XL-15104-028, and the sequence scanner v1.0. The obtained sequence was searched using the Basic Local Alignment Search Tool (BLAST) and compared with sequences of other closely related species retrieved from the GenBank database (http://www.ncbi.nlm.nih.gov/BLAST/) followed by alignment using the MEGA software version 5.

### 7.2.11 Phylogenetic Analysis of Bacterial Isolate

The evolutionary relationship of the newly isolated potential strains from agricultural soil was compared with closely related *Bacillus* species using 16S rRNA, and a phylogenetic tree was constructed. Phylogenetic analysis of the 16S rRNA sequences of the selected mosquitocidal bacteria was performed with MEGA 5. The evolutionary history was inferred using the UPGMA method (Sneath and Sokal, 1973). The evolutionary distances were computed using the maximum composite likelihood method (Tamura et al., 2004) and are expressed as the units of the number of base substitutions per site. Finally, the evolutionary analysis was conducted in MEGA 5 (Tamura et al., 2007).

## 7.3 STATISTICAL ANALYSIS

The percentage of larval mortality was calculated using the Abbott formula (1925). The dose–response data were subjected to probit regression analysis (Finney, 1971) for calculating $LC_{50}$, $LC_{90}$, and 95% fiducial limits of upper confidence limit and lower confidence limit, and chi-square values were calculated using the software developed by Reddy et al. (1992). Results with $p < 0.05$ were considered to be statistically significant.

## 7.4 RESULTS

Entomopathogenic bacteria used to control mosquitoes are favorable and environment friendly alternative to synthetic pesticides. However, in view of the current situation of rapid development of resistance to *B. sphaericus* and *B. thuringenesis*, a determined need for finding new microbial agents from the natural environment is inevitable. In the present study, 15 soil samples were collected for isolate native bacterium. The isolated bacterium showed gram-positive rods, endospore (subterminal)-bearing motile bacillus, older cultures may stain pink, due to the deterioration of cell walls. Colonies were minute, circular, shiny, and pale yellow and became irregular (Figure 7.1). However, the isolate was positive for production of

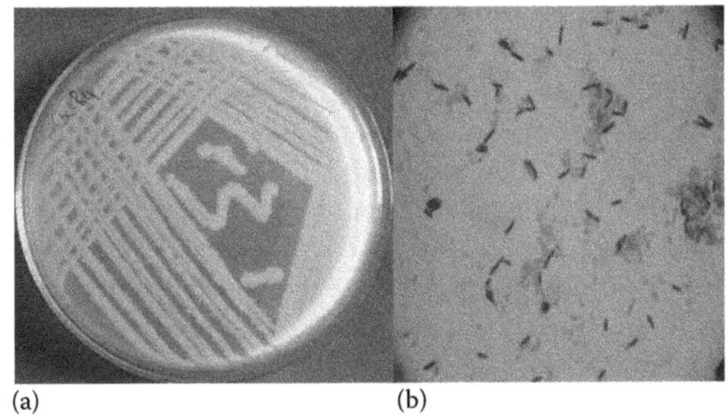


**FIGURE 7.1** (a) Colony morphology of *B. cereus*. (b) Gram staining of isolated *B. cereus* (gram-positive rods).

catalase, oxidase, Voges-Proskauer, and citrate tests. Negative results for indole test. The best growth of isolate was noted in 2%, 4% (W/V) NaCl followed by the two other concentrations, that is, 6% and 8% (W/V). Growth was completely absent under anaerobic conditions. Based on the results obtained, the isolate belonged to the genus *Bacillus* sp.

The preliminary larvicidal activities of *B. cereus* were tested against mosquito larvae (*Ae. aegypti*, *An. stephensi*, and *Cx. quinquefasciatus*). The results depicted the isolate as having mosquitocidal property. DNA fragment (1450 bp) was amplified from the 16S rRNA (Figure 7.2) and analyzed for sequencing pattern. The comparison of the generated bacterial 16S rRNA sequences was analyzed using the standard nucleotide–nucleotide BLAST program (National Center for Biotechnology Information, NCBI). BLAST analysis of the isolate showed 98% similarity with *B. cereus* and the sequence was deposited in GenBank (KR150253.1). This isolate has been

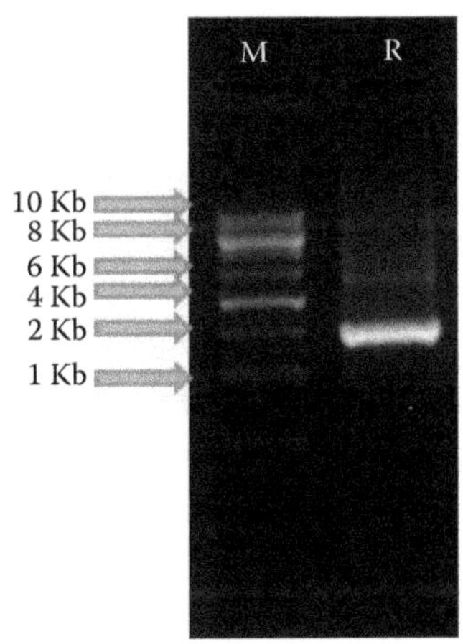


**FIGURE 7.2** PCR amplification (16s rRNA) of isolated genomic DNA from *B. cereus* M-marker.

identified as new, and its respective 16S rRNA sequences have been submitted to NCBI (accession number: KR061331.1). The evolutionary history was inferred using the maximum likelihood method based on the Tamura-Nei model. The tree is drawn to scale; with branch length measured in the number of substitutions per site. The analysis involved nine nucleotide sequences (Figure 7.3). All positions containing gaps and missing data were eliminated. There were a total of 723 positions in the final data sets. A neighbor-joining (NJ) tree of *Bacillus* 16S rRNA sequences including 10 *Bacillus* species type strains were clustered of the isolates belonging to the previously identified species to the corresponding species. We found that the *Bacillus* isolates were distributed among three clades of the NJ tree. In addition, *Bacillus* species revealed in the present study were separated into each known type strain of *Bacillus*.

The larvicidal activity of *B. cereus* extracts showed better activity against all mosquitoes. Larvae of *Ae. aegypti* were found to be highly susceptible to *B. cereus* metabolites with an $LC_{50}$ of third and fourth instar larvae (52.694 and 51.119, respectively) and an $LC_{90}$ of 58.257 and 56.513 mg/ml, respectively, after 24 h. *Cx. quinquefasciatus* third and fourth instar larvae had an $LC_{50}$ value of 55.044 and 58.515, respectively, and an $LC_{90}$ value of 60.858 and 64.702, respectively. The *An. stephensi* third and fourth instar larvae had an $LC_{50}$ value of 53.041 and 56.336, respectively, and an $LC_{90}$ of 58.674 and 62.297 mg/l, respectively. After treatment of *B. cereus* extract tested against *Ae. aegypti* and *An. stephensi*, larval survival rate was significantly reduced and *Cx. quinquefasciatus* mortality rates were slower than compared with other mosquitoes, but larvae suffered severe deformities. The sublethal effects on later larval instars were correlated with reduced survival of the early instars. Third instar larvae were most susceptible in bioassay at the optimized concentration. At the 1 mg/l concentration of *B. cereus* metabolite showed activity against *Ae. aegypti* larvae, the mortality rate was faster compared with other mosquitoes tested, with the mortality occurred within 24 h. At the concentrations of 1 mg/l metabolites over 90% of the mortality was observed in third and fourth instar of three selected mosquitoes (Table 7.2 and Figure 7.4).

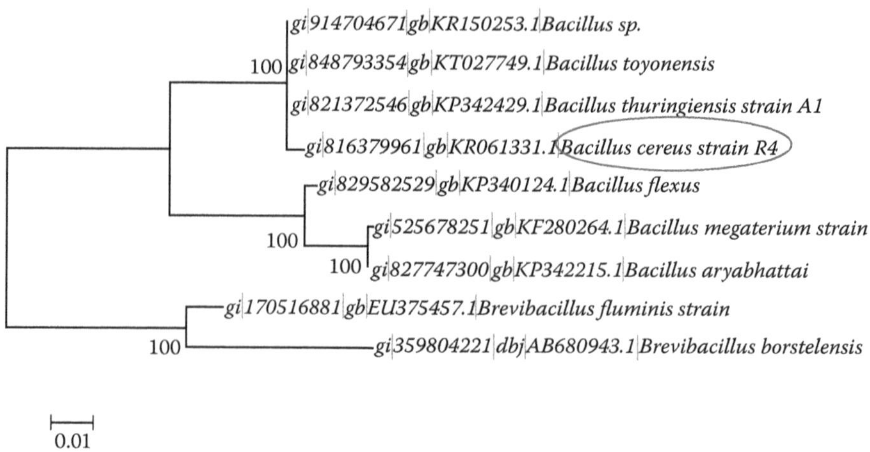


**FIGURE 7.3** Phylogenetic relationships of the isolates with other bacteria.

**TABLE 7.2**
**Effect of the Crude Extracts (1 mg/ml) of *B. cereus* against the Third and Fourth Instar Larval Stages of Three Different Mosquito Species**

| Mosquito Species | Larvae Stages | $LC_{50}$ | $LC_{90}$ | $\chi^2$ | df | Significance ($p < 0.05$) |
|---|---|---|---|---|---|---|
| *Ae. aegypti* | III | 52.694 | 58.257 | 2.420 | 4 | 0.659* |
| | IV | 51.119 | 56.513 | 0.231 | 4 | 0.994* |
| *An. stephensi* | III | 53.041 | 58.674 | 2.284 | 4 | 0.684* |
| | IV | 56.336 | 62.297 | 0.397 | 4 | 0.983* |
| *Cx. quinquefasciatus* | III | 55.044 | 60.858 | 0.439 | 4 | 0.979* |
| | IV | 58.515 | 64.702 | 0.118 | 4 | 0.998* |

*Notes:* Control (chlorine-free water with DMSO), nil mortality. $LC_{50}$, lethal concentration that kills 50% of the exposed larvae; $LC_{90}$, lethal concentration that kills 90% of the exposed larvae; df, degree of freedom; $\chi^2$, chi-square values are significant at $p < 0.05$ levels.
Mean value of three replicates.
*$p < 0.05$

**FIGURE 7.4** Larvicidal activity of *B. cereus* metabolite against *Ae. aegypti* larvae.

## 7.5 DISCUSSION

*Bacillus* is a gram-positive, rod-shaped, aerobic or facultative anaerobic spore-forming bacterium. The genus *Bacillus* is phenotypically heterogeneous, with its members exhibiting an extremely wide range of nutritional requirements, growth conditions, metabolic diversity, and DNA base composition (Claus and Berkeley, 1986). In addition, the results of 16S rRNA sequence analysis reconfirm the insufficient defined genera on the basis of phenotypic criteria (Woese, 1987).

In recent years, there has been a considerable amount of research focused on identifying potential mosquitocidal microbes from natural sources for the control of disease-transmitting mosquito vectors. In the present study, agriculture soil samples (from Salem district, Tamil Nadu, India) were taken for identifying the potential larvicide bacterial isolates. *B. cereus* (NCBI KR061331.1) extracts were identified for the first time for potent mosquitocidal agent. 16S rRNA sequences are the best indicator of phylogenetic relationships among bacteria at the intra- and interspecies levels. Identification of *B. cereus* using 16S rRNA sequencing was supported by many researchers (Zhang et al., 2012; Mani et al., 2015). The phylogenetic analysis of this isolate revealed its homology with other *Bacillus* species. These results are in agreement with earlier studies on the phylogenetic relationship of mosquitocidal bacteria (Zahner et al., 1999; Freitas et al., 2008).

In the present study, the relative trends on the mosquito toxicity of *B. cereus* have been reported as *Ae. aegypti* ≥ *An. stephensi* ≥ *Cx. quinquefasciatus*. Similarly, a few researchers reported *B. cereus* is a natural facultative mosquito pathogen (Krattiger, 1997; Plearnpis et al., 2001; Chatterjee et al., 2008). Insecticidal activity of spores of *B. cereus* against *Ae. aegypti* has been already reported by Tyrell et al. (1981). Significant larval reduction was observed using *B. cereus* against *An. subpictus* larvae (Chatterjee et al., 2010). This relative trend on toxicity was in evidence with Bti (Federici et al., 2006). *B. thuringiensis* was identified as an entomopathogen and acted as lethal to many species of lepidopterans, dipterans in particular, and is routinely used worldwide in microbial pesticide formulations (de Maagd et al., 2001).

The varying degree of biolarvicidal agents were reported by many researchers of varied range of effective $LC_{50}$ and $LC_{90}$ values against mosquito vectors. This observation constituted a probable breakthrough for the management of problematic insects and would have offered the lead of reducing the occurrence of resistance to the latest insecticides. The larvicidal activities of microbial metabolites are influenced by the specificity of the metabolites, physiological condition of the host, resistance mechanism, cuticle, and epicuticle of the insects (Weiser et al., 1992). Geetha et al. (2011) worked on the identification of *B. amyloliquefaciens* and exhibits better larvicidal and pupicidal activity against selected mosquitoes. Kovendan et al. (2011) reported an excellent larvicidal of *B. sphaericus* for first to fourth instar larvae of the malarial vector *An. stephensi* (0.051%, 0.057%, 0.062%, and 0.066%). The marine actinobacterial (EA) extracts reported that the highest larval mortality was found in extract of LK-3 ($LC_{50}$ = 108.08 and $LC_{90}$ = 609.15 ppm/l against early fourth instar larvae of *Cx. gelidus* and $LC_{50}$ = 146.24 and $LC_{90}$ = 762.69 ppm/l) against *Cx. tritaeniorhynchus* after 24 h of exposure (Karthik et al., 2011). This study clearly shows *B. cereus*, which possesses an ability to be a mosquitocidal agent, is target-specific and does not exhibit any effect over nontargeted organisms in the environment.

## 7.6 SUMMARY AND CONCLUSION

In this study, entomopathogenic bacteria were isolated from agricultural soils of Salem, Tamil Nadu, India. Based on the morphological, physiological, biochemical, and molecular characters, the isolate was identified as *Bacillus cereus* (KR06133.1). Phylogenetic analysis of the 16S rDNA gene sequence of *B. cereus* isolate showed similarity among the closely related *Bacillus* strains. The broth culture of *B. cereus* were extracted by soaking in an orbital shaker for 3 days with the n-hexane:ethyl acetate (1:1) solvents. The solvent extract was removed, dried properly, dissolved in DMSO and used for larvicidal bioassay. Toxicity assay was performed using *B. cereus* extracts against selected mosquito larvae and the results revealed the *Aedes aegypti* (a vector for dengue fever) is more susceptible than the other two species (*Anopheles stephensi* and *Culex quinquefasciatus*). The $LC_{50}$ and $LC_{90}$ values for third and fourth instar larvae of *Ae. aegypti* were 51.119 and 52.694, and 56.513 and 58.257 mg/ml, respectively. No effect of *B. cereus* was found on nontargeted organisms. However, *B. cereus* metabolites resulted in more than 90% mortality of third and fourth instar larvae of *Ae. aegypti*, *An. stephensi*, and *Cx. quinquefasciatus* at the desired concentration (1 mg/ml). Based on the findings of the present study, the bacterial metabolites can be used as ecofriendly, safe, and sustainable insecticides to overcome vector-borne diseases in the near future.

## ACKNOWLEDGMENTS

The first author express heartfelt thanks to Periyar University for released grant under University Research Fellowship Scheme (Ref No: PU/A&A-3/URF/2014) and CSIR, New Delhi awarded Senior Research Fellowship (Ref. Lr.09/810 (0024)/ 2016-EMR-I, Date: 30-03-2017) to carry out further study. We acknowledge the Department of Biotechnology, School of Biosciences, Periyar University for providing the necessary infrastructural facility to successfully carry out this research work, and we also extend our thanks to Xcelris Lab, Ahmedabad, India, for molecular identification of bacterial species.

## REFERENCES

Abbott, W. S. 1925. A method of computing the effectiveness of an insecticide. *J. Econ. Entom.*, 18, 256–267.

Amalraj, D., and Das, P. K. 1998. Estimation of predation by the larvae of *Toxorhynchites splendens* on the aquatic stages of *Aedes aegypti*. *Southeast Asian J. Trop. Med. Public Health*, 29, 177–183.

Amer, A., and Mehlhorn, H. 2006a. Larvicidal effects of various essential oils against *Aedes*, *Anopheles*, and *Culex* larvae (Diptera, Culicidae). *Parasitol. Res.*, 99, 466–472.

Amer, A., and Mehlhorn, H. 2006b. Repellency effect of forty-one essential oils against *Aedes*, *Anopheles* and *Culex* mosquitoes. *Parasitol. Res.*, 99, 478–490.

Armengol, G., Hernandez, J., Velez, J. G., and Orduz, S. 2006. Long-lasting effects of a *Bacillus thuringiensis serovar israelensis* experimental tablet formulation for *Aedes aegypti* (Diptera: Culicidae) control. *J. Econ. Entomol.*, 99, 1590–1595.

Balaraman, K., Balasubramanian, M., and Jambulingam, P. 1983. Field trial of *Bacillus thuringiensis* H-14 (VCRC B-17) against *Culex* and *Anopheles* larvae. *Ind. J. Med. Res.*, 77, 38–43.

Balaraman, K., Bheema, R. U. S., and Rajagopalan, P. K. 1979. Bacterial pathogens of mosquito larvae—*Bacillus alvei* (Cheshire and Cheyene) and *Bacillus brevis* (Migula)—Isolated in Pondicherry. *Ind. J. Med. Res.*, 70, 615–619.

Balaraman, K., Gunasekaran, K., Pillai, P. K., and Manonmani, A. M. 1987. Field trial with different formulations of *Bacillus sphaericus* for mosquito control. *Ind. J Med Res.*, 85, 620–625.

Baruah, I., Laishram, S. K. S., Das, S. C., and Kalita, J. 2008. Larvicidal activity of indigenous entomopathogenic soil bacteria from northeastern region of India. *J. Cell. Tissue Res.*, 8, 1271–1274.

Ben-Dov, E. 2014. *Bacillus thuringiensis* subsp. *israelensis* and its dipteran-specific toxins. *Toxins*, 6., 1222–1243.

Benelli, G., Lo Iacono, A., Canale, A., and Mehlhorn, H. 2016. Mosquito vectors and the spread of cancer: An overlooked connection? *Parasitol. Res.*, doi:10.1007/s00436-016-5037-y.

Benelli, G., and Mehlhorn, H. G. 2016. Declining malaria, rising dengue and Zika virus: Insights for mosquito vector control. *Parasitol. Res.*, doi:10.1007/s00436-016-4971-z.

Benelli, G., Murugan, K., Panneerselvam, C., Madhiyazhagan, P., Conti, B., and Nicoletti, M., 2015. Old ingredients for a new recipe? Neem cake, a low-cost botanical by-product in the fight against mosquito-borne diseases. *Parasitol Res.*, 114, 391–397.

Blackmore, M. S., 1994. Mermithid parasitism of adult mosquitoes in Sweden. *Am. Midl. Nat.*, 132, 192–198.

Blanford, S., Chan, B. H., Jenkins, N., Sim, D., Turner, R. J., Read, A. F., and Thomas, M. B. 2005. Fungal pathogen reduces potential for malaria transmission. *Science*, 308, 1531–1533.

Chandrahas, R. K., and Rajagopalan, P. K. 1979. Observations on mosquito breeding and the natural parasitism of larvae by a fungus *Coelomomyces* and a mermithid nematode *Romanomermis* in paddy fields in Pondicherry. *Ind. J. Med. Res.*, 69, 63–67.

Chatterjee, S., Subhra Ghosh, T., and Das, S. 2010. Virulence of *Bacillus cereus* as natural facultative pathogen of *Anopheles subpictus* Grassi (Diptera: Culicidae) larvae in submerged rice-fields and shallow ponds. *Afr. J. Biotech.*, 41, 6983–6987.

Chatterjee, S. N., Das, D., and Dangar, T. K. 2008. Efficacy of *Bacillus sphaericus* against *Aedes (Stegomyia) aegypti*, *Mansonia Indiana* (Edward) and *Culex vishnui* under laboratory and field conditions. *Entomon-Trivandrum*, 33, 181–188.

Chobu, M., Nkwengulila, G., Mahande, A. M., Mwang'onde, B. J., and Kweka, E. J. 2015. Direct and indirect effect of predators on *Anopheles gambiae* sensu stricto. *Acta Trop.*, 142, 131–137.

Clarridge, J. E. 2004. Impact of 16S rRNA gene sequence analysis for identification of bacteria on clinical microbiology and infectious diseases. *Clin. Microbiol. Rev.*, 17, 840–862.

Claus, D., and Berkeley, C. W. 1986. The genus *Bacillus*. In: *Bergey's Manual of Systematic Bacteriology*, vol 2, P. H. A. Sneath (Ed.), 1105–1139. Williams, Wilkins, Baltimore.

Claus, M. 1992. A standardized gram staining procedure. *World J. Microbiol. Biotech* 8:451–452.

Darriet, F., and Hougard, J. M. 2002. An isolate of *Bacillus circulans* toxic to mosquito larvae. *J. Am. Mosq. Control. Assoc.*, 18, 65–67.

Das, K., and Mukherjee, A. K. 2006. Assessment of mosquito larvicidal potency of cyclic lipopeptides produced by *Bacillus subtilis* strains. *Acta Trop.*, 97, 168–173.

Das, M. K., and Prasad, R. N. 1991. Evaluation of mosquito fish *Gambusia affinis* in the control of mosquito breeding in rice fields. *Ind. J. Malariol.*, 28, 171–177.

de Maagd, R. A., Bravo, A., and Crickmore, N. 2001. How *Bacillus thuringiensis* has evolved specific toxins to colonize the insect world. *Trends Genet.*, 17, 193–199.

Dhanasekaran, D., and Thangaraj, R. 2013. Evaluation of larvicidal activity of biogenic nanoparticles against filariasis causing *Culex* mosquito vector. *Asian Pacific J. Trop Dis.*, 3, 174–179.

Dhanasekaran, D., and Thangaraj, R. 2014. Microbial secondary metabolites are an alternative approaches against insect vector to prevent zoonotic diseases. *Asian. Pac. J. Trop. Dis.*, 4, 253–261.

Dhiman, R. C., Pahwa, S., Dhillon, G. P., and Dash, A. P. 2010. Climate change and threat of vector-borne diseases in India: Are we prepared? *Parasitol. Res.*, 106, 763–773.

Dinesh, D., Murugan, K., Madhiyazhagan, P., Panneerselvam, C., Nicoletti, M., Jiang, W., Benelli, G., Chandramohan, B., and Suresh, U. 2015. Mosquitocidal and antibacterial activity of green-synthesized silver nanoparticles from *Aloe vera* extracts: Towards an effective tool against the malaria vector Anopheles stephensi? *Parasitol. Res.*, 114, 1519–1529.

Federici, B. A., Park, H. W., Bideshi, D. K., Wirth, M. C., and Johnson, J. J. 2003. Recombinant bacteria for mosquito control. *J. Exp. Biol.*, 206, 3877–3885.

Federici B. A., Park, H. W., and Sakano, Y. 2006. Insectidal protien crystals of *Bacillus thuringiensis*. In: *Microbiology Monographs Series, Inclusions in Prokaryotes*, vol. 515, J. M. Shively (Ed.), 195–236. Springer, Berlin.

Feitelson, J. S., Payne, J., and Kim, L. 1992. *Bacillus thuringiensis*: Insects and beyond. *Bio/ Technol.*, 10, 271–275.

Finney, D. J. 1971. *Probit Analysis*, 3rd ed. Cambridge University Press, Cambridge.

Freitas, D. B., Reis, M. P., Bittencourt, C. I. L., Costa, P. S., Assis, P. S., Chartone-souza, E., and Nascimento, A. M. A. 2008. Genotypic and phenotypic diversity of *Bacillus* species isolated from steel plant waste. *BMC Res. Notes*, 1, 92–103.

Gajanana, A., Tewari, S. C., Reuben, R., and Rajagopalan, P. K. 1979. Partial suppression of malaria parasites in *Aedes aegypti* and *Anopheles stephensi* doubly infected with Nosema algerae and Plasmodium. *Ind. J. Med. Res.*, 70, 417.

Geetha, I., Manonmani, A. M., Prabakaran, G. 2011. *Bacillus amyloliquefaciens*: A mosquitocidal bacterium from mangrove forests of Andaman & Nicobar islands, India. *Acta Trop.*, 120, 155–159.

Geetha, I., Prabakaran, G., Paily, K. P., Manonmani, A. M., and Balaraman, K. 2007. Characterization of three mosquitocidal *Bacillus* strains isolated from mangrove forest. *Biol. Control*, 42, 34–40.

Gopalan, S. S., and Das, A. 2009. Household economic impact of an emerging disease in terms of catastrophic out of pocket health care expenditure and loss of productivity: Investigation of an outbreak of chikungunya in Orissa. *India. J. Vector Borne. Dis.*, 46, 57–64.

Govindarajan, M. 2010. Chemical composition and larvicidal activity of leaf essential oil from *Clausena anisata* (Willd.) Hook. *f. ex Benth (Rutaceae) against three mosquito species. Asian Pacific J. Trop. Med.*, 3, 874–877.

Gratz, N. G., and Pal, R. 1988. Malaria vector control: Larviciding. In: *Malaria: Principles and Practice of Malariology*, W. H. Wernsdorfer and I. McGregor (Eds.), 1213–1226. Churchill Livingstone, Edinburgh, UK.

Grewal, P. S., De Nardo, E. A. B., and Aguillera, M. M. 2001. Entomopathogenic nematodes: Potential for exploration and use in South America. *Neotrop. Entomol.*, 30, 191–205.

Griffin, L. F., and Knight, J. M. 2012. A review of the role of fish as biological control agents of disease vector mosquitoes in mangrove forests: Reducing human health risks while reducing environmental risk. *Wetl. Ecol. Manag.*, 20, 243–252.

Grimont, F., and Grimont, P. A. D. 1995. Determination of rRNA gene restriction patterns. In: *Diagnostic Bacteriology Protocols*, J. Howard and D. M. Whitcombe (Eds.), 149–164. Humana Press, Totowa, NJ.

Guzman, D., and Axtell, R. 1986. Effect of nutrient concentration in culturing three isolates of the mosquito fungal pathogen, *Lagenidium giganteum* (Oomycetes: Lagenidiales), on sunflower seed extract. *J. Am. Mosq. Control Assoc.*, 2, 196–200.

Harve, G., and Kamath, V. 2004. Larvicidal activity of plant extracts used alone and in combination with known synthetic larvicidal agents against *Aedes aegypti*. *Ind. J. Exp. Biol.*, 42, 1216–1219.

Hussain, A. A., Mostafa, S. A., Ghazal, S. A., and Ibrahim, S. Y. 2002. Studies on antifungal antibiotic and bioinsecticidal activities of some actinomycete isolates. *Afr. J. Mycol Biotechnol.*, 1, 374–381.

Kalfon, A., Charles, J. F., Bourgouin, C., and de Barjac H. 1984. Sporulation of *Bacillus sphaericus* 2297: An electron microscope study of crystal like inclusion, biogenesis and toxicity to mosquito larvae. *J. Gen. Microbiol.*, 130, 893–900.

Kali, A. 2015. Microbial entomopathogens in control of mosquito borne diseases. *Int. J. Pharm. Bio. Sci.*, 6, 232–237.

Kamaraj, C., Rahuman, A. A., and Bagavan, A. 2008. Screening for antifeedant and larvicidal activity of plant extracts against *Helicoverpa armigera* (Hübner), *Sylepta derogata* (F.) and *Anopheles stephensi* (Liston). *Parasitol Res.*, 103, 1361–1368.

Karthik, L., Gaurav, K., Bhaskara Rao, K. V., Rajakumar, G., and Rahuman, A. A. 2011. Larvicidal, repellent and ovicidal activity of marine actinobacteria extracts against *Culex tritaeniorhynchus* and *Culex gelidus*. *Parasitol. Res.*, 108, 1447–1455.

Karunamoorthi, K. 2012. Global malaria burden: Socialomics implications. *J. Socialomics*, 1, e108.

Khyami-Horani, H., Katbeh-Bader, A., and Mohsen, Z. H. 1999. Isolation of endospore-forming bacilli toxic to *Culiseta longiareolata* (Diptera: Culicidae) in Jordan. *Lett. Appl. Microbiol.*, 128, 57–60.

Kovendan, K., Murugan, K., Vincent, S., and Kamalakannan, S., 2011. Larvicidal efficacy of *Jatropha curcas* and bacterial insecticide, *Bacillus thuringiensis*, against lymphatic filarial vector, *Culex quinquefasciatus* Say (Diptera: Culicidae). *Parasitol. Res.*, doi:10.1007/s00436-011-2368-6.

Krattiger, F. 1997. Insect resistant crops: A case study of *Bacillus thuringiensis* (BT) and its transfer to developing countries. *ISAAA Briefs*, 2, 42.

Kudo, R. 1929. Studies on microsporidia parasitic in mosquitoes. VII. Notes on microsporidia of some Indian mosquitoes. *Arch. Protistenkd.*, 67, 1.

Kweka, E. J., Zhou, G., Gilbreath, T. M. 3rd, Afrane, Y., Nyindo, M., Githeko, A. K., and Yan, G. 2011. Predation efficiency of *Anopheles gambiae* larvae by aquatic predators in Western Kenya Highlands. *Parasit. Vectors*, 4, 128.

Lin, H. F., Chen, T. H, and Liu, S. H. 2011. The antifungal mechanism of *Bacillus subtilis* against *Pestalotiopsis eugeniae* and its development for commercial applications against wax apple infection. *Afr. J. Microbiol. Res.*, 5, 1723–1728.

Mani, C., Thitugnanasambantham, K., Sundarapandian, S., and Poopathi, S. 2015. Identification and characterization of a novel marine *Bacillus cereus* VCRC-B540 for mosquito control. *Biocontrol.*, 60, 71–79. Doi: 10.1007/s10526-014-9605-8.

Medeiros, F. P., Santos, M. A., Regis, L., Rios, E. M., and Rolim, N. P. J. 2005. Development of a *Bacillus sphaericus* tablet formulation and its evaluation as a larvicide in the biological control of *Culex quinquefasciatus*. *Mem. Inst. Oswaldo. Cruz.*, 100, 431–434.

Mitchell, F. L., and Smith, J. W. 1985. Pathology and bioassays of the lesser cornstalk borer (*Elasmo palpalpus lignosellus*) entomopoxvirus. *J. Invertebr. Pathol.*, 45, 75–80.

Mohanty, S. S., and Prakash, S. 2003. Laboratory evaluation of *Trichophyton ajelloi*, a fungal pathogen of *Anopheles stephensi* and *Culex quinquefasciatus*. *J. Am. Mosq. Control Assoc.*, 16, 254–257.

Muspratt, J. 1945. Observation on the larvae of tree-hole breeding Culicini (Diptera: Culicidae) and two of their parasites. *J. Entomol. Soc. S. Africa*, 8, 13–20.

Najitha Banu, A., and Balasubramanian, C. 2014. Myco-synthesis of silver nanoparticles using *Beauveria bassiana* against dengue vector *Aedes aegypti* (diptera: culicidae). *Parasitol Res.*, 113, 2869–2877.

Najitha Banu, A., Balasubramanian, C., and Vinayaga Moorthi, P. 2014. Biosynthesis of silver nanoparticles using *Bacillus thuringiensis* against dengue vector, *Aedes aegypti* (Diptera: Culicidae). *Parasitol. Res.*, 113, 311–316.

Patil, C. D., Patil, S. V., Salunke, B. K., and Salunkhe, R. B. 2011. Prodigiosin produced by *Serratia marcescens* NMCC46 as a mosquito larvicidal agent against *Aedes aegypti* and *Anopheles stephensi*. *Parasitol. Res.*, 109, 1179–1187.

Patil, S. V., Patil, C. D., Salunke, B. K., and Salunkhe, R. B. 2010. Larvicidal efficacy of six plants against two mosquito species *Aedes aegypti* and *Anopheles stephensi*. *Trop. Biomed.*, 27, 360–365.

Paupy, C., Delatte, H., Bagny, L., Corbel, V., and Fontenille, D. 2009. *Aedes albopictus*, an arbovirus vector: From the darkness to light. *Microb. Infect.*, 11, 1177–1185.

Petersen, J. J. 1985. Nematodes as biological control agents: Part I. Mermithidae. *Adv. Parasit.*, 24, 307–346.

Philip, R. S. 2001. Current status of malaria and potential for control. *Clin. Microbiol. Rev.*, 14, 208–226.

Plearnpis, L., Haruyuki, A., and Sakol, P. 2001. Isolation of bacterial strains colonizable in mosquito larval guts as novel host cells for mosquito control. *J. Biosci. Bioeng.*, 92, 342–345.

Poinar, G. O. 1979. *Nematodes for biological control of insects*. CRC Press, Boca Raton, FL.

Poinar, G. O. 2001. Nematoda and nematomorpha. In: *Ecology and Classification of North American Freshwater Invertebrates*, 2nd ed., J. H. Thorp and A. P. Covich (Eds.), 255–295. Academic, San Diego, California.

Porter, A. G., Davidson, E. W., and Liu, J. W. 1993. Mosquitocidal toxins of bacilli and their genetic manipulation for effective biological control of mosquitoes. *Microbiol. Rev.*, 57, 838–861.

Prasad, A., Kumar, D., Sharma, E., and Mathur, P. 2014. Isolation, identification and characterization of bacterial isolates from soil samples and their possible role in the management of malaria vector *Anopheles stephensi*. *Int. J. Recent Scient. Res.*, 5, 2332–2336.

Prescott, L., Harley, J., and Klein, D. 1996. *Microbiology*, 3rd ed., 811–877. WCB Publishers, Chicago, IL.

Priyanka, and Prakash, S. 2003. Laboratory efficacy tests for fungal metabolites of *Chrysosporium tropicum* against *Culex quinquefasciatus*. *J. Am. Mosq. Control Assoc.*, 19, 404–407.

Priyanka, Srivastava, J. N., and Prakash, S. 2001. *Chrysosporium tropicum* efficacy against *Anopheles stephensi* larvae in the laboratory. *J. Am. Mosq. Control Assoc.*, 17, 127–130.

Ragavendran, C., and Natarajan, D. 2015. Insecticidal potency of *Aspergillus terreus* against larvae and pupae of three mosquito species *Anopheles stephensi*, *Culex quinquefasciatus*, and *Aedes aegypti*. *Environ. Sci. Pollut. Res.*, 22, 17224–17237.

Ragavendran, C., Dubey, N. K., and Natarajan, D. 2017. *Beauveria bassiana* (Clavicipitaceae): A potent fungal agent for controlling mosquito vectors of *Anopheles stephensi*, *Culex quinquefasciatus* and *Aedes aegypti* (Diptera: Culicidae). *RSC Adv.*, 7, 3838.

Rahuman, A., Bagavan, A., Kamaraj, A., Saravanan, C., Zahir, E., and Elango, G. 2009. Efficacy of larvicidal botanical extracts against *Culex quinquefasciatus* Say (Diptera: Culicidae). *Parasitol. Res.*, 104, 1365–1372.

Reddy, P. J., Krishna, D., Murthy, U. S., and Jamil, K. 1992. A microcomputer FORTRAN program for rapid determination of lethal concentration of biocides in mosquito control. *Comput. Appl. Biosci.*, 83, 209–213.

Roberts, D. W., and Strand, A. 1977. Pathogens of medically important arthropods. *Bull. WHO*, 55, (Suppl. I): 419.

Rodriguez, J., Garcia, I. G., Diaz, M., Avila, I. G., and Sanchez, J. E. 2003. Pathogenic effect of the parasite nematode *Strelkovimermmis spiculatus* (Nematoda Mermithidae) in larvae of mosquito *Culex quinquefasciatus* (Diptera: Culicidae) under laboratory conditions in Cuba. *J. Med. Ent.*, 55, 124–125.

Rupp, H. R. 1996. Adverse assessments of *Gambusia affinis*: An alternate view for mosquito control practitioners. *J. Am. Mosq. Control. Assoc.*, 12, 155–159; discussion 160–166.

Sambrook, J., Fritsch, E. F., and Maniatis, T. 1989. *Molecular Cloning: A Laboratory Manual*. Cold Spring Harbor Laboratory Press, Cold Spring Harbor, NY.

Sanad, M. M., Shamseldean, M. S., Elgindi, A. E., and Gaugler, R. 2013. Host penetration and emergence patterns of the mosquito-parasitic mermithids *Romanomermis iyengari* and *Strelkovimermis spiculatus* (Nematoda: Mermithidae). *J. Nematol.*, 45, 30–38.

Scholte, E. J., Knols, B. G., Samson, R. A., and Takken, W. 2004. Entomopathogenic fungi for mosquito control: A review. *J. Insect Sci.*, 4, 19.

Sen, P. 1941. On the microsporidian infesting some *Anophelines* of India. *J. Malaria Inst.*, 4, 257.

Shaalan, E. A. S., Canyonb, D., Younesc, M. W. F., Abdel-Wahaba, H., and Mansoura, A. H. 2005. A review of botanical phytochemicals with mosquitocidal potential. *Environ. Inst.*, 3, 1149–1166.

Shida, O., Takagi, H., Kadowaki, K., Yano, H., and Komagata, F., 1996. Differentiation of species in the *Bacillus brevis* group and the *Bacillus aneurinolyticus* group based on the electrophoretic whole-cell protein pattern. *Antonie Van Leeuwenhoek*, 70, 31–39.

Singh, R. K., Dhiman, R. C., and Singh, S. P. 2003. Laboratory studies on the predatory potential of dragonfly nymphs on mosquito larvae. *Commun. Dis.*, 35, 96–101.

Sneath, P. H. A. 1986. Endospore forming Gram positive rods and cocci. In: *Bergey's Manual of Systematic Bacteriology*, vol. 2, P. H. A. Sneath, N. Mair, M. Sharpe, and J. Holt (Eds.), 1104–1207. Williams, Wilkins, Baltimore, MD.

Sneath, P. H. A., and Sokal, R. R. 1973. *Numerical Taxonomy*. Freeman, San Francisco, CA.

Soni, N., and Prakash, S. 2012. Efficacy of fungus mediated silver and gold nanoparticles against *Aedes aegypti* larvae. *Parasitol Res.*, 110, 175–184.

Sosan, M. B., Adewoyin, F. B., and Adewunmi, C. O. 2001. Larvicidal properties of three indigenous plant oils on the mosquito *Aedes aegypti. Nigerian J. Nat. Prod. Med.*, 5, 30–33.

Su, X., Zuo, F., Guo, Q., Huang, J., and Chen, T. X. 2001. A report on a mosquito killing fungus, *Pythium carolimanium. Fungal Divers.*, 7, 129–113.

Sujitha, V., Murugan, K., Paulpandi, M., Panneerselvam, C., Suresh, U., Roni, M., Nicoletti, M., Higuchi, A., Madhiyazhagan, P., Subramaniam, J., Dinesh, D., Vadivalagan, C., Chandramohan, B., Abdullah, A., Murugan, A., and Benelli, G. 2015. Green-synthesized silver nanoparticles as a novel control tool against dengue virus (DEN-2) and its primary vector *Aedes aegypti. Parasitol. Res.*, 114, 3315–3325.

Tamura, K., Dudley, J., Nei, M., and Kumar, S. 2007. MEGA4: Molecular Evolutionary Genetics Analysis (MEGA) software version 4.0. *Mol. Biol. Evol.*, 24, 1596–1599.

Tamura, K., Nei, M., and Kumar, S. 2004. Prospects for inferring very large phylogenies by using the neighbor-joining method. *Proc. Natl. Acad. Sci. USA*, 101, 11030–11035.

Trevors, J. T., Barkey, T., and Bourquin, A. W. 1987. Gene transfer among bacteria in soil and aquatic environments: A review. *Can. J. Microbiol.*, 33, 191–198.

Tyrell, D. J., Lee, A., and Bulla, J. R. 1981. Characterization of spore coat proteins of *Bacillus thuringiensis* and *Bacillus cereus. Comp Biochem Physiol Part B: Biochem Mol Biol.*, 70 (3), 535–539.

Van Dam, A. R., and Walton, W. E. 2007. Comparison of mosquito control provided by the Arroyo chub (*Gila orcutti*) and the mosquito fish (*Gambusia affinis*). *J. Am. Mosq. Control Assoc.*, 23, 430–441.

Weisberg, W. G., Barns, S. M., Pelletier, D. A., and Lane, D. J. 1991. 16s ribosomal DNA amplification for Phylogenetic study. *J. Bacteriol.*, 173, 697–703.

Weiser, J., Matha, V., Zizka, Z., and Jegorov, A. 1992. Ultrastructural changesin *Culex pipiens* larvae treated with tolypin, the insecticidal metabolite of *Tolypocladium inflatum* Gams (Deuteromycetes). *Cytobios*, 69, 179–186.

Wirth, M. C., Yang, Y., Walton, W. E., and Federici, B. A. 2001. *Evaluation of alternative resistance management strategies for Bacillus sphaericus. Mosquito Control Research, Annual Report.* Division of Agriculture and Natural Resources, University of California.

Woese, C. R. 1987. Bacterial evolution. *Microbiol. Rev.*, 51, 221–271.

World Health Organization (WHO). 2003. *Guidelines for integrated vector management.* WHO Regional Office for Africa, Harare, Zimbabwe.

World Health Organization (WHO). 2007. *Global plan to combat neglected tropical diseases 2008–2015. WHO/CDS/NTD/2007.40.* World Health Organization, Geneva.

World Health Organization (WHO). 2012. *Handbook for integrated vector management.* World Health Organization, Geneva.

World Health Organization (WHO). 2014. *Lymphatic filariasis. Fact sheet no. 102.* World Health Organization, Geneva.

World Health Organization (WHO). 2015 *Dengue and severe dengue. Fact sheet no. 117.* World Health Organization, Geneva.

Zahner, V., Rabinovitch, L., Suffys, P., and Momen, H. 1999. Genotypic diversity among *Brevibacillus laterosporus* strains. *Appl. Environ. Microbiol.*, 65, 5182–5185.

Zhang, Y., Fan, T., Jia, W., Zhang, W., Liu, Q., Li, B., and Zhang, L. 2012. Identification and characterization of a *Bacillus subtilis* strain TSO6 as bio-control agent of strawberry replant disease (*Fusarium* and *Verticillium* wilts). *Afr. J. Biotech.*, 11, 570–580.

# 8 Natural Weapons Used against Dengue Vector Mosquito, *Aedes aegypti*

*Nilüfer Orhan and Didem Deliorman Orhan*

## CONTENTS



## 8.1 INTRODUCTION

According to data from the World Health Organization (WHO), more than 1 million people worldwide die each year from vector-borne diseases. These infections transmit to humans or from animals to humans by vectors, such as mosquitoes, ticks, triatomine bugs, sandflies, and blackflies. Therefore, vector control plays a vitally important role to prevent the spread of these diseases. The conventional methods used for vector control are habitat control (removing stagnant water and empty cans), reducing contact with vectors (using bed nets, adding window screens to homes, or wearing protective clothing), chemical control (long-lasting insecticides, larvicides, rodenticides, repellents, and indoor residual spraying) and biological control (bacterial toxins or botanical compounds) (Crop Life International, 2017; European Center for Disease Prevention and Control, 2017).

Chikungunya, dengue fever, Rift Valley fever, yellow fever, and Zika are transmitted by *Aedes* sp.; Japanese encephalitis, lymphatic filariasis, and West Nile fever by *Culex* sp.; and Malaria by *Anopheles* sp. (WHO, 2016).

Today, among the vector-borne diseases, there is no vaccine for dengue fever; hence prevention of the dengue virus (DENV) infection depends upon reducing the risk of exposure to insects or animals that are vectors of diseases and development of environmental management programs.

Currently, approximately 390 million people worldwide are infected with DENV. The clinical manifestations of 96 million of these patients are compatible with dengue fever disease. In the world, nearly 20,000 people die every year due to this disease. Today, there is no effective antiviral drug treatment of this disease. Early diagnosis and hospital care is crucial for saving the life of the patients.

Dengue fever is a vector-borne disease caused by antigenically related four dengue virus (DENV-1, -2, -3, and -4) serotypes belonging to the genus Flavivirus in the Flaviviridae family and are transmitted by infected mosquitoes such as *Aedes aegypti*. Other vector mosquito species, such as *Ae. albopictus* can transmit at least 22 various viruses, including yellow fever virus, West Nile virus, Japanese encephalitis virus, Rift Valley fever virus, equine encephalitis viruses, and Zika virus (Abd Kadir et al., 2013). *Ae. japonicus*, the third invasive mosquito species reported in Europe, has been identified as a competent vector of West Nile virus by experimental studies, but also for La Crosse virus and Japanese encephalitis virus. This mosquito is also capable of transmitting chikungunya and dengue, but less successfully when compared to the two aforementioned species (Sánchez-Trinidad et al., 2014). The full life cycle of dengue fever virus involves the role of mosquitoes as a transmitter and humans as the main victim and source of infection.

In this chapter, we focus on effective microorganism-derived compounds, plant extracts, and natural compounds against the dengue vector mosquito, *Ae. aegypti*, and dengue fever virus.

## 8.2 DENGUE VIRUS INFECTION

After the bite of a Flavivirus-infected *Ae. aegypti* mosquito, the DENV enters the bloodstream of the person. Symptoms begin to appear within 4 to 6 days, and may last up to 3 to 14 days. If the infected person is bitten by a mosquito within approximately 6 days, dengue virus may pass to this biting mosquito. Symptoms of typical dengue often begin with a sudden onset of high fever (39°C–40°C). Then, frequent headaches, retro-orbital pain, nausea, vomiting, and skin rashes are observed in patients. A small proportion of patients are accompanied by bleeding from the nose, gums, and under the skin to classical dengue symptoms, and this development of the disease is defined as "severe dengue" (Ghosh and Basu, 2008; Behnam et al., 2016).

## 8.3 TREATMENT PROTOCOLS FOR DENGUE FEVER

At the present, no spesific antiviral drugs are available for the treatment of dengue fever. Generally, supportive treatments against serious complications, such as dehydration, liver and kidney failure, and fever, are used in medical centers to increase patients' survival rate (WHO, 1997). Secondary bacterial infections can be treated with appropriate antibiotics. The patients should be hospitalized under observation.

The protocol that can be applied for the treatment of yellow fever disease according to the WHO (1997) is as follows:

1. Adequate and appropriate fluid administration against loss of plasma volume
2. An electrolyte replacement solution treatment for diarrhea
3. A single dose of chloral hydrate (12.5–50 mg/kg) for agitated children
4. Blood transfusion should be made if there is bleeding in a patient with shock syndrome.

Vaccination, the eradication of *Aedes* mosquitoes, and protection from mosquito bites are measures taken for yellow fever prevention. Hence, there is an urgent need to discover new antiflaviviral compounds and mosqouitocidal agents.

## 8.4 BIOLOGICAL CONTROL OF DENGUE VECTOR AND ANTIFLAVIVIRAL AGENTS

Since ancient times, microorganisms have been utilized in production of alcoholic beverages, as food, or traditional folk medicine. Currently, the microorganisms (especially fungi and bacteria) play a crucial role in the production of fermented food, beverages, biodiesel, antibiotics, vaccines, insulin, diagnostic kits, and bioalcohol. Therefore, these microorganisms can also be useful in the discovery of novel drugs having antiflaviviral and/or mosqouitocidal effect. For this aim, we reviewed available literature data on biologically active compounds of microbial origin having aforementioned biological activities.

### 8.4.1 Bacteria and Fungi (Bacteriophyta and Mycophyta)

Geneticin produced by the bacterium *Micromonospora rhodorangea* is an aminoglycoside antibiotic. Zhang et al. (2009) displayed that geneticin-inhibited DENV-2 induced cytopathology, viral titers, RNA replication, and translation in baby hamster kidney cells in a dose-dependent manner (Zhang et al., 2009).

Kaptein et al. (2010) tested the inhibitory effect on the dengue virus replication cycle of some derivatives of doxorubicin (SA-3, SA-11, and SA-17) produced by *Streptomyces peucetius* subsp. *caesius* ATCC 27952. SA-17 [$EC_{50}$: 0.34 ± 0.20 μg/ml (0.52 ± 0.31 μM)] was found to be a selective inhibitor of DENV serotype 2 RNA replication 8.1). Also, it is observed that this derivate is significantly less cytostatic than other derivatives (SA-3 and SA-11). On the other hand, SA-17 [$EC_{50}$: 3.1 ± 1.0 μg/ml (4.8 ± 1.5 μM)] moderately inhibited yellow fever virus 17D RNA replication.

A promising candidate, Narasin, is considered as a therapeutic anti-DENV agent for treatment of yellow fever. It is a polyether ionophore antibiotic obtained from fermentation of *Streptomyces aureofaciens* strain and was found to be effective against four serotypes of DENV by blocking of viral protein expression (Low et al., 2011).

Ivermectin, an FDA-approved antihelminthic macrocyclic lactone, is a chemical derivative obtained by the fermentation of the soil actinomycete, *Streptomyces avermitilis*. The effect of ivermectin (Figure 8.1) on *in vitro* replication of yellow

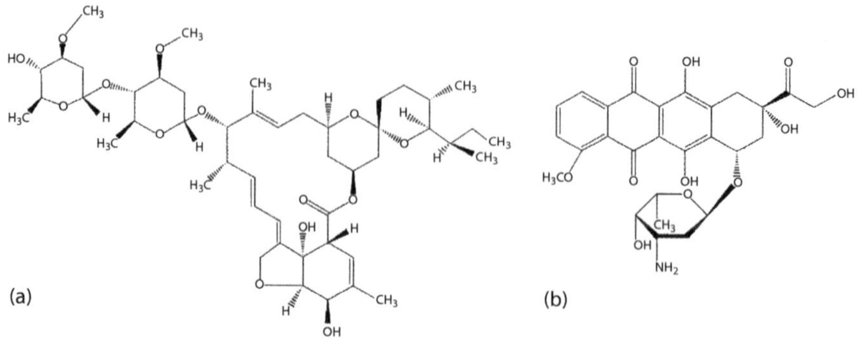


**FIGURE 8.1** The structures of (a) doxorubicin and (b) ivermectin.

fever virus (YFV), dengue viruses (DENV), and West Nile virus (WNV) was studied using helicase/enzymatic assays by Mastrangelo et al. (2012). In this report, ivermectin ($EC_{50}$: >1 μM) was found to selectively inhibit nonstructural protein 3 helicase activity playing an important role in viral replication of DENV.

*Bacillus alvei*, *B. thuringiensis* subsp. *kurstaki* (protoxin 65 kDA), *B. thuringiensis* subsp. *jegathesan* (protoxins 81, 70–72, 65, 37, 26, and 16 kDa), *B. brevis*, *B. circulans*, *Clostridium bifermentas* ser. Malaysia (protoxins 66, 18, and 16 kDa), *Pseudomonas fluorescens* and *Brevibacillus laterosporus* could be utilized for biological control of *Ae. aegypti* mosquito due to their larvicidal activities. Monnerat et al. (2004) reported that *B. sphaericus* (protoxins 100, 51, 42, and 32 kDA) isolated from the soil sample have toxic activity against *Ae. aegypti*. In the market, there are products (Aquabee®, Bactimos®, Gnatrol®, LarvX®, Teknar®, Vectobac®, Skeetal®, Mosquito Attack®) obtained from the microorganism *B. thuringiensis* var. *israelensis* (protoxins 134, 128, 78, 72, and 27 kDa) having insecticide activity against larvae of *Ae. aegypti* (Porter, 1996; Prasad et al., 2012).

Prabakaran et al. (2015) examined the pupicidal efficacy of this microorganism against pupae of *Ae. aegypti*, after identification of insecticidal activity of *Pseudomonas fluorescens* Migula (VCRC B426). Consequently, a metabolite having strong pupicidal activity ($LC_{50}$: 0.003 mg/100 ml) from *P. fluorescens* (VCRC B426) was defined as a di-rhamnolipid by different NMR and MS techniques (Figure 8.2) (Prabakaran et al., 2015).

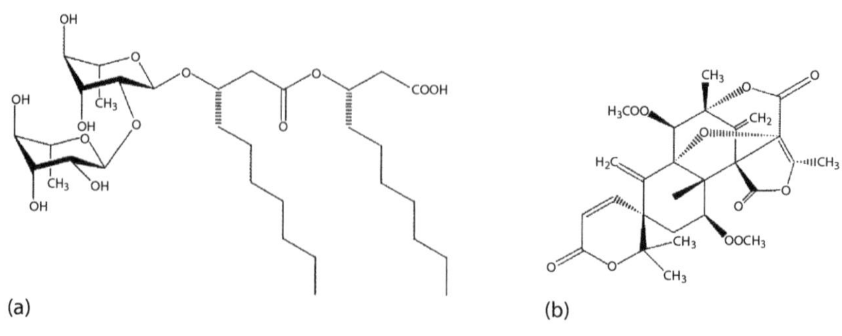


**FIGURE 8.2** The structure of (a) di-rhamnolipid and (b) dehydroaustin.

Gram-negative entomopathogenic bacterium *Photorhabdus asymbiotica* is a pathogen of both insects and humans (Costa et al., 2009). Ahantarig et al. (2009) studied larvicidal activity of *Photorhabdus* insect-related proteins (Pir) from *P. asymbiotica* against *Ae. aegypti*. Consequently, a 100% mortality rate of *Ae. aegypti* mosquitoes was observed on both day 1 and 2 for PirAB among the insect-related proteins tested.

*Pseudomonas aeruginosa* strains isolated from rhizosphere soil of the agricultural areas in Tamil Nadu (India) were evaluated for the control of dengue vector *Ae. aegypti*. Lyophilized exotoxins obtained from different *P. aeruginosa* isolates (KUN1, KUN2, KUN3, KUN4, and KUBS) were tested against the fourth instar larvae of *Ae. aegypti*. It is observed that petroleum ether extract of KUN2 isolate (100% mortality at 100 μg/ml at 24 h) possessed the maximum larvicidal activity (Lalithambika and Vani, 2016).

Mosquitocidal bacteria have many advantages:

- They are nonhazardous for human and animals.
- Their toxicities are minimal.
- Their recycling potential is high.
- They are friendly with the environment.
- The development of resistance against these insecticides is an extremely slow process.
- They are effective against the larvae even at very low dose.

Mazet and Vey (1995) isolated a toxic protein, hirsutellin A, from *Hirsutella thompsonii* (the mite fungal pathogen) using ammonium sulfate precipitation and different chromatographic techniques. The highest larval mortality of hirsutellin A (20 μg/ml concentration) having a molecular mass of 15 kDa and an isoelectric point of 10.5 was observed 100% at 72 h postinoculation.

Ninety-three isolates of six entomopathogenic fungi species (*Metarhizium anisopliae*, *Beauveria bassiana*, *Isaria fumosorosea*, *I. farinosa*, *I. flavovirescens*, and *Lecanicillium* sp.) were screened for effect on the survival rate of *Ae. aegypti* mosquito. *B. bassiana* FI-277 and FI-278 (soil samples) were considered to be the most virulent two Australian isolates against *Ae. aegypti* mosquito at 20°C, 26°C, and 32°C (Darbro et al., 2011).

Geris et al. (2008) studied the larvicidal effects of seven secondary fungal metabolites (preaustinoid, preaustinoid B, preaustinoid A2, dehydroaustin [Figure 8.2], neoastin, and acetoxydehydroaustin) isolated from *Penicillium* sp. against the third instar larvae of *Ae. aegypti*. According to $LC_{50}$ values, it was found that dehydroaustin ($LC_{50}$: 2.9 ppm) and acetoxydehydroaustin ($LC_{50}$: 7.3 ppm) displayed strong larvicidal activity against the third instar larvae of *Ae. aegypti*. Dehydroaustin killed 100% of *Ae. aegypti* larvae, after 24 h of exposure.

The efficacy of the entomopathogenic fungus *Metarhizium anisoplae* was investigated, as well as the time course of conidial adhesion and germination on both larvae and adult *Ae. aegypti* by Vieira et al. (2013). *Ae. aegypti* larvae were infected by conidial suspensions of *M. anisoplae*. It was noticed conidial germination and penetration on the thorax of adult *Ae. aegypti* between 24 and 48 hours. The outcomes were considered that this entomopathogenic fungus will be a hopeful natural insecticide for microbial control of adult *Ae. aegypti*.

### 8.4.2 Lichens (Lichenes)

Nanayakkara et al. (2005) studied the methanol and dichlorometan extracts of 48 lichenes collected from Sri Lanka against the second instar larvae of *Ae. aegypti*. The dichlorometan extracts of *Cladonia coniocraea*, *Parmeliella* sp., *Everniastrum* sp., *Parmelina tiliacea*, *Parmotrema chinense*, *P. crinitum*, *P. tinctorium*, *Rimelia reticulata*, *Dirinaria applanta*, *Heterodermia leucomelos*, *Roccella montagnei*, *Stereocaulan* sp., *Myriotrema* sp., *Usnea* sp. and methanol extracts of *Hypogymnia* sp., *Heterodermia leucomelos*, *Lepraria atrotomentosa* showed high activity (91%–100%) against *Ae. aegypti* larvae.

Methyl-5-chloro-3-formyl-2,4-dihydroxy-6-methylbenzoate,methyl-orcinolcarboxylate, zeorin, atranorin, and chloroatranorin were isolated from the hexane and dichlorometan extracts of the lichen *Heterodermia microphylla* by using different chromatographic techniques. All of these compounds showed moderate larvicidal activity against second instar larvae of the *Ae. aegypti* mosquito (90%–100% mortality at 100 ppm after 24 and 48 h) (Bombuwela et al., 2008).

The chloroform extract of *Cladonia substellata* at concentrations ranging from 1 to 15 ppm was tested against *Ae. aegypti* larvae. Usnic acid was isolated from chloroform extract ($LC_{50}$: 7.70 ppm), which was effective against the larvae, using chromatographic techniques. Usnic acid ($LC_{50}$: 6.61 ppm) was considered to be the active compound of *C. substellata* chloroform extract for larvicidal effect (Bomfim et al., 2009).

Methanol extract (1 mg/ml) of *Everniastrum cirrhatum* samples collected from tree barks in the Bhadra wildlife sanctuary in Karnataka, India, produced 100% mortality against the second instar *Ae. aegypti* larvae. Thin-layer chromatography (TLC) analysis showed that larvicidal effect of the methanol extract may be due to the presence of atranorin, salazinic acid, and protolichesterinic acid (Swathi et al., 2010).

2-hydroxy-4-methoxy-6-propyl-methyl benzoate ($LC_{50}$: 4.85 μg/ml) and (+)-usnic acid ($LC_{50}$: 4.48 μg/ml) isolated from the methanol extract of the lichen *Ramalina usnea* by Moreira et al. (2016) displayed larvicidal activity against third instar larvae of the *Ae. aegypti* mosquito.

### 8.4.3 Algae

Toxicity of hydrophilic and lipophilic cell-free extracts of 76 planktonic cyanobacteria strains was performed using larvae (late third or early fourth instar) of vector *Ae. aegypti*. The fractions containing hepatotoxin and anatoxin-a (neurotoxin) isolated from the active aqueous extracts caused toxic effects in the larvae of *Ae. aegypti* (Kiviranta et al., 1993).

The cyanobacterium *Oscillatoria agardhii* 27 was determined to be highly toxic to fourth ($LC_{50}$: 8.7 μg live cells/ml) and second instar ($LC_{50}$: 6.1 μg live cells/ml) larval stages of *Ae. aegypti*. Blue-green algae live naturally in natural mosquito habitats and propagate under appropriate conditions. These organisms could be used as food for mosquito larvae. Therefore, blue-green algae have significant advantages in biological control of mosquitoes as they can be consumed such as food for mosquitoes (Kiviranta and Abdel-Hameed, 1994).

Methanol extract of the *Westiellopsis* sp. (blue-green algae) was screened for its larvicidal activity against *Culex quinquefasciatus*, *Cx. tritaeniorhynchus*, *Ae. aegypti*, and *Anopheles stephensi*. This blue-green algae extract showed significant larvicidal activity at $LC_{50}$ values of 3.95 mg/ml after 48 h. *Westiellopsis* sp. may be one of the promising cyanobacteria species for the mosquito control program (Rao et al., 1999).

Ahmad et al. (2001) investigated the effects of four microalgal chlorophytes (*Chlorella vulgaris*, *Scenedesmus quadricauda*, *Chlorococcum* sp., and *Ankistrodesmus convolutus*) on larvicidal mortality, larval development, and adult body size of *Ae. aegypti*. In conclusion, body size reduction and wing length shortening in larvae fed with *S. quadricauda*, *C. vulgaris*, and *Chlorococcum* sp. was observed when compared with the controls. Additionally, the mortality rate of larvae fed with the same chlorophytes at the end of the 6 days was found between 84% and 100%. This report suggests that *S. quadricauda*, *C. vulgaris*, and *Chlorococcum* sp. may be promising larvicidal agents (Ahmad et al., 2001).

Methanol extracts of 20 marine algae collected from the southwest coast of India were investigated for larvicidal activity against the second and third instar larvae of the vector *Ae. aegypti*. Among the samples tested, *Lobophora variegata* showed the highest larvicidal activity with an $LD_{50}$ value of 70.38 μg/ml, followed by *Spatoglossum asperum* ($LD_{50}$: 81.28 μg/ml) > *Stoechospermum marginatum* ($LD_{50}$: 82.95 μg/ml) > *Sargassum wightii* ($LD_{50}$: 84.82 μg/ml) > *Acrosiphonia orientalis* ($LD_{50}$: 86.13 μg/ml) > *Centroceras clavulatum* ($LD_{50}$: 91.54 μg/ml) > *Padina tetrastromanica* ($LD_{50}$: 97.41 μg/ml) (Manilal et al., 2011).

Bantoto and Dy (2013) compared the larvicidal activities of the crude extracts of brown algae *Padina minor* and *Dictyota linearis* collected from the Philippines against *Ae. aegypti* mosquito. In this report, *P. minor* ($LC_{50}$: 50.8 mg/ml) exhibited more markedly larvicidal activity than that of *D. linearis* ($LC_{50}$: 60 mg/ml). Interestingly, both algae extracts prolonged the larval stage of *Ae. aegypti* mosquito. The findings revealed that *P. minor* and *D. linearis* possess larvicidal activity against fourth instar *Ae. aegypti* larvae.

Marine algae (*Ulva lactuca*, *Padina gymnospora*, *Sargassum vulgare*, *Hypnea musciformis*, and *Digenea simplex*) collected from the northeastern region of Brazil were examined with respect to their larvicidal activity against *Ae. aegypti*. Dichloromethane, methanol, ethanol, and water extracts from these algae were prepared in order to determine the concentrations required to kill $LC_{10}$, $LC_{50}$, and $LC_{90}$ of the larvae. Among the extracts tested, the dichlomethane extracts of *H. musciformis* and *P. gymnospora* exhibited strong larvicidal activity. Afterward, the dichloramethane extracts were fractionated with hexane and chloroform and also tested with the same procedure. In conclusion, hexane (17.230 μg/ml) and chloroform (29.018 μg/ml) fractions of *P. gymnospora* displayed markedly larvicidal activity (Guedes et al., 2014).

The lethal effect of fungus *Metarhizium anisopleae* was investigated on the fourth instar larvae of *Ae. aegypti*. According to the results, it was assumed that this fungus gradually decreases the survival rate of *Ae. aegypti* larvae. Furthermore, *M. anisopleae* was thought to be safe for humans and other animals (Priyadarshini and Lekeshmanaswamy, 2014).

Salvador-Neto et al. (2016) investigated the effect of the red algae *Laurencia dendroidea* on the second instar *Ae. aegypti* larvae. The dichloromethane extract of the red algae at 5 ppm concentration caused the death of the larvae. Two halogenated sesquiterpenoids, namely, (–)-elatol and (+)-obtusol isolated from the active extract, were tested on the same experimental set up. (+)-Obtusol displayed higher toxic activity with a $LC_{50}$ value of 3.5 ppm when compared to (–)-elatol. In order to explain the mechanism responsible for the larvicidal activity, the effects of the compounds were evaluated on the intestinal epithelium of larvae with histological analysis. As a result, it has been proposed that (+)-obtusol may contribute to larvicidal activity by increasing reactive oxygen species in the midgut of the larvae. Other algae reported to be effective against *Ae. aegypti* mosquito are given in Table 8.1.

### 8.4.4 Marine Sponges

The methanol extracts of sponges (*Psammaplysilla purpurea*, *Dendrilla nigra*, *Petrosia testudinaria*, *P. similes*, *Haliclona Pigmentifera*, *Ircinia fusca*, *Sigmadocia fibulata*, *S. pumila*, *Clathria reinwardti*, *Spirastrella inconstans*, *Fasciospongia cavernosa*, *F. chondroides*, *Callyspongia diffusa*, *Spongia officinalis* var. *ceylonensis*, *Dysidea herbacea*, *Cinachyra cavernosa*, and *Heteronema erecta*) collected from Palk Bay and Gulf of Mannar waters of India were screened for larvicidal properties against the fourth instar larvae of *Ae. aegypti* by Rao et al. (2008). The extracts of *P. purpurea* and *H. cribricutis* exhibited the highest larvicidal activity against larvae with $LC_{50}$ values < 50 ppm. The rest of the extracts displayed $LC_{50}$ values at <100 ppm.

Larvicidal effect of the mixture of chloroform with methanol extracts prepared from *Clathria gorgonoids*, *C. reinwardti*, *Callyspongia diffusa*, *Haliclona pigmentifera*, *Sigmadocia carnosa*, *S. fibulata*, *Petrosia similes*, *Dendrilla nigra*, *Ircinia fusca*, *Spirastrella inconstans*, and *Ircinia campana* collected from the Cape Comorin coasts of the Indian Ocean was examined on the fifth stadium instar larvae of *Ae. aegypti* by Sujatha and Joseph (2011). The extracts of *Clathria gorgonoids* ($LC_{50}$: 25.90 ppm) and *Callyspongia diffusa* ($LC_{50}$: 33.46 ppm) were found to be the most toxic to *Ae. aegypti* larvae.

The larvicidal and ovicidal effects of hexane, ethyl acetate, and methanol extracts obtained from marine sponge *Cliona celata* were evaluated against *Ae. aegypti* mosquito by Reegan et al. (2013). Methanol extract of *C. celata* exerted the highest larvicidal activity at 500 ppm. Concentrations causing 50% and 90% mortality were calculated as 158.40 and 780.16 ppm, respectively. Additionally, 72% ovicidal activity at 500 ppm concentration of methanol extract was observed against fresh eggs of *Ae. aegypti* after a 120 h posttreatment period.

**TABLE 8.1**
**Insecticidal Algae Species Effective on Controlling and Reducing the Population of Dengue Vector *Aedes aegypti***

| Type of Algae | Species | Active Extract/Principle | Reference |
|---|---|---|---|
| Blue-green algae | *Anabaena circinalis* | Microcystins | Marten (2007) |
| Blue-green algae | *Oscillatoria agardhii* | Microcystins | Marten (2007) |
| Blue-green algae | *Westielopsis* sp. | Methanol extract | Marten (2007) |
| Green algae | *Acrosiphonia orientalis* | Methanol | Yu et al. (2014) |
| Green algae | *Caulerpa racemosa* | 3:1 Ethanol:water | Yu et al. (2014) |
| Green algae | *Caulerpa peltata* | Petroleum ether:acetone | Yu et al. (2014) |
| | | Acetone | Navarro et al. (2013) |
| Green algae | *Caulerpa racemosa* | Ethanol | Navarro et al. (2013) |
| | | Acetone | Navarro et al. (2013) |
| Green algae | *Caulerpa taxifolia* | Ethanol | Navarro et al. (2013) |
| Green algae | *Caulerpa scalpelliformis* | Petroleum ether:acetone | Yu et al. (2014) |
| | | Acetone | Yu et al. (2014) |
| | | 1:9 Methanol:acetone | Yu et al. (2014) |
| | | 2:8 Methanol:acetone | Yu et al. (2014) |
| | | 3:7 Methanol:acetone | Yu et al. (2014) |
| | | 3:1 Ethanol:water | Yu et al. (2014) |
| Green algae | *Chaetomorpha antennina* | Phenolic acids | Navarro et al. (2013) |
| Green algae | *Chlorella ellipsoidea* | Ether extract | Marten (2007) |
| Green algae | *Chlorella vulgaris* | – | Marten (2007) |
| Green algae | *Chlorococcum* UMACC 218, 213 | – | Ahmad et al. (2004) |
| Green algae | *Enteromorpha intestinalis* | 3:1 Ethanol:water | Yu et al. (2014) |
| | | Petroleum ether:acetone | Yu et al. (2014) |
| | | Ethanol | Navarro et al. (2013) |
| Green algae | *Enteromorpha clathrata* | Petroleum ether:acetone | Yu et al. (2014) |
| | | Acetone | Yu et al. (2014) |
| Green algae | *Rhizoclonium hieroglyphicum* | Ether extract | Marten (2007) |
| Green algae | *Scenedesmus* UMACC 101 | – | Ahmad et al. (2004) |

*(Continued)*

**TABLE 8.1 (CONTINUED)**
**Insecticidal Algae Species Effective on Controlling and Reducing the Population of Dengue Vector *Aedes aegypti***

| Type of Algae | Species | Active Extract/Principle | Reference |
|---|---|---|---|
| Green algae | *Ulva lactuca* | 3:1 Ethanol:water | Yu et al. (2014) |
| | | Acetone | Navarro et al. (2013) |
| | | Acetone | Yu et al. (2014) |
| Brown algae | *Dictyota dichotoma* | 1:9 Methanol:acetone | Navarro et al. (2013) |
| | | 2:8 Methanol:acetone | Navarro et al. (2013) |
| | | 3:1 Ethanol:water | Navarro et al. (2013) |
| | | Ethanol | Navarro et al. (2013) |
| | | Acetone | Navarro et al. (2013) |
| Brown algae | *Dictyota dichotoma* var. *velutricata* | Ethanol | Navarro et al. (2013) |
| Brown algae | *Dictyota linearis* | 95% Ethanol | Yu et al. (2014) |
| | | Ethanol | Bantoto et al. (2013) |
| Brown algae | *Lobophora variegata* | Methanol | Yu et al. (2014) |
| Brown algae | *Padina minor* | 95% Ethanol | Yu et al. (2014) |
| Brown algae | *Padina tetrastromatica* | Methanol | Yu et al. (2014) |
| Brown algae | *Spatoglossum asperum* | Methanol | Yu et al. (2014) |
| | | Ethanol | Navarro et al. (2013) |
| Brown algae | *Sargassum myriocystum* | 3:1 Ethanol:water | Yu et al. (2014) |
| Brown algae | *Sargassum wightii* | Methanol | Yu et al. (2014) |
| | | Ethanol | Navarro et al. (2013) |
| Brown algae | *Sargassum variegatum* | Ethanol | Navarro et al. (2013) |
| Brown algae | *Sargassum swartzii* | Ethanol | Navarro et al. (2013) |
| Brown algae | *Sargassum tenerrimum* | Ethanol | Navarro et al. (2013) |

(*Continued*)

**TABLE 8.1 (CONTINUED)**
**Insecticidal Algae Species Effective on Controlling and Reducing the Population of Dengue Vector *Aedes aegypti***

| Type of Algae | Species | Active Extract/Principle | Reference |
|---|---|---|---|
| Brown algae | *Stoechospermum marginatum* | Methanol | Yu et al. (2014) |
| | | Ethanol | Navarro et al. (2013) |
| Brown algae | *Stokeyia indica* | Ethanol | Navarro et al. (2013) |
| Brown algae | *Turbinaria decurrens* | 3:1 Ethanol:water | Yu et al. (2014) |
| Red algae | *Centroceras clavulatum* | Methanol | Yu et al. (2014) |
| Red algae | *Cryptonemia crenulata* | Aqueous/DL-hybrid galactan C2S-3 | Talarico et al. (2007) |
| Red algae | *Gracilaria corticata* | 3:1 Ethanol:water | Yu et al. (2014) |
| Red algae | *Laurencia dendroidea* | 2:1 Dichloromethane: methanol/elatol (sesquiterpene) | Yu et al. (2014) |
| Red algae | *Melanothamnus somalensis* | Ethanol | Navarro et al. (2013) |
| Red algae | *Meristiella gelidium* | Aqueous/iota-, kappa-, nu-hybrid carrageenans | De S.F.-Tischer et al. (2006) |

## 8.5 ANTIFLAVIVIRAL (ANTI-DENGUE) AND MOSQUITOCIDAL PLANTS AND THEIR ACTIVE PRINCIPLES

### 8.5.1 Antiflaviviral (Anti-Dengue) Plants and Their Active Principles

Medicinal plants have been used throughout the world for health care and treatment of various diseases for centuries. Several of these may have been used to treat viral infections in the past; however, the first recognized interest in their development of antiviral agent began in the mid-20th century (Chantrill et al., 1952). Many studies reported that medicinal plant extracts significantly inhibited the replication of various viruses. Most of these studies have utilized either aqueous or alcoholic extracts of medicinal plants, and limited efforts have been directed toward the identification of active principles exhibiting the antiviral activity. Additionally, novel studies pointing out the antiviral potential of plant extracts against viral strains resistant to classical antiviral agents (Serkedjieva, 2003; Tolo et al., 2006) have challenged the modern drug discovery practices, and deem a careful look toward discovering antiviral constituents in the plants. Medicinal plants that have anti-dengue activity are summarized in Table 8.2. Their active part, extract, and/or principles are also mentioned.

**TABLE 8.2**
**Anti-Dengue Plant Extracts and Their Active Principles**

| Species | Family | Plant Part | Active Extract/Principle | Reference |
|---|---|---|---|---|
| *Alternanthera philoxeroides* | Amaranthaceae | Whole plant | Petroleum ether | Jiang et al. (2005) |
| *Anacolosa pervilleana* | Olacaceae | Leaf | Ethylacetate/acetylenic acids | Bourjot et al. (2012a) |
| *Andrographis paniculata* | Acanthaceae | Aerial part | Methanol | Tang et al. (2012) |
| *Artemisia douglasiana* | Asteraceae | Leaf | Essential oil | Garcia et al. (2003) |
| *Artemisia vulgaris* | Asteraceae | – | Essential oil | Meneses et al. (2009) |
| *Azadirachta indica* | Meliaceae | Leaf | Aqueous | Parida et al. (2002) |
| *Boesenbergia rotunda* | Zingiberaceae | Rhizome | Cyclohexenyl chalchone derivatives (4-hydroxy panduratin A, panduratin A, pinostrobin, cardamonin) | Kiat et al. (2006a) |
| *Castaspermum australae* | Fabaceae | Seed | Alkaloid (castanospermine) | Whitby et al. (2005) |
| *Cissampelos pareira* | Menispermaceae | Aerial parts | Methanol | Bhatnagar et al. (2010) |
| *Cladogynos orientalis* | Euphorbiaceae | Whole plant | Dichloromethane and ethanol | Klawikkan et al. (2011) |
| *Cryptocarya chartacea* | Lauraceae | Stem bark | Dialkylated flavanones (chartaceones C–F) | Allard et al. (2011) |
| *Curcuma longa* | Zingiberaceae | Rhizome | Methanol | Kiat et al. (2006b) |
| *Curcuma mangga* | Zingiberaceae | Rhizome | Methanol, hexane | Kiat et al. (2006b) |
| *Distictella elongata* | Bignoniaceae | Leaf | Ethanol | Simoes et al. (2011) |
| *Eupatorium patens* | Asteraceae | Leaf, flower, fruit | Essential oil | Garcia et al. (2003) |
| *Flacourtia ramontchi* | Salicaceae | Stem bark | Ethyl acetate/betulinic acid 3β-caffeate, flacourtosides | Bourjot et al. (2012b) |
| *Flagellaria indica* | Flagellariaceae | Whole plant | Dichloromethane and ethanol | Klawikkan et al. (2011) |
| *Gastrodia elata* | Orchidaceae | – | Sulfated α-D-glucan | Tong et al. (2010) |

(*Continued*)

**TABLE 8.2 (CONTINUED)**
**Anti-Dengue Plant Extracts and Their Active Principles**

| Species | Family | Plant Part | Active Extract/Principle | Reference |
|---|---|---|---|---|
| *Hippophae rhamnoides* | Elaeagnaceae | Leaf | Aqueous ethanol | Jain et al. (2008) |
| *Houttuynia cordata* | Saururaceae | Whole plant, stem, leaf | Dichloromethane and ethanol; aqueous | Klawikkan et al. (2011), Leardkamolkarn et al. (2012) |
| *Kaempferia parviflora* | Zingiberaceae | Stem, leaf | – | Phurimsak and Leardkamolkarn (2005) |
| *Leucaena leucocephala* | Fabaceae | Seed | Sulfated galactomannans | Ono et al. (2003) |
| *Lippia alba* | Verbenaceae | – | Essential oil | Meneses et al. (2009), Ocazionez et al. (2010) |
| *Lippia citriodora* | Verbenaceae | – | Essential oil | Ocazionez et al. (2010) |
| *Lippia origanoides* | Verbenaceae | – | Essential oil | Meneses et al. (2009) |
| *Mimosa scabrella* | Fabaceae | Seed | Sulphated galactomannans | Ono et al. (2003) |
| *Momordica charantia* | Cucurbitaceae | Aerial part | Methanol | Tang et al. (2012) |
| *Myrtopsis corymbosa* | Rutaceae | Bark, leaf | Ethylacetate | Coulerie et al. (2013) |
| *Origanum vulgare* | Lamiaceae | – | Essential oil | Meneses et al. (2009) |
| *Piper retrofractum* | Piperaceae | Whole plant | Dichloromethane and ethanol | Klawikkan et al. (2011) |
| *Psidium guajava* | Myrtaceae | Leaf | – | Abd Kadir et al. (2013) |
| *Quercus lusitanica* | Fagaceae | Seed | Methanol | Muliawan et al. (2006) |
| *Rhizophora apiculata* | Rhizophoraceae | Whole plant | Dichloromethane and ethanol | Klawikkan et al. (2011) |
| *Tephrosia madrensis* | Fabaceae | – | Flavones (glabrin, 7-O-methyl-glabranine) | Sanchez et al. (2000) |
| *Trigonostemon cherrieri* | Euphorbiaceae | Bark, wood | Ethylacetate/diterpenoid (trigocherriolide A) | Allard et al. (2012) |

*(Continued)*

**TABLE 8.2 (CONTINUED)**
**Anti-Dengue Plant Extracts and Their Active Principles**

| Species | Family | Plant Part | Active Extract/Principle | Reference |
|---|---|---|---|---|
| *Uncaria tomentosa* | Rubiaceae | Stem bark | Ethanol, oxindole alkaloids | Reis et al. (2008) |
| *Zingiber zerumbet* | Zingiberaceae | Rhizome | Methanol | Kiat et al. (2006b) |

The mechanism of antiviral activity of plants may vary among different viruses. They may inhibit the viral adhesion, binding of the virus to the receptor of the host cell, internalization of the virus, and the postentry process of viral replication and show direct virucidal effect against the virus strain. Besides a broad-spectrum antiviral effect, plant extracts boost inherent antiviral defense of the body and support the immune system by their immunomodulatory effect. These effects could be associated with a single phytochemical or a number of different plant constituents. There have been numerous reports on the antiviral activity of various secondary metabolites of medicinal plants having anti-dengue activity and they showed a wide range of activity against DENV. The isolated and identified active compounds belong to different chemical classes such as alkaloids, flavonoids, quinones, and sulfated polysaccharides. We compiled the anti-dengue natural compounds derived from plants and animals in Table 8.3 and mentioned their mechanism of actions.

### 8.5.2 Mosquitocidal Plants and Their Active Principles Used against *Aedes aegypti*

Due to vector-borne diseases and serious harms caused by mosquitoes, controlling them has been important and become the subject of many studies over the past decades. The technique used to control mosquitoes depends on the larval stages (egg, larvae, pupae, and adult) of the target. In recent years, use of many of the former synthetic insecticides in mosquito control programs has been limited. It is because of the lack of new insecticides, high cost of synthetic insecticides, concern for environmental sustainability, harmful effect on human health and other nontarget populations, their non biodegradable nature, higher rate of biological magnification through the ecosystem, and increasing insecticide resistance on a global scale (Ghosh et al., 2012).

Phytochemicals with mosquitocidal activity are recognized as potent alternative insecticides to replace synthetic insecticides in mosquito control programs due to their excellent larvicidal, pupicidal, and adulticidal properties. These phytochemicals have been projected as weapons in future mosquito control programs as they are shown to function as general toxicants, growth and reproductive inhibitors, repellents, and oviposition-deterrents (Panneerselvam et al., 2012).

The effects of medicinal plants on controlling and reducing the population of dengue vector *Ae. aegypti* are summarized in Table 8.4.

**TABLE 8.3**
**Anti-Dengue Natural Compounds Derived from Plants and Animals**

| Class | Compound | Mechanism | Reference |
|---|---|---|---|
| Alkaloid | Emetin | Inhibits the postentry process of viral replication | Low et al. (2009) |
| | Lycorine | Inhibits replication of virus | Zou et al. (2009) |
| | Palmatine | Inhibits replication of virus (inhibit NS2B-NS3 protease) | Jia et al. (2010) |
| Aminosteroid | Squalamine | Inhibits viral adhesion | Ozaslan (2012) |
| Flavonoid | Baicalein | Direct virucidal activity against dengue virus type-2 | Zandi et al. (2012) |
| | Baicalin | Inhibits adsorption and replication of virus | Moghaddam et al. (2014) |
| | Fisetin | Inhibits replication of virus | Zandi et al. (2011b) |
| | Naringenin | Direct virucidal activity against Dengue virus type-2 | Zandi et al. (2011b) |
| | Quercetin | Inhibits replication of virus | Zandi et al. (2011a) |
| | Rutin | Inhibits replication of virus | Silva et al. (2011) |
| Phenolic acid | Zosteric acid | Inhibits binding of the virus to the receptor of the host cell | Rees et al. (2008) |
| Polycyclic quinone | Gymnochrome D | Inhibits replication of virus | Laille et al. (1998) |
| | Isogymnochrome D | Inhibits replication of virus | Laille et al. (1998) |
| Saponin | Glycyrrhizin | Inhibits replication of virus | Crance et al. (2003) |
| Sulfate polysaccharide | λ-Carrageenan | Inhibits virus adsorption and internalization | Talarico et al. (2016) |
| | Fucoidan | Inhibits binding of the virus to the receptor of the host cell | Hidari et al. (2008) |
| | DL-Galactan hybrids | Inhibits binding of the virus to the receptor of the host cell | Pujol et al. (2002) |

## 8.6 CONCLUSION

Vector-borne diseases continue to create serious public health threats in Southeast Asia and Western Pacific regions. Mosquitoes, sandflies, ticks, fleas, triatomine bugs, tsetse flies, blackflies, and aquatic snails carry sicknesses easily transmitted to people. Among these vectors, a kind of the mosquito, also known as *Aedes aegypti*,

**TABLE 8.4**
**Effect of Plant Extracts on Controlling and Reducing the Population of Dengue Vector *Aedes aegypti***

| Species | Family | Plant Part | Active Extract/ Principle | Effect | Reference |
|---|---|---|---|---|---|
| *Acalypha alnifolia* | Euphorbiaceae | Leaf | Methanol | Larvicidal | Kovendan et al. (2012) |
| *Ageratina adenophora* | Asteraceae | Twigs | Acetone | Larvicidal | Raj Mohan and Ramaswamy (2007) |
| *Ageratum conyzoides* | Asteraceae | Whole plant | Essential oil | Larvicidal | de Mendonça et al. (2005) |
| *Agremone mexicana* | Papaveraceae | Seed | Acetone | Larvicidal | Ruiz-Guerrero et al. (2015) |
| *Anacardium occidentalis* | Anacardiaceae | Nut | Oil | Larvicidal | De Mendonça et al. (2005) |
| *Annona crassiflora* | Annonaceae | Root wood, root bark, stem | Ethanol | Larvicidal | Omena et al. (2007) |
| *Annona glabra* | Annonaceae | Seed | Ethanol | Larvicidal | Omena et al. (2007) |
| *Annona muricata* | Annonaceae | Root | Ethanol | Larvicidal | Omena et al. (2007) |
| *Annona squamosa* | Annonaceae | Root, leaf | Ethanol | Larvicidal | Omena et al. (2007) |
| *Apium graveolens* | Apiaceae | Seed | Ethanol, essential oil | Larvicidal, repellent | Choochate et al. (2004), Pitasawat et al. (2007) |
| *Artemisia nilagirica* | Asteraceae | Leaf | Methanol | Adulticidal, larvicidal, pupicidal, repellent | Panneerselvam et al. (2012) |
| *Atlantia monophylla* | Rutaceae | Leaf | Methanol | Larvicidal, pupicidal | Sivagnaname and Kalyanasundaram (2004) |
| *Azadirachta indica* | Meliaceae | Leaf | Ethanol | Larvicidal | Mgbemena (2010) |
| *Boenninghausenia albiflora* | Rutaceae | Leaf | Petroleum ether | Larvicidal | Alam et al. (2011) |
| *Bowdichia virgilioides* | Fabaceae | Heartwood | Cyclohexane/ isoflavonoid (maackiain) | Larvicidal | Bezerra-Silva et al. (2015) |

(*Continued*)

**TABLE 8.4 (CONTINUED)**
**Effect of Plant Extracts on Controlling and Reducing the Population of Dengue Vector *Aedes aegypti***

| Species | Family | Plant Part | Active Extract/ Principle | Effect | Reference |
|---|---|---|---|---|---|
| *Cadaba indica* | Capparaceae | Leaf | Ethanol | Larvicidal | Kalimuthu et al. (2012) |
| *Carapa guianensis* | Meliaceae | Seed | Oil | Larvicidal | De Mendonça et al. (2005) |
| *Carum carvi* | Apiaceae | Seed | Essential oil | Larvicidal | Pitasawat et al. (2007) |
| *Cassia obtusifolia* | Fabaceae | Seed | Methanol, chloroform/ anthraquinone (emodin) | Larvicidal | Jang et al. (2002), Yang et al. (2003) |
| *Cassia tora* | Fabaceae | Seed | Methanol | Larvicidal | Jang et al. (2002) |
| *Catharanthus roseus* | Apocgynaceae | Leaf | Acetone, petroleum ether | Larvicidal | Remia and Logaswamy (2010), Alam et al. (2011) |
| *Centella asiatica* | Apiaceae | Leaf | Diethyl ether, hexane | Larvicidal | Nair et al. (2014) |
| *Cinnamomum cassia* | Lauraceae | Bark | Essential oil | Larvicidal | Zhu et al. (2008) |
| *Citrullus colocynthis* | Cucurbitaceae | Leaf | Petroleum ether | Larvicidal | Rahuman and Venkatesan (2008) |
| *Citrullus vulgaris* | Cucurbitaceae | Leaf | Benzene | Larvicidal, ovicidal | Mullai et al. (2008) |
| *Citrus grandis* | Rutaceae | Bark | Methanol | Larvicidal | Gutierrez et al. (2014) |
| *Citrus limetta* | Rutaceae | Fruit exocarp | Hexane, petroleum ether | Larvicidal | Kumar et al. (2012) |
| *Clausena dentata* | Rutaceae | Leaf | Essential oil | Repellent | Rajkumar and Jebanesan (2010) |
| *Clerodendrum inerme* | Lamiaceae | Leaf | Aqueous, petroleum ether | Larvicidal | Patil et al. (2013), Yankanchi et al. (2014) |
| *Clitoria ternatea* | Fabaceae | Seed | Methanol | Larvicidal | Mathew et al. (2009) |
| *Coccinia indica* | Cucurbitaceae | Leaf | Methanol | Larvicidal | Rahuman et al. (2008) |

(*Continued*)

**TABLE 8.4 (CONTINUED)**
**Effect of Plant Extracts on Controlling and Reducing the Population of Dengue Vector *Aedes aegypti***

| Species | Family | Plant Part | Active Extract/ Principle | Effect | Reference |
|---|---|---|---|---|---|
| *Copaifera langsdorffii* | Fabaceae | – | Oil | Larvicidal | de Mendonça et al. (2005) |
| *Cucumus sativus* | Cucurbitaceae | Leaf | Methanol | Larvicidal | Rahuman et al. (2008) |
| *Curcuma aromatica* | Zingiberaceae | Rhizome | Hexane, essential oil | Adulticidal, larvicidal, repellent | Choochate et al. (2005) |
| *Curcuma longa* | Zingiberaceae | Rhizome | Essential oil | Larvicidal | Zhu et al. (2008) |
| *Curcuma zedoaria* | Zingiberaceae | Rhizome | Essential oil | Larvicidal | Pitasawat et al. (2007) |
| *Cybistax antisyphilitica* | Bignoniaceae | Stem, wood | Hexane/ naphtaquinone (lapachol) | Larvicidal | Rodrigues et al. (2005) |
| *Cymbopogon citratus* | Rutaceae | Leaf | Ethanol | Larvicidal | Mgbemena (2010) |
| *Cymbopogon winterianus* | Poaceae | Whole plant | Essential oil | Larvicidal | de Mendonça et al. (2005) |
| *Delonix elata* | Fabaceae | Leaf, seed | Methanol | Larvicidal, ovicidal | Marimuthu et al. (2012) |
| *Erythrina mulungu* | Fabaceae | Stem bark | Ethanol | Larvicidal | Omena et al. (2007) |
| *Eucalyptus citriodora* | Myrtaceae | Leaf | Hexane, essential oil | Larvicidal | Singh et al. (2007), Zhu et al. (2008) |
| *Eucalyptus globulus* | Myrtaceae | Leaf | Hexane | Larvicidal | Nair et al. (2014) |
| *Eupatorium odoratum* | Asteraceae | Leaf | Petroleum ether | Larvicidal | Alam et al. (2011) |
| *Feronia limonia* | Rutaceae | Leaf | Fatty acid (palmitic acid) | Larvicidal | Rahuman et al. (2000) |
| *Foeniculum vulgare* | Apiaceae | Seed | Essential oil | Larvicidal | Pitasawat et al. (2007) |
| *Gluta renghas* | Anacardiaceae | Bark, leaf | Methanol | Larvicidal | Yousaf et al. (2015) |
| *Lantana camara* | Verbenaceae | Leaf | Acetone | Larvicidal | Remia et al. (2010) |
| *Magnifera indica* | Anacardiaceae | Bark, leaf | Methanol | Larvicidal | Yousaf and Zuharah (2015) |

(*Continued*)

**TABLE 8.4 (CONTINUED)**
**Effect of Plant Extracts on Controlling and Reducing the Population of Dengue Vector *Aedes aegypti***

| Species | Family | Plant Part | Active Extract/ Principle | Effect | Reference |
|---|---|---|---|---|---|
| *Mentha piperita* | Lamiaceae | Leaf | Essential oil | Larvicidal | Geetha and Roy (2014) |
| *Millettia dura* | Fabaceae | Seed | Chloroform/ rotenoids (deguelin, tephrosin) | Larvicidal | Yenesew et al. (2003) |
| *Millingtonia hortensis* | Bignoniaceae | Leaf | Acetone | Larvicidal | Kaushik and Saini (2008) |
| *Momordica charantia* | Cucurbitaceae | Fruit, leaf | Hexane, methanol | Larvicidal | Singh et al. (2006), Rahuman et al. (2008) |
| *Mundula sericea* | Fabaceae | Root bark | Hexane, methanol | Larvicidal | Langat et al. (2012) |
| *Nyctanthes arbor-tristis* | Oleaceae | Leaf | Petroleum ether | Larvicidal | Alam et al. (2011) |
| *Ocimum canum* | Lamiaceae | Leaf | Methanol | Adulticidal, larvicidal | Murugan et al. (2016) |
| *Ocimum gratissimium* | Lamiaceae | Leaf | Ethanol | Larvicidal | Mgbemena (2010) |
| *Ocimum sanctum* | Lamiaceae | Leaf | Acetone, ethylacetate | Larvicidal | Anees (2008) |
| *Piper aduncum* | Piperaceae | Leaf | Hexane, dillapiole, essential oil | Larvicidal | Marques et al. (2015) |
| *Piper auritum* | Piperaceae | Leaf | Essential oil | Larvicidal | Marques et al. (2015) |
| *Piper betle* | Piperaceae | Leaf | Essential oil | Larvicidal | Marques et al. (2015) |
| *Piper fimbriulatum* | Piperaceae | Leaf | – | Larvicidal | Marques et al. (2015) |
| *Piper gaudichaudianum* | Piperaceae | Leaf | Essential oil | Larvicidal | Marques et al. (2015) |
| *Piper hostmannianum* | Piperaceae | Leaf | Essential oil | Larvicidal | Marques et al. (2015) |
| *Piper jacquemontianum* | Piperaceae | Leaf | – | Larvicidal | Marques et al. (2015) |
| *Piper klotzschianum* | Piperaceae | Seed, root | Essential oil | Larvicidal | Marques et al. (2015) |

*(Continued)*

## TABLE 8.4 (CONTINUED)
## Effect of Plant Extracts on Controlling and Reducing the Population of Dengue Vector *Aedes aegypti*

| Species | Family | Plant Part | Active Extract/ Principle | Effect | Reference |
|---|---|---|---|---|---|
| *Piper longum* | Piperaceae | Fruit exocarp, fruit | Ethanol, methanol, hexane, essential oil | Larvicidal | Chaithong et al. (2006), Marques et al. (2015) |
| *Piper marginatum* | Piperaceae | Fruit | Essential oil | Larvicidal | Marques et al. (2015) |
| *Piper nigrum* | Piperaceae | Fruit, root | Ethanol, methanol, piperidine alkaloids (piperine), N-isobutylamide alkaloids, essential oil, ethyl acetate, hexane, chloroform | Larvicidal | Marques et al. (2015) |
| *Piper peltata* | Piperaceae | Root | – | Larvicidal | Marques et al. (2015) |
| *Piper permucronatum* | Piperaceae | Leaf | Essential oil | Larvicidal | Marques et al. (2015) |
| *Piper retrofractum* | Piperaceae | Fruit | Aqueous | Larvicidal | Chansang et al. (2005) |
| *Piper ribesoides* | Piperaceae | Fruit exocarp | Ethanol | Larvicidal | Chaithong et al. (2006) |
| *Piper sarmentosum* | Piperaceae | Fruit exocarp, fruit | Ethanol, essential oil | Larvicidal | Chaithong et al. (2006), Marques et al. (2015) |
| *Piper tuberculatum* | Piperaceae | Leaf | Essential oil | Larvicidal | Marques et al. (2015) |
| *Piper variabile* | Piperaceae | Leaf | – | Larvicidal | Marques et al. (2015) |
| *Pseudosmodingium perniciosum* | Anacardiaceae | Stem bark | Hexane | Larvicidal | Ruiz-Guerrero et al. (2015) |
| *Pterodon polygalaeflorus* | Fabaceae | Seed | Ethanol | Larvicidal | Omena et al. (2007) |

(*Continued*)

**TABLE 8.4 (CONTINUED)**
**Effect of Plant Extracts on Controlling and Reducing the Population of Dengue Vector *Aedes aegypti***

| Species | Family | Plant Part | Active Extract/ Principle | Effect | Reference |
|---|---|---|---|---|---|
| *Rhizophora mucronata* | Rhizophoraceae | Bark, pith, stem wood | Ethanol | Larvicidal | Kabaru and Gichia (2001) |
| *Santalum album* | Santalaceae | Wood | Essential oil | Larvicidal | Zhu et al. (2008) |
| *Solanum nigrescens* | Solanaceae | Root | Aqueous | Larvicidal | Reyes-Villanueva et al. (2008) |
| *Solanum nigrum* | Solanaceae | Dried fruit | Hexane, aqueous | Larvicidal | Raghavendra et al. (2009) |
| *Solanum villosum* | Solanaceae | Fruit, leaf | Chloroform: methanol, aqueous/ protein | Larvicidal | Chowdhury et al. (2008a, 2008b) |
| *Spondias mombin* | Anacardiaceae | Leaf | Dichloromethane | Adulticidal | Ajaegbu et al. (2016) |
| *Tinospora rumphii* | Menispermaceae | Leaf, stem | Methanol | Larvicidal | Gutierrez et al. (2014) |
| *Trichosanthes anguina* | Cucurbitaceae | Leaf | Acetone | Larvicidal | Rahuman et al. (2008) |
| *Valeriana hardwickii* | Valerianaceae | Leaf | Petroleum ether | Larvicidal | Alam et al. (2011) |
| *Vitex negundo* | Lamiaceae | Leaf | Petroleum ether | Larvicidal | Yankanchi et al. (2014) |
| *Zanthoxylum limonella* | Rutaceae | Fruit | Essential oil | Larvicidal | Pitasawat et al. (2007) |

carries the dengue virus (DENV) belong to the genus Flavivirus, family Flaviviridae. DENV passes to humans by infected-mosquito biting. Unfortunately, this disease does not currently have a radical drug treatment. Therefore, numerous researchers around the world have been focusing on the discovery of effective and safe antiflaviviral and/or anti-dengue drug candidates from plant and microorganism sources.

The results of the studies have pointed out that compounds isolated, especially from microorganisms, are highly promising for the treatment of DENV. Geneticin, narasin, some doxorubicin derivatives, ivermectin, dehydroaustin, acetoxydehydroaustin, *Photorhabdus* insect-related proteins, hirsutellin, and several protoxins having larvicidal activity against *Ae. aegypti* have been isolated from different microorganisms (e.g., *Bacillus sphaericus*, *B. alvei*, *B. thuringiensis* var. *israelensis*, *B. thuringiensis* subsp. *kurstaki*, *B. thuringiensis* subsp. *jegathesan*, *B. brevis*, *B. circulans*, *Brevibacillus laterosporus*, *Clostridium bifermentas* ser. Malaysia,

*Hirsutella thompsonii*, *Micromonospora rhodorangea*, *Penicillium* sp., *Photorhabdus asymbiotica*, *Pseudomonas fluorescens*, *Streptomyces peucetius* subsp. *caesius*, and *Str. avermitilis*).

Furthermore, phytochemicals with mosquitocidal activity are recognized as potent alternative insecticides to replace synthetic insecticides in mosquito control programs due to their excellent larvicidal, pupicidal, and adulticidal properties. Therefore, in this chapter, we also reviewed important studies on plants, lichens, algae, and natural compounds having antiflaviviral and larvicidal activity.

## REFERENCES

Abd Kadir, S. L., H. Yaakob, and R. Mohamed Zulkifli, 2013. Potential anti-dengue medicinal plants: A review. *J Nat Med.* 67: 677–689.

Ahantarig, A., N. Chantawat, N. R. Waterfield, R. Ffrench-Constant, and P. Kittayapong, 2009. PirAB toxin from *Photorhabdus asymbiotica* as a larvicide against Dengue vectors. *Appl Environ Microbiol.* 75(13): 4627–4629.

Ahmad, R., W. L. Chu, H. L. Lee, and S. M. Phang, 2001. Effect of four chlorophytes on larval survival, development and adult body size of the mosquito *Aedes aegypti*. *J Appl Phycol.* 13(4): 369–374.

Ahmad, R., W. L. Chu, Z. Ismail, H. L. Lee, and S. M. Phang, 2004. Effect of ten chlorophytes on larval survival, development and adult body size of the mosquito *Aedes aegypti*. *Southeast Asian J Trop Med Public Health.* 35(1): 79–87.

Ajaegbu, E. E., S. P. Danga, I. U. Chijoke, and F. B. Okoye, 2016. Mosquito adulticidal activity of the leaf extracts of *Spondias mombin* L. against *Aedes aegypti* L. and isolation of active principles. *J Vector Borne Dis.* 53(1): 17–22.

Alam, M. F., M. S. Mohammed, A. K. Chopra, and V. K. Dua, 2011. Toxicological properties of several medicinal plants from the Himalayas (India) against vectors of malaria, filariasis and dengue. *Trop Biomed.* 28(2): 343–350.

Allard, P. M., E. T. Dau, C. Eydoux, J. C. Guillemot, V. Dumontet, C. Poullain, B. Canard, F. Guéritte, and M. Litaudon, 2011. Alkylated flavanones from the bark of *Cryptocarya chartacea* as dengue virus NS5 polymerase inhibitors. *J Nat Prod.* 74 (11): 2446–2453.

Allard, P. M., P. Leyssen, M. T. Martin, M. Bourjot, V. Dumontet, C. Eydoux, J. C. Guillemot, B. Canard, C. Poullain, F. Guéritte, and M. Litaudon, 2012. Antiviral chlorinated daphnane diterpenoid orthoesters from the bark and wood of *Trigonostemon cherrieri*. *Phytochemistry*. 84: 160–168.

Anees, A. M. 2008. Larvicidal activity of *Ocimum sanctum* Linn. (Labiatae) against *Aedes aegypti* (L.) and *Culex quinquefasciatus* (Say). *Parasitol Res.* 101: 1451–1453.

Bantoto, V., and D. Dy, 2013. The larvicidal activity of brown algae *Padina minor* (Yamada 1925) and *Dicyota linearis* (Greville 1830) against the dengue vector, *Aedes aegypti* (Linn 1762) (Diptera: Culicidae). *J Vector Borne Dis.* 50(1): 68–70.

Behnam, M. A., C. Nitsche, V. Boldescu, and C. D. Klein, 2016. The medicinal chemistry of dengue virus. *J Med Chem.* 59(12): 5622–5649.

Bezerra-Silva, P. C., J. C. Santos, G. K. N. Santos, K. A. Dutra, A. L. Santana, C. A. Maranhão, M. S. Nascimento, D. M. A. F. Navarro, and L. W. Bieber, 2015. Extract of *Bowdichia virgilioides* and maackiain as larvicidal agent against *Aedes aegypti* mosquito. *Exp Parasitol.* 153: 160–164.

Bhatnagar, P. K. et al., 2010. Anti dengue activity of *Cissampelos pareira* extracts, Patent application: WO 2010-IB50299 20100123.

Bombuwela, K., S. Kathirgamanathar, V. Thadhani, R. G. U. Jayalal, N. K. B. Adikaram, D. S. A. Wijesundara, R. Andersen, P. Wolseley, and V. Karunaratne, 2008. Chemistry of

*Heterodermia microphylla*, a lichen new to Sri Lanka. *J Natn Sci Foundation Sri Lanka.* 36(3): 251–252.

Bomfim, R. R., A. A. S. Araújo, S. Cuadros-Orellana, M. G. D. Melo, L. J. Quintans-Júnior, and S. C. H. Cavalcanti, 2009. Larvicidal activity of *Cladonia substellata* extract and usnic acid against *Aedes aegypti* and *Artemia salina. Lat Am J Pharm.* 28(4): 580–584.

Bourjot, M., P. Leyssen, C. Eydoux, J. C. Guillemot, B. Canard, P. Rasoanaivo, F. Guéritte, and M. Litaudon, 2012a. Chemical constituents of *Anacolosa pervilleana* and their antiviral activities. *Fitoterapia.* 83(6): 1076–1080.

Bourjot, M., P. Leyssen, C. Eydoux, J. C. Guillemot, B. Canard, P. Rasoanaivo, F. Guéritte, and M. Litaudon, 2012b. Flacourtosides A-F, phenolic glycosides isolated from *Flacourtia ramontchi. J Nat Prod.* 75(4): 752–758.

Chaithong, U., W. Choochote, K. Kamsuk, A. Jitpakdi, P. Tippawangkosol, D. Chaiyasit, D. Champakaew, B. Tuetun, and B. Pitasawat, 2006. Larvicidal effect of pepper plants on *Aedes aegypti* (L.) (Diptera: Culicidae). *J Vector Ecol.* 31: 138–144.

Chansang, U., N. S. Zahiri, J. Bansiddhi, T. Boonruad, P. Thongsrirak, J. Mingmuang, N. Benjapong, and M. S. Mulla, 2005. Mosquito larvicidal activity of aqueous extracts of long pepper (*Piper retrofractum* Vahl) from Thailand. *J Vector Ecol.* 30: 195–200.

Chantrill, B. H., C. E. Coulthard, L. Dickinson, G. W. Inkley, W. Morris, and A. H. Pyle, 1952. The action of plant extracts on a bacteriophage of Pseudomonas pyocyanea and on inflienza A virus. *J Gen Microbiol.* 6(1–2): 74–84.

Choochate, W., D. Chaiyasit, D. Kanjanapothi, E. Rattanachanpichai, A. Jitpakdi, B. Tuetun, and B. Pitasawat, 2005. Chemical composition and anti-mosquito potential of rhizome extract and volatile oil derived from *Curcuma aromatica* against *Aedes aegypti* (Diptera: Culicidae). *J Vector Ecol.* 30(2): 302–309.

Choochate, W., B. Tuetun, D. Kanjanapothi, E. Rattanachanpichai, U. Chaithong, P. Chaiwong, A. Jitpakdi, P. Tippawangkosol, D. Riyong, and B. Pitasawat, 2004. Potential of crude seed extract of celery, *Apium graveolens* L., against the mosquito *Aedes aegypti* (L.) (Diptera: Culicidae). *J Vector Ecol.* 29: 340–346.

Chowdhury, N., A. Ghosh, and G. Chandra, 2008a. Mosquito larvicidal activities of *Solanum villosum* berry extract against the Dengue vector *Stegomyia aegypti. BMC Complement Altern Med.* 8: 10.

Chowdhury, N., S. Laskar, and G. Chandra, 2008b. Mosquito larvicidal and antimicrobial activity of protein of *Solanum villosum* leaves. *BMC Complement Altern Med.* 8: 62.

Costa, S. C. P., P. A. Girard, and R. Zumbihl, 2009. The emerging human pathogen *Photorhabdus asymbiotica* is a facultative intracellular bacterium and induces apoptosis of macrophage-like cells. *Infect Immun.* 77(3): 1022–1030.

Coulerie, P., A. Maciuk, N. Lebouvier, E. Hnawia, J.-C. Guillemot, B. Canard, B. Figadère, and M. Nour, 2013. Phytochemical Study of *Myrtopsis corymbosa*, Perspectives for anti-dengue natural compound research. *Rec Nat Prod.* 7(3): 250–253.

Crance, J. M., N. Scaramozzino, A. Jouan, and D. Garin, 2003. Interferon, ribavirin, 6-azauridine and glycyrrhizin: Antiviral compounds active against pathogenic flaviviruses. *Antiviral Res.* 58(1): 73–79.

Crop Life International, 2017. Vector control. http://croplife.org/wp-content/uploads/pdf_files /Vector-Control-fact-sheet.pdf.

Darbro, J. M., R. I. Graham, B. H. Kay, P. A. Ryan, and M. B. Thomas, 2011. Evaluation of entomopathogenic fungi as potential biological control agents of the dengue mosquito, *Aedes aegypti* (Diptera: Culicidae). *Biocontrol Sci Techn.* 21(9): 1027–1047.

De Mendonça, F. A. C., K. F. S. da Silva, K. K. dos Santos, K. A. L. Ribeiro Jr., and A. E. G. Sant'Ana, 2005. Activities of some Brazilian plants against larvae of the mosquito *Aedes aegypti. Fitoterapia.* 76: 629–636.

De S. F.-Tischer, P. C., L. B. Talarico, M. D. Noseda, S. M. P. B. Guimaraes, E. B. Damonte, and M. E. R. Duarte, 2006. Chemical structure and antiviral activity of carrageenans from Meristiella gelidium against herpes simplex and dengue virus. *Carbohydr Polym.* 63: 459–465.

European Center for Disease Prevention and Control, 2017. Vector borne diseases. http://ecdc.europa.eu/en/healthtopics/climate_change/health_effects/Pages/vector_borne_diseases.aspx.

Garcia, C. C., L. Talarico, N. Almeida, S. Colombres, C. Duschatzky, and E. B. Damonte, 2003. Virucidal activity of essential oils from aromatic plants of San Luis, Argentina. *Phytother Res.* 17(9): 1073–1075.

Geetha, R. V., and A. Roy, 2014. Essential oil repellents—A short review. *Int J Drug Dev & Res.* 6(2): 20–27.

Geris, R., E. Rodrigues-Fo, H. H. Garcia da Silva, and I. Garcia da Silva, 2008. Larvicidal effects of fungal meroterpenoids in the control of *Aedes aegypti* L., the main vector of dengue and yellow fever. *Chem Biodivers.* 5(2): 341–345.

Ghosh, A., N. Chowdhury, and G. Chandra, 2012. Plant extracts as potential mosquito larvicides. *Indian J Med Res.* 135: 581–598.

Ghosh, D., and A. Basu, 2008. Present perspectives on flaviviral chemotherapy. *Drug Discov Today.* 13(13–14): 619–624.

Guedes, E. A. C., C. M. Carvalho, K. A. L. Ribeiro Jr., T. F. L. Ribeiro, L. D. Barros, M. R. F. Lima, F. B. P. Moura, and A. E. G. Sant'Ana, 2014. Larvicidal activity against *Aedes aegypti* and molluscicidal activity against *Biomphalaria glabrata* of Brazilian marine algae. *J Parasitol Res.* 2014: Article ID 501328.

Gutierrez, P. M. Jr., A. N. Antepuesto, B. A. L. Eugenio, and M. F. L. Santos, 2014. Larvicidal activity of selected plant extracts against the Dengue vector *Aedes aegypti* mosquito. *Int Res J Biological Sci.* 3(4): 23–32.

Hidari, K. I., N. Takahashi, M. Arihara, M. Nagaoka, K. Morita and T. Suzuki, 2008. Structure and anti-dengue virus activity of sulfated polysaccharide from a marine alga. *Biochem Biophys Res Commun.* 376(1): 91–95.

Jain, M., L. Ganju, A. Katiyal, Y. Padwad, K.P. Mishra, S. Chanda, D. Karan, K. M. S. Yogendra, and R. C. Sawhney, 2008. Effect of *Hippophae rhamnoides* leaf extracts against Dengue virus infection in human blood-derived macrophages. *Phytomedicine.* 15: 793–799.

Jang, Y. S., B. R. Baek, Y. C. Yang, M. K. Kim, and H. S. Lee, 2002. Larvicidal activity of leguminous seeds and grains against *Aedes aegypti* and *Culex pipiens pallens. J Am Mosq Control Assoc.* 18: 210–213.

Jia, F., G. Zou, J. Fan, and Z. Yuan, 2010. Identification of palmatine as an inhibitor of West Nile virus. *Arch Virol.* 155(8): 1325–1329.

Jiang, W. L., X. L. Luo, and S. J. Kuang, 2005. Effects of *Alternanthera philoxeroides* Griseb. against dengue virus *in vitro. J First Mil Med Univ.* 25(4): 454–456.

Kabaru, J. M., and L. Gichia, 2001. Insecticidal activity of extracts derived from different parts of the mangrove tree *Rhizophora mucronata* (Rhizophoraceae) Lam. against three arthropods. *Afr J Sci Tech.* 2: 44–49.

Kalimuthu, K., K. Murugan, C. Panneerselvam, and J.-S. Hwang, 2012. Mosquito larvicidal activity of *Cadaba indica* Lam. leaf extracts against the Dengue vector, *Aedes aegypti. Asian J Plant Sci Res.* 2(5): 633–637.

Kaptein, S. J., T. De Burghgraeve, M. Froeyen, B. Pastorino, M. M. Alen, J. A. Mondotte, P. Herdewijn, M. Jacobs, X. de Lamballerie, D. Schols, A. V. Gamarnik, F. Sztaricskai, and J. Neyts, 2010. A derivate of the antibiotic doxorubicin is a selective inhibitor of dengue and yellow fever virus replication *in vitro*. *Antimicrob Agents Chemother.* 54 (12): 5269–5280.

Kaushik, R., and P. Saini, 2008. Larvicidal activity of leaf extract of *Millingtonia hortensis* (Family: Bignoniaceae) against *Anopheles stephensi*, *Culex quinquefasciatus* and *Aedes aegypti. J Vector Borne Dis.* 45: 66–69.

Kiat, T. S., R. Pippen, R. Yusof, H. Ibrahim, N. Khalid, and N. A. Rahman, 2006a. Inhibitory activity of cyclohexenyl chalcone derivatives and flavonoids of fingerroot, *Boesenbergia rotunda* (L.), towards dengue-2 virus NS3 protease. *Bioorg Med Chem Lett.* 16(12): 3337–3340.

Kiat, T.S., R. Pippen, R. Yusof, N. A. Rahman, H. Ibrahim, and N. Khalid 2006b. Screening of selected Zingiberaceae extracts for dengue-2 virus protease inhibitory activities. *SAJ.* 3: 1–7.

Kiviranta, J., and A. Abdel-Hameed, 1994. Toxicity of the blue-green alga *Oscillatoria agardhii* to the mosquito *Aedes aegypti* and the shrimp *Artemia salina. World J Microbiol Biotechnol.* 10(5): 517–520.

Kiviranta, J., A. Abdel-Hameed, K. Sivonen, S. I. Niemelä, and G. Carlberg, 1993. Toxicity of cyanobacteria to mosquito larvae-screening of active G. compounds. *Environ Toxicol Water Quality.* 8: 63–71.

Klawikkan, N., V. Nukoolkarn, N. Jirakanjanakir, S. Yoksan, C. Wiwat, and K. Thirapanmethee, 2011. Effect of Thai medicinal plant extracts against Dengue virus *in vitro. MU J Pharm.* 38(1–2): 13–18.

Kovendan, K., K. Murugan, and S. Vincent, 2012. Evaluation of larvicidal activity of *Acalypha alnifolia* Klein ex Willd. (Euphorbiaceae) leaf extract against the malarial vector, *Anopheles stephensi*, dengue vector, *Aedes aegypti* and *Bancroftian filariasis* vector, *Culex quinquefasciatus* (Diptera: Culicidae). *Parasitol Res.* 110: 571–581.

Kumar, S., M. Warikoo, M. Mishra, A. Seth, and N. Wahab, 2012. Larvicidal efficacy of the *Citrus limetta* peel extracts against Indian strains of *Anopheles stephensi* Liston and *Aedes aegypti* L. *Parasitol Res.* 111(1): 173–178.

Laille, M., F. Gerald, and C. Debitus, 1998. *In vitro* antiviral activity on dengue virus of marine natural products. *Cell Mol Life Sci.* 54(2): 167–170.

Lalithambika, B., and C. Vani, 2016. *Pseudomonas aeruginosa* KUN2, extracellular toxins—A potential source for the control of dengue vector. *J Vector Borne Dis.* 53: 105–111.

Langat, B. K., D. K. Siele, C. Wainaina, C. Mwandawiro, J. Ondicho, W. K. Tonui, C. Anjili, L. N. Ireri, and C. K. Mutai, 2012. Larvicidal effect of *Mundulea sericea* (Leguminosaea) plant extract against *Aedes aegypti* (L.) (Diptera: Culicidae). *Afr J Pharmacol Ther.* 1(3): 106–109.

Leardkamolkarn, V., W. Srigulpanit, C. Phurimsak, S. Kumkate, L. Himakoun, and B. Sripanidkulchai, 2012. The inhibitory actions of *Houttuynia cordata* aqueous extract on Dengue virus and Dengue-infected cells. *J Food Biochem.* 26: 86–92.

Low, J. S. Y., K. C. Chen, K. X. Wu, M. M. L. Ng, and J. J. H. Chu, 2009. Antiviral activity of emetine dihydrochloride against Dengue virus infection. *J Antivir Antiretrovir.* 1: 062–071.

Low, J. S. Y., K. X. Wu, K. C. Chen, M. M. L. Ng, and J. J. H. Chu, 2011. Narasin, a novel antiviral compound that blocks dengue virus protein expression. *Antivir Ther.* 16(8): 1203–1218.

Manilal, A., N. Thajuddin, J. Selvin, A. Idhayadhulla, R. Surendra Kumar, and S. Sujith, 2011. *In vitro* mosquito larvicidal activity of marine algae against the human vectors, *Culex quinquefasciatus* (Say) and *Aedes aegypti* (Linnaeus) (Diptera: Culicidae). *Int J Zool Res.* 7(3): 272–278.

Marimuthu, G., S. Rajamohan, R. Mohan, and Y. Krishnamoorthy, 2012. Larvicidal and ovicidal properties of leaf and seed extracts of *Delonix elata* (L.) Gamble (Family: Fabaceae) against malaria (*Anopheles stephensi* Liston) and dengue (*Aedes aegypti* Linn.) (Diptera: Culicidae) vector mosquitoes. *Parasitol Res.* 111(1): 65–77.

Marten, G. G. 2007. Larvicidal algae. In Tom Floore (ed.), *Biorational Control of Mosquitoes, American Mosquito Control Association Bulletin* No. 7: 179–185.

Marques, A. M., and M. A. C. Kaplan, 2015. Active metabolites of the genus *Piper* against *Aedes aegypti*: natural alternative sources for dengue vector control. *Univ Sci.* 20(1): 61–82.

Mastrangelo, E., M. Pezzullo, T. De Burghgraeve, S. Kaptein, B. Pastorino, K. Dallmeier, X. de Lamballerie, J. Neyts, A. M, Hanson, D. N. Frick, M. Bolognesi, and M. Milani, 2012. Ivermectin is a potent inhibitor of flavivirus replication specifically targeting NS3 helicase activity: New prospects for an old drug. *J Antimicrob Chemother.* 67: 1884–1894.

Mathew, N., M. G. Anitha, T. S. Bala, S. M. Sivakumar, R. Narmadha, and M. Kalyanasundaram, 2009. Larvicidal activity of *Saraca indica*, *Nyctanthes arbor-tristis*, and *Clitoria ternatea* extracts against three mosquito vector species. *Parasitol Res.* 104 (5): 1017–1025.

Mazet, I., and A. Vey, 1995. Hirsutellin a, a toxic protein produced *in vitro* by *Hirsutella thompsonii. Microbiology.* 141: 1343–1348.

Meneses, R., R. E. Ocazionez, J. R. Martinez, and E. E. Stashenko, 2009. Inhibitory effect of essential oils obtained from plants grown in Colombia on yellow fever virus replication *in vitro. Ann Clin Microbiol Antimicrob.* 8: 8.

Mgbemena, I. C. 2010. Comparative evaluation of larvicidal potentials of three plant extracts on *Aedes aegypti. J Am Sci.* 6: 435–440.

Moghaddam, E., B. T. Teoh, S. S. Sam, R. Lani, P. Hassandarvish, Z. Chik, A. Yueh, S. Abubakar, and K. Zandi, 2014. Baicalin, a metabolite of baicalein with antiviral activity against dengue virus. *Sci Rep.* 4: 5452.

Monnerat, R., S. da Silva, D. S. Dias, E. S. Martins, L. B. Praça, G. W. Jones, C. M. Soares, J. M. C. de Souza Dias, and C. Berry, 2004. Screening of Brazilian *Bacillus Sphaericus* strains for high toxicity against *Culex quinquefasciatus* and Aedes aegypti. *JEN.* 128(7): 469–473.

Moreiraa, A. S. N., R. O. S. Fernandes, F. J. A. Lemos, R. Braz-Filho, and I. J. C. Vieira, 2016. Larvicidal activity of *Ramalina usnea* lichen against *Aedes aegypti. Rev Bras Farmacogn.* 26: 530–532.

Muliawan, S. Y., L. S. Kit, S. Devi, O. Hashim, and R. Yusof, 2006. Inhibitory potential of *Quercus lusitanica* extract on Dengue virus type 2 replication. *Southeast Asian J Trop Med Public Health.* 37(9): 132–135.

Mullai, K., A. Jebanesan, and T. Pushpanathan, 2008. Effect of bioactive fractions of *Citrullus vulgaris* Schrad. leaf extract against *Anopheles stephensi* and *Aedes aegypti. Parasitol Res.* 102: 951–955.

Murugan, J. M., G. Ramkumar, and M. S. Shivakumar, 2016. Insecticidal potential of *Ocimum canum* plant extracts against *Anopheles stephensi*, *Aedes aegypti* and *Culex quinquefasciatus* larval and adult mosquitoes (Diptera: Culicidae). *Nat Prod Res.* 30(10): 1193–1196.

Nair, S. S., V. Shetty, and N. J. Shetty, 2014. Relative toxicity of leaf extracts of *Eucalyptus globulus* and *Centella asiatica* against mosquito vectors *Aedes aegypti* and *Anopheles stephensi. J Insects.* 2014: Article ID 985463.

Nanayakkara, C., K. Bombuwala, S. Kathirgamanathar, N. K. B. Adikaram, D. S. A. Wijesundara, G. N. Hariharan, P. Wolseleys, and V. Karunaratne, 2005. Effect of some lichen extracts from Sri Lanka on larvae of *Aedes aegypti* and the fungus *Cladosporium cladosporioides. J Natn Sci Foundation Sri Lanka.* 33(2): 147–149.

Navarro, D. M. A. F., P. C. B. Silva, M. F. R. Silva, T. H. Napoleao, and P. M. G. Paiva, 2013. Larvicidal activity of plant and algae extracts, essential oils and isolated chemical constituents against *Aedes aegypti. NPJ.* 3: 268–291.

Ocazionez, R. E., R. Meneses, F. A. Torres, and E. Stashenko, 2010. Virucidal activity of Colombian *Lippia* essential oils on Dengue virus replication *in vitro. Mem Inst Oswaldo Cruz.* 105(3): 304–309.

Omena, de C., D. M. A. F. Navarro, J. E. Paula de, J. S. M. R. Ferreira de Lima, and A. E. G. Sant'Ana, 2007. Larvicidal activities against *Aedes aegypti* of Brazilian medicinal plants. *Bioresour Technol.* 98: 2549–2556.

Ono, L., W. Wollinger, I. M. Rocco, T. L. M. Coimbra, P. A. J. Gorin, and M. R. Sierakowski, 2003. *In vitro* and *in vivo* antiviral properties of sulfated galactomannans against yellow fever virus (BeH111 strain) and dengue 1 virus (Hawaii strain). *Antivir Res.* 60: 201–208.

Ozaslan, M. 2012. Squalamine: May be an effective viral control. *Int J Virol.* 8(3): 285–287.

Panneerselvam, C., K. Murugan, K. Kovendan, and P. Mahesh Kumar, 2012. Mosquito larvicidal, pupicidal, adulticidal, and repellent activity of *Artemisia nilagirica* (Family: Compositae) against *Anopheles stephensi* and *Aedes aegypti. Parasitol Res.* 111(6): 2241–2251.

Parida, M. M., C. Upadhyay, G. Pandya, and A. M. Jana, 2002. Inhibitory potential of neem *Azidarachta indica* Juss) leaves on Dengue virus type-2 replication. *J Ethnopharmacol.* 79: 273–278.

Patil, P. B., V. L. Kallapur, and S. N. Holihosur, 2013. Evaluation of *Clerodendron inerme* Gaertn. plant extract against *Aedes aegypti* L. mosquito. *Int J Nat Prod Res.* 2(2): 36–38.

Phurimsak, C., and V. Leardkamolkarn, 2005. Screening for antiviral effect of Thai herbs; *Kaempferia parviflora, Ellipeiopsis cherrevensis* and *Stemona tuberosa* against dengue virus type-2. 31st Congress on Science and Technology of Thailand, Suranaree University of Technology, October 18–20.

Pitasawat, B., D. Champakaew, W. Choochote, A. Jitpakdi, U. Chaithong, D. Kanjanapothi, E. Rattanachanpichai, P. Tippawangkosol, D. Riyong, B. Tuetun, and D. Chaiyasit, 2007. Aromatic plant-derived essential oil: An alternative larvicide for mosquito control. *Fitoterapia.* 78: 205–210.

Porter, A. G., 1996. Mosquitocidal toxins, genes and bacteria: The hit squad. *Parasitol Today.* 12(5): 175–178.

Prabakaran, G., S. L. Hoti, H. Surya Prakash Rao, and S. Vijjapu, 2015. Di-rhamnolipid is a mosquito pupicidal metabolite from *Pseudomonas fluorescens* (VCRC B426). *Acta Trop.* 148: 24–31.

Prasad, A., D. Kumar, M. Srivastava, E. Sharma, and P. Mathur, 2012. Soil bacteria and their possible role in mosquito control: A short review. *World J Envir Biosci.* 2(1): 40–48.

Priyadarshini, T., M. Lekeshmanaswamy, 2014. Larvicidal effect of fungus *Metarhizium anisopleae* on *Aedes aegypti. SIRJ HMS.* 1(1): 26–30.

Pujol, C. A., J. M. Estevez, M. J. Carlucci, M. Ciancia, A. S. Cerezo, and E. B. Damonte 2002. Novel DL-galactan hybrids from the red seaweed *Gymnogongrus torulosus* are potent inhibitors of herpes simplex virus and dengue virus. *Antivir Chem Chemother.* 13(2): 83–89.

Raghavendra, K., S. P. Singh, S. K. Subbarao, and A. P. Dash, 2009. Laboratory studies on mosquito larvicidal efficacy of aqueous & hexane extracts of dried fruit of *Solanum nigrum* Linn. *Indian J Med Res.* 130: 74–77.

Rahuman, A. A., G. Gopalakrishnan, S. Ghouse, S. Arumugam, and B. Himalayan, 2000. Effect of *Feronia limonia* on mosquito Larvae. *Fitoterapia.* 71: 553–555.

Rahuman, A. A., and P. Venkatesan, 2008. Larvicidal efficacy of five cucurbitaceous plant leaf extracts against mosquito species. *Parasitol Res.* 103: 133–139.

Raj Mohan, D., M. Ramaswamy, 2007. Evaluation of larvicidal activity of the leaf extract of a weed plant, *Ageratina adenophora*, against two important species of mosquitoes, *Aedes aegypti* and *Culex quinquefasciatus. Afr J Biotech.* 6: 631–638.

Rajkumar, S., and A. Jebanesan, 2010. Prevention of Dengue fever through plant based mosquito repellent *Clausena dentata* (Willd.) M. Roem (Family: Rutaceae) essential oil against *Aedes aegypti* L. (Diptera: Culicidae) mosquito. *Eur Rev Med Pharmacol Sci.* 14: 231–234.

Rao, D. R., C. Thangavel, L. Kabilan, S. Suguna, T. R., Mani and S. Shanmugasundaram, 1999. Larvicidal properties of the cyanobacterium *Westiellopsis* sp. (blue-green algae) against mosquito vectors. *Trans R Soc Trop Med Hyg.* 93(3): 232.

Rao, J. V., P. K. Usman, and J. B. Kumar, 2008. Larvicidal and insecticidal properties of some marine sponges collected in Palk Bay and Gulf of Mannar waters. *Afr J Biotechnol.* 7(2): 109–113.

Reegan, A. D., A. V. Kinsalin, M. G. Paulraj, and S. Ignacimuthu, 2013. Larvicidal, ovicidal, and repellent activities of marine sponge *Cliona celata* (Grant) extracts against *Culex quinquefasciatus* Say and *Aedes aegypti* L. (Diptera: Culicidae). *SRN Entomol.* 2013: Article ID 315389.

Rees, C. R., J. M. Costin, R. C. Fink, M. McMichael, K. A. Fontaine, S. Isern, and S. F. Michael, 2008. *In vitro* inhibition of dengue virus entry by p-sulfoxy-cinnamic acid and structurally related combinatorial chemistries. *Antiviral Res.* 80(2): 135–142.

Reis, S. R. I. N., L. M. M. Valente, A. L. Sampaio, A. C. Siani, M. Gandini, E. L. Azeredo, L. A. D'Avila, J. L. Mazzei, M. G. M. Henriques, and C. F. Kubelka, 2008. Immunomodulating and antiviral activities of *Uncaria tomentosa* on human monocytes infected with Dengue Virus-2. *Int Immunopharmacol.* 8(3): 468–476.

Remia, K. M., and S. Logaswamy. 2010. Larvicidal efficacy of leaf extract of two botanicals against the mosquito vector *Aedes aegypti* (Diptera: Culicidae). *IJNPR.* 1 (2): 208–212.

Reyes-Vilanueva, F., O.J. Gonzales-Gaona, and M. A. Rodriguez-Perez, 2008. Larvicidal effect of medicinal plants against *Aedes aegypti* (L.) (Diptera: Culicidae) in Mexico. *Bioassay.* 3: 7.

Rodrigues, A. M. S., J. E. Paula, F. Roblot, A. Fournet, and L. S. Espíndola, 2005. Larvicidal activity of *Cybistax antisyphilitica* against *Aedes aegypti* larvae. *Fitoterapia.* 76: 755—757.

Ruiz-Guerrero, R., M. A. Rodriguez-Perez, and M. Norzagaray-Campos, 2015. Toxicity of Mexican native plant extracts against larvae of *Aedes aegypti* (Diptera: Culicidae). *Asian Pac J Trop Biomed.* 5(4): 287–291.

Salvador-Neto, O., S. A. Gomes, A. R. Soares, F. L. Machado, R. I. Samuels, R. Nunes da Fonseca, J. Souza-Menezes, J. L. Moraes, E. Campos, F. B. Mury, and J. R. Silva, 2016. Larvicidal potential of the halogenated sesquiterpene (+)-obtusol, isolated from the alga *Laurencia dendroidea* J. Agardh (Ceramiales: Rhodomelaceae), against the Dengue vector mosquito *Aedes aegypti* (Linnaeus) (Diptera: Culicidae). *Mar Drugs.* 14 (2): E20.

Sanchez, I., F. G. Garibay, J. Taboada, and B. H. Ruiz, 2000. Antiviral effect of flavonoids on the Dengue virus. *Phytother Res.* 14: 89–92.

Sánchez-Trinidad, A., F. Ordoñez-Sánchez, M. T. Valdes-Perezgasga, F. J. Sánchez-Ramos, T. J. Zavortink, A. J. Cortés-Guzmánand, and A. I. Ortega-Morales, 2014. Geographical distribution of the *Aedes triseriatus* group (Diptera: Culicidae) in Mexico. *J Vector Ecol.* 39(1): 134–137.

Serkedjieva, J., 2003. Influenza virus variants with reduced susceptibility to inhibition by a polyphenol extract from *Geranium sanguineum* L. *Pharmazie.* 58(1): 53–57.

Silva, A. R. A., S. M. Morais, M. M. M. Marques, D. M. Lima, S. C. C. Santos, R. R. Almeida, I. G. P. Vieira, and M. I. F. Guedes, 2011. Antiviral activities of extracts and phenolic components of two *Spondias* species against dengue virus. *J Venom Anim Toxins Incl Trop Dis.* 17(4): 406–413.

Simoes, L. R., G. M. Maciel, G. C. Brandão, E. G. Kroon, R. O. Castilho, and A. B. Oliveira, 2011. Antiviral activity of *Distictella elongata* (Vahl Urb. (Bignoniaceae)), a potentially useful source of anti-dengue drugs from the state of Minas Gerais, Brazil. *Lett Appl Microbiol.* 53(6):602–607.

Singh, R. K., R. C. Dhiman, and P. K. Mittal, 2006. Mosquito larvicidal properties of *Momordica charantia* Linn (Family: Cucurbitaceae). *J Vect Borne Dis.* 43: 88—91.

Singh, R. K., R. C. Dhiman, and P. K. Mittal, 2007. Studies on mosquito larvicidal properties of *Eucalyptus citriodora* Hook (Family Myrtaceae). *J Comm Dis.* 39: 233–236.

Sivagnaname, S., and M. Kalyanasundaram, 2004. Laboratory evaluation of methanolic extract of *Atlanta monophylla* (Family: Rutaceae) against immature stage of mosquitoes and nontarget organisms. *Mem Inst Oswaldo Cruz.* 99: 115–118.

Sujatha, S., and B. Joseph, 2011. Effect of Few marine sponges and its biological activity against *Aedes aegypti* Linn. *Musca domestica (Linnaeus, 1758) (Diptera: Culicidae). JFAS.* 6(2): 170–177.

Swathi, D., Y. Suchitha, T. R. Prashith Kekuda, T. M. Venugopal, K. S. Vinayaka, N. Mallikarjun, and H. L. Raghavendra, 2010. Antimicrobial, anthelmintic and insecticidal activity of a Macrolichen *Everniastrum cirrhatum* (Fr.) Hale. *Int J Drug Dev & Res.* 2(4): 780–789.

Talarico, L. B., and E. B. Damonte, 2016. Characterization of *in vitro* dengue virus resistance to carrageenan. *J Med. Virol.* 88(7): 1120–1129.

Talarico, L. B., M. E. Duarte, R. G. Zibetti, M. D. Noseda, and E. B. Damonte, 2007. An algal-derived DL-galactan hybrid is an efficient preventing agent for *in vitro* dengue virus infection. *Planta Med.* 73(14): 1464–1468.

Tang, L. I. C., A. P. K. Ling, R. Y. Koh, S. M. Chye, and K. G. L. Voon, 2012. Screening of anti-dengue activity in methanolic extracts of medicinal plants. *BMC Complement Altern Med.* 12: 3.

Tolo, F. M., G. M. Rukunga, F. W. Muli, E. N. Njagi, W. Njue, K. Kumon, G. M. Mungai, C. N. Muthaura, J. M. Muli, L. K. Keter, E. Oishi, and M. W. Kofi-Tsekpo, 2006. Antiviral activity of the extracts of a Kenyan medicinal plant *Carissa edulis* against herpes simplex virus. *J Ethnopharmacol.* 104(1/2): 92–99.

Tong, X., H. Q. X. Zhang, L. Shi, G. Wang, F. Ji, H. Ding, W. Tang, K. Ding, and J. Zuo, 2010. WSS45, a sulfated α-D-glucan, strongly interferes with Dengue 2 virus infection *in vitro. Acta Pharmacol Sin.* 31(5): 585–592.

Vieira, L. P., A. R. Paula, C. O. Paula, R. A. DaMatta, and R. I. Samuels, 2013. Infection of *Aedes aegypti* (Diptera: Culicidae) larvae and adults by the entomopathogenic fungus *Metarhizium anisopliae* (Metschn.) Sorokin. *Br Microbiol Res J.* 3(3): 309–317.

Whitby, K., T. C. Pierson, B. Geiss, K. Lane, M. Engle, Y. Zhou, R. W. Doms, and M. S. Diamond, 2005. Castanospermine, a potent inhibitor of Dengue virus infection in vitro and in vivo. *J Virol.* 79(14): 8698–8706.

World Health Organization (WHO), 2016. *Vector-borne diseases.* http://www.who.int/mediacentre/factsheets/fs387/en/.

World Health Organization (WHO), 1997. Treatment. In *Dengue Haemorrhagic Fever: Diagnosis, Treatment, Prevention and Control.* 2nd edition, chap, 3. Geneva. http://www.who.int/csr/resources/publications/dengue/024-33.pdf?ua=1.

Yang, Y. C., M. Y. Lim, and H. S. Lee, 2003. Emodin isolated from *Cassia obtusifolia* (Leguminosae) seed shows larvicidal activity against three mosquito species. *J Agric Food Chem.* 51: 7629–7631.

Yankanchi, S. R., O. V. Yadav, and G. S. Jadhav, 2014. Synergistic and individual efficacy of certain plant extracts against Dengue vector mosquito, *Aedes aegypti. J Biopest.* 7(1): 22–28.

Yenesew, A., S. Derese, J. O. Midiwo, M. Heydenreich, and M. J. Peter, 2003. Effect of rotenoids from the seeds of *Millettia dura* on larvae of *Aedes aegypti. Pest Manage Sci.* 59: 1159–1161.

Yousaf, A., and W. F. Zuharah, 2015. Lethal response of the Dengue vectors to the plant extracts from family Anacardiaceae. *Asian Pac J Trop Biomed.* 5(10): 812–818.

Yu, K. X., I. Jantan, R. Ahmad, and C. L. Wong, 2014. The major bioactive components of seaweeds and their mosquitocidal potential. *Parasitol Res.* 113(9): 3121–3141.

Zandi, K., B. T. Teoh, S. S. Sam, P. F. Wong, M. R. Mustafa, and S. AbuBakar, 2011a. Antiviral activity of four types of bioflavonoid against dengue virus type-2. *Virol J.* 8: 560.

Zandi, K., B. T. Teoh, S. S. Sam, P. F. Wong, M. R. Mustafa, and S. AbuBakar, 2011b. *In vitro* antiviral activity of fisetin, rutin and naringenin against Dengue virus type-2. *J Med Plant Res.* 5(23): 5534–5539.

Zandi, K., B. T. Teoh, S. S. Sam, P. F. Wong, M. R. Mustafa, and S. Abubakar, 2012. Novel antiviral activity of baicalein against dengue virus. *BMC Complement Altern Med.* 12: 214.

Zhang, X. G., P. W. Mason, E. J. Dubovi, X. Xu, N. Bourne, R. W. Renshaw, T. M. Block, and A. V. Birk, 2009. Antiviral activity of geneticin against dengue virus. *Antiviral Res.* 83 (1): 21–27.

Zhu, J., X. Zeng, M. O'Neal, G. Schultz, B. Tucker, J. Coats, L. Bartholomay, and R. D. Xue, 2008. Mosquito larvicidal activity of botanical-based mosquito repellents. *J Am Mosq Control Assoc.* 24(1): 161–168.

Zou, G., F. Puig-Basagoiti, B. Zhang, M. Qing, L. Chen, K. W. Pankiewicz, K. Felczak, Z. Yuan, and P. Y. Shi, 2009. A single-amino acid substitution in West Nile virus 2K peptide between NS4A and NS4B confers resistance to lycorine, a flavivirus inhibitor. *Virology.* 384: 242–252.

# 9 Paratransgenesis Involving Microbes

## *A New Avenue for the Control of Vector-Borne Diseases*

*Vitthalrao B. Khyade and Brij Kishore Tyagi*

**CONTENTS**



## 9.1 INTRODUCTION

Insect-borne diseases cause significant human morbidity and mortality (Kirchhoff, 2006; Moncayo, 2003; Planning Commission, Government of India, 1996; Tanowitz et al., 2011; Thakur and Kumar, 1992; WHO, 2009, 2010, 2014). Insect-borne diseases are responsible for severely affecting human life around the world, causing significant morbidity and mortality (Miller et al., 2006; Moncayo and Yanine, 2006; Travi et al., 1990). Malaria alone is responsible for 1 million to 2 million deaths annually, and approximately 300 million are at risk of becoming infected. The lack of effective vaccines to control the spread of a significant number of vector-borne diseases (VBDs) makes control mainly dependent on insecticides (Gurtler et al., 1994, 2004; Subhadra et al., 2009; Sundar et al., 2006, 2009, 2010, 2011). However, the appearance of insecticide resistance requires the development of new strategies to reduce pathogen transmission in the field (Cecere et al., 2004; Hill et al., 2005; Picollo et al., 2005). Current control and preventive methods against vector-borne diseases rely mainly on insecticides (Kishore et al., 2004). The emergence of insecticide resistance in many disease vectors highlights the necessity to develop new strategies to control these insects. Novel approaches against vector-borne diseases include transgenesis and paratransgenesis to reduce vector competence. Vector transgenesis and paratransgenesis are novel strategies that aim at reducing insect vectorial capacity, or seek to eliminate transmission of pathogens such as *Plasmodium* sp., *Trypanosoma* sp., and dengue virus (Mereschkowsky, 1905, 1910; Rassi et al., 2010; Ren et al., 2008). Vector transgenesis relies on direct genetic manipulation of disease vectors making them incapable of functioning as vectors of a given pathogen. Paratransgenesis focuses on utilizing genetically modified insect symbionts to express molecules within the vector that are deleterious to pathogens they transmit. Despite the many successes achieved in developing such techniques in the last several years, many significant barriers remain and need to be overcome prior to any of these approaches becoming a reality. In the process of transgenesis, an exogenous gene (also called a transgene) is introduced into a living organism. This results in a change in the genetic makeup of the organism, and it will exhibit a new property and transmit that property to its offspring. The process of transgenesis can be facilitated by liposomes, enzymes, plasmid vectors, viral vectors, pronuclear injection, protoplast fusion, and ballistic DNA injection. Transgenesis can occur in nature (DiCarlo et al., 2013). This transgenic organism incorporates the genetic code and it is able to express the foreign genes. Paratransgenesis is the process that attempts to eliminate a pathogen from vector populations through the transgenesis of a symbiont of the vector. The goal of this technique is to control vector-borne diseases. The very first step in paratransgenesis is to identify proteins that prevent the vector species from transmitting the pathogen. Then, genes that are coded for these proteins are introduced into the symbiont. This enables the vector to achieve the expected protein sequence. And the last step in the paratransgenic approach is to introduce these transgenic symbionts into vector populations in the wild. This strategy is as good as a Trojan horse. The Trojan horse appears harmless, but is, in fact, malicious. In the paratransgenic strategy, the native microbial flora in the body of vectors are isolated and used for *in vitro* genetic transformation. The first example of this technique used in *Rhodnius prolixus*, which is associated with the symbiont *Rhodococcus rhodnii*. *R. prolixus* is an important insect vector of Chagas disease that is caused by *Trypanosoma cruzi* (Fieck et al., 2010; Figueiro et al., 1995; Schmunis, 2007; Schmunis and Yadon, 2010). The strategy was to engineer *R. rhodnii* to express proteins such as cecropin A that are toxic to *T. cruzi* or that block the transmission of *T. cruzi* (Durvasula et al., 1997, 1999, 2008; Maguire et al., 1987; Markiv et al., 2011; Matthews et al., 2011). Attempts are also made in tsetse flies using bacteria (Aksoy et al., 2008; De Vooght et al., 2012) and in malaria mosquitoes using fungi (Fang et al., 2011), viruses (Ren et al., 2008), and bacteria (Rodrigues et al., 2008). For the successful control of vector borne diseases, through the paratransgenesis, there need to handle the matter carefully. The symbiotic bacteria can be easily grown *in vitro*. They can be genetically modified, such as through transformation with a plasmid containing the desired gene. The engineered symbiont is stable and safe. The association between vector and symbiont cannot be attenuated. Field delivery should be easily handled.

## 9.2 ENDOSYMBIONTS FOR THE VECTOR CONTROL

The endosymbiont or endobiont is any organism that lives within the body or cells of another organism, forming an endosymbiosis (Dasch et al., 1984; Hillesland et al., 2008). Examples are nitrogen-fixing bacteria (called rhizobia), which live in root nodules on legume roots, single-cell algae inside reef-building corals, and bacterial endosymbionts that provide essential nutrients to about 10% to 15% of insects (Margulis and Chapman, 2009). Many instances of endosymbiosis are obligate; that is, either the endosymbiont or the host cannot survive without the other, such as the gutless marine worms of the genus *Riftia*, which get nutrition from their endosymbiotic bacteria. The most common examples of obligate endosymbioses are mitochondria and chloroplasts. Some human parasites (e.g., *Wuchereria bancrofti* and *Mansonella perstans*) thrive in their intermediate insect hosts because of an obligate endosymbiosis with *Wolbachia* spp. They can both be eliminated from said hosts by treatments that target this bacterium. However, not all endosymbioses are obligate. Also, some endosymbioses can be harmful to either of the organisms involved. Two major types of organelles in eukaryotic cells, mitochondria, and plastids such as chloroplasts, originated as bacterial endosymbionts. Such symbiogenesis was first suggested in 1905, and articulated by the Russian botanist Konstantin Mereschkowski in 1910. It was reintroduced by Lynn Margulis in the 1960s.

There are number of symbionts in the various arthropods that are transmitted to their young ones by various ways and means. The significantly present *Wolbachia* organisms of *Aedes* and *Culex* spp. mosquitoes are transmitted in a vertical manner via gametes to their next generations (Townson, 2002). The young ones of tsetse flies get some of the symbiotic organisms in the secretion of milk glands during their intrauterine life (Aksoy et al., 1997, 2008). The *Rhodococcus* is a genus of aerobic, nonsporulating, nonmotile gram-positive bacteria closely related to *Mycobacterium* and *Corynebacterium* (van der Geize and Dijkhuizen, 2004; Burkovski, 2008). Although a few species are pathogenic, most are benign, and have been found to thrive in a broad range of environments, including soil, water, and eukaryotic cells. Fully sequenced in October 2006, the genome is known to be 9.7 megabase pairs long and 67% G/C (McLeod et al., 2006). The triatomine bug (*Rhodinus prolixus*) is the vector for Chagas disease. The mid gut of this triatomine bug is with the *Rhodococcus rhodhini* symbionts (Cooper et al., 1993). These symbionts get transmitted from adult to the progenies through coprophagy (Beard et al. 1992, 1993, 1998; Baines, 1956; Bestetti and Muccillo, 1997). Understanding of the biology of the transmission of symbionts from the adult vectors to progenies is imperative for the symbiont-based control approach to be effectively used in the field conditions. Perturbation of the interaction between endosymbiont, vector, and parasite is the ultimate idea of the symbiont-based approach to control vector and vector-borne diseases (VVBDs). Arthropod vectors benefit from the symbiosis and augment their functional capabilities to facilitate their expansion into novel niches. The best example is *Wigglesworthia glossinidia*, an intracellular endosymbiont of *Glossina* fly, which synthesizes a plethora of vitamin biosynthetic products that may promote host reproduction. *Wolbachia* organism's biosynthesis of riboflavin and flavin adenine dinucleotide (FAD) is essential for the *Brugia malayi* coenzymes and cytochromes. The presence of symbionts and the pathogenic organisms together in the vector and the possible role of the interaction between the two in favor of disease transmission had been reported among the bacterial gut flora the *S. marcescens* chitinase, which has the ability to digest the peritrophic membrane of mosquito (Huber et al., 1991). *Plasmodium* ookinetes produce chitinase to invade the midgut epithelium of mosquito. The competence of the vector to transmit a parasite/pathogen requires a molecule(s) within the vector that interacts with the parasite/pathogen. Proteolytic lectin present in the midgut of tsetse fly is stated to have trypanosome-transforming activity (Amin et al., 2006; Volf et al., 2002). Through the creation of a paratransgenic tsetse fly with the phenotype to check the activity of proteolytic lectin of the fly would limit the transmission of African trypanosomiasis (Barrett and Croft, 2012; Hale and Hancock, 2007; Hancock and Scott, 2000).

## 9.3 BACTERIAL ENDOSYMBIONTS IN INSECTS

Scientists classify insect endosymbionts in two broad categories: primary and secondary. Primary endosymbionts (sometimes referred to as P-endosymbionts) have been associated with their insect hosts for millions of years (from 10 to several hundred million years in some cases). They form obligate associations (see later) and display cospeciation with their insect hosts. Secondary endosymbionts exhibit a more recently developed association, are sometimes horizontally transferred between hosts, live in the hemolymph of the insects (not specialized bacteriocytes, see later), and are not obligate. Among primary endosymbionts of insects, the best studied are the pea aphid (*Acyrthosiphon pisum*) and its endosymbiont *Buchnera* sp. APS (Douglas, 1998), the tsetse fly *Glossina morsitans morsitans* and its endosymbiont *Wigglesworthia glossinidia brevipalpis*, and the endosymbiotic protists in lower termites. As with endosymbiosis in other insects, the symbiosis is obligate in that neither the bacteria nor the insect is viable without the other. Scientists have been unable to cultivate the bacteria in lab conditions outside of the insect. With special nutritionally enhanced diets, the insects can survive but are unhealthy, and at best survive only a few generations. In some insect groups, these endosymbionts live in specialized insect cells called bacteriocytes (also called *mycetocytes*), and are maternally transmitted (i.e., the mother transmits her endosymbionts to her offspring). In some cases, the bacteria are transmitted in the egg, as in *Buchnera*; in others like *Wigglesworthia*, they are transmitted via milk to the developing insect embryo. In termites, the endosymbionts reside within the hindguts and are transmitted through trophallaxis among colony members. The primary endosymbionts are thought to help the host either by providing nutrients that the host cannot obtain itself or by metabolizing insect waste products into safer forms. For example, the putative primary role of *Buchnera* is to synthesize essential amino acids that the aphid cannot acquire from its natural diet of plant sap. Likewise, the primary role of *Wigglesworthia*, it is presumed, is to synthesize vitamins that the tsetse fly does not get from the blood that it eats. In lower termites, the endosymbiotic protists play a major role in the digestion of lignocellulosic materials that constitute a bulk of the termite's diet. Bacteria benefit from the reduced exposure to predators and competition from other bacterial species, the ample supply of nutrients, and relative environmental stability inside the host. Genome sequencing reveals that obligate bacterial endosymbionts of insects have among the smallest of known bacterial genomes and have lost many genes that are commonly found in closely related bacteria (Ganz, 2003). Several theories have been put forth to explain the loss of genes. It is presumed that some of these genes are not needed in the environment of the host insect cell. A complementary theory suggests that the relatively small numbers of bacteria inside each insect decrease the efficiency of natural selection in "purging" deleterious mutations and small mutations from the population, resulting in a loss of genes over many millions of years. Research in which a parallel phylogeny of bacteria and insects was inferred supports the belief that the primary endosymbionts are transferred only vertically (i.e., from the mother) and not horizontally (i.e., by escaping the host and entering a new host). Attacking obligate bacterial endosymbionts may present a way to control their insect hosts, many of which are pests or carriers of human disease. For example, aphids are crop pests and the tsetse fly carries the organism *Trypanosoma brucei* that causes African sleeping sickness. Other motivations for their study is to understand symbiosis and to understand how bacteria with severely depleted genomes are able to survive, thus improving our knowledge of genetics and molecular biology. Less is known about secondary endosymbionts. The pea aphid (*Acyrthosiphon pisum*) is known to contain at least three secondary endosymbionts: *Hamiltonella defensa*, *Regiella insecticola*, and *Serratia symbiotica*. *H. defensa* aids in defending the insect from parasitoids. *Sodalis glossinidius* is a secondary endosymbiont of tsetse flies that lives inter- and intracellularly in various host tissues, including the midgut and hemolymph. Phylogenetic studies have not indicated a correlation between evolution of *Sodalis* and tsetse (Aksoy et al., 1995). Unlike tsetse's P-symbiont *Wigglesworthia*, *Sodalis* has been cultured *in vitro* (Welburn et al., 1987).

## 9.4 STRATEGIES OF CONTROLLING VECTOR-BORNE DISEASES THROUGH MICROBIALS

The endosymbionts of arthropod vectors can be targeted and/or manipulated in different ways to control the VVBDs of man and animals (Yang et al., 2002). The attempt of vector control can be achieved in three different ways by utilizing *Wolbachia* by chemotherapeutic, immunological, and *Wolbachia* cytoplasmic incompatibility (CI)-based approaches. The chemotherapeutic approach exploits the endosymbionts of arthropods vectors as a chemotherapeutic target with the aim to disturb the symbiosis sterility was observed in healthy tsetse flies fed with tetracycline (2500 μg/ml) due to damage to the mycetome bacterial endosymbionts (Nogge, 1976). Antibiotics (tetracycline group) against the endosymbionts of nematode parasites when administered to the mammalian host disrupt the symbiosis and lead to growth retardation and/or infertility of the parasite. The immunization of animals with the whole killed endosymbionts or purified antigens or recombinant antigens of the endosymbionts would render them immune to tick vectors. The concealed antigens or the midgut antigens of the blood-feeding arthropods like ticks, Glossina, and mosquitoes are the potential vaccine targets being exploited by researchers for the control of vectors (Willadsen et al., 1995). Instead of targeting the host (vector) antigens, the endosymbionts could be targeted to disturb the symbiotic relationship between the vector and the symbiont. Following ingestion of the blood from immunized animals, these antibodies together with other components of the immune system will destroy the symbionts inside the vector, leading either to death or to disruption of normal gut physiology of the tick and reduce growth and egg-laying ability (Zasloff, 2002).

In the *Wolbachia* CI-based approach, *Wolbachia* infections in arthropods can manipulate reproduction of their hosts in a variety of ways, including induced parthenogenesis, male killing, parthenogenesis, and CI (Huber et al., 1991). The *Wolbachia* of mosquito species is an extensively studied symbiont and the same can be said for the widely explored *Wolbachia* of certain other arthropod pests or vectors. The CI phenomenon of the *Wolbachia* of mosquito species is presently being explored as a means to control the mosquito population. The *Wolbachia* CI is a phenomenon in which mating between a *Wolbachia*-infected male insect and female insect of the same species without *Wolbachia* infection (Unidirectional CI) and mating between insects of the same species with different *Wolbachia* strain infection (bidirectional CI) result in embryonic mortality (Bourtzis et al., 1998). The reciprocal mating (infected female × uninfected male) and mating between infected individuals are fully compatible.

Although the mechanism of CI has not been elucidated at the molecular level, presently it is explained with two terminologies: modification and rescue. Modification is the process in which the *Wolbachia* modifies the sperm of the infected male during spermatogenesis by an unknown process. The modified mature sperm is devoid of *Wolbachia*. If a modified sperm enters an incompatible egg (uninfected or infected with different strain), a delay in break down of the nuclear membrane of pronuclei of sperm results in mitotic asynchrony (Tram and Sullivan, 2002). Among the incompatible crosses, bidirectional CI could be exploited to control the mosquito population, which in turn limits the transmission of mosquito-borne diseases. The genetic transformation of commensal or symbiotic bacteria of the arthropod vector alters the vector's ability to transmit pathogen. It is an alternative means of blocking the transmission of VBDs. The midgut bacteria of arthropod vectors can be engineered to express and secrete effector proteins, which block the parasite invasion or kill the parasite in the midgut or hemolymph or reproductive tract. The arthropod vectors that harbor the genetically transformed endosymbionts are called paratransgenic vectors. For this strategy to be used in vector-borne disease control, bacteria that survive inside the vector's body have to be identified. The endosymbionts of arthropod vectors can be cultured and genetically transformed to express the effector gene inside the vector in such a way that the gene product kills the parasite/pathogen that the vector transmits.

Isolation and characterization of the endosymbionts present in midgut, hemolymph, and reproductive tracts is the very first step toward the paratransgenic approach. In this strategy, the normal symbiont population of the arthropod vector can be replaced with genetically modified symbionts, resulting in a population of arthropod vectors refractory to the particular vector-borne parasite/pathogen. This strategy has shown promise in controlling the transmission of *Trypanosoma cruzi* by *Rhodnius prolixus*. The genetically transformed *Rhodococcus rhodnii* was orally delivered into the asymbiotic first instar nymph in such a manner to express an antimicrobial peptide L-cecropin A inside the gut lumen, which conferred resistance status to the paratransgenics (Durvasula et al., 1997; Beard et al., 2002). The *Wolbachia* might be used to prevent the transovarian transmission of arboviruses, namely, dengue virus, La Crosse virus, and Rift Valley fever virus. With the paratransgenic approach, African trypanosomiasis can be controlled using the tsetse symbiont. Here the genes are not inserted to the tsetse chromosome but instead to the tsetse symbiont *Sodalis*. The *Sodalis* is well suited for paratransgenesis. This is because it resides in the gut in close proximity to pathogenic trypanosomes; a system for culturing *Sodalis in vitro* has been developed (Welburn et al., 1987) and can be used in conjunction with standard molecular biology techniques to insert and express foreign genes of interest in this bacterium (Beard, 1993). *Sodalis* is highly resistant to many trypanocidal peptides (Hao et al., 2001; Hu and Aksoy, 2005); recombinant *Sodalis* (recSodalis) can be reintroduced into tsetse by thoracic microinjection and passed on to future progeny where they successfully express the marker gene product. Last, its genome is completely sequenced and annotated, and this information will serve as a valuable resource that can be exploited to improve the efficiency of this expression system. Because these commensal bacteria live naturally in close proximity to where trypanosomes develop and replicate in the tsetse midgut, expression of trypanocidal products in *Sodalis* has the potential to block parasite development in the fly. Though this technology is useful in many aspects, it has some disadvantages, namely, socioethical issues, fitness of the genetically modified (GM) symbiont, alteration of the ecosystem, horizontal gene transfer among compatible bacteria, transfer of GM symbiont to nontarget arthropods and subsequently getting into the food chain.

The endosymbionts of arthropod have small genome and/or more psuedogenes, which indicate the long-term coevolution with loss of considerable size of their genome and functional genes by conserving the genes essential to adapt to the arthropod microenvironment. This genome erosion and adaptive degeneration phenomenon questions the stability of the transgene/effector gene in the long run. The advancements in the biotechnology will find safer solutions to the aforementioned problems in the near future.

Whether transformed symbionts can replace nontransformed symbionts in natural insect populations can be said with certainty, despite of the early success with the transformation of insect vector symbionts. Symbionts perhaps have no fitness load on insect hosts and are capable of being transmitted via transovarian or lateral transmission. Thus, a strong gene drive system can potentiate the effectiveness of paratransgenesis. An example of such a gene drive system can be *Wolbachia* endosymbionts. Although there has been great advances in the development of stable lines of genetically modified disease vectors (Ito, 2002; Lobo, 2002; Perera et al., 2002), there are many challenges in to the application of transgenesis to control vector-borne diseases outside the laboratory. Future studies mimicking field conditions likely will uncover the importance of fitness to the establishment of transgenic mosquitoes in natural habitats. India, which suffers from dengue, chikungunya, malaria, and filariasis, needs a program on a *Wolbachia* CI and paratransgenic based-approach to counter the aforementioned VBDs. The main constraint to the establishment of an efficient transgenic vector approach is the scarcity of a transgene that effectively reduces pathogen load. Again, further studies utilizing quantitative trait locus (QTL) mapping, reverse genetics, gene knockdown, or other techniques to identify traits associated with vector competence can reveal candidate genes that, when targeted, may effectively block pathogen development and transmission.

Conclusively enough, a variety of very effective methods have been employed for suppressing arthropod vector populations, including the application of biological control agents and the elimination of breeding sites, with a continuing and heavy reliance on the use of chemical insecticides (Hurwitz et al., 2011). However, the development of insecticide resistance by vector insects, the cost of developing and registering new insecticidal compounds, and the increase in legislation to combat the detrimental effects of insecticidal residues on the environment have emphasized the need to assess alternative strategies for vector control that are cost effective and safer. Endosymbionts of the arthropod vector are identified as a potential source for the control of VVBDs, and symbiont-based approaches are considered cost effective. Bioprospecting of molecules is involved. The strategy for the control of vector-borne diseases should be paratransgenic. Symbiotic or commensal microbial pathogens should be transformed to express products of genes that interfere with the transmission of pathogens. That is to say the genetically modified microbes are to be reintroduced back to the insect (vector). This method decreases the ability of the vector to transmit the pathogen into the human body. The paratransgenic method, if utilized efficiently, can reduce, for example, *Trypanosoma cruzi*, the causative agent of Chagas disease, transmitted by the triatomine bug *Rhodnius prolixus*; and the parasite *Leishmania donovani*, the causative agent of visceral leishmaniasis, transmitted by the sand fly *Phlebotomus argentipes* (Dhiman et al., 2003; Feliciangeli et al., 1999; Ferro et al., 1997; Ghosh KN and Bhattacharya A. 1991; Ghosh et al., 1999; Guerin et al., 2002; Hanson et al., 1961; Hati et al., 1983; Ilango et al., 1994; Kesari et al., 2000; Mukhopadhyay et al., 1996; Pandya and Niyogi, 1980; Shortt et al., 1930; Singh et al., 2001; Smith et al., 1936). There are many effector biomolecules (e.g., antimicrobial peptides and specific single chain antibodies) currently being explored for the activities of an “antiparasite” in the systems to control vector-borne diseases.

## REFERENCES

Aksoy S, Chen X, Hypsa V. 1997. Phylogeny and potential transmission routes of midgut-associated endosymbiots of tsetse (Diptera: Glossinadae). *Insect Mol Biol.* 6(2): 183–190.

Aksoy S, Pourhosseini A, Chow A. 1995. Mycetome endosymbionts of tsetse flies constitute a distinct lineage related to Enterobacteriaceae. *Insect Mol Biol.* 4: 15–22.

Aksoy S, Weiss B, Attardo G. 2008. Paratransgenesis applied for control of tsetse transmitted sleeping sickness. *Adv Exper Med Biol.* 627: 35–48. doi:10.1007/978-0-387-78225-6_3. PMID 18510012.

Amin DN, Kamitha SG, Muluvi GM, Machuka J, Hammok BD, Osir EO. 2006. Glossina proteolytic lectin does not require a carbohydrate moiety for enzymatic or trypanosome transforming activity. *J Med Entomol.* 43(2): 301–308.

Baines S. 1956. The role of the symbiotic bacteria in the nutrition of Rhodnius prolixus. *J Exp Biol.* 33: 533–541.

Barrett, MP, Croft SL. 2012. Management of trypanosomiasis and leishmaniasis. *Brit Med Bull.* 104: 175–196. doi:10.1093/bmb/lds031. PMC 3530408. PMID 23137768.

Beard CB, Cordon-Rosales C, Durvasula RV. 2002. Bacterial symbionts of the triatominae and their potential use in control of Chagas disease transmission. *Annu Rev Entomol.* 47: 123–141.

Beard CB, Durvsula RV, Richards FF. 1998. Bacterial symbiosis in arthropods and the control of disease transmission. *Emerg Infect Dis.* 4(4): 581–591.

Beard CB, Mason PW, Aksoy S, Tesh RB, Richards FF. 1992. Transformation of an insect symbiont and expression of a foreign gene in the Chagas' disease vector *Rhodnius prolixus*. *Am J Trop Med Hyg.* 46: 195–200.

Beard CB, O'Neill SL, Mason P, Mandelco L, Woese SR, Tesh RB, Richards FF, Aksoy S. 1993. Genetic transformation and phylogeny of bacterial symbionts from tsetse. *Insect Mol Biol.* 1: 123–131.

Bestetti RB, Muccillo G. 1997. Clinical course of Chagas' heart disease: A comparison with dilated cardiomyopathy. *Int J Cardiol.* 60: 187–193.

Bourtzis K, Dobson SL, Braig HR, O'Neill SL. 1998. Rescuing Wolbachia has been over looked. *Nature* 391: 852–853.

Burkovski A (editor). 2008. *Corynebacteria: Genomics and Molecular Biology*. Norfolk, UK: Caister Academic Press.

Cecere MC, Vazquez-Prokopec GM, Gurtler RE, Kitron U. 2004. Spatio-temporal analysis of reinfestation by Triatoma infestans (Hemiptera: Reduviidae) following insecticide spraying in a rural community in northwestern Argentina. *Am J Trop Med Hyg.* 71: 803–810.

Cooper R, de Jesus AR, Cross GA. 1993. Deletion of an immunodominant *Trypanosoma cruzi* surface glycoprotein disrupts flagellum-cell adhesion. *J Cell Biol.* 122: 149–156.

Dasch GA, Weiss E, Chang S. 1984. *Endosymbionts in Insects*. Baltimore, MD: Williams & Wilkins.

De Vooght L, Caljon G, Stijlemans B, De Baetselier P, Coosemans M, Van den Abbeele J. 2012. Expression and extracellular release of a functional anti-trypanosome Nanobody® in *Sodalis glossinidius*, a bacterial symbiont of the tsetse fly. *Microb Cell Fact.* 11: 23. doi:10.1186/1475-2859-11-23. PMID 22335892.

Dhiman RC, Raghavendra K, Kumar V, Kesari S, Kishore K. 2003. Susceptibility status of *Phlebotomus argentipes* to insecticides in districts Vaishaii and Patna (Bihar). *J Commun Dis.* 35: 49–51.

Dhiman RC, Shetty PS, Dhanda V. 1983. Breeding habitats of phlebotomine sandflies in Bihar, India. *Indian J Med Res.* 77: 29–32.

DiCarlo JE, Norville JE, Mali P, Rios X, Aach J, Church GM. 2013. Genome engineering in Saccharomyces cerevisiae using CRISPR-Cas systems. *Nucleic Acids Res.* 41(7): 4336–4343. doi:10.1093/nar/gkt135. PMC 3627607. PMID 23460208.

Douglas AE. 1998. Nutritional interactions in insect-microbial symbioses: Aphids and their symbiotic bacteria Buchnera. *Ann Rev Entomol.* 43: 17–38. doi:10.1146/annurev.ento.43.1.17. ISSN 0066-4170. PMID 15012383.

Durvasula RV, Gumbs A, Panackal A, Kruglov O, Aksoy S, Merrifield BR, Richards FF, Beard C. 1997. Prevention of insect borne diseases: An approach using transgenic symbiotic bacteria. *Proc Natl Acad Sci USA.* 94: 3274–3278.

Durvasula RV, Gumbs A, Panackal A, Kruglov O, Taneja J, Kang AS, Cordon-Rosales C, Richards FF, Whitham RG, Beard CB. 1999. Expression of a functional antibody fragment in the gut of *Rhodnius prolixus* via transgenic bacterial symbiont *Rhodococcus rhodnii*. *Med Vet Entomol.* 13: 115–119.

Durvasula RV, Kroger A, Goodwin M, Panackal A, Kruglov O, Taneja J, Gumbs A, Richards FF, Beard CB, Cordon-Rosales C. 1999. Strategy for introduction of foreign genes into field populations of Chagas disease vectors. *Ann Entomol Soc Am.* 92: 937–943.

Durvasula RV, Sundaram RK, Kirsch P, Hurwitz I, Crawford CV, Dotson E, Beard CB. 2008. Genetic transformation of a Corynebacterial symbiont from the Chagas disease vector *Triatoma infestans*. *Exp Parasitol.* 119: 94–98.

Fang W, Vega-Rodríguez J, Ghosh AK, Jacobs-Lorena M, Kang A, St. Leger RJ. 2011. Development of transgenic fungi that kill human malaria parasites in mosquitoes. *Science.* 331(6020): 1074–1077. doi:10.1126/science.1199115. PMID 21350178.

Feliciangeli MD, Rodriguez N, De Guglielmo Z, Rodriguez A. 1999. The re-emergence of American visceral leishmaniasis in an old focus in Venezuela. II. Vectors and parasites. *Parasite.* 6: 113–120.

Ferro C, Pardo R, Torres M, Morrison AC. 1997. Larval microhabitats of *Lutzomyia longipalpis* (Diptera: Psychodidae) in an endemic focus of visceral leishmaniasis in Colombia. *J Med Entomol.* 34: 719–728.

Fieck A, Hurwitz I, Kang AS, Durvasula R. 2010. *Trypanosoma cruzi*: Synergistic cytotoxicity of multiple amphipathic anti-microbial peptides to *T. cruzi* and potential bacterial hosts. *Exp Parasitol.* 125: 342–347.

Figueiro AR, Nunes ZG, Silvia AAL, Giordano-Dias CMG, Conra JR, Hofer E. 1995. Isolation of microorganisms of triatomines maintained in artificial and sylvatic conditions. *Mem Inst Oswaldo Cruz.* 90: 228.

Ganz T. 2003. Defensins: Antimicrobial peptides of innate immunity. *Nat Rev Immunol.* 3: 710–720.

Ghosh KN, Bhattacharya A. 1991. Breeding places of Phlebotomus argentipes Annandale and Brunetti (Diptera: Psychodidae) in West Bengal, India. *Parassitologia.* 33: 267–272.

Ghosh KN, Mukhopadhyay JM, Guzman H, Tesh RB, Munstermann LE. 1999. Interspecific hybridization and genetic variability of Phlebotomus sandflies. *Med Vet Entomol.* 13: 78–88.

Guerin PJ, Olliaro P, Sundar S, Boelaert M, Croft SL, Desjeux P, Wasunna MK, Bryceson AD. 2002. Visceral leishmaniasis: Current status of control, diagnosis, and treatment, and a proposed research and development agenda. *Lancet.* 2: 494–501.

Gurtler RE, Canale DM, Spillmann C, Stariolo R, Salomon OD, Blanco S, Segura EL. 2004. Effectiveness of residual spraying of peridomestic ecotopes with deltamethrin and permethrin on Triatoma infestans in rural western Argentina: A district-wide randomized trial. *Bull WHO.* 82: 196–205.

Gurtler RE, Petersen RM, Cecere MC, Schweigmann NJ, Chuit R, Gualtieri JM, Wisnivesky-Colli C. 1994. Chagas disease in north-west Argentina: Risk of domestic reinfestation by *Triatoma infestans* after a single community-wide application of deltamethrin. *Trans R Soc Trop Med Hyg.* 88: 27–30.

Hale JD, Hancock RE. 2007. Alternative mechanisms of action of cationic antimicrobial peptides on bacteria. *Expert Rev Anti Infect Ther.* 5: 951–959.

Hancock RE, Scott MG. 2000. The role of antimicrobial peptides in animal defenses. *Proc Natl Acad Sci USA.* 97: 8856–8861.

Hanson WJ. 1961. The breeding places of *Phlebotomus* in Panama (Diptera: Psychodidae). *Ann Entomol Soc Am.* 54: 317–322.

Hao Z, Kasumba I, Lehane MJ, Gibson WC, Kwon J, Aksoy S. 2001. Tsetse immune responses and trypanosome transmission: Implications for the development of tsetse-based strategies to reduce trypanosomiasis. *Proc Natl Acad Sci USA.* 98(22): 12648–12653.

Hati AK. 1983. Current status of leishmaniasis—Vector biology. In *Proceedings of Indo-UK Workshop on Leishmaniasis*, RC Mahajan (editor). New Delhi, India: Indian Council of Medical Research, pp. 84–91.

Hill CA, Kafatos FC, Stansfield SK, Collins FH. 2005. Arthropod-borne diseases: Vector control in the genomics era. *Nat Rev Microbiol.* 3(3): 262–268.

Hillesland H, Read A, Subhadra B, Hurwitz I, McKelvey R, Ghosh K, Das P, Durvasula R. 2008. Identification of aerobic gut bacteria from the kala azar vector, Phlebotomus argentipes: A platform for potential paratransgenic manipulation of sand flies. *Am J Trop Med Hyg.* 79: 881–886.

Hu Y, Aksoy S. 2005. An antimicrobial peptide with trypanocidal activity characterized from *Glossina morsitans morsitans. Insect Biochem Mol Biol.* 35: 105–115.

Huber M, Cabib E, Miller LH. 1991. Malaria parasite chitinase and penetration of the mosquito peritrophic membrane. *Proc Natl Acad Sci USA.* 88: 2807–2810.

Hurwitz I, Hillesland H, Fieck A, Das P, Durvasula R. 2011. The paratransgenic sand fly: A platform for control of Leishmania transmission. *Parasit Vectors.* 4: e82.

Ilango K, Dhanda V, Srinivasan R, Sadanand AV, Lane RP. 1994. Phlebotomine sandflies (Diptera: Psychodidae) of Tamil Nadu and Pondicherry, southern India, in relation to visceral leishmaniasis. *Ann Trop Med Parasitol.* 88: 413–431.

Ito J. 2002. Transgenic anopheline mosquitoes impaired in transmission of a malaria parasite. *Nature.* 417: 452–455.

Kesari S, Kishore K, Palit A, Kumar V, Roy MS, Sivakumar S, Kar SK. 2000. An entomological field evaluation of larval biology of sandfly in Kala-azar endemic focus of Bihar—Exploration of larval control tool. *J Commun Dis.* 32: 284–288.

Kirchhoff LV. 2006. American trypanosomiasis (Chagas' disease) In *Tropical Infectious Diseases: Principles, Pathogens & Practice*, R Guerrant, DH Walker, PF Weller (editors). Philadelphia: Churchill Livingstone, p. 1082.

Kishore K, Kumar V, Kesari S, Bhattacharya SK, Das P. 2004. Susceptibility of Phlebotomus argentipes against DDT in endemic Districts of North Bihar, India. *J Commun Dis.* 36: 41–44.

Lobo NF. 2002. Germ line transformation of the yellow fever mosquito, Aedes aegypti, mediated by transpositional insertion of a piggyBac vector. *Insect Mol Biol.* 11: 133–139.

Maguire JH, Hoff R, Sherlock I, Guimaraes AC, Sleigh AC, Ramos NB, Mott KE, Weller TH. 1987. Cardiac morbidity and mortality due to Chagas' disease: Prospective electrocardiographic study of a Brazilian community. *Circulation.* 75: 1140–1145.

Margulis L, Chapman MJ. 2009. *Kingdoms & Domains an Illustrated Guide to the Phyla of Life on Earth* (4th ed.). Amsterdam: Academic Press/Elsevier, p. 493.

Markiv A, Anani B, Durvasula RV, Kang AS. 2011. Module based antibody engineering: A novel synthetic REDantibody. *J Immunol Methods.* 364: 40–49.

Matthews S, Sreehari Rao V, Durvasula RV. 2011. Modeling horizontal gene transfer (HGT) in the gut of the Chagas disease vector Rhodnius prolixus. *Parasit Vectors.* 4: e77.

McLeod MP, Warren RL, Hsiao WW, Araki N, Myhre M, Fernandes C, Miyazawa D, Wong W, Lillquist AL, Wang D et al. (2006). The complete genome of *Rhodococcus* sp. RHA1 provides insights into a catabolic powerhouse. *PNAS.* 103(42): 15582–15587. doi:10.1073/pnas.0607048103. PMC 1622865. PMID 17030794.

Mereschkowsky C. 1905. Uber Natur und Ursprung der Chromatophoren im Pflanzenreiche. *Biol Centralbl.* 25: 593–604.

Mereschkowsky K. 1910. Theorie der zwei Plasmaarten als Grundlage der Symbiogenesis, einer neuen Lehre von der Ent-stehung der Organismen. *Biol Centralbl.* 30: 353–367.

Miller T, Lauzon C, Lampe D, Durvasula R, Matthews S. 2006. Paratransgenesis applied to insect-transmitted disease: The Pierce's Disease Case. In *Insect Symbiosis*, T Miller, K Bourtzis (editors). London: Taylor & Francis.

Moncayo A. 2003. Chagas disease: Current epidemiological trends after the interruption of vectorial and transfusional transmission in the Southern Cone countries. *Mem Inst Oswaldo Cruz.* 98: 577–591.

Moncayo A, Ortiz Yanine MI. 2006. An update on Chagas disease (human American trypanosomiasis). *Ann Trop Med Parasitol.* 100: 663–677.

Mukhopadhyay AK, Hati AK, Chakraborty S, Saxena NB. 1996. Effect of DDT on Phlebotomus sandflies in Kala-Azar endemic foci in West Bengal. *J Commun Dis.* 28: 171–175.

Nogge G, 1976. Sterility in tsetse flies (Glossinia morsitans Westwood) caused by loss of symbionts. *Experientia.* 32: 995–996.

Pandya AP, Niyogi AK. 1980. Ecological study on immature stages of phlebotomid sandflies in Gujarat. *Indian J Med Res.* 72: 355–358.

Perera OP, Harrell IR, Handler AM. 2002. Germ-line transformation of the South American malaria vector, Anopheles albimanus, with a piggyBac/EGFP transposon vector is routine and highly efficient. *Insect Mol Biol.* 11: 291–297.

Picollo MI, Vassena C, Santo Orihuela P, Barrios S, Zaidemberg M, Zerba E. 2005. High resistance to pyrethroid insecticides associated with ineffective field treatments in *Triatoma infestans* (Hemiptera: Reduviidae) from Northern Argentina. *J Med Entomol.* 42: 637–642.

Planning Commission, Government of India. 1996. Kala azar. In 9th Five Year Plan, Sec 3.4.89.

Rassi AJr, Rassi A, Marin-Neto JA. 2010. Chagas disease. *Lancet.* 375: 1388–1402.

Ren X, Hoiczyk E, Rasgon JL. 2008. Viral paratransgenesis in the malaria vector *Anopheles gambiae*. *PLOS Pathogens.* 4(8): e1000135. doi:10.1371/journal.ppat.1000135. PMC 2500179. PMID 18725926.

Rodrigues FG, Santos MN, de Carvalho TX, Rocha BC, Riehle MA, Pimenta PF, Abraham EG, Jacobs-Lorena M, Alves de Brito CF, Moreira LA. 2008. Expression of a mutated phospholipase A2 in transgenic *Aedes fluviatilis* mosquitoes impacts *Plasmodium gallinaceum* development. *Insect Mol Biol.* 17(2): 175–183. doi:10.1111/j.1365-2583.2008.00791.x. PMID 18353106.

Schmunis GA. 2007. Epidemiology of Chagas disease in non-endemic countries: The role of international migration. *Mem Inst Oswaldo Cruz.* 102(Suppl 1): 75–85.

Schmunis GA, Yadon ZE. 2010. Chagas disease: A Latin American health problem becoming a world health problem. *Acta Trop.* 115: 14–21.

Shortt HE, Smith ROA, Swaminath CS. 1930. The breeding nature of Phlebotomus argentipes Ann & Brun. *Bull Entomol Res.* 21: 269–271

Singh R, Das RK, Sharma SK. 2001. Resistance of sandflies to DDT in Kala-azar endemic districts of Bihar, India. *Bull WHO.* 79: 793.

Smith ROA, Mukherjee S, Lal D. 1936. Bionomics of P. argentipes. Part III. The breeding sites of P. argentipes and an attempt to control these insects by antilarval measures. *Indian J Med Res.* 24: 557–562.

Subhadra B, Hurwitz I, Fieck A, Subba Rao G, Subba Rao DV, Durvasula RV. 2009. Development of paratransgenic artemia as a platform for control of infectious diseases in shrimp agriculture. *J Appl Microbiol.* 108: 831–840.

Sundar S, Agrawal N, Arora R, Agarwal D, Rai M, Chakravarty J. 2009. Short-course paromomycin treatment of visceral leishmaniasis in India: 14-day vs 21-day treatment. *Clin Infect Dis.* 49: 914–918.

Sundar S, Chakravarty J, Agarwal D, Rai M, Murray HW. 2010. Single-dose liposomal amphotericin B for visceral leishmaniasis in India. *N Engl J Med.* 362: 504–512.

Sundar S, Jha TK, Thakur CP, Bhattacharya SK, Rai M. 2006. Oral miltefosine for the treatment of Indian visceral leishmaniasis. *Trans R SocTrop Med Hyg.* 100: S26–33.

Sundar S, Sinha PK, Rai M, Verma DK, Nawin K, Alam S, Chakravarty J, Vaillant M, Verma N, Pandey K et al. 2011. Comparison of short-course multidrug treatment with standard therapy for visceral leishmaniasis in India: An open-label, non-inferiority, randomised controlled trial. *Lancet.* 377: 477–486.

Tanowitz HB, Weiss LM, Montgomery SP. 2011. Chagas disease has now gone global. *PLoS Negl Trop Dis.* 5: e1136.

Thakur CP, Kumar K. 1992. Post kala-azar dermal leishmaniasis: A neglected aspect of kala-azar control programmes. *Ann Trop Med Parasitol.* 86: 355–359.

Tram U, Sullivan W. 2002. Role of delayed nuclear envelope breakdown and mitosis in *Wolbachia* induced cytoplasmic incompatibility. *Science.* 296: 1124–1126.

Townson H, 2002. Wolbachia as a potential tool for suppressing filarial transmission. *Ann. Trop. Med. Parasitol.* 96(Suppl. 2): 117–127.

Travi BL, Velez ID, Brutus L, Segura I, Jaramillo C, Montoya J. 1990. Lutzomyia evansi, an alternate vector of *Leishmania chagasi* in a Colombian focus of visceral leishmaniasis. *Trans R Soc Trop Med Hyg.* 84: 676–677.

van der Geize R, Dijkhuizen L. 2004. Harnessing the catabolic diversity of Rhodococci for environmental and biotechnological applications. *Microbiology.* 7(3): 255–261. doi:10.1016/j.mib.2004.04.001. PMID 15196492.
Volf P, Kiewegova A, Nemec A. 2002. Bacterial colonisation in the gut of *Phlebotomus duboseqi* (Diptera: Psychodidae): Transtadial passage and the role of female diet. *Folia Parasitol (Praha).* 49: 73–77.
Welburn SC, Maudlin I, Ellis DS. 1987. In vitro cultivation of rickettsia-like-organisms from Glossina spp. *Ann. Trop. Med. Parasitol.* 81(3): 331–335.
Willadsen P, Bird P, Cobon GS, Hungerford J. 1995. Commercialization of a recombinant vaccine against Boophilus microplus. *Parasitology* 110: 43–50.
World Health Organization (WHO). 2009. Chagas disease: Control and elimination: Report by the Secretariat. Sixty-second World Health Assembly, A62/17. Geneva: WHO.
World Health Organization (WHO). 2010. Control of the Leishmaniases. *WHO Tech Rep Ser.* 949: 22–26.
World Health Organization (WHO). 2014. *Leishmaniasis. Fact sheet no. 375.* Geneva: WHO.
Yang D, Biragyn A, Kwak LW, Oppenheim JJ. 2002. Mammalian defensins in immunity: More than just microbicidal. *Trends Immunol.* 23: 291–296.
Zasloff M. 2002. Antimicrobial peptides of multicellular organisms. *Nature.* 415: 389–395.

# 10 Microbial Control of Vector-Borne Diseases in Thailand: A Review

*Viroj Wiwanitkit*

## CONTENTS



## 10.1 INTRODUCTION

Vector-borne infectious diseases is an important group of infections that is a global public health issue (Sedda et al., 2014). Zika virus is the best example of a problematic vector-borne infectious disease. According to the definition, a vector-borne infectious disease is any disease that has the vector as an intermediate carrier for carrying to the host (hereby human beings) or transmission of disease. There are many medically important vectors such as the mosquito and fly. The abundance of vectors can be seen around the world, especially in tropical zones.

There are many vector-borne diseases that are still prevalent in the present day. As already mentioned, those diseases are usually prevalent in tropical zones and are an important focus in tropical medicine and public health. Examples of well-known vector borne diseases are dengue, malaria, filariasis, and Japanese encephalitis. The most recent important infection is the emerging Zika virus, which became a global issue in 2016 and can cause abnormality in newborns. The transmission of the aforementioned infections causes deaths around the world, and the high morbidity and mortality lead public health and medical personnel to focus on vector-borne diseases.

Early diagnosis and prompt treatment is the common goal in medical management of vector-borne infectious diseases (Adelman and Tu, 2016). Concerning public health, how to effectively control and prevent a disease is the big issue. Since vector-borne diseases relate to the vector, which is usually the insect, they are very difficult to control and prevent. Ideally, the control or prevention of vector-borne infectious diseases can be of these activities: (a) control of the pathogen (which is usually a parasite or virus), (b) control of the environment, (c) control of the vector, and (d) control of the host (human beings or animals). Controlling for the pathogen is difficult. As a microbe, manipulation of controlling the pathogen seems to be extremely hard. It requires a sophisticated genetic editing technique that is impossible at present. It is also impossible at present the control of the environment. Global warming has significantly increased the epidemics of vector-borne infectious diseases (McMichael, 2015) and it is not possible to change the weather. However, it is possible to modify the microenvironment of the vector. Good examples are "reducing the availability of larval habitats (habitat modification and habitat manipulation), or by adding substances to standing water that either kill or inhibit the development of larvae (larviciding)" (Tusting et al., 2013). For control of the host, health education might be a useful tool to modify human behavior. But it is not possible to use the knowledge–attitude–practice concept to modify animal behavior. In addition, not all human beings can display the desirable preventive behavior to vector-borne diseases using a common educational tool and model. For vector control, it is similarly hard but it is the main focus and practice in public health. Since the vector is the main part of vector-borne infectious disease transmission, it is simply said that if we can get rid of the vector, then there will be no vector-borne infectious disease. However, this is not realistic. There are many billions insects are around the world and they all have potential to carry and transmit vector-borne infectious diseases.

There are several means for controlling vectors and many tools have been used for a long time. Examples include using insecticide/chemicals and bed nets. Biological control is an interesting means of vector control (Weiser, 1975). In fact, biological control of vectors is not a new concept that relies on biological things (Benelli et al., 2016). It has been proposed for many decades. The important advantage of biological control is due to its ecofriendly nature. Benelli et al. (2016) noted that "biocontrol strategies aim to be sustainable and target a range of different mosquito species to reduce the current reliance on insecticide-based mosquito control." At first, biological control usually relied on the use of a macrobiological agent for controlling (Chapman, 1974). A predator of the vector is a good example. The use of predator of vector larvae for controlling of larvae is classical practice In addition to the use of a predator (Figure 10.1), the use of a macrobiological agent should be mentioned (Chapman et al., 1972). For example, there are many aquatic insects that are predators of mosquito vector larvae. García Avila et al. (1996) proposed that species that are effective for control of mosquito vector larvae included "effective bioregulators of mosquito larvae such as *Ischnura ramburii* (Selys), *Ceratura capreola* (Hagen), *Erithrodoplax cleopatra Ris*, *Aeschna elsia Calvert*, *Tropisternus dorsalis Brullé*, *Thermonectes circunscripta*

(a)

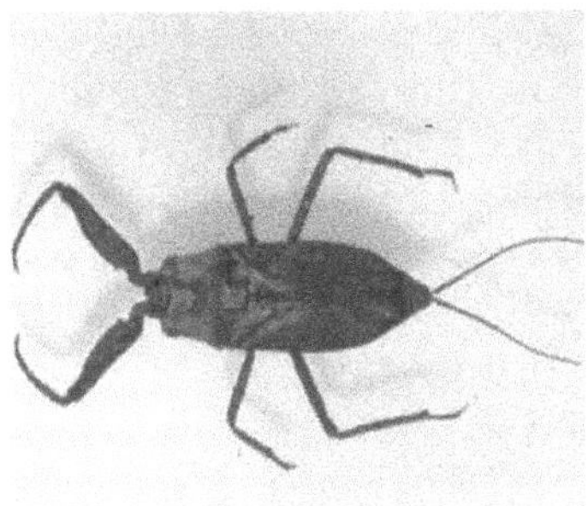

(b)

**FIGURE 10.1** Some macrobiological agents that can be useful in control of vector-borne infectious diseases: (a) The guppy is known to control mosquito larvae. (b) Nepidae is a predator that eat mosquito larvae.

*Latr, Belostoma asiaticum* (Mayr), and *Buenoa absidata Truxal.*" A more recognized predator of vector-borne infectious disease is the larvivorous fish (Walshe et al., 2013). Several fish species are presently used for control of vector. Nevertheless, Walshe et al. (2013) noted that "research could also usefully examine the effects on native fish and other nontarget species."

Similar to any animal, the vector of a vector-borne infectious disease can also have a disease. The pathogen that can cause the disease to the vector can be useful in vector control (Chapman et al., 1972). This is an interesting concept of using a microbiological agent for controlling. Several microbe species are studied and the interesting species that are widely used at present is the *Wolbachia* species (Mustafa et al., 2016; Wiwanitkit, 2001). Mustafa et al. (2016) mentioned that the *Wolbachia* species acted as a "Trojan Horse" and could prevent transmission of vector-borne infectious disease by "transmission by either directly blocking the virus or by decreasing the lifespan of the vector." The use of the pathogenic microbe for controlling and preventing vector-borne infectious diseases is the new hope for management of the problem. Nevertheless, it is still in the beginning phase and requires more study and development.

Hence, the use of microbial controls for vector-borne diseases is very interesting. It is presently applied in several areas around the world. In the tropical world where tropical vector-borne infectious diseases are highly prevalent, the use of biological controls is very interesting. At present, there are many ongoing studies on this topic. In this chapter, the author will briefly summarize and discuss specific reports on microbial control of important vector-borne diseases (such as malaria, dengue, and filariasis) from Thailand, a tropical country in Southeast Asia.

## 10.2 BIOLOGICAL METHODS FOR CONTROLLING VECTOR-BORNE DISEASES IN THAILAND

### 10.2.1 Application of Predators of the Vector

#### 10.2.1.1 Using Aquatic Insects

There are many local species of aquatic insects in Thailand that could naturally control mosquito vector larvae. However, there is still no report on using aquatic insects for control of vector-borne infectious diseases in Thailand. Of interest, the local people know that aquatic insects can be useful for getting rid of vector larvae. Nevertheless, to control an aquatic insect seems to be hard and there is no research group in Thailand that focuses on using aquatic insects as a biological control mean for vector-borne infectious diseases.

#### 10.2.1.2 Using Larvivorous Fish

Similar to aquatic insects, there are many species of larvivorous fish in Thailand. Local people usually feed the larvivorous fish in their houses aiming at control of mosquito vectors. Nevertheless, there are some interesting publications reporting the effectiveness of using larvivorous fish in control of vector-borne infectious diseases especially for dengue. For example, Swaddiwudhipong et al. (1992) reported on the "effect of health education on community participation in control of dengue hemorrhagic fever in an urban area of Thailand." Swaddiwudhipong et al. concluded that "the introduction of larvivorous fish may be an effective method of larval control." However, the success of the technique is still not confirmed. The effectiveness of using larvivorous fish to control vector-borne infectious diseases in Thailand is raised by Mulla et al. (1997). Mulla et al. mentioned that the "presence of larvivorous fish did not seem to influence larval populations because there were heavy populations of mosquito larvae present."

### 10.2.2 Application of Pathogens of Vectors

#### 10.2.2.1 Using *Wolbachia* Species

*Wolbachia* species are "maternally inherited bacteria that induce cytoplasmic incompatibility in mosquitoes" and can use "these patterns of sterility to spread themselves through populations" (Sinkins, 2004). Hence, *Wolbachia* becomes the new focus in research on biological control of vector-borne infectious diseases. The important reports are usually on filariasis by a research group from Chulalongkorn University, Bangkok (Sanprasert et al., 2010; Sungpradit et al., 2012). For clinical implications in disease control and management, Sungpradit et al. (2012) reported that "doxycycline and rifampicin have an effect on the obligate intracellular gram-negative bacteria, *Wolbachia*, which appears to exert an influence on filarial nematode embryonic and larval development, adult female fertility, and filarial survival." Nevertheless, most studies in Thailand are usually on the epidemiology of *Wolbachia*-infected vectors. There is still no report on a new intervention or trial using *Wolbachia* as a tool for disease control.

#### 10.2.2.2 Using *Bacillus* Species

Different from *Wolbachia* species, there are some reports on using *Bacillus* species as tools for management and control of vector-borne infectious disease in Thailand. For example, Mulla et al. (1999) studied and confirmed the efficacy of new formulations of the microbial larvicide *B. sphaericus* against polluted water mosquitoes in Thailand. Mulla et al. (2003) also preformed another study to "find practical strategies for controlling resistant mosquitoes and more importantly to develop resistance management strategies that would prevent or delay development of resistance" and found "field evidence for delay or prevention of resistance to microbial agents in larval *Cx. quinquefasciatus* by using mixtures of *Bacillus thuringiensis var. israelensis* (Bti) and Bsph." In another report, Pantuwatana et al. (1989) studied the long residual activity of *B. sphaericus* 1593 against *Cx. quinquefasciatus* larvae in artificial pools and found that *B. sphaericus* 1593 could recycle in their study.

#### 10.2.2.3 Using *Penicillium* Species

The use of *Penicillium* species is an interesting new avenue in vector control research. In 1998, da Costa et al. first proposed the possible pathogenic action of Penicillium species against the medially important vectors and there has been further research on this specific issue. There is an interesting report on this topic from Thailand. Maketon et al. (2014) reported their study on using 20 local isolates of entomopathogenic fungi in Thailand for control of the larvae and adults of *Cx. quinquefasciatus*. Maketon et al. (2014) found that *P. citrinum* CM-010 could decrease survival of *Cx. quinquefasciatus* larvae in their study.

## 10.3 REPORTS OF USING BIOLOGICAL CONTROLS ON VECTOR-BORNE DISEASES IN THAILAND

### 10.3.1 Dengue

Thailand is located in Southeast Asia where dengue is extremely prevalent. Dengue is an arboviral infection caused by the dengue virus. This virus is transmitted by *Aedes species* mosquito. The disease is classified as a tropical infection presenting with acute febrile illness with common clinical triad (hemconcentration, atypical lymphocytosis, and thrombocytopenia). In Thailand, there have been several outbreaks affecting thousands of local people each year. Many thousands of patients of all age groups are hospitalized with dengue during the endemic rainy season in Thailand. The control of dengue is important among local public health policies (Bhumiratana et al., 2014). There are many interesting reports on biological methods for control of dengue in Thailand, as shown in Table 10.1. However, some medical

**TABLE 10.1**
**Reports on Biological Methods for Control of Dengue in Thailand**

| Reference | Details |
|---|---|
| Choochote et al. (2004) | Choochote et al. (2004) reported the study on potential of crude seed extract of celery, *Apium graveolens L.*, against the mosquito vector of dengue and proposed that the extract can be a new biological substance for control of dengue. |
| Chowanadisai et al. (1987) | Chowanadisai et al. (1987) performed a study to determine the efficacy of B.t. H-14 against *Ae. aegypti* in domestic containers under field conditions in a low-income residential area covering 137 houses in Bangkok, the capital of Thailand, and found "high reduction in adult densities during epidemic season for ten weeks." |
| Dieng et al. (2016) | Dieng et al. (2016) mentioned that "tea and its waste affect most key components of *Ae. albopictus* vectorial capacity and may be useful for dengue control." Although this is not a direct report on biological control of dengue, it is an interesting report on using a naturally derived biological substance for control of dengue. |
| Phuanukoonnon et al. (2005) | Phuanukoonnon et al. (2005) studied the effectiveness of dengue control practices in household water containers in Northeast Thailand and found that "keeping fish" was the most effective method. |

personnel have poor knowledge and misunderstand that the basic technique such as application of larvivorous fish in homes can effectively control malaria vector, which is generally not seen in house but in forest streams.

### 10.3.2 Malaria

Malaria is a disease caused by the protozoa *Plasmodium* species. This vector borne infectious disease can be seen in both animals and human beings. Malaria is a kind of blood infection that can result in acute febrile illness. Severe infection (such as cerebral malaria) can be seen and it can be fatal. The disease is classified as a mosquito-borne infectious disease and the important mosquito vector for malaria is the *Anopheles* species. Malaria is common in tropical countries. In Thailand, malaria is still prevalent in many areas especially along the borders with other countries in Indochina (Cambodia, Laos, and Myanmar). There are some interesting reports on biological methods for control of malaria in Thailand as shown in Table 10.2. However, since the malarial vector usually lives in forested areas and the visiting these areas is the main risk factor for getting disease, the attempt to control the vector is not as much a priority compared to dengue, which is present in crowded cities.

### 10.3.3 Filariasis

Filariasis is the disease that is caused by the nematode parasite. The infection is transmitted by several species of mosquito vectors. It is classified as a blood infection. The pathogenic parasite usually lives in the lymphatic system and causes pathologies. The common clinical feature is elephantiasis, which is the most common

**TABLE 10.2**
**Reports on Biological Methods for Control of Malaria in Thailand**

| Reference | Details |
|---|---|
| Kitthawee et al. (1993) | Kitthawee et al. (1993) reported the situation on abundance of malarial mosquito vector in gem pits in southern Tha Mai District, Chanthaburi Province, Thailand, which is the border province to Cambodia. Kitthawee et al. found that "variations in [*Anopheles*] *dirus* density and occurrence were related to predators populations, ie Notonectidae and fish" |
| Pitasawat et al. (2007) | Pitasawat et al. (2007) reported that essential oil from mullilam (*Zanthoxylum limonella*) was highly effective for *Anopheles dirus* larvicidal activity. Although this is not a direct report on biological control of malaria, it is an interesting report on using a naturally derived biological substance for control of malaria. |

**TABLE 10.3**
**Reports on Biological Methods for Control of Filariasis in Thailand**

| Reference | Details |
|---|---|
| Bay and Self (1972) | Bay and Self (1972) reported on observations of the guppy in *Culex pipiens fatigans* breeding sites in Bangkok, Rangoon, and Taipei. |
| Changbunjong et al. (2010) | Changbunjong et al. (2010) studied on the effect of crude extract of *Solanum xanthocarpum* against *Culex quinquefasciatus* and concluded that "a preparation of ingredients from this plant may be used as a biological larvicide." Although this is not a direct report on biological control of filariasis, it is an interesting report on using a naturally derived biological substance for control of filariasis. |

kind of infectious-induced disability. In Thailand, this infection is still prevalent in border areas to Myanmar and Malaysia. The local public health policies against this infection can still be seen and the vector control is one of the important policies.

There are some interesting reports on biological methods for control of filariasis in Thailand as shown in Table 10.3.

### 10.3.4 Japanese Encephalitis

Japanese encephalitis is another arboviral infection that is common in Southeast Asian countries including to Thailand. It is another disease that can be transmitted by mosquito. The main mosquito vector is *Culex* species. Due to the endemic status of this infection, the disease is of focused by local public health policies. Different from other aforementioned vector-borne infectious diseases, there is already an effective method for prevention and control of Japanese encephalitis (Wiwanitkit, 2007, 2009). The vaccination against Japanese encephalitis is routinely used in Thailand. The use of biological controls against this infection is still locally used but there is no report on its efficacy and effectiveness.

## 10.4 CONCLUSION

As a tropical country in Southeast Asia, vector-borne infectious diseases is common in Thailand. The control of the endemic tropical vector-borne infectious diseases by biological controls of vectors can be seen in Thailand. There are many reports of using management techniques for many of these diseases such as dengue, malaria, and filariasis. The success is usually mentioned in research reports. Nevertheless, the exact usefulness and success in actuality is limited. Also, the studies are mainly on diseases affecting human beings. The diseases of animals is usually neglected. There is still the need for further studies and development of biological controls of vector-borne infectious diseases with a hope that it can be the way for successful management of vector-borne infectious diseases in Thailand.

## REFERENCES

Adelman, Z. N., Tu, Z., 2016. Control of mosquito-borne infectious diseases: Sex and Gene Drive. *Trends Parasitol* 32: 219–29.

Bay, E. C., Self, L. S., 1972. Observations of the guppy, *Poecilia reticulata* Peters, in *Culex pipiens* fatigans breeding sites in Bangkok, Rangoon, and Taipei. *Bull World Health Organ* 4: 407–16.

Benelli, G., Jeffries, C. L., Walker, T., 2016. Biological control of mosquito vectors: Past, present, and future. *Insects* 7(4). pii: E52.

Bhumiratana, A., Intarapuk, A., Chujun, S., Kaewwaen, W., Sorosjinda-Nunthawarasilp, P., Koyadun, S., 2014. Thailand momentum on policy and practice in local legislation on dengue vector control. *Interdiscip Perspect Infect Dis* 2014:217237.

Changbunjong, T., Wongwit, W., Leemingsawat, S., Tongtokit, Y., Deesin, V., 2010. Effect of crude extract of *Solanum xanthocarpum* against snails and mosquito larvae. *Southeast Asian J Trop Med Public Health* 41: 320–5.

Chapman, H. C., 1974. Biological control of mosquito larvae. *Annu Rev Entomol* 19: 33–59.

Chapman, H. C., Petersen, J. J., Fukuda, T., 1972. Predators and pathogens for mosquito control. *Am J Trop Med Hyg* 21:777–81.

Choochote, W., Tuetun, B., Kanjanapothi, D., Rattanachanpichai, E., Chaithong, U., Chaiwong, P., Jitpakdi, A., Tippawangkosol, P., Riyong, D., Pitasawat, B., 2004. Potential of crude seed extract of celery, *Apium graveolens* L., against the mosquito *Aedes aegypti* (L.) (Diptera: Culicidae). *J Vector Ecol* 29: 340–6.

Chowanadisai, L., Thanasripukdikul, S., Phanthumachinda, B., 1987. Field trials of granular formulation of *Bacillus thuringiensis* H-14 against *Aedes aegypti* DHF vector of dengue hemorrhagic fever in Thailand. *Commun Dis J* 13: 193–201.

da Costa, G. L., de Moraes, A. M., de Oliveira, P. C., 1998. Pathogenic action of *Penicillium* species on mosquito vectors of human tropical diseases. *J Basic Microbiol* 38: 337–41.

Dieng, H., Tan Yusop, N. S., Kamal, N. N., Ahmad, A. H., Ghani, I. A., Abang, F., Satho, T., Ahmad H, Zuharah, W. F., Majid, A. H., Morales, R. E., Morales, N. P., Hipolito, C. N., Noweg, G. T., 2016. Exposure of a dengue vector to tea and its waste: Survival, developmental consequences, and significance for pest management. *J Agric Food Chem* 64: 3485–91.

García Avila, I., Vivar Guzmán, R., Quezada Márquez, J., Huamán Mayta, P., 1996. Aquatic insects that are bioregulators of mosquito larvae present in Pantanos de Villa, Lima, Peru. *Rev Cubana Med Trop* 48: 227–8.

Kitthawee, S., Edman, J. D., Upatham, E. S., 1993. Mosquito larvae and associated macroorganisms occurring in gem pits in southern Tha Mai District, Chanthaburi Province, Thailand. *Southeast Asian J Trop Med Public Health* 24: 143–51.

Maketon, M., Amnuaykanjanasin, A., Kaysorngup, A., 2014. A rapid knockdown effect of Penicillium citrinum for control of the mosquito *Culex quinquefasciatus* in Thailand. *World J Microbiol Biotechnol* 30: 727–36.

McMichael, A. J., 2015. Extreme weather events and infectious disease outbreaks. *Virulence* 6: 543–7.

Mulla, M. S., Rodcharoen, J., Ngamsuk, W., Tawatsin, A., Pan-Urai, P., Thavara, U., 1997. Field trials with *Bacillus sphaericus* formulations against polluted water mosquitoes in a suburban area of Bangkok, Thailand. *J Am Mosq Control Assoc* 13(4): 297–304.

Mulla, M. S., Thavara, U., Tawatsin, A., Chomposri, J., Su, T., 2003. Emergence of resistance and resistance management in field populations of tropical *Culex quinquefasciatus* to the microbial control agent *Bacillus sphaericus*. *J Am Mosq Control Assoc* 19: 39–46.

Mulla, M. S., Su, T., Thavara, U., Tawatsin, A., Ngamsuk, W., Pan-Urai, P., 1999. Efficacy of new formulations of the microbial larvicide *Bacillus sphaericus* against polluted water mosquitoes in Thailand. *J Vector Ecol* 24: 99–110.

Mustafa, M. S., Rastogi, V., Gupta, R. K., Jain, S., Singh, P. M., Gupta, A., 2016. *Wolbachia*: The selfish Trojan Horse in dengue control. *Med J Armed Forces India* 72: 373–6.

Pantuwatana, S., Maneeroj, R., Upatham, E. S., 1989. Long residual activity of *Bacillus sphaericus* 1593 against *Culex quinquefasciatus* larvae in artificial pools. Southeast *Asian J Trop Med Public Health* 20: 421–7.

Phuanukoonnon, S., Mueller, I., Bryan J. H., 2005. Effectiveness of dengue control practices in household water containers in Northeast Thailand. *Trop Med Int Health* 10: 755–63.

Pitasawat, B., Champakaew, D., Choochote, W., Jitpakdi, A., Chaithong, U., Kanjanapothi, D., Rattanachanpichai, E., Tippawangkosol, P., Riyong, D., Tuetun, B., Chaiyasit, D., 2007. Aromatic plant-derived essential oil: An alternative larvicide for mosquito control. *Fitoterapia* 7: 205–10.

Sanprasert, V., Sujariyakul, A., Nuchprayoon S., 2010. A single dose of doxycycline in combination with diethylcarbamazine for treatment of bancroftian filariasis. *Southeast Asian J Trop Med Public Health* 41: 800–12.

Sedda, L., Morley, D. W., Braks, M. A., De Simone, L., Benz, D., Rogers, D. J., 2014. Risk assessment of vector-borne diseases for public health governance. *Public Health* 128: 1049–58.

Sinkins, S.P., 2004. *Wolbachia* and cytoplasmic incompatibility in mosquitoes. *Insect Biochem Mol Biol* 34: 723–9.

Sungpradit, S., Chatsuwan, T., Nuchprayoon, S., 2012. Susceptibility of *Wolbachia*, an endosymbiont of Brugia malayi microfilariae, to doxycycline determined by quantitative PCR assay. *Southeast Asian J Trop Med Public Health* 43: 841–50.

Swaddiwudhipong, W., Chaovakiratipong, C., Nguntra, P., Koonchote, S., Khumklam, P., Lerdlukanavonge, P., 1992. Effect of health education on community participation in control of dengue hemorrhagic fever in an urban area of Thailand. *Southeast Asian J Trop Med Public Health* 23: 200–6.

Tusting, L. S., Thwing, J., Sinclair, D., Fillinger, U., Gimnig, J., Bonner, K. E., Bottomley, C., Lindsay, S. W., 2013. Mosquito larval source management for controlling malaria. *Cochrane Database Syst Rev* 8: CD008923. doi: 10.1002/14651858.CD008923.pub2.

Walshe, D. P., Garner, P., Abdel-Hameed Adeel, A. A., Pyke, G. H., Burkot, T., 2013. Larvivorous fish for preventing malaria transmission. *Cochrane Database Syst Rev* 12: CD008090.

Weiser, J., 1975. Significant recent advances in biological control of vector insects. *Adv Vet Sci Comp Med* 19: 47–72.

Wiwanitkit, V., 2001. Wolbachia infection in filarial nematodes, new trend in the study on macrofilaricides. *Songklanagarind Med J* 19: 235–9.

Wiwanitkit, V., 2007. Vaccination against mosquito borne viral infections: Current status. *Iran J Immunol* 4: 186–96.

Wiwanitkit, V., 2009. Vaccination for tropical mosquito borne encephalitis. *Acta Neurol Taiwan* 18: 60–3.

# 11 Biocontrol Potential of Symbiotic Bacterium *Wolbachia* against Mosquito-Borne Diseases

*Saikat Mondal and Bhaskar Biswas*

## CONTENTS



## 11.1 INTRODUCTION

Mosquito-borne diseases are responsible for significant human morbidity and mortality. It acts as a vector of a broad range of harmful parasitic and viral diseases both in humans and other animals. Mosquitoes are the causative agents of various types of diseases, such as, malaria, dengue, filariasis, chikungunya, and West Nile fever.

Rapid urbanization, human travel, and failures of preventative measures result in rapid enhancement in annual incidences of mosquito-borne diseases including dengue fever, yellow fever, Japanese encephalitis, West Nile virus, chikungunya, and lymphatic filariasis (Adams and Kapan, 2009; Chen et al., 2010; Gould et al., 2008; Gubler, 2002). Malaria, caused by infection of the protozoan parasite *Plasmodium*, results in an estimated 1 million to 2 million deaths per year, led to a dramatic impact on health and socioeconomic development in affected areas (World Health Organization [WHO], 2008). Dengue is the most important and fatal arboviral disease in humans; 40% of the population of the world in more than 100 countries is at risk of infection and an estimated 50 million to 100 million cases occur annually (Guzman and Kouri, 2002; Kyle and Harris, 2008; WHO, 2009). Current control and preventive methods against mosquito-borne (vector) diseases depends mainly on insecticides. The use of insecticides to target mosquitoes as a means of disease control can be effective, but is often prohibitively expensive, unsustainable, and environmentally undesirable. Furthermore, repeated exposure of mosquitoes to insecticides has allowed the development of insecticide resistance in mosquitoes. The unjudicial use of chemical insecticide has led to bioaccumulation, biomagnification, and environmental degradation in the ecosystems. The emergence of insecticide resistance in many disease vectors highlights the necessity to develop new and environment-friendly strategies to control these insects.

*Wolbachia* is a intracellular gram-negative endosymbiotic bacterium found in arthropods and nematodes, and is estimated to infect more than 65% of insect species and a large number of other taxa, including freshwater crustaceans, arachnids, and nematodes (Hilgenboecker et al., 2008). *Wolbachia* is able to induce a number of reproductive modifications that helps its spread and maintenance in natural populations. One of the most reproductive manipulations induced by *Wolbachia* is cytoplasmic incompatibility (CI), which has received a great and considerable attention to control insect vectors and diseases.

Beside CI, the presence of *Wolbachia* in insects has been reported to cause numerous other phenotypic changes in reproduction, including feminization, male killing, and parthenogenesis. All these phenotypes benefit the spread and maintenance of these endosymbiotic bacteria. *Wolbachia* strains can invade and sustain themselves in mosquito populations and reduce the mosquito adult life span, affect mosquito reproduction, and interfere with pathogen replication and interfere with parasites and viruses. Recent studies revealed that *Wolbachia* is able to interact with a wide range of pathogens in transfected mosquitoes including malarian parasite, dengue, and chikungunya viruses (Brelsfoard and Dobson, 2011).

A recent study stated that *Plasmodium falciparum* development in *Anopheles gambiae* is suppressed transiently as a result of *Wolbachia* infection. This reproductive parasite is known to indirectly support and upregulate the insect-host immune system and suppress the pathogen (Pinto et al., 2012) from the release of relatively small numbers of individuals. *Wolbachia* provides a biological method to manipulate mosquito populations and reduce disease transmission (Iturbe et al., 2011). A lot of extensive research and recent findings have open a new arena to control diseases that are transmitted mainly by mosquitoes. In this respect *Wolbachia* has the benefit of being more environment-friendly than insecticide-based approaches and provides a novel strategy for mosquito control.

## 11.2 BRIEF HISTORY ABOUT WOLCHIA

The intracellular bacteria was first reported within the reproductive tissues of the mosquito *Culex pipiens* by Hertig and Wolbach in 1924, and these rickettsiae were subsequently named *Wolbachia pipientis.* In the mid-20th century, Ghelelovitch (1952) and Laven (1967) found that there were some incompatibility within *Culex* mosquitoes when certain intraspecific crosses were done. Laven established that the incompatibility factor had a cytoplasmic inheritance pattern (i.e., inheritance through females only) and named this phenomenon cytoplasmic incompatibility. Since then, many different strains of *Wolbachia* have been discovered in a wide range of invertebrate hosts, including many filarial nematode and arthropod species.

## 11.3 TAXONOMY AND PHYLOGENY OF THE *WOLBACHIA*

During their 100-million-year interaction with their hosts, the obligate intracellular bacteria *Wolbachia* has evolved as a reproductive parasite that belong to the class Alphaproteobacteria, order Rickettsiales, and family Rickettsiaceae. It is considered as most common endosymbiotic bacteria that infect millions of insect species from all around the globe (Hilgenboecker et al., 2008; Jeyaprakash and Hoy, 2000). The closest bacteria to the *Wolbachia* is a group of rickettsiae that include *Ehrlichia equii, Ehrlichia canis, Cowdria ruminata*, and *Anaplasma marginale*.

Moreover, some recent literature reported that infectious strains of *Wolbachia* can further be classified into three more supergroups, A–K super orders (Bordenstein et al., 2009; Ros et al., 2009; Zhou et al., 1998) as shown in Table 11.1.

**TABLE 11.1**
***Wolbachia* Infections into 11 Supergroups**

| Supergroup | Host Name | Reference |
|---|---|---|
| A, B, E | Arthropods | Vandekerckhove et al., 1999; Lo et al., 2002, 2007 |
| C, D | Filarial nematodes | Lo et al., 2007 |
| F | Both arthropods and nematodes | Campbell et al., 1992; Rasgon and Scott, 2004 |
| G | Spiders | Goodacre et al., 2006 |
| H | Termites | Bordenstein et al., 2009 |
| I | Siphonaptera | Ros et al., 2009 |
| J | Spirurida, Nematoda | Ros et al., 2009 |
| K | Prostigmata, Acarina | Ros et al., 2009 |

## 11.4 CYTOPLASMIC INCOMPATIBILITY

Cytoplasmic incompatibility (CI) is the most abundant *Wolbachia*-induced phenomena that alters the host's reproduction. CI is a mating incompatibility induced by *Wolbachia*. It is called cytoplasmic because *Wolbachia* is transmitted from the mother through the cytoplasm of the egg to the offspring. There are two basic forms: unidirectional and bidirectional. If a *Wolbachia*-infected male crosses with an uninfected female, this causes unidirectional CI. When both the male and female carry opposite strains of contrastive to each other are commonly known as bidirectional CI (Telschow et al., 2005) (Figure 11.1).

Unidirectional CI is most frequent and usually occurs between males infected with a single strain or female without any strain of *Wolbachia*. While the bidirectional CI have been reported only in the condition when both partners are infected with the same *Wolbachia* strains. Incompatibility apparently involves a two-component system; bacterial "modification" of sperm and a bacterial "rescue" in the fertilized egg. According to this model, bacteria present in the testes modify the developing sperm (possibly via chromatin binding proteins). The same bacterial strain must then be present in the egg to rescue this modification. If rescue does not occur, then incompatibility occurs between the egg and sperm.

### 11.4.1 Outcome of Cytoplasmic Incompatibility

In CI embryos, the first mitotic division following fertilization is severely disrupted. Cytogenetic mechanisms of CI have been studied in *C. pipiens, Drosophila simulans*, and *Nasonia vitripennis*. In all cases, CI is associated with early mitotic defects in the fertilized egg. The first cytological studies to describe this phenotype were performed in the parasitic wasp, *Nasonia vitripennis*, reporting paternal "tangled chromatin" next to the properly condensed female chromosomes during the metaphase of the first mitotic division (Ryan and Saul, 1968). CI embryos also exhibit abnormal centrosome behavior. Immediately following fertilization, the sperm basal body transforms into a bonafide centrosome (Ghelelovitch, 1952). The nascent centrosome duplicates and during pronuclear apposition, the daughter centrosomes migrate away from one another to form the poles of the first mitotic spindle. In *D. simulans* CI embryos, the cytoplasm contains a large excess of centrosomes unassociated with the maternal and

| | **Mating** | | | **Progeny** |
|---|---|---|---|---|
| Uninfected female | x | *Wolbachia*-infected male | = | x |
| *Wolbachia*-infected female | x | Uninfected male | = | *Wolbachia* infected |
| Uninfected female | x | Uninfected male | = | Uninfected |
| *Wolbachia*-infected female | x | *Wolbachia*-infected male | = | *Wolbachia* infected |

**FIGURE 11.1** *Wolbachia*-induced cytoplasmic incompatibility in mosquitoes. No progeny is born when a *Wolbachia*-infected male mosquito mates with an uninfected female mosquito. *Wolbachia*-infected females produce infected progeny in all matings allowing the infection to rapidly spread through mosquito populations.

paternal pronuclei (Callaini et al., 1997). Excess centrosomes have also been found in embryos lacking specific mitotic kinases (Renault et al., 2003), suggesting that *Wolbachia*-induced CI may disrupt mitotic kinases that coordinate the cytoplasmic-driven centrosome replication cycle with the nuclear-driven mitotic cycle.

In mosquitoes, *Wolbachia* strains can induce CI, which results in the generation of unviable offspring when an uninfected female mates with a *Wolbachia*-infected male (Werren et al., 2008). In contrast, *Wolbachia*-infected females can produce viable progeny when they mate with both infected and uninfected males, resulting in a reproductive advantage over uninfected females. The CI phenotype allows the maternally transmitted *Wolbachia* to efficiently invade host populations without being infectious or moving horizontally between individuals (Werren et al., 2008). CI can also result when mating occurs between mosquitoes with two different incompatible *Wolbachia* strains. This CI can occur bidirectionally if crosses of both males and females with each different *Wolbachia* strain are incompatible, resulting in embryonic death in both crosses (Min and Benzer, 1997).

## 11.5 TRANSINFECTION OF *WOLBACHIA* INTO MOSQUITOES

Stable infection (transinfection) to target mosquito species is the first criteria for potential use of *Wolbachia*-mediated mosquito biocontrol. *Stegomyia aegypti*, the principle vector of dengue virus (DENV) was the first target species that was selected for *Wolbachia* infection. Although *Wolbachia* infections are relatively common in mosquitoes (Kittayapong et al., 2000; Ricci et al., 2002) including *Culex pipiens* (Yen and Barr, 1973), *C. quinquefasciatus*, *Aedes fluviatilis* (Moreira et al., 2009a), and *Ae. albopictus* (Sinkins et al., 1995), the main vectors for dengue fever (*Ae. aegypti*) and malaria (*Anopheles* spp.) are not naturally infected by *Wolbachia*. Approaches that use *Wolbachia* for the control of diseases transmitted by uninfected, naive insects rely on the successful establishment of stable *Wolbachia* infections, usually by embryonic microinjection of *Wolbachia*-infected cytoplasm or *Wolbachia* purified from infected insect hosts. To create stable transinfected lines, embryo injections must target the region near the pole cells in preblastoderm embryos to incorporate *Wolbachia* into the developing germline and favor the transmission of *Wolbachia* to offspring. Several *Wolbachia* strains have been transferred across sometimes phylogenetically distant insects (Table 11.2) and, importantly, the phenotypes induced by these strains in their native hosts are generally also expressed in the newly infected hosts. Progress in transferring *Wolbachia* between mosquitoes species face intense hindrance due to the challenging nature of mosquito microinjection. The transfer of *Wolbachia* into mosquito embryos is intrinsically more difficult in comparison to the transfer into *Drosophila*, because mosquito embryos are less responsive to inoculation, especially *Anopheles* embryos. Several factors such as the age at the time of injection and desiccation level are especially critical for both injectability and survival of *Anopheles* embryos.

*Wolbachia* transinfection experiments are more likely to be successful when the donor and recipient organisms are closely related. In line with this, the transfer of *w*MelPop from its natural host, *Drosophila melanogaster*, into the dengue fever vector *Ae. aegypti* was achieved in laboratory after *Wolbachia* was first maintained by

**TABLE 11.2**
***Wolbachia* Induced Phenotypes in the Transinfected Mosquitoes**

| Strain | Original Host | Transinfected Host | Phenotype in New Host | Reference |
|---|---|---|---|---|
| *w*Mel | *Drosophila melanogaster*, then *Aedes albopictus* cell line | *Aedes aegypti* | CI, DENV interference | Walker et al., 2011 |
| *w*MelPop-CLA | *Drosophila melanogaster*, then *Aedes albopictus* cell line | *Aedes aegypti* | CI, life shortening, blood-feeding alteration, bendy proboscis, *Plasmodium*, DENV and CHIKV interference, increased metabolism, increased activity, inhibition of filarial nematodes | Evans et al., 2009; Kambris et al., 2009; McMeniman et al., 2009; Moreira et al., 2009a,b; Turley et al., 2009 |
| *w*AlbB | *Aedes albopictus* | *Aedes aegypti* | CI, DENV interference | Bian et al., 2010; Xi et al., 2005 |
| *w*AlbA, B | *Aedes albopictus* | *Aedes aegypti* | Partial CI | Ruang-Areerate and Kittayapong, 2006 |
| *w*Ri | *Drosophila simulans* | *Aedes albopictus* | CI | Xi et al., 2008 |
| *w*MelPop | *Drosophila melanogaster* | *Aedes albopictus* | CI, life shortening, embryo mortality | Suh et al., 2009 |
| *w*Pip | *Culex pipiens* | *Aedes albopictus* | CI, lower hatch rate, reduced fecundity | Calvitti et al., 2010 |

continuous passage in *Ae. albopictus in vitro* cell culture for almost 4 years (McMeniman et al., 2008). *Wolbachia* adapted to a mosquito intracellular environment, facilitating transfection *in vivo*. After microinjection of thousands of *Ae. aegypti* embryos, two stable *w*MelPop-CLA (cell-line-adapted) lines with maternal transmission rates of approximately 100% were generated (McMeniman et al., 2009). *w*MelPop-CLA-infected mosquitoes showed an approximately 50% reduction in adult life span, compared with their uninfected counterparts (McMeniman et al., 2009). The halving of adult mosquito life span and the high *Wolbachia* maternal transmission rates were also maintained in more genetically diverse outbred mosquitoes, and larval nutrition did not affect the life-shortening ability of the *w*MelPop-CLA strain (Yeap et al., 2010). The *w*MelPop-CLA infection is widespread in *Ae. aegypti* tissues, with high bacterial densities in the head (brain and ommatidia), thorax (salivary glands, muscle), and abdomen (fat tissue, reproductive tissues, and Malpighian tubules) (Moreira et al., 2009a).

Wide distribution across tissues has been found in other transinfected mosquitoes, such as *Ae. aegypti* infected with the *w*AlbB strain from *Ae. albopictus* (Bian et al., 2010). By using quantitative polymerase chain reaction (PCR) and Western blot analyses, this strain was also found in reproductive tissues, midgut, muscles, and heads in both native *Ae. albopictus* (Dobson et al., 1999) and the transfected *Ae. aegypti* (Bian et al., 2010), although the densities are lower than those found in *w*MelPop-CLA-infected *Ae. aegypti.*

## 11.6 PATHOGENIC INTERFERENCE

A key element in the use of *Wolbachia* for the control of mosquito-borne disease is its ability to induce pathogenic interference. The presence of *Wolbachia* interferes with a wider range of pathogens in transinfected mosquitoes including nematodes and bacteria (Kambris et al., 2009), viruses such as DENV and chikungunya virus (CHIKV) (Bian et al., 2010; Moreira et al., 2009a), as well as the avian and rodent malaria parasites *Plasmodium gallinaceum* (Moreira et al., 2009a) and *P. berghei* (Kambris et al., 2010). Natural *Wolbachia* strains that infect mosquitoes have also been shown to induce resistance to viruses, as in *C. quinquefasciatus* mosquitoes that are resistant to West Nile virus (Glaser and Meola, 2010).

Recent study suggests that pathogen interference is partly mediated by the induction of antimicrobial peptides and preactivation of the innate immune response in the insect (Kambris et al., 2009, 2010). The presence of *w*MelPop-CLA *Wolbachia* in *Ae. aegypti* induced the expression of several immune effector molecules, including cecropin, defensin, thioester-containing proteins, and C-type lectins, diptericin, GNBPB1, SPZ1A, and cactus (Moreira et al., 2009a). Whole-genome arrays showed that the presence of *w*MelPop in mosquitoes upregulated the host immune gene expression, resulting in inhibition of pathogen replication or multiplication. The upregulated six immune genes observed in *w*MelPop-infected *An. gambiae* are LRIM1, TEP1, CEC1, DEF1, CTL4, and CLIPB3 (Moreira et al., 2009a). *w*MelPop strain was transiently injected into adult *An. gambiae* and several immune genes were upregulated, (Kambris et al., 2009), resulting in the inhibition of *Plasmodium* development (Kambris et al., 2010). Some of these genes, in particular *LRIM1* and *TEP1*, have been found to inhibit *Plasmodium* development by interfering with the opsonization pathway (Povelones et al., 2009). Kokoza and colleagues (2010), who showed that the coexpression of two antimicrobial peptides (cecropin A and defensin A) in transgenic *Ae. aegypti* induces resistance to infection with the bacterial pathogen *Pseudomonas aeruginosa*, and reduces the number of *P. gallinaceum* oocysts, completely blocking transmission of this avian malaria parasite to naive chickens. Furthermore, *Wolbachia* infections in *Ae. aegypti* and *Drosophila* have been shown to increase hemolymph melanization (Thomas et al., 2010), a key constituent of the insect innate immune system that is involved in the encapsulation of foreign bodies and parasites (Carton and Nappi, 1997; Theopold et al., 2004).

Pathogen interference mediated by *Wolbachia* is mainly due to an upregulation of the host immune system. In mosquito cell lines, *Wolbachia* infection provides protection against DENV, even though these cells lack tissue-specific immunity (Frentiu et al., 2010). Similarly, the *Wolbachia* strain *w*Ri does not induce an immune response in its native *D. simulans* hosts (Bourtzis et al., 2000), and yet confers protection against *Drosophila* C virus (DCV) in these flies (Osborne et al., 2009). However, *w*Ri induces an immune response in transinfected *D. melanogaster* cell lines (Xi et al., 2008), and mosquitoes artificially infected with *Wolbachia* also show upregulation of the immune system (Kambris et al., 2009, 2010; Moreira et al., 2009a). Given that immune response to *Wolbachia* infection has been observed only in hosts artificially infected with new strains, such responses might not be due to the *Wolbachia* infection, but instead to an unusual effect of the new host–symbiont combination, such as elevated, unnatural *Wolbachia* densities (McGraw et al., 2002).

Pathogenic interference mediated by *Wolbachia* has some important features:

1. The strength of inhibition depends on the density of *Wolbachia.*
2. Pathogenic interference can only occur locally, meaning that *Wolbachia* has to be present in the same tissue in which the pathogen resides in order to induce an antipathogen effect.
3. *Wolbachia* mediated pathogenic interference is broad-spectrum, with resistance reported against dengue, chikungunya, yellow fever, West Nile virus, plasmodium parasites, and filarial worms.
4. *Wolbachia*-mediated pathogen mosquitoes to be released into the field.

## 11.7 POTENTIAL USE OF *WOLBACHIA* FOR ARBOVIRAL DISEASES

There are more than 80 different arboviruses that cause human diseases that are mostly transmitted by several species of Culicine mosquitoes. Most human cases result from infection with dengue virus (DENV) and other closely related Flaviviruses. The genus Flavivirus also includes West Nile virus (WNV), yellow fever virus (YFV), Zika virus (ZIKV), and Japanese encephalitis virus (JEV) (Table 11.3).

*Wolbachia*-mediated antiviral protection was initially found in *Drosophila melanogaster* (Hedges et al., 2008; Teixeira et al., 2008). *Wolbachia*-infected flies were protected from virus infection compared to paired groups of flies cured of their *Wolbachia* infection. In the presence of *Wolbachia* there is a significant delay in the mortality induced by the pathogenic RNA viruses *Drosophila* C virus (DCV; Dicistroviridae), Cricket paralysis virus (CrPV; Dicistroviridae), and Flock House virus (FHV; Nodaviridae) (Hedges et al., 2008; Teixeira et al., 2008).

Several studies indicate that the life-shortening strain of *Wolbachia w*Melpop has the ability to protect against RNA virus infection by means of delaying the mortality of flies infected with a range of pathogenic viruses in Drosophila. When this strain of *Wolbachia* was introduced artificially into the dengue virus mosquito vector *Ae. aegypti*, there was remarkable reduction in the proliferation of dengue

**TABLE 11.3**
**Arboviruses and Their Common Mosquito vectors**

| Examples of Arboviruses | Common Vectors | Reference |
|---|---|---|
| Dengue virus | *Aedes aegypti*, *Aedes albopictus* | Ros et al., 2009; Weaver and Reisen, 2010 |
| Japanese encephalitis virus | *Culex* spp. | |
| St. Louis encephalitis Virus | *Culex* spp. | |
| West Nile virus | *Culex* spp. | |
| Yellow fever virus | *Aedes* spp. | |
| Chikungunya virus | *Aedes albopictus*, *Aedes aegypti* | Ros et al., 2009; Weaver and Reisen, 2010 |
| O'nyong nyong virus | *Anopheles* spp. | |
| Semliki Forest virus | *Aedes* spp. | |
| Venezuelan equine encephalitis virus | *Aedes* spp., *Culex* spp. | |
| La Crosse virus | *Aedes triseriatus* | Ros et al., 2009; Weaver and Reisen, 2010 |
| Rift Valley fever virus | *Aedes* spp., *Culex* spp. | |

virus and also reduced adult life span, compared with uninfected mosquito population (Min and Benzer, 1997; Osborne et al., 2009).

*Wolbachia*-infected *Ae. aegypti* are also likely to have the added benefit of reducing transmission of additional arboviruses vectored by this mosquito species, including CHIKV (Moreira et al., 2009a) and YFV (Van den Hurk et al., 2012) and potentially ZIKV (Table 11.4). JEV is predominantly transmitted by *Cx. tritaeniorhynchus* mosquitoes, and a *Wolbachia*-based biocontrol strategy has the potential to reduce transmission if stably infected lines can be generated (Jeffries and Walker, 2015). JEV is part of the same genus as DENV (Flavivirus), so *Wolbachia* strains would likely provide similar inhibitory effects on transinfected mosquitoes. *Wolbachia* can inhibit virus transmission to new the host by altering the extrinsic incubation period of DENV. *Wolbachia*-mediated viral blockage and the extant of inhibition depend on the nature of *Wolbachia* strain and viruses. In Drosophila, *Wolbachia* only inhibit RNA viruses. DNA viruses are not affected by *Wolbachia* in the host cell (Teixeira et al., 2008).

Protective *Wolbachia* are found at higher density than nonprotective strains in *D. simulans* and protection is lost if the *Wolbachia* density is experimentally decreased (Osborne et al., 2012). *Wolbachia* density tends to be positively correlated with the strength of antiviral protection both in *Drosophila* and transinfected *St. aegypti*.

**TABLE 11.4**
**Protection from Virus in Mosquitoes Naturally or Artificially Infected with *Wolbachia***

| Host Species | Mode of *Wolbachia* Infection | *Wolbachia* Strain | Virus | Antiviral Effect | Reference |
|---|---|---|---|---|---|
| *Culex quinquefasciatus* | Natural | *w*Pip | WNV | Reduced virus load and transmission | Glaser and Meola, 2010 |
| *Culex pipiens* | Natural | Not typed | WNV | No effect | Micieli and Glaser, 2014 |
| *Culex tarsalis* | Transient transinfection | *w*AlbB | WNV | Enhanced infection rate | Dodson et al., 2014 |
| *Aedes albopictus* | Natural | *w*AlbA and *w*AlbB | DENV | No effect | Lu et al., 2012 |
| | Natural | *w*AlbA and *w*AlbB | DENV | No effect on virus load, reduced dissemination | Jaenike, 2009 |
| | | | CHIKV | No effect | Suh et al., 2009 |
| | Introgressed | *w*AlbA and *w*AlbB | CHIKV | No effect | Blagrove et al., 2013 |
| | Stable transinfection | *w*Mel | DENV | Reduced transmission | Blagrove et al., 2012 |
| | | | CHIKV | Reduced transmission | Blagrove et al., 2013 |
| *Aedes aegypti* | Stable transinfection | *w*MelPop | DENV | Reduced infection rate, virus load, and transmission | Moreira et al., 2009a |
| | | | CHKV | Reduced infection rate, virus load | Moreira et al., 2009a |
| | | | WNV | Reduced infection rate, virus load, and transmission | Hussain et al., 2013 |
| | | | YFV | Reduced infection rate and virus load | Van de Hurk et al., 2012 |
| | | *w*Mel | DENV | Reduced virus load, dissemination and transmission | Goodacre et al., 2006 |
| | | | CHKV | Reduced virus load and transmission | Van den Hurk et al., 2012 |
| | | | WNV | Delayed virus accumulation, reduced transmission | Hussain et al., 2013 |
| | | | YFV | Reduced virus load | Van den Hurk et al., 2012 |
| | | *w*AlbB | DENV | Reduced infection rate, virus load, and transmission | Bian et al., 2010 |

In mosquitoes and mosquito cells, *Wolbachia* density has also been implicated in antiviral effects (Bian et al., 2010; Frentiu et al., 2010; Lu et al., 2012). In general, resident *Wolbachia* strains in wild populations do not appear to grow to such high densities as transinfected strains and therefore have less impact on arboviral vector competence. For example, a high-density *w*Pip strain in a laboratory colonized line of *Cx. quinquefasciatus* was observed to show resistance to WNV, compared to a *w*Pip cleared line (Blagrove et al., 2012). Recent studies indicate that the removal of resident *Wolbachia* strains from *Stegomyia albopicta*, followed by transinfection of the *Drosophila w*Mel strain highlights strong inhibition of both DENV and CHIKV (Blagrove et al., 2012). Therefore, the transinfection of mosquito species with *Drosophila* and *Wolbachia* strains would likely demonstrate strong inhibitory effects on arbovirus transmission.

The importance of *Wolbachia* density to the outcome of infection may be at a cellular or tissue level. In addition to localization in the germline tissues, *Wolbachia* has been identified in a number of somatic tissue types across insect species (Bordenstein and Wernegreen, 2009; Dobson et al., 1999; Frydman et al., 2006). Both host and *Wolbachia* factors influenced *Wolbachia* distribution (Jaenike, 2009) and density can vary across tissues within a host (Osborne et al., 2012). There is some evidence to suggest that the outcome of infection is determined by the presence and density of *Wolbachia* in cells/tissues that are the site of arbovirus replication (Frentiu et al., 2010; Moreira et al., 2009a).

The role of *Wolbachia* density in antiviral effects suggests there may be competition for resources between the virus, host, and *Wolbachia* or remodeling of the host cell environment, and there is some evidence for this competition (Hilgenboecker et al., 2008; Osborne et al., 2009; Rainey et al., 2014). Another potential mechanism for antiviral effects is immune stimulation by the presence *of Wolbachia*; in particular this has been implicated in mosquitoes transinfected with *Wolbachia* (Bian et al., 2010; Kambris et al., 2010).

Recent studies revealed that reactive oxygen species (ROS)-dependent activation of the Toll pathway induced by *Wolbachia* was able to control dengue virus in *Ae. aegypti*. Microarray and real-time PCR analyses revealed that when *Ae. aegypti* was infected with the *Wolbachia w*AlbB strain, the Toll pathway and antioxidant genes were induced in the mosquito. This bacterium elevated the expression of NADPH oxidase and of ROS. In turn, ROS upregulation resulted in activation of the Toll pathway, which mediated both the antioxidant and production of two antimicrobial peptides: cecropins and defensins. So *Wolbachia* is able to boost immunity and can block dengue virus proliferation in a newly acquired host (Pan et al., 2012). In addition, a role for microRNAs in controlling *Wolbachia*-mediated antiviral effects has also been proposed (Asgari, 2014, 2015).

## 11.8 WOLBACHIA BACTERIA IN FILARIAL IMMUNITY AND DISEASE

Endosymbiotic *Wolbachia* bacteria play an important role in the biology of filarial nematodes. Phylogenetic analysis and the effects of antibiotic therapy on embryogenesis, development and viability show that *Wolbachia* appears to have evolved an essential mutualistic association with its filarial hosts (Taylor et al., 2000). The presence of large numbers of endosymbionts throughout all stages of the pathogenic filariae of humans suggest that the host will be exposed to *Wolbachia* following death of the parasite or through the release of bacterial products.

Recent study suggests that *Wolbachia* initiates acute inflammatory responses associated with death of parasites, leading to acute filarial lymphangitis and adverse reactions to antifilarial chemotherapy. They infect Lymphatic filarial nematodes with endosymbiotic *Wolbachia* bacteria and find that lipopolysaccharide (LPS) from these bacteria is the major activator of innate inflammatory responses induced directly by the parasite (Taylor et al., 2001).

## 11.9 EFFECT OF *WOLBACHIA* ON *PLASMODIUM*

Several studies present evidence that *Wolbachia* is likely to provide some protection against human malaria *Plasmodium* parasites if stable transinfection of *Anopheles* is achieved (Table 11.5). The effect of the life shortening *Wolbachia* strain *w*MelPop-CLA on *P. gallinaceum* was tested as this species of malaria parasite is known to be able to infect *Ae. Aegypti* mosquitoes in the laboratory. The *P. gallinaceum* oocyst load was reduced by *w*MelPop-CLA-infected *Ae. aegypti* mosquitoes by triggering the coexpression of two antimicrobial peptides ceropin A and defensin A, compared to *Wolbachia*-uninfected mosquitoes 7 days after feeding on an infected chicken (Kokoza et al., 2010; Moreira et al., 2009a). In *An. gambiae* females transiently infected with *w*MelPop using adult injection, mean *Plasmodium berghei* levels were reduced by 75% to 84% (Kambris et al., 2010). When *w*MelPop transinfects into *An. gambiae*, several immune genes are activated, resulting in inhibition of *Plasmodium* parasite development (Kambris et al., 2009). Several recent studies in *Drosophila* revealed that the immune genes involved in inhibition of *Plasmodium* are LRIM1 and TEP1, and they are supposed to inhibit the human malarial parasite *Plasmodium*, if stable infection is transmitted vertically to *Anopheles* (Hughes et al., 2011). The density and tissue distribution of *Wolbachia* infections in mosquitoes is an important factor in inhibiting *Plasmodium* (Jin et al., 2009). Although this combination of vector–parasite does not occur in nature, these results do highlight the ability of *Wolbachia* to significantly reduce the levels of malaria parasites in *Anopheles* mosquitoes.

**TABLE 11.5**
**Pathogen Protection in the *Wolbachia* Induced Mosquitoes**

| *Wolbachia* Strain | Mosquito Species | Pathogen Inhibited |
|---|---|---|
| *w*MelPop-CLA | *Anopheles gambie* | *Plasmodium berghei* (Kambris et al., 2010) |
| | | *Plasmodium falciparum* (Hughes et al., 2011) |
| *w*MelPop-CLA | *Aedes aegypti* | *Plasmodium gallinaceum* (Moreira et al., 2009a) |

## 11.10 GENOMIC INSIGHTS INTO *WOLBACHIA*–HOST INTERACTIONS

The molecular mechanisms behind the reproductive phenotypes and pathogen interference of *Wolbachia* remain mostly unknown, but recent progress on of wolbachian genomics is able to inform about molecular interactions between *Wolbachia* and its host. The entire genomes of arthropod and nematode-infecting *Wolbachia* strains have been sequenced, and alleles of selected genes have been sequenced from a large number of *Wolbachia* strains.

*Wolbachia* have genome sizes ranging from 1.2 to 1.6 Mb and 0.9 to 1.1 Mb (Foster et al., 2005). Particular effort has been devoted to elucidating the mechanisms for CI, the main driving force for *Wolbachia* population invasion in transinfected mosquitoes. Among the effectors proposed to induce CI in *Drosophila* and mosquitoes, ankyrin repeat proteins, WO phage, and *Wolbachia* surface protein (WSP) have received special attention.

### 11.10.1 Ankyrin Repeat Proteins

The *wMel* genome contains 23 proteins with ankyrin repeats (Ank proteins) (Wu et al., 2004). Eight of these proteins are encoded in phage regions. A similarly high number of Ank proteins was found in other arthropod-infecting *Wolbachia* strains: 34 in *wAna* (Fenn and Blaxter, 2006) and 54 in *wPip* (Duron et al., 2007). The symbiotic *Wolbachia* strain of the nematode *Brugia, wBm*, has only 9 such proteins (Foster et al., 2005). This correlation suggests that Ank proteins are instrumental in the parasitic lifestyle of arthropod-borne *Wolbachia.*

Ank proteins are of great interest in the analysis of host–pathogen relationships because they are involved in a diversity of protein–protein interactions (Mosavi et al., 2004). Most Ank proteins are found in eukaryotes, but they occur in prokaryotes as well. Viruses and bacteria are thought to have acquired them by horizontal gene transfer (Bork, 1993). ANK repeats are found in proteins involved in cell-cycle regulation, in transcriptional regulators, toxins, and cyclin-dependent kinase inhibitors (Al-Khodor et al., 2010; Bork, 1993). Since *Wolbachia*-induced CI causes an asynchrony in Cdk1 activation in female and male pronuclei (Tram and Sullivan, 2002), one intriguing idea is that *Wolbachia* Ank proteins may influence cell-cycle timing during CI expression.

### 11.10.2 WO Phage

*Wolbachia* strains that infect arthropods contain a varying number of WO prophages (Masui et al., 2000) in contrast to the strictly mutualistic *Wolbachia* strains in filarial nematodes that do not contain phage DNA. The sequenced *w*Mel and and *w*Riverside *Wolbachia* strains from *Drosophila melanogaster* and *D. simulans* harbor two divergent prophage families, the active WO-B and the dormant WO-A. Approximately 90% of all *Wolbachia* strains examined to date carry WO-B bacteriophages (Bordenstein et al., 2004).

The WO *Wolbachia* bacteriophage, which infects up to 90% of insect *Wolbachia* strains (Bordenstein and Wernegreen, 2004; Gavotte et al., 2007) has also been suggested to induce CI in insects, particularly as viral filtrates obtained from *Wolbachia* extracts are able to induce CI in *Nasonia* wasps (Williams et al., 1993). A series of WO-encoded genes, such as the virulence factor *VrlC*, have been implicated in *Wolbachia* pathogenicity (Kent and Bordenstein, 2010), due to the correlation between sequence variation in these genes and the induction of CI in *C. pipiens* mosquitoes (Duron et al., 2006). However, no correlation was found between the presence of phage capsid genes and the induction of CI in *Culex* mosquitoes (Gavotte et al., 2007).

### 11.10.3 *Wolbachia* Surface Protein

Wsp (*Wolbachia* surface protein), a low molecular weight protein of approximately 22 kDa, was originally identified as an abundant *Wolbachia*-derived protein present in fly ovaries and testis. Wsp contains a conserved gram-negative secretion signal sequence. Sequence analysis reveals Wsp exists as a single-copy gene and shares homology with *Rickettsiae* outer-membrane proteins (Braig et al., 1998).

Wsp may function to inhibit apoptosis of host cells. Polymorphonuclear (PMN) cells in mammals are the first responders against invading bacteria and act through bacterial phagocytosis. Many bacteria defend themselves against this innate immune response by inhibiting apoptosis of the phagocytosing host cells (Power et al., 2004). Wsp also has been shown to function as a ligand of TLR2, a receptor that controls apoptosis of PMNs and other types of host cells (Francois et al., 2005).

## 11.11 *WOLBACHIA* AND HOST MEMBRANE

Like other obligate intracellular bacteria, *Wolbachia* reside within a membrane compartment (Avakyan and Popov, 1984). Several electron microscopies from *Drosophila* embryos show *Wolbachia* to be surrounded by three membrane layers, the outermost layer probably derived from membraneous organelles of the host cell (Dobson, 2003; O'Neill et al., 1997; Wright and Barr, 1981).

In some cases, *Wolbachia* appear to be surrounded by multiple layers of host membrane but which host organelle or organelles serve as the source of this membrane is not clear yet. *Wolbachia* face some interesting challenges, such as targeting of membrane-encased bacteria to degradative lysosomal compartments and retaining the ability to access food, replicate at a sufficient rate, and maintain a presence in the germline while encased within a host-derived membrane compartment. The first challenge is overcome by changing the complement of protein markers associated with their surrounding host membranes such that they do not precisely resemble any endogenous host organelle. The other challenges are overcome by interactions between host proteins and *Wolbachia*-derived outer membrane proteins. Fusion with host organelles would provide *Wolbachia*-encompassing compartments with infusions of nutrients as well as new membrane to coat the *Wolbachia* as they replicate within the compartment. It has been also suggested that host proteins and/or *Wolbachia* factors secreted to the exterior surface of the membrane compartment probably help *Wolbachia* interaction with microtubule motor proteins. This would be inherently useful to *Wolbachia*, as their association with microtubules and microtubule motors appears strongly linked to their maternal transmission.

## 11.12 *WOLBACHIA* INVASION OF MOSQUITO POPULATIONS

Successful *Wolbachia*-based biocontrol would also require invasion of wild mosquito populations. *Wolbachia*-infected females must vertically transmit the bacteria to their progeny at a high frequency. All three transinfected *Wolbachia* strains (*w*AlbB, *w*Mel, and *w*MelPop) show maternal transmission rates close to 100% and induce high levels of CI in *St. aegypti* (Xi et al., 2005). The ability of transinfected Wolbachia to successfully invade wild *St. aegypti* mosquito populations will depend on a balance between negative selection imposed by fitness costs of the bacteria on the mosquito and positive selection associated with CI induction. The *w*MelPop strain results in significant fitness costs (including impacts on adult longevity and fecundity), so it was considered inappropriate for initial test releases into wild mosquito populations. Invasion will also depend on different fitness attributes such as the ability of females carrying the *Wolbachia* strain to feed and locate breeding sites, and the ability of larvae with the *Wolbachia* strain to compete against other conspecific larvae and other species.

## 11.13 FITNESS OF *WOLBACHIA*-INFECTED MOSQUITOES

Fitness cost of infection should be lesser than the fitness advantage, provided by the CI for spreading of *Wolbachia*-infected mosquitoes and invasion. Fitness effects induced by the *w*MelPop-CLA infection in *Ae. aegypti* include an increase of metabolic rate and activity in the mosquito (Evans et al., 2009), reduction in life span, and a fecundity cost. Another significant effect of *w*MelPop-CLA infection in *Ae. aegypti* is the reduction of egg survival during periods of embryonic quiescence (McMeniman and O'Neill, 2010).

*w*MelPop-CLA-infected mosquitoes that feed on human host has been tasted regarding their different abilities such as the volume of blood they have ingested, their ability to probe successfully, and other aspects of their biting behavior (Moreira et al., 2009b; Turley et al., 2009). *Wolbachia* does not affect the response time of mosquitoes to humans, but it reduces the number and size of blood meals taken by the mosquito. *w*MelPop-CLA *Wolbachia* also is able to induce behavioral changes in old mosquitoes, termed "shaky" or "bendy," in which the proboscis bends and is unable to penetrate the skin; 65% of 35-day-old insects showed the bendy phenotype (Turley et al., 2009). *Wolbachia*-infected *Ae. aegypti* produce smaller volumes of saliva, which contain the same levels of the anti-platelet-aggregation enzyme apyrase as uninfected mosquitoes (Moreira et al., 2009b). The impact of *Wolbachia* on the immature stage has been reported in various hosts (McMeniman and O'Neill, 2010). In *Aedes* mosquitoes, a negative impact of *Wolbachia* infection on larval survival and development time has been observed (McMeniman et al., 2008).

In the parasitic filarial nematode species responsible for elephantiasis, such as *Brugia malayi* and *Wuchereria bancrofti*, *Wolbachia* has become an obligate endosymbiont and provides the host with chemicals necessary to its reproduction and survival (Foster et al., 2005). Elimination of the *Wolbachia* symbionts through antibiotic treatment therefore prevents reproduction of the nematode and eventually results in its premature death.

## 11.14 CONCLUSIONS AND PERSPECTIVES

Previously it was believed that *Wolbachia* species were maternally inheritable symbionts, that is *Wolbachia* cause infection by means of vertical transfer (from egg cytoplasm to offsprings) only. However, recent study highlights that *Wolbachia* could also cause infection through horizontal transfer from infected uninfected species. It can also be cultured outside the cell (Rasgon et al., 2006) that have great potential to transfer infection and live outside the host cell (culture cell) over a reasonable period of time (Dobson et al., 2003). This discovery of *in vitro* production of *Wolbachia* opens various strategies to control the mosquito vector of various fatal human diseases, including Dengue, chikungunya, and *Plasmodium* (Moreira et al., 2009a).

Due to its prevalent phenotypic behavior, diverse host range, and potential for rapid host speciation and *Wolbachia* has a great ability to alter host embryonic progress and mitotic developments. *Wolbachia* becomes a novel and promising natural microbiocontrol agent to induce cytoplasmic incompatibility, killing of male embryos, parthenogenesis, and feminization. It also has autonomous transgenic capability, survives on several host species, and the ability to modify the host reproductive systems. Theoretical and experimental analysis show that it can act as a microbial natural enemy with multiple ways of infectious phenotypes to control the targeted pest population and can be efficiently used as a biocontrol agent in pest management strategies.

*Wolbachia* has been placed at the frontline of new approaches targeting mosquito-borne diseases in an environmentally friendly manner due to the ability of some *Wolbachia* strains to invade a mosquito population through the induction of CI, interfere with pathogen replication, and also reduce the life span of the mosquito. The release of *Wolbachia*-infected, dengue-refractory mosquitoes in Australia in 2011 will create the basis for the implementation of this strategy in Vietnam, Thailand, and possibly other countries in which the magnitude of dengue fever is higher, and the ecology and breeding habits of mosquitoes—particularly in large urban areas—make control challenging. This *Wolbachia*-based controlling approach could be applied to invasive mosquito species, such as *Ae. albopictus*, which has recently become a secondary dengue fever vector in Asia and has rapidly spread in the United States and large parts of Europe and Africa (Knudsen et al., 1996; Lambrechts et al., 2010). *Ae. albopictus* is also a vector for CHIKV in these countries (Bonilauri et al., 2008; Pages et al., 2009).

The main focus for the *Wolbachia*-based biocontrol approach is stable trans-infection and invasion. The major goal is to transfer *Wolbachia* into anopheline mosquitoes, the most common vectors of human malaria. The ability to maintain *Wolbachia* in *Anopheles* cell lines (McMeniman et al., 2008; Rasgon et al., 2006) and the successful establishment of transient somatic *w*MelPop infections in *Anopheles* mosquitoes (Jin et al., 2009; Kambris et al., 2010) activates mosquito immune response and therefore inhibit development of *Plasmodium* (Kambris et al., 2010) are promising advances that suggest stable transinfection might be possible. To summarize, a *Wolbachia*-based biocontrol approach will provide environmentally benign and effective long-term control options for mosquito-borne diseases and also highlights a novel strategy for controlling and manipulating mosquito populations and reducing transmission of various diseases that are responsible for human health hazards and burdens.

## LIST OF ABBREVIATIONS

| | |
|---|---|
| **CHIKV** | chikungunya virus |
| **CI** | cytoplasmic incompatibility |
| **CLA** | cell-line adapted |
| **DCV** | *Drosophila* C virus |
| **DENV** | dengue virus |
| **LRIM1** | leucine-rich repeat immune protein |
| **TEP1** | thioester-containing protein |

## REFERENCES

Adams B, Kapan DD. 2009. Man bites mosquito: Understanding the contribution of human movement to vector-borne disease dynamics. *PLoS ONE.* 4: e6763.

Al-Khodor S, Price CT, Kalia A, Abu Kwaik Y. 2010. Functional diversity of ankyrin repeats in microbial proteins. *Trends Microbiol.* 18: 132–139.

Asgari S. 2014. Role of microRNAs in arbovirus/vector interactions. *Viruses.* 6: 3514–3534.

Asgari S. 2015. Regulatory role of cellular and viral microRNAs in insect-virus interactions. *Curr Opin Insect Sci.* 8: 104–110.

Avakyan AA, Popov VL. 1984. Rickettsiaceae and Chlamydiaceae: Comparative electron microscopic studies. *Acta Virol.* 28: 159–173.

Bian G, Xu Y, Lu P, Xie Y, Xi Z. 2010. The endosymbiotic bacterium *Wolbachia* induces resistance to dengue virus in *Aedes aegypti. PLoS Pathog.* 6: e1000833.

Blagrove MS, Arias-Goeta C, Failloux AB, Sinkins SP. 2012. *Wolbachia* strain *w*Mel induces cytoplasmic incompatibility and blocks dengue transmission in *Aedes albopictus. Proc Natl Acad Sci USA.* 109: 255–260.

Blagrove MS, Arias-Goeta C, Di Genua C, Failloux AB, Sinkins SP. 2013. A *Wolbachia* wMel transinfection in *Aedes albopictus* is not detrimental to host fitness and inhibits Chikungunya virus. *PloS Negl Trop Dis.* 27: e2152.

Bonilauri P, Bellini R, Calzolari M, Angelini R, Venturi L et al. 2008. Chikungunya virus in *Aedes albopictus*, Italy. *Emerg Infect Dis.* 14: 852–854.

Bordenstein SR, Paraskevopoulos C, Hotopp JC, Sapountzis P, Lo N et al. 2009. Parasitism and mutualism in *Wolbachia*: What the phylogenomic trees can and cannot say. *Mol Biol Evol.* 26: 231–241.

Bordenstein SR, Wernegreen JJ. 2004. Bacteriophage flux in endosymbionts (*Wolbachia*): Infection frequency, lateral transfer, and recombination rates. *Mol Biol Evol.* 21: 1981–1991.

Bork P. 1993. Hundreds of ankyrin-like repeats in functionally diverse proteins: Mobile modules that cross phyla horizontally? *Proteins.* 17: 363–374.

Bourtzis K, Pettigrew MM, O'Neill SL. 2000. *Wolbachia* neither induces nor suppresses transcripts encoding antimicrobial peptides. *Insect Mol Biol.* 9: 635–639.

Braig HR, Zhou W, Dobson SL, O'Neill SL. 1998. Cloning and characterization of a gene encoding the major surface protein of the bacterial endosymbiont *Wolbachia pipientis*. *J Bacteriol.* 180: 2373–2378.

Brelsfoard CL, Dobson SL. 2011. An update on the utility of *Wolbachia* for controlling insect vectors and disease transmission. *Asian Pac J Mol Biol Biotechnol.* 19(3): 85–92.

Callaini G, Dallai R, Riparbelli MG. 1997. *Wolbachia*-induced delay of paternal chromatin condensation does not prevent maternal chromosomes from entering anaphase in incompatible crosses of *Drosophila simulans*. *J Cell Sci.* 110(Pt 2): 271–280.

Calvitti M, Moretti R, Lampazzi E, Bellini R, Dobson SL. 2010. Characterization of a new *Aedes albopictus* (Diptera: Culicidae)-*Wolbachia pipientis* (Rickettsiales: *Rickettsiaceae*) symbiotic association generated by artificial transfer of the *w*Pip strain from *Culex pipiens* (Diptera: *Culicidae*). *J Med Entomol.* 47: 179–187.

Campbell BC, Bragg TS, Turner CE. 1992. Phylogeny of symbiotic bacteria of four weevil species (Coleoptera: Curculionidae) based on analysis of 16S ribosomal DNA. *Insect Biochem Mol Biol.* 22: 415–421.

Carton Y, Nappi AJ. 1997. *Drosophila* cellular immunity against parasitoids. *Parasitol Today.* 13: 218–227.

Chen LH, Wilson ME. 2010. Dengue and chikungunya infections in travelers. *Curr Opin Infect Dis.* 23: 438–444.

Dodson BL, Hughes GL, Paul O, Matacchiero AC, Kramer LD et al. 2014. *Wolbachia* enhances West Nile virus (WNV) infection in the mosquito *Culex tarsalis*. *PLoS Negl Trop Dis.* 8: e2965.

Dobson SL, Bourtzis K, Braig HR, Jones BF, Zhou W et al. 1999. *Wolbachia* infections are distributed throughout insect somatic and germ line tissues. *Insect Biochem Mol Biol.* 29: 153–160.

Dobson SL. 2003. *Wolbachia pipientis*: Impotent by association. In *Insect Symbiosis*, eds. K Bourtzis, TA Miller, pp. 199–215. New York: CRC Press.

Duron O, Bernard C, Unal S, Berthomieu A, Berticat C et al. 2006. Tracking factors modulating cytoplasmic incompatibilities in the mosquito *Culex pipiens*. *Mol Ecol.* 15: 3061–3071.

Duron O, Fort P, Weill M. 2007. Influence of aging on cytoplasmic incompatibility, sperm modification and *Wolbachia* density in *Culex pipiens* mosquitoes. *Heredity.* 98: 368–74.

Evans O, Caragata EP, McMeniman CJ, Woolfit M, Green DC et al. 2009. Increased locomotor activity and metabolism of *Aedes aegypti* infected with a life-shortening strain of *Wolbachia pipientis*. *J Exp Biol.* 212: 1436–1441.

Fenn K, Blaxter M. 2006. *Wolbachia* genomes: Revealing the biology of parasitism and mutualism. *Trends Parasitol.* 22: 60–65.

Foster J, Ganatra M, Kamal I, Ware J, Makarova K et al. 2005. The *Wolbachia* genome of *Brugia malayi*: Endosymbiont evolution within a human pathogenic nematode. *PLoS Biol.* 3: e121.

Francois S, El Benna J, Dang PM, Pedruzzi E, Gougerot-Pocidalo MA et al. 2005. Inhibition of neutrophil apoptosis by TLR agonists in whole blood: Involvement of the phosphoinositide 3-kinase/Akt and NF-kappaB signaling pathways, leading to increased levels of Mcl-1, A1, and phosphorylated Bad. *J Immunol.* 174: 3633–3642.

Frentiu FD, Robinson J, Young PR, McGraw EA, O'Neill SL. 2010. *Wolbachia*-mediated resistance to dengue virus infection and death at the cellular level. *PLoS ONE.* 5: e13398.

Frydman HM, Li JM, Robson DN, Wieschaus E. 2006. Somatic stem cell niche tropism in Wolbachia. *Nature.* 441: 509–512.

Gavotte L, Henri H, Stouthamer R, Charif D, Charlat S et al. 2007. A survey of the bacteriophage WO in the endosymbiotic bacteria *Wolbachia*. *Mol Biol Evol.* 24: 427–435.

Ghelelovitch S. 1952. Genetic determinism of sterility in the cross-breeding of various strains of *Culex autogenicus* Roubaud. *C R Hebd Seances Acad Sci.* 234: 2386–2388.

Glaser RL, Meola MA. 2010. The native *Wolbachia* endosymbionts of *Drosophila melanogaster* and *Culex quinquefasciatus* increase host resistance to West Nile virus infection. *PLoS ONE.* 5: e11977.

Goodacre SL, Martin OY, Thomas CG, Hewitt GM. 2006. *Wolbachia* and other endosymbiont infections in spiders. *Mol Ecol.* 15: 517–527.

Gould EA, Solomon T. 2008. Pathogenic flaviviruses. *Lancet.* 371: 500–509.

Gubler DJ. 2002. Epidemic dengue/dengue hemorrhagic fever as a public health, social and economic problem in the 21st century. *Trends Microbiol.* 10: 100–103.

Guzman MG, Kouri G. 2002. Dengue: An update. *Lancet Infect Dis.* 2: 33–42.

Hedges LM, Brownlie JC, O'Neill SL, Johnson KN. 2008. *Wolbachia* and virus protection in insects. *Science.* 322(5905): 702.

Hertig M, Wolbach SB. 1924. Studies of *Rickettsia*-like microorganisms in insects. *J Med Res.* 44: 329–374.

Hilgenboecker K, Hammerstein P, Schlattmann P, Telschow A, Werren JH. 2008. How many species are infected with *Wolbachia*?—A statistical analysis of current data. *FEMS Microbiol Lett.* 281: 215–220.

Hughes GL, Koga R, Xue P, Fukatsu T, Rasgon JL. 2011. *Wolbachia* infections are virulent and inhibit the human malaria parasite *Plasmodium falciparum* in *Anopheles gambiae*. *PLoS Pathogens.* May; 7(5): e1002043.

Hussain M, Lu GJ, Torres S, Edmonds JH, Kay BH et al. 2013. Effect of *Wolbachia* on replication of West Nile Virus in a mosquito cell line and adult mosquitoes. *J Virol.* 87: 851–858.

Iturbe-Ormaetxe I, Woolfit M, Rances E, Duplouy A, O'Neill SL. 2011. A simple protocol to obtain highly pure *Wolbachia* endosymbiont DNA for genome sequencing. *J Microbiol Methods.* 84: 134–136.

Jaenike J. 2009. Coupled population dynamics of endosymbionts within and between hosts. *Oikos.* 118: 353–362.

Jeffries CL, Walker T. 2015. The potential use of *Wolbachia*-based mosquito biocontrol strategies for Japanese encephalitis. *PLoS Negl Trop Dis.* 9: e0003576.

Jeyaprakash A, Hoy MA. 2000. Long PCRimproves *Wolbachia* DNAamplification: Wsp sequences found in 76% of sixty-three arthropod species. *Insect Molec Biol.* 9: 393–405.

Jin CY, Ren XX, Rasgon JL. 2009. The virulent *Wolbachia* strain wMelPop efficiently establishes somatic infections in the malaria vector *Anopheles gambiae*. *Appl Environ Microbiol.* 75(10): 3373–3376.

Kambris Z, Blagborough AM, Pinto SB, Blagrove MS, Godfray HC et al. 2010. *Wolbachia* stimulates immune gene expression and inhibits plasmodium development in *Anopheles gambiae*. *PLoS Pathog.* 6: e1001143.

Kambris Z, Cook PE, Phuc HK, Sinkins SP. 2009. Immune activation by life-shortening *Wolbachia* and reduced filarial competence in mosquitoes. *Science.* 326: 134–136.

Kent BN, Bordenstein SR. 2010. Phage WO of *Wolbachia*: Lambda of the endosymbiont world. *Trends Microbiol.* 18: 173–181.

Kittayapong P, Baisley KJ, Baimai V, O'Neill SL. 2000. Distribution and diversity of *Wolbachia* infections in Southeast Asian mosquitoes (Diptera: Culicidae). *J Med Entomol.* 37: 340–345.

Knudsen AB, Romi R, Majori G. 1996. Occurrence and spread in Italy of *Aedes albopictus*, with implications for its introduction into other parts of Europe. *J Am Mosq Control Assoc.* 12: 177–183.

Kokoza V, Ahmed A, Woon Shin S, Okafor N, Zou Z et al. 2010. Blocking of *Plasmodium* transmission by cooperative action of Cecropin A and Defensin Ain transgenic *Aedes aegypti* mosquitoes. *Proc Natl Acad Sci USA.* 107: 8111–8116.

Kyle JL, Harris E. 2008. Global spread and persistence of dengue. *Annu Rev Microbiol.* 62: 71–92.

Lambrechts L, Scott TW, Gubler DJ. 2010. Consequences of the expanding global distribution of *Aedes albopictus* for dengue virus transmission. *PLoS Negl Trop Dis.* 4: e646.

Laven H. 1967. Eradication of *Culex pipiens* fatigans through cytoplasmic incompatibility. *Nature.* 216: 383–384.

Lo N, Casiraghi M, Salati E, Bazzocchi C, Bandi C. 2002. How many *Wolbachia* supergroups exist? *Mol Biol Evol.* 19: 341–346.

Lo N, Paraskevopoulos C, Bourtzis K, O'Neill S, Werren J et al. 2007. Taxonomic status of the intracellular bacterium *Wolbachia pipientis. Int J Syst Evol Microbiol.* 57: 654–657.

Lu P, Bian GW, Pan XL, Xi ZY. 2012. *Wolbachia* induces density-dependent inhibition to Dengue virus in mosquito cells. *PLoS Negl Trop Dis.* 6(7): e1754.

Masui S, Kamoda S, Sasaki T, Ishikawa H. 2000. Distribution and evolution of bacteriophage WO in *Wolbachia*, the endosymbiont causing sexual alterations in arthropods. *J Mol Evol.* 51: 491–497.

McGraw EA, Merritt DJ, Droller JN, O'Neill SL. 2002. *Wolbachia* density and virulence attenuation after transfer into a novel host. *Proc Natl Acad Sci USA.* 99: 2918–2923

McMeniman CJ, Lane RV, Cass BN, Fong AW, Sidhu M et al. 2009. Stable introduction of a life-shortening *Wolbachia* infection into the mosquito *Aedes aegypti. Science.* 323: 141–144.

McMeniman CJ, Lane AM, Fong AW, Voronin DA, Iturbe-Ormaetxe I et al. 2008. Host adaptation of a *Wolbachia* strain after long-term serial passage in mosquito cell lines. *Appl Environ Microbiol.* 74: 6963–6969.

McMeniman CJ, O'Neill SL. 2010. Avirulent *Wolbachia* infection decreases the viability of the dengue vector *Aedes aegypti* during periods of embryonic quiescence. *PLoS Negl Trop Dis.* 4: e74.

Micieli MV, Glaser RL. 2014. Somatic *Wolbachia* (Rickettsiales: Rickettsiaceae) levels in *Culex quinquefasciatus* and *Culex pipiens* (Diptera: Culicidae) and resistance to West Nile virus infection. *J Med Entomol.* 51: 189–199.

Min KT, Benzer S. 1997. Wolbachia, normally a symbionts of Drosophila, can be virulent, causing degeneration and early death. *Proc Natl Acad Sci USA.* 94(20): 10792–10796.

Moreira LA, Iturbe-Ormaetxe I, Jeffery JA, Lu GJ, Pyke AT et al. 2009a. A *Wolbachia* symbiont in Aedes aegypti limits infection with dengue, Chikungunya, and *Plasmodium. Cell.* 139: 1268–1278.

Moreira LA, Saig E, Turley AP, Ribeiro JM, O'Neill SL et al. 2009b. Human probing behavior of *Aedes aegypti* when infected with a life-shortening strain of *Wolbachia. PLoS Negl Trop Dis.* 3: e568.

Mosavi LK, Cammett TJ, Desrosiers DC, Peng ZY. 2004. The ankyrin repeat as molecular architecture for protein recognition. *Protein Sci.* 13: 1435–1448.

O'Neill SL, Pettigrew MM, Sinkins SP, Braig HR, Andreadis TG et al. 1997. In vitro cultivation of *Wolbachia pipientis* in an *Aedes albopictus* cell line. *Insect Mol Biol.* 6: 33–39.

Osborne SE, Iturbe-Ormaetxe I, Brownlie JC, O'Neill SL, Johnson KN. 2012. Antiviral protection and the importance of *Wolbachia* density and tissue tropism in *Drosophila simulans. Appl Environ Microbiol.* 78: 6922–6929.

Osborne SE, Leong YS, O'Neill SL, Johnson KN. 2009. Variation in antiviral protection mediated by different *Wolbachia* strains in *Drosophila simulans. PLoS Pathog.* 5: e1000656.

Pages F, Peyrefitte CN, Mve MT, Jarjaval F, Brisse S et al. 2009. *Aedes albopictus* mosquito: The main vector of the 2007 Chikungunya outbreak in Gabon. *PLoS ONE.* 4: e4.

Pan XL, Zhou GL, Wu JH, Bian GW, Lu P et al. 2012. *Wolbachia* induces reactive oxygen species (ROS)-dependent activation of the Toll pathway to control dengue virus in the mosquito *Aedes aegypti. Proc Natl Acad Sci USA.* 109: E23–E31.

Pinto SB, Mariconti M, Bazzocchi C, Bandi C, Sinkins SP. 2012. *Wolbachia* surface protein induces innate immune responses in mosquito cells. *BMC Microbiol.* 12 (Suppl 1): S11.

Power CP, Wang JH, Manning B, Kell MR, Aherne NJ et al. 2004. Bacterial lipoprotein delays apoptosis in human neutrophils through inhibition of caspase-3 activity: Regulatory roles for CD14 and TLR-2. *J Immunol.* 173: 5229–5237.

Povelones M, Waterhouse RM, Kafatos FC, Christophides GK. 2009. Leucine-rich repeat protein complex activates mosquito complement in defense against *Plasmodium* parasites. *Science.* 324: 258–261.

Rainey SM, Shah P, Kohl A, Dietrich I. 2014. Understanding the *Wolbachia*-mediated inhibition of arboviruses in mosquitoes: Progress and challenges. *J Gen Virol.* 95: 517–530.

Rasgon JL, Gamston CE, Ren X. 2006. Survival of *Wolbachia* pipientis in cell-free medium. *Appl Environ Microbiol.* 72: 6934–6937.

Rasgon JL, Scott TW. 2004. Phylogenetic characterization of *Wolbachia* symbionts infecting Cimex lectularius L. and *Oeciacus vicarius* (Hemiptera: Cimicidae). *J Med Entomol.* 41: 1175–1178.

Renault AD, Zhang XH, Alphey LS, Frenz LM, Glover DM et al. 2003. Giant nuclei is essential in the cell cycle transition from meiosis to mitosis. *Development.* 130: 2997–3005.

Ricci I, Cancrini G, Gabrielli S, D'Amelio S, Favi G. 2002. Searching for *Wolbachia* (Rickettsiales: *Rickettsiceae*) in mosquitoes (Diptera: *Culicidae*): Large polymerase chain reaction survey and new identifications. *J Med Entomol.* 39: 562–567.

Ros VI, Fleming VM, Feil EJ, Breeuwer JA. 2009. How diverse is the genus *Wolbachia*? Multiple-gene sequencing reveals a putatively new *Wolbachia* supergroup recovered from spider mites (Acari: Tetranychidae). *Appl Envir Microbiol.* 75: 1036–1043.

Ruang-Areerate T, Kittayapong P. 2006. *Wolbachia* transinfection in *Aedes aegypti*: A potential gene driver of dengue vectors. *Proc Natl Acad Sci USA.* 103: 12534–12539.

Ryan SL, Saul GB 2nd. 1968. Post-fertilization effect of incompatibility factors in Mormoniella. *Mol Gen Genet.* 103: 29–36.

Sinkins SP, Braig HR, O'Neill SL. 1995. *Wolbachia pipientis*: Bacterial density and unidirectional cytoplasmic incompatibility between infected populations of *Aedes albopictus*. *Exp Parasitol.* 81: 284–291.

Suh E, Mercer DR, Fu Y, Dobson SL. 2009. Pathogenicity of life-shortening *Wolbachia* in *Aedes albopictus* after transfer from *Drosophila melanogaster*. *Appl Environ Microbiol.* 75: 7783–7788.

Taylor MJ, Bandi C, Hoerauf AM et al. 2000. *Wolbachia* bacteria of filarial nematodes: A target for control? *Parasitol Today.* 16: 179–180.

Taylor MJ, Cross HF, Ford L, Makunde WH, Prasad GBKS et al. 2001. *Wolbachia* bacteria in filarial immunity and disease. *Parasite Immunol.* 23: 401–409.

Teixeira L, Ferreira A, Ashburner M. 2008. The bacterial symbiont *Wolbachia* induces resistance to RNA viral infections in *Drosophila melanogaster*. *PLoS Biol.* 6(12): 2.

Telschow A, Hammerstein P, Werren JH, Gavrilets S. 2005. The effect of *Wolbachia* Versus genetic incompatibilities on reinforcement and speciation. *Evolution.* 59: 1607–1619.

Theopold U, Schmidt O, Soderhall K, Dushay MS. 2004. Coagulation in arthropods: Defence, wound closure and healing. *Trends Immunol.* 25: 289–294.

Thomas P, Kenny N, Eyles D, Moreira LA, O'Neill SL et al. 2010 Infection with the *w*Mel and *w*MelPop strains of *Wolbachia* leads to higher levels of melanization in the hemolymph of *Drosophila melanogaster*, *Drosophila simulans* and *Aedes aegypti*. *Dev Comp Immunol.* 35: 360–365.

Tram U, Sullivan W. 2002. Role of delayed nuclear envelope breakdown and mitosis in *Wolbachia* induced cytoplasmic incompatibility. *Science.* 296: 1124–1126.

Turley AP, Moreira LA, O'Neill SL, McGraw EA. 2009. *Wolbachia* infection reduces blood-feeding success in the dengue fever mosquito, *Aedes aegypti. PLoS Negl Trop Dis.* 3: e516.

Van den Hurk AF, Hall-Mendelin S, Pyke AT, Frentiu FD, McElroy K et al. 2012. Impact of *Wolbachia* on infection with Chikungunya and Yellow Fever viruses in the mosquito vector *Aedes aegypti. PLoS Negl Trop Dis.* 6(11): e1892.

Vandekerckhove TT, Watteyne S, Willems A, Swings JG, Mertens J, Gillis M. 1999. Phylogenetic analysis of the 16S rDNA of the cytoplasmic bacterium *Wolbachia* from the novel host *Folsomia candida* (Hexapoda, Collembola) and its implications for *Wolbachial* taxonomy. *FEMS Microbiol Lett.* 180: 279–286.

Walker T, Johnson PH, Moreira LA, Iturbe-Ormaetxe I, Frentiu FD et al. 2011. The wMel *Wolbachia* strain blocks dengue and invades caged *Aedes aegypti* populations. *Nature.* 476: 450–453.

Weaver SC, Reisen WK. 2010. Present and future arboviral threats. *Antivir Res.* 85: 328–345.

Werren JH, Baldo L, Clark ME. 2008. *Wolbachia*: Master manipulators of invertebrate biology. *Nat Rev Microbiol.* 6: 741–751.

World Health Organization (WHO). 2008. *World malaria report*. Geneva, Switzerland: World Health Organization.

World Health Organization (WHO). 2009. *Dengue and dengue hemorrhagic fever. Fact sheet*, vol 117. Geneva, Switzerland: World Health Organization.

Williams EH, Fields S, Saul GB 2nd. 1993. Transfer of incompatibility factors between stocks of *Nasonia* (=Mormoniella) *vitripennis*. *J Invertebr Pathol.* 61: 206–210.

Wright JD, Barr AR. 1981. *Wolbachia* and the normal and incompatible eggs of *Aedes polynesiensis* (Diptera: Culicidae). *J Invertebr Pathol.* 38: 409–418.

Wu M, Sun LV, Vamathevan J, Riegler M, Deboy R et al. 2004. Phylogenomics of the reproductive parasite *Wolbachia pipientis wMel*: A streamlined genome overrun by mobile genetic elements. *PLoS Biol.* 2: e69.

Xi Z, Gavotte L, Xie Y, Dobson SL. 2008. Genome-wide analysis of the interaction between the endosymbiotic bacterium *Wolbachia* and its *Drosophila* host. *BMC Genomics.* 9: 1.

Xi Z, Khoo CC, Dobson SL. 2005. *Wolbachia* establishment and invasion in an *Aedes aegypti* laboratory population. *Science.* 14: 326–328.

Yeap HL, Mee P, Walker T, Weeks AR, O'Neill SL et al. 2010. Dynamics of the "popcorn" *Wolbachia* infection in outbred *Aedes aegypti* informs prospects for mosquito vector control. *Genetics.* 187: 583–595.

Yen JH, Barr AR. 1973. The etiological agent of cytoplasmic incompatibility in *Culex pipiens. J Invertebr Pathol.* 22: 242–250.

Zhou W, Rousset F, O'Neill S. 1998. Phylogeny and PCR–based classification of *Wolbachia* strains using wsp gene sequences. *Proc Royal Soc London B: Biol Sci.* 265: 509–515.

# 12 *Wolbachia*
## *An Effective Biocontrol Agent*

*Alwarappan Sankaranarayanan, Natarajan Amaresan, and Hardik Naik Jinal*

**CONTENTS**



## 12.1 INTRODUCTION

Vector-borne diseases are responsible for human morbidity and mortality, and economic loss of agricultural crops throughout the world (Brelsfoard and Dobson, 2009). A diversified group of vectors carry important pathogens responsible for different diseases in humans. As per a recent report, the important diseases caused by vector include malaria, dengue, chikungunya, yellow fever, leishmaniasis, and sleeping sickness (Saldana et al., 2017). Vector-borne disease control strategies mainly focus on eradicating the spreading of disease through control of vectors. However, insecticide-resistant vectors and inefficiency in the control of vectors through traditional practices has led to research for an effective control protocol especially through biological agents. The pandemic Zika virus, reemerging yellow fever virus in Africa, and *Leishmania* in Middle East highlight the need for alternate tools to control vector-borne diseases in an effective manner (Al-Salem et al., 2016; Barrett, 2016; Weaver et al., 2016).

*Wolbachia* is a gram-negative bacteria, more specifically, an obligate intracellular parasite, that belongs to the order of Rickettsiales. *Wolbachia* was first reported in the tissues of the mosquito *Cule pipiens* (Hertig and Wolbach, 1924), and the genus of bacteria was named Wolbach by Hertig in lieu of his research collaborator (Hertig, 1936). Based on 16S rDNA (partial small subunit ribosomal DNA, groEl (heat shock protein gene), wsp (outer surface coat protein gene) (Baldo et al., 2006; Paraskevopoulos et al., 2006), and ftsZ (cell division gene), *Wolbachia* have been primarily classified into eight supergroups; three more supergroups were recently added and renamed as supergroup A–K (Zhou et al., 1998; Bordenstein et al., 2009; Ros et al., 2009; Salunke et al., 2010; Ali et al., 2016). *Wolbachia* infects a wide range of hosts, especially arthropods including mosquitoes, ticks, flies, and nematodes. It can live inside of the host at various organs, especially ovaries and testes, and is involved in the alteration of the reproduction behavior of the host; hence it may be considered as a biological control (Sinkins et al., 1997). *Wolbachia* produce many reproductive ill effects on its host (Jayaprakash and Hoy, 2000; Werren and Windsor, 2000). Apart from the reproductive organs, *Wolbachia* invades hemolymph, glands, and nervous tissues (Huigens, 2003). Another advantage of *Wolbachia* is its representation in more than 16% of insect species belonging to major insect orders (Werren et al., 1995). In Canada, infections were observed in 47% of arthropod populations. Unfortunately, only 19% and 1% of peer-reviewed research articles concerning *Wolbachia* are available in journals specific to arthropod and biocontrol, respectively (Floate et al., 2006). This reveals a need for more concentration on *Wolbachia*-based biocontrol research studies.

## 12.2 *WOLBACHIA* AS A BIOCONTROL AGENT

Undoubtedly, *Wolbachia* is an important biocontrol agent. Basically, *Wolbachia* are obligate endosymbionts. There are four main reasons why *Wolbachia* sp. is considered as an effective biocontrol agent:

1. Alter the reproductive ability of the host (cytoplasmic incompatibility, CI)
2. Feminization
3. Parthenogenesis induction (PI)
4. Male killing (MK)

The details of these reasons will be briefly discussed later. Among all the biotic hosts, *Wolbachia* is predominant in arthropod vectors. Among the 105 species tested belonging to 10 orders, *Wolbachia* was found in 46% of arthropods (Floate et al., 2006) (Table 12.1). Other surveys reported 76% (Stevens et al., 2001) and 40% of the insect population (Zug and Hammerstein, 2012; Hilgenboecker et al., 2008). Among the arthropod group, *Wolbachia* sp. represented in insects, mites, spiders, and terrestrial isopods, as well as filarial nematodes (Werren, 1997; Zhou et al., 1998; Lo et al., 2002; Gotoh et al., 2003; Cordaux et al., 2004; Goodacre et al., 2006; Hilgenboecker et al., 2008; Werren et al., 2008).

Based on the outcome of earlier research, biologists strongly believed *Wolbachia* sp. vertically transferred from egg cytoplasm to offspring. But later, the possibility for horizontal transfer from infected to uninfected host was revealed (Kawai et al., 2009), cultured outside the cell (Rasgon et al., 2006), and live outside the host cell for a short time period (Dobson et al., 2002). It leads to a positive direction to control the insect vector of various human disease by *Wolbachia* sp. (Moreira et al., 2009). Another prominent feature for *Wolbachia* is the global distribution of the organism. Based on

**TABLE 12.1**
**Biocontrol Efficiency of *Wolbachia* in Arthropod Species**

| Order | Family | Species | Biocontrol Efficiency |
|---|---|---|---|
| Acari | Phytoseiidae | *Amblyseius cucumeris* | – |
| | | *Phytoseiulus persimilis* | – |
| | Tetranychidae | *Tetranychus urticae* | – |
| Coleoptera | Chrysomelidae | *Aphthona lacertosa* | – |
| | | *Aphthona nigriscutis* | – |
| | Coccinellidae | *Cryptolaemus montrouzieri* | – |
| | | *Galerucella calmariensis* | – |
| | Curculionidae | *Ceutorhynchus assimilis* | + |
| | | *Mecinus janthinus* | – |
| | | *Mogulones cruciger* | – |
| Hymenoptera | Encyrtidae | *Tachinophaegus zealandicus* | + |
| | Eucoilidae | *Kleiodotoma* sp. | + |
| | Ichneumonidae | *Phygadewon prob. Fumator* | – |
| | Megachilidae | *Megachile rotundata* | – |
| | Pteromalidae | *Muscidifurax raptor* | – |
| | | *Muscidifurax raptorellus* | – |
| | | *Muscidifurax uniraptor* | + |
| | | *Muscidifurax zaraptor* | – |
| | | *Nasonia vitripennis* | + |
| | | *Pachycrepoideus vindemiae* | + |
| | | *Spalangia cameroni* | + |
| | | *Splangia endius* | + |
| | Scolytidae | *Dendroctomus pondervsae* | – |
| | Tenebrionidae | *Tenebrio molitor* | – |
| | | *Triboliium castaneum* | – |
| Dictyoptera | Blaberidae | *Blaberus giganteus* | – |
| | | *Gromphadorlina portentosa* | – |
| Diptera | Caliphoridae | *Pollenia rudis* | + |
| | | *Protocalliphora sialia* | + |
| | | *Protophormia terraenovae* | – |
| | Muscidae | *Haematobia irritans* | + |
| | | *Hydrotaea aenescens* | – |
| | | *Musca autumnalis* | – |
| | | *Spalangia geminia* | + |
| | | *Splangia nigra* | + |
| | | *Splangia nigroaenea* | + |

(*Continued*)

## TABLE 12.1 (CONTINUED)
## Biocontrol Efficiency of *Wolbachia* in Arthropod Species

| Order | Family | Species | Biocontrol Efficiency |
|---|---|---|---|
| | | *Trichomalopsis sarcophagae* | + |
| | | *Trichomalopsis viridescens* | – |
| | | *Urolepsis rufipes* | + |
| | Trichogrammatidae | *Trichogrammabrassicae* | + |
| | | *Trichogramma cacoeciae* | + |
| | | *Trichogramma chilonis* | + |
| | | *Trichogramma leptoparameron* | + |
| | | *Trichogramma minutum* | + |
| | | *Trichogramma pintoi* | + |
| | | *Trichogramma platneri* | + |
| | | *Trichogramma sibericum* | + |
| Lepidoptera | Lasiocampidae | *Malacosoma disstria* | – |
| | | *Musca domestica* | – |
| | Lymantriidae | *Stomoxys calcitrans* | – |
| | Oestridae | *Hypoderma lineatum* | – |
| | Sarcophagidae | *Ravinia querula* | + |
| | | *Sarcophaga bullata* | – |
| | Scathophagidae | *Scatophaga stercoraria* | – |
| | | *Scatophaga furcata* | – |
| | Sepsidae | *Sepsis* sp. | – |
| Hemiptera | Cimicidae | *Cimex lectularius* | + |
| | Miridae | *Lygus borealis* | – |
| | | *Lygus elisus* | – |
| | | *Lygus keltoni* | – |
| | | *Lygus lineolaris* | – |
| | Pemphigidae | *Pemphigus betae* | – |
| | | *Pemphigus populicaulis* | – |
| | | *Pemphigus populiglobuli* | – |
| | | *Pemphigus populiramorulum* | – |
| | | *Lymantria dispar* | – |
| | | *Orgyia antique* | – |
| | | *Orgyia leucostigma* | – |
| | | *Orgyia pseudotsugata* | – |
| | Noctuidae | *Lambdina fiscellaria fiscellaria* | – |

(*Continued*)

**TABLE 12.1 (CONTINUED)**
**Biocontrol Efficiency of *Wolbachia* in Arthropod Species**

| Order | Family | Species | Biocontrol Efficiency |
|---|---|---|---|
| | | *Trichoplusia ni* | – |
| | Pyralidae | *Diorcyctria abietivorella* | – |
| | Saturniidae | *Bombyx mori* | – |
| | Tortricidae | *Choristoneura conflictana* | – |
| | | *Choristoneura fumiferana* | – |
| | | *Choristoneura occidentalis* | – |
| | | *Choristoneura pinus* | – |
| | | *Choristoneura rosaceana* | – |
| Orthoptera | Grylidae | *Gryllus bimaculatus* | + |
| | | *Gryllus pennsylvanicus* | + |
| | | *Gryllus rubens* | + |
| | | *Teleogryllus oceanicus* | + |
| Phthiraptera | Haematopinidae | *Haematopinus suis* | + |
| Hymenoptera | Braconidae | *Aphaereta pallipes* | + |
| | | *Glyptapanteles* sp. | + |
| | | *Microplitis mediator* | – |
| | | *Macrocentrus linearis* | – |
| | | *Peristenus digoneutus* | + |
| | | *Peristeneus stygicus* | + |
| | Cephidae | *Cephus cinctus* | – |
| | Chalcididae | *Brachymeria podagrica* | + |
| | | *Dirhinus himalayanus* | + |
| | Diapriidae | *Trichopria nigra* | + |
| | Diprionidae | *Gilpinia hercyniae* | + |
| | Linognathidae | *Linognathus africanus* | + |
| | | *Linognathus setosus* | + |
| | | *Linognathus vituli* | + |
| | Pediculidae | *Pediculus humanus capitis* | + |
| | | *Pediculus humanus humanus* | + |
| | | *Pthirus pubis* | + |
| | Polyplacidae | *Polypax serrata* | + |
| | Trichodectidae | *Bovicola bovis* | + |
| Siphonaptera | Pulicidae | *Ctenocephalides felis* | + |

*Source:* Data from Floate, K. D., Kyei-Poku, G. K., and Coghlin, P. C., 2006, Overview and relevance of *Wolbachia* bacteria in biocontrol research, *Biocontrol Sci. Technol.*, 16(8), 767–788.

*Note:* +, present; –, absent.

**TABLE 12.2**
***Wolbachia*-Based Research Work Carried Out with Various Hosts in Different Countries**

| Host | Country | Reference |
|---|---|---|
| Arthropod | Belgium | Vandekerckhove et al., 1999 |
| Filarial nematodes | Australia | Lo et al., 2007 |
| Spiders | England | Goodacre et al., 2006 |
| Termites | United States | Bordenstein et al., 2009 |
| Siphonaptera | Europe | Ros et al., 2009 |
| Spirurida | Europe | Ros et al., 2009 |
| Prostigmata | Europe | Ros et al., 2009 |
| Nematoda | Europe | Ros et al., 2009 |
| Acarina | Europe | Ros et al., 2009 |
| Coleoptera | United Kingdom | Majerus and Zakharov, 2000 |
| Diptera | United States | Dyer and Jaenike, 2004 |
| | Thailand and China | Wiwatanaratanabutr and Zhang, 2016 |
| | Malaysia | Afizah et al., 2015; Joanne et al., 2015 |
| | Brazil | Dutra et al., 2016 |
| Lepidoptera | Uganda | Jiggins et al., 1998 |
| | Uganda | Jiggins et al., 2000 |
| | Japan | Fuji et al., 2001 |
| | Uganda | Jiggins et al., 2001 |
| | Fiji | Dyson et al., 2002 |
| Arachnida | Republic of Panama | Zeh et al., 2005 |
| | Republic of Panama | Koop et al., 2009 |
| Copepod | Thailand | Wiwatanaratanabutr and Grandjean, 2016 |
| Cladocerans | Thailand | Wiwatanaratanabutr, 2013 |
| Sand flea | Brazil | Heukelbach et al., 2004 |
| Fruit fly | Austria and Italy | Zabalou et al., 2004 |

the existing literature and scientific reports, different aspects of research work on *Wolbachia* were reported in almost all the countries with a few exceptions. This clearly reveals that climatic and other parameters are not barriers for this organism. Details of the representation of *Wolbachia*-based research work carried out in various countries are presented in Table 12.2.

### 12.2.1 Cytoplasmic Incompatibility (CI)

Cytoplasmic incompatibility (CI) is a major tool used by *Wolbachia* in reproductive manipulation of insects. *Wolbachia* sp. lives in the cytoplasm of the host cell, entered through the cytoplasm of the maternal line (vertical transmission) and between insect species (through horizontal transmission) (Ahmed et al., 2015; Jiggins and Hurst, 2011;

O'Neill et al., 1992; Werren et al., 1995; Vavre et al., 1999). Cytoplasmic-mediated incompatibility is used to control viral infection spread by mosquitos. In an *in vitro* study, after transinfection of *Wolbachia* into *Aedes aegypti*, carrying the dengue virus (DENV), the dengue virus population was significantly reduced (Joubert et al., 2016). There are three types of CI based on the effect of *Wolbachia* on host. They are

1. Female mortality CI type
2. Male development CI type
3. Intermediate CI type

In the female mortality CI type, with the diploid species (Yenn and Bar, 1971) and haplo-diploid species (Breeuwer, 1997; Vavre et al., 2000) the fertilized eggs die or they may be developed into males (male development CI type) (Breeuwer and Werren, 1990; Giordano et al., 1997). The intermediate CI type was observed in Parasitoid wasp *Leptopillina heterotoma*. In this type, some fertilized eggs of the single female died; others developed into males (Vavre et al., 2001).

CI increases the *Wolbachia*-infected individual population. CI-related things were well documented in *Wolbachia*-infected hosts in many species (Serbus et al., 2008). CI not only controls the dengue virus in the *Ae. aegypti* population (Moreira et al., 2009; Bian et al., 2010; Walker et al., 2011), it also reduces the mosquito population (Laven, 1967; O'Connor et al., 2012).

If sperm modified by *Wolbachia* of infected males are not rescued by an uninfected egg, then incompatibility occurs. This is called unidirectional incompatibility. In bidirectional incompatibility, different strains of *Wolbachia* of an infected male and female, have a the different modified-rescue system (Figure 12.1) (Stouthamer et al., 1999). Several researchers have reported CI as the reason for high embryonic death and fewer progenies (Breeuwer and Werren, 1990; O'Neill and Karr, 1990; Bourtzis and O'Neill, 1998; Bourtzis and Miller, 2014). According to the findings of Tram and his research team (2003), a factor produced by *Wolbachia* in infected males hinders the embryogenesis in the fertilized egg, and the rescue occurs if the female partner is infected by the same strain of *Wolbachia*. These types of biological tools may be harnessed to control the insect populations (Tagami et al., 2006). In diploid species, abnormalities in mitosis and disruption in embryogenesis have been reported (Lassy and Karr, 1996).

Two biochemical models have been proposed: one is the infected male produces a product that disrupts the sperm and egg processing, and the other one is bacteria in a male acts as a "hurdle" for the normal processing of sperm with the egg (Kose and Karr, 1995; Reed and Werren, 1995; Lassy and Karr, 1996). The biochemical mechanism behind CI is still unclear (Ju et al., 2017); a histone binding protein may play a role (Werren, 1997). Behind the success of incompatibility are certain factors like bacterial strain and its density, and the host genotype also plays a role. To support this view, CI and non-CI-inducing strains have been reported (Giordano et al., 1995; Hoffman et al., 1996; Rousset et al., 1992).

However, recent research suspects that in *Laodelphax striatellus*, an important rice pest in the testes of male, the iLvE gene mediates branched chain amino acid (BCAA) in the reproductive processes. However further detailed study is needed

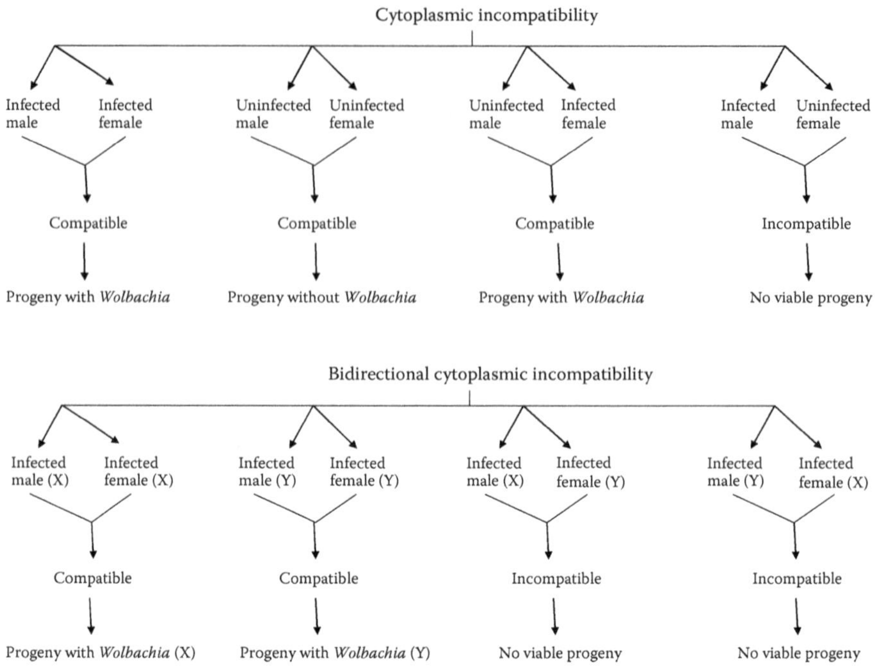


**FIGURE 12.1** Cytoplasmic incompatibility in host by *Wolbachia*.

(Ju et al., 2017). A recent study analyzed the proteome of a mosquito host cell response against *Wolbachia*. The proteins suppress the host DNA replication, transcription, and translation, suggesting that *Wolbachia* suppressed the host cell growth and proliferation (Baldridge et al., 2017).

In another point of view, the host genes JhI-26 and HIRA are involved in *Wolbachia*-mediated CI. However, the exact mechanism is not known in the sperm modification that makes the embryo nonviable (Zheng et al., 2011; Liu et al., 2014). The host methylation and chromatin structure (Tram and Sullivan, 2002; Tram et al., 2006; Landmann et al., 2009; Pinto et al., 2013) may be modified due to the presence of *Wolbachia* in leafhopper and mosquitoes responsible for the CI (Negri et al., 2009; Ye et al., 2013).

### 12.2.2 Feminization

Due to the presence of *Wolbachia*, sometimes the male host is forced to develop into a functional female; this is is called feminization, or alteration of sex. This is mainly due to hormonal abnormalities, especially in the androgenic gland by the presence of *Wolbachia* in the host (Correa and Ballard, 2016). The feminization concept was analyzed in the terrestrial isopod *Armadillidium vulgare* (Rousset et al., 1992; Cordaux et al., 2012). The importance of feminization was insisted by Negri (2012). Except for CI, the rest of the effect of *Wolbachia* on the host was interconnected with feminization. Feminization leads to an imbalanced sex ratio (Rigaud et al., 1999). Due to feminization, male killing, and parthenogenesis induction, the male–female ratio was highly disturbed in host biotic communities. The male individual became increasingly rare, raising questions of reproductive success (Charlat et al., 2003). In lepidopteran and hemipteran insects, feminization was induced by *Wolbachia* and spiders (Negri, 2012; Curry et al., 2015). *Eurema hecabe*, a butterfly, has demonstrated that the *Wolbachia* feminizing effect was observed during the larval developmental period (Narita et al., 2007). Hurst (1992) observed that feminization, male killing, and parthenogenesis were together responsible for the increasing host infection.

*Wolbachia* in the host inhibited the development of the androgenic gland responsible for the secretion of hormone and blocked the receptor site, which leads to the male converting into a functional female (Rigaud, 1997). The feminization was also observed in marine amphipod crustaceans, terrestrial isopod crustaceans, lepidopteran insects (Stouthamer et al., 1999) and two species of isopod (Juchalt et al., 1994). In *Wolbachia*-infected isopod crustaceans, females were more prevalent than males due to feminization reported by Bouchon and colleagues (1998). Androgenic gland induced the sexual differentiation due to the presence of *Wolbachia* sp. in *Armadilldium vulgare* (Martin et al., 1990). The androgenic hormone played an active role in the intersex determination and *Wolbachia* mediated the hormonal impact on the host. And in another view, feminization results in the loss of the male-determining chromosome and the presence of sex-suppressing genes that acts as the sex-determining mechanism (Werren, 1997). Next to CI, researchers on *Wolbachia* concentrated more on feminization (Rigaud et al., 1999; Rousset et al., 1992; Cordaux et al., 2012; Wiwatanaratanabutr, 2013).

### 12.2.3 Parthenogenesis Induction (PI)

*Wolbachia*-induced parthenogenesis induction (PI) (thelytoky) is not as common a phenomena as CI. Parthenogenesis, a manipulation of the host reproduction, causes the infected female to produce infected female offspring (Weeks and Breeuwer, 2001). This is due to changes in the first mitotic division (Stouthamer and Kazmer, 1994); the anaphase was aborted and hence a diploid nucleus formed from the unfertilized egg. *Wolbachia*-mediated PI has been reported in the Hymenoptera order (Zchori-Fein et al., 1992; Stouthamer, 1997; Plantard et al., 1998) and in a species of thrips (Arakaki et al., 2001). Stouthamer and colleagues (1999) reported two reasons in Hymenoptera for *Wolbachia*-induced PI: one on haplo-diploid sex determination and the other diploidy restored through gamete duplication. In *Wolbachia*-induced PI, the host species reproduced through haplo-diploid mode (Wrensch and Ebbert, 1993). Diplo-diploid sex determination was not vulnerable; the reason is unknown (Huigens, 2003). In PI-induced *Wolbachia*, both the vertical and horizontal transfer were observed, though vertical is main (Huigens et al., 2000). Concisely, males developed from haploid eggs and females from diploid eggs. Existing reports revealed two views on PI in Hymenoptera. First, *Wolbachia* infections interfere with the separation of chromosomes in the anaphase of the first mitotic division (Stouthamer and Kazmer, 1994; Pannebakker et al., 2004). In the Netherlands, six species belonging to the genus *Bryobia* (mites) were studied for PI (Weeks and Breeuwer, 2001). PI was observed in haplo-diploid insects infected with *Wolbachia*.

### 12.2.4 Male Killing (MK)

Male killing (MK) is the killing of males during embryonic development (Floate et al., 2006). Due to MK, the male–female sex ratio is distorted and leads to the biocontrol of host. MK of *Wolbachia* has been reported in several hosts, including the flour beetles, ladybird beetles, Nymphalid beetles, and *Drosophila* (Hurst et al., 1999, 2000; Fialho and Stevens, 2000). However, MK is not a common as common as CI in connection with biocontrol of hosts. Compared to *Wolbachia* supergroup A, B is highly involved in MK; the reason is unknown. Apart from *Wolbachia*, other bacteria in the genera *Rickettsia*, *Spiroplasma*, *Flavobacteria* *γ-Proteobacteria* are also involved in MK (Hackett et al., 1986; Werren et al., 1986, 1994; Hurst et al., 1996, 1997; Jiggins et al., 2000). It is a cytological-mediated cause; however, the detail mechanism is unexplored so far (Stevens et al., 2001). A research study concerning the interaction between *Wolbachia* and *Drosophila bifasciata* concluded that there is a cell-mediated cause behind MK. This research study revealed that during the early development of the male embryo in *D. bifasciata*, improper chromatid segregation, unusual chromosome bridging, abnormal mitotic spindle, and loss of their centrosome leads to complication in early nuclear division, causing defects in the cellular blastoderm leading to the death of male embryo (Riparbelli et al., 2012). In other research, a genetically mediated cause behind MK was found. The male killing *Spiroplasma* in *Drosophila willistoni* can target the genetic-based components of male-specific sex determination pathway (Veneti et al., 2005).

In *Acraea encedon*, the female-biased sex ratios were mediated through meiotic drive (Jiggins et al., 1998); however, it is curable by treating with antibiotics. The MK efficiency of the host by *Wolbachia* has been proven in several studies. In *A. encedon*, 86% of wild females had the efficiency of MK (Jiggins et al., 1998). However as per earlier reports, the *Wolbachia*-infected females can be cured with antibiotics (i.e., rifampicin and sulphamethoxazole) (Hurst et al., 1999). But in contradiction, in a research study carried out in Fiji, the antibiotic treatment failed to cure the female broods in the butterfly community *Hypolimnas bolina*. The reason is unknown; however, it may be due to antibiotic resistance of the *Wolbachia* or the host did not responded to the antibiotics (Dyson et al., 2002). However this study proved *Wolbachia* is responsible for MK in *H. bolina* in Fiji. The existing research report revealed the polymerase chain reaction (PCR) assay was used to ensure the presence of *Wolbachia* in MK female insects. With the aid of specific primers, based on *Wolbachia* 16S rDNA (Werren et al., 1995), the *fts*Z gene (Werren et al., 1995), and the *wsp* gene (Zhou et al., 1998), the appropriate size of the PCR products were amplified. In *A. encedon* and *Adalia bipunctata*, the presence of *Wolbachia* alone associated with the hosts was proven by PCR experiment, in which the 16S rDNA of eubacteria were amplified by specific primers. The results revealed the presence of *Wolbachia* in both hosts (Hurst et al., 1999).

## 12.3 RECENT DEVELOPMENTS IN *WOLBACHIA*-BASED RESEARCH

In a recent study conducted in China on *Trichogramma dendrolimi* (Hymenoptera: Trichogrammatidae), parasitoid wasps were widely used as a biocontrol against Asian corn borer *Ostrinia furnacalis* (Lepidoptera: Pyralidae). In this study *Wolbachia*-infected and -uninfected *T. dendrolimi* are compared against *O. Dendrolimi* at different temperatures. *Wolbachia* responded well to the temperature change (Dong et al., 2017). *Wolbachia* strains combat the dispersal of *Ae. aegypti*, a promising tool for dengue suppression in urban areas of Australia. This spatial study, analyzed with a statistical model, yielded a promising result (Schmidt et al., 2017). In a recent study and first report from Korea on *Wolbachia* revealed 99% of collected mosquito *Aedes albopictus* was found to be infected with *w*AlbA and *w*AlbB, and the *Wolbachia*-based WSP gene in these populations by PCR assay was identified (Park et al., 2016). In a recent study in the United States, high larval density populations infected with *Wolbachia* on *Ae. aegypti* reduced the populations more effectively than the low larval densities and adult populations of *Ae. aegypti* (Suh et al., 2017). In a recent research study conducted in Japan, *Wolbachia pipientisis*-infected flour moth *Ephestia kuehniella* suffers deleterious impacts on both life history and reproductive traits, hence *Wolbachia* may be considered as an effective pest control (Sumida et al., 2017). A recent study from Singapore reported a *w*Mel strain of *Wolbachia*-infected in *Ae. aegypti* reduced replication and dissemination of dengue virus, Zika virus, chikungunya virus, and yellow fever virus. Interestingly, there was no detectable chikungunya and Zika in the salivary glands of *Ae. aegypti* at 14 days postinfectious blood meal (Tan et al., 2017). A recent research work focused on the pathogen blocking and probable mechanism behind the *Wolbachia*-infected *Ae. aegypti* (Terradas and McGraw, 2017). This study revealed the immune priming and competition between symbiont and pathogen for host resources. Further, *Wolbachia* increases the immune gene expression and blocks the pathogens in the host. In addition to immune regulators, iron metabolism and reactive oxygen species (ROS) production may be behind the biocontrol activity. Researchers have also suggested double infection (two *Wolbachia* infecting one individual) may enhance pathogen blocking for a prolonged period.

## 12.4 CONCLUSION

Based on the various research reports in different perspectives, there is a strong belief that *Wolbachia* may be considered an effective biocontrol agent to control vector-mediated diseases. However, the exact mechanisms of CI, MK, PI, and feminization are still unknown based on the existing research study reports. However, once the unknown becomes known, *Wolbachia* can be used more efficiently as a biocontrol tool. Though, the recent research concepts concentrated on molecular-based mechanisms behind the interaction of *Wolbachia* and hosts, gene-mediate studies, genome analysis, amino acid biosynthesis concepts, pathogen-blocking mechanisms in the host, and PCR-based studies have yielded hopes to use *Wolbachia* as an effective biocontrol agent in the near future.

## ACKNOWLEDGMENTS

We gratefully acknowledge all the authorities of Uka Tarsadia University for their constant encouragement and support during writing of the article. We are also thankful to the editors for their help, support, and critical suggestions in the draft version of the article. We also thank Kevin D. Float for permission us to use the data from his published research article.

## REFERENCES


Afizah, N. A., Roziah, A., Nazni, W. A. and Lee, H. L. 2015. Detection of *Wolbachia* from field collected *Aedes albopictus* Skuse in Malaysia. *Indian J. Med. Res.*, 142, 205–210.

Ahmed, M. Z., Li, S. J., Xue, X., Yin, X. J., Ren, S. X., Jiggins, F. M., Greeff, J. M. and Qiu, B. L. 2015. The intracellular bacterium *Wolbachia* uses parasitoid wasps as phoretic vectors for efficient horizontal transmission. *PLoS Pathog.*, 11(2), e1004672. https://doi.org/10.1371/journal.ppat.1004672.

Ali, H., Hou, Y., Tang, B., Shi, Z., Huang, B., Muhammad, A. and N. B. Sanda 2016. A way of reproductive manipulation and Biology of *Wolbachia pipientis*. *J. Exper. Biol. Agr. Sci.*, 4(2), 156–168.

Al-Salem W. S., Pigott, D. M., Subramaniam, K., Haines, L. R., Kelly-Hope, L., Molyneux D. H., Hay, S. I. and Serrano, A. A. 2016. Cutaneous leishmaniasis and conflict in Syria. *Emerging Infect Dis.*, 16(22), 931–933. doi:http://dx.doi.org/10.3201/eid2205.160042.

Arakakki, N., Miyoshi, T., and Noda, H. 2001. *Wolbachia* mediated parthenogenesis in the predatory thrips *Franklintothrips vespiformis* (Thysanoptera: Insecta). *Proc. R. Soc. Lond. B*, 268, 1011–1016.

Baldo, L, Hotopp, J. C., Jolley, K. A., Bordenstein, S. R., Biber, S. A., Choudhury, R. R., Hayashi, C., Maiden, M. C., Tettelin, H. and Werren, J. H. 2006. Multilocus sequence typing system for the endosymbiont *Wolbachia pipientis*. *Appl. Environ. Microbiol.*, 72, 7098–7110.

Baldridge, G., Higgins, L., Witthuhn, B., Markowski, T., Baldridge, A., Armien, A. and Falloon, A., 2017. Proteomic analysis of a mosquito host cell response to persistent *Wolbachia* infection. *Res. Microbiol.*, in press, doi:10.1016/j.resmic.2017.04.005.

Barrett, A. D. T. 2016. Yellow fever in Angola and beyond—The problem of vaccine supply and demand. *N. Engl. J. Med.*, 375, 301–303.

Bian, G., Xu, Y., Lu, P., Xie, Y. and Xi, Z. 2010. The endosymbiotic bacterium Wolbachia induces resistance to dengue virus in *Aedes aegypti*. *PLoS Pathogens*, 6(4), e1000833 https://doi.org/10.1371/journal.ppat.1000833.

Bordenstein, S. R., Paraskevopoulos, C., Hotopp, J. C., Sapountzis, P., Lo, N., Bandi, C., Tettelin, H., Werren, J. H. and Bourtzis, K. 2009. Parasitism and mutualism in *Wolbachia*: What the phylogenomic trees can and cannot say. *Mol. Biol. Evol.*, 26, 231–241.

Bouchon, D., Rigaud, T. and Juchault, P. 1998. Evidence for widespread *Wolbachia* infection in isopod crustaceans: Molecular identification and host feminization. *Proc. R. Soc. London Ser. B*, 265, 1081–1090.

Bourtzis K. and Miller, T. A. (eds). 2006. *Insect symbiosis*. Vol. 2. Boca Raton, FL: CRC Press.

Bourtzis, K. and O'Neill, S. 1998. "*Wolbachia*" infections and arthropod reproduction. *Bioscience*, 48, 287–293.

Breeuwer, J. A. J. 1997. *Wolbachia* and cytoplasmic incompatibility in the spider mites *Tetranychus urticae* and *T. turkestani*. *Heredity*, 79, 41–17.

Breeuwer, J. A. J. and Werren, J. H. 1990. Microorganisms associated with chromosome destruction and reproductive isolation between two insect species. *Nature*, 346, 558–560.

Brelsfoard, C. L. and Dobson, S. L. 2009. *Wolbachia* based strategies to control insect pests and disease vectors. *AS Pac J Mol. Biol. Biotechnol.*, 17(3), 55–63.

Charlat, S., Hurst, G. D. D. and Merçot, H. 2003. Evolutionary consequences of *Wolbachia* infections. *Trends Genet.*, 19, 217–223.

Cordaux, R., Michel-Salzat, A., Frelon-Raimond, M., Rigaud, T. and Bouchon, D. 2004. Evidence for a new feminizing *Wolbachia* strain in the isopod *Armadillidium vulgare*: Evolutionary implications. *Heredity*, 93, 78–84.

Cordaux, R., Pichon, S., Hatira, H. B., Doublet, V., Greve, P., Marcade, I., Varnier, C. B., Grosset, C. S., Cheikhrouha, F. C. and Bouchon, D. 2012. Widespread *Wolbachia* infection in terrestrial isopods and other crustaceans. *Zookeys*, 176, 123–131.

Correa, C. C. and Ballard, J. W. O. 2016. *Wolbachia* associations with insects: Winning or losing against a master manipulator. *Front. Ecol. Evol.*, 3, 153, 1–18.

Curry, M. M., Paliulis, L. V., Welch, K. D., Harwood, J. D., and White, J. A. 2015. Multiple endosymbiont infections and reproductive manipulations in alinyphiid spider population. *Heredity*, 115, 146–152.

Dobson, S. L, Marsland E. J., Veneti, Z., Bourtzis, K. and O'Neill, S. L. 2002. Characterization of *Wolbachia* host cell range via the in vitro establishment of infections. *Appl. Environ. Microbiol.*, 68, 656–660.

Dong, H., Liu, Q., Xie, L., Cong, B. and Wang, H. 2017. Functional response of *Wolbachia* infected and uninfected *Trichogramma dendrolini* Matsumura (Hymenoptera: Trichogrammatidae) to Asian corn borer *Ostrinia furnacalis* Guenee (Lepidoptera: Pyralidae) eggs. *J. Asia-Pac. Entomol.*, in press, doi:10.1016/j.aspen.2017.05.001.

Dutra, H. L. C., Rocha, M. V., Dias, F. B. S., Mansur, S. B., Caragata, E. P. and Moreira, L. A. 2016. *Wolbachia* blocks currently circulating Zika virus isolates in Brazilian *Aedes aegypti* mosquitoes. *Cell Host Microbe*, 19, 1–4.

Dyer, K. A. and Jaenike, J. 2004. Evolutionarily stable infection by a male-killing endosymbiont in *Drosophila innubila* molecular evidence from the host and parasite genomes. *Genetics*, 168, 1443–1455.

Dyson, E. A., Kamath, M. K. and Hurst, G. D. D. 2002. *Wolbahia* infection associated with all-female broods in *Hypolimnas bolina* (Lepidoptera: Nymphalidae): Evidence for horizontal transmission of a butterfly male killer. *Heredity*, 88, 166–171.

Fialho, R. F. and Stevens, L. 2000. Male-killing *Wolbachia* in a flour beetle. *Proc. R. Soc. London Ser. B*, 267, 1469–73.

Floate, K. D., Kyei-Poku, G. K. and Coghlin, P. C. 2006. Overview and relevance of *Wolbachia* bacteria in biocontrol research. *Biocontrol Sci. Technol.*, 16(8), 767–788.

Fuji, Y., Kageyama, D., Hoshizaki, S., Ishikawa, H. and Sasaki, T. 2001. Transfection of *Wolbachia* in Lepidoptera: The feminizer of the adzuki bean borer *Ostrinia scapulalis* causes male killing in the Mediterranean flour moth *Ephestia kuehniella*. *Proc. R. Soc. London B Biol. Sci.*, 268, 855–859.

Giordano, R., O'Neill, S. L. and Robertson, H. M. 1995. *Wolbachia* infections and the expression of cytoplasmic incompatibility in *Drosophila sechellia* and *D. mauritiana*. *Genetics*, 140, 1307–1317.

Giordano, R., Jackson, J. J. and Robertson, H. M. 1997. The role of *Wolbachia* bacteria in reproductive incompatibilities and hybrid zones of *Diabrotica* beetles and *Gryllus* crickets. *Proc. Natl. Acad. Sci. U.S.A.* 94, 11439–11444.

Goodacre, S. L., Martin, O. Y., Thomas, C. G. and Hewitt, G. M. 2006. *Wolbachia* and other endosymbiont infections in spiders. *Mol. Ecol.* 15, 517–527.

Gotoh, T., Noda, H. and Hong, X. Y. 2003. *Wolbachia* distribution and cytoplasmic incompatibility based on a survey of 42 spider mite species (Acari: Tetranychidae) in Japan. *Heredity*, 91, 208–216.

Hackett, K. J., Lynn, D. E., Williamson, D. L., Ginsberg, A. S. and Whitcomb, R. F. 1986. Cultivation of the *Drosophila* sex-ratio spiroplasm. *Science*, 232, 1253–1255.

Hertig, M. 1936. The rickettsia, *Wolbachia pipientis* (gen. et sp. n.) and associated inclusions of the mosquito, *Culex pipiens*. *Parasitology*, 28, 453–486.

Hertig, M. and Wolbach, S. B. 1924. Studies on rickettsia-like microorganisms in insects. *J Med. Res.*, 44, 329–374.

Heukelbach, J., Bonow, I., Witt, L., Feldmeier, H. and Fischer, P. 2004. High infection rate of *Wolbachia* endobacteria in the sand flea *Tunga penetrans* from Brazil. *Acta Trop.*, 92, 225–230.

Hilgenboecker, K., Hammerstein, P., Schlattmann, P., Telschow, A. and Werren, J. H. 2008. How many species are infected with *Wolbachia*?—A statistical analysis of current data. *FEMS Microbiol. Lett.*, 281, 215–220.

Hoffmann, A. A, Clancy, D. J. and Duncan, J. 1996. Naturally-occurring *Wolbachia* infection in *Drosophila simulans* that does not cause cytoplasmic incompatibility. *Heredity*, 76, 1–8.

Huigens, M. E., Luck, R. F., Klaassen, R. H. G., Maas, M. F. P. M., Timmermans, M. J. T. N. and Stouthamer, R. 2000. Infectious Parthenogenesis. *Nature*, 405, 178–179.

Huigens, M. E. 2003. *On the evolution of Wolbachia induced parthenogenesis in Trichogramma wasps*. Thesis, Wageningen University, The Netherlands.

Hurst, G. D. D., Hammarton, T. C., Bandi, C., Majerus, T. M. O., Bertrand, D. and Majerus, M. E. N. 1997. The diversity of inherited parasites of insects: The male-killing agent of the lady bird beetle *Coleomegilla magulata* is a member of the Flavobacteria. *Genet. Res.*, 70, 1–6.

Hurst, G. D. D., Hammarton, T. C., Obrycki, J. J., Majerus, T. M. O., Walker, L. E., Bertrand, D. and Majerus, M. E. N. 1996. Male killing bacterium in a fifth lady bird beetle, *Coleomegilla magulata* (Coleoptera: Coccinellidae). *Heredity*, 77, 177–185.

Hurst G. D. D., Jiggins, F. J., von der Schulenburg, J. H. G., Bertrand, D., West, S. A., Goriacheva, I. I., Zakharov, I. A., Werren, J. H., Stouthamer, R. and Majerus, M. E. N. 1999. Male killing *Wolbachia* in two species of insect. *Proc. R. Soc. London Ser. B*, 266, 735–740.

Hurst, G. D. D., Johnson, A. P., von der Schulenburg, J. H. G. and Fuyama, Y. 2000. Male-killing *Wolbachia* in *Drosophila*: A temperature sensitive trait with a threshold bacterial density. *Genetics*, 156, 699–709.

Hurst, L. D. 1992. Intra genomic conflict as an evolutionary force. *Proc. R. Soc. Lond. B Biol. Sci.*, 248, 135–140.

Jayaprakash, A. and Hoy, M. A. 2000. Long PCR improves *Wolbachia* DNA amplification: *wsp* sequences found in 76% of sixty three arthropod species. *Insect. Mol. Biol.* 9, 393–405.

Jiggins, F. M. and Hurst, G. D. 2011. Microbiology. *Rapid insect evolution by symbiont transfer. Science*, 332, 185–186.

Jiggins, F. M., Hurst, G. D. and Majerus, M. E. 1998. Sex ratio distortion in *Acraea encedon* (Lepidoptera: Nymphalidae) is caused by a male-killing bacterium. *Heredity*, 81, 87–91.

Jiggins, F. M., Hurst, G. D. and Majerus, M. E. 2000. Sex-ratio-distorting *Wolbachia* causes sex-role reversal in its butterfly host. *Proc. R. Soc. London B Biol. Sci.*, 267, 69–73.

Jiggins, F. M., Hurst, G. D., Schulenburg, J. H. G. and Majerus, M. E. 2001. Two male-killing *Wolbachia* strains coexist within a population of the butterfly *Acraea encedon*. *Heredity*, 86, 161–166.

Joanne, S., Vythilingam, I., Yugavathy, N., Leong, C., Wong, M. and AbuBaker, S. 2015. Distribution and dynamics of *Wolbachia* infection in Malaysia *Aedes albopictus*. *Acta Trop.*, 148, 38–45.

Joubert, D. A., Walker, T., Carrington, L. B., De Bruyne, J. T., Kien, D. H. T., Hoang, N. L. T., Chau, N. V. V., Ormaetxe, I. I., Simmons, C. P. and O'Neill, S. L. 2016. Establishment of a *Wolbachia* superinfection in *Aedes aegypti* mosquitoes as a potential approach for future resistance management. *PLoS Pathog.*, 12(2), e1005434. https://doi.org/10.1371/journal.ppat.1005434.

Ju, J. F., Hoffman, A. A., Zhang, Y. K., Duan, X., Guo, Y., Gong, J. T., Zhu, W. C. and Hong, X. Y. 2017. *Wolbachia* induced loss of male fertility is likely related to branch chain amino acid biosynthesis and iLvE in *Laodelphax striatellus*. *Insect Biochem. Mol. Biol.*, 85, 11–20.

Juchault, P., Frelon, M., Bouchon, D. and Rigaud, T. 1994. New evidence for feminizing bacteria in terrestrial isopods: Evolutionary implications. *C.R. Acad. Sci., Ser. III*, 317, 225–230.

Kawai, S., Matsumoto, Y., Gotoh, T. and Noda, H. 2009. Transinfection of *Wolbachia* in planthoppers: Nymphal injection of cultured *Wolbachia* and infection dynamics. *Environ. Entomol.*, 38, 1626–1633.

Koop, J. L., Zeh, D. W., Bonilla, M. M. and Zeh, J. A. 2009. Reproductive compensation favours male-killing *Wolbachia* in a live-bearing host. *Proc. R. Soc. London B Biol. Sci.*, 276, 4021–4028.

Kose, H., Karr, T. L. 1995. Organization of *Wolbachia pipientis* in the *Drosophila* fertilized egg and embryo revealed by an anti-*Wolbachia* monoclonal antibody. *Mech. Dev.*, 51, 275–288.

Landmann, F., Orsi, G. A., Loppin, B., and Sullivan, W. 2009. *Wolbachia*-mediated cytoplasmic incompatibility is associated with impaired histone deposition in the male pronucleus. *PLoS Pathog.*, 5(3), e1000343. https://doi.org/10.1371/journal.ppat.1000343.

Lassy, C. W. and Karr, T. L. 1996. Cytological analysis of fertilization and early embryonic development in incompatible crosses of *Drosophila simulans*. *Mech. Dev.*, 57(1), 47–58.

Laven, H. 1967. Eradication of Culex pipiens fatigans through cytoplasmic incompatibility. *Nature*, 216, 383–384.

Liu, C., Wang, J. L., Zheng, Y., Xiong, E. J., Li, J. J., Yuan, L. L., Yu, X. Q. and Wang, Y. F. 2014. *Wolbachia* induced paternal defect in *Drosophila* is likely by interaction with the Juvenile hormone pathway. *Insect Biochem. Mol. Biol.*, 49, 49–58.

Lo, N., Casiraghi, M., Salati, E., Bazzocchi, C. and Bandi, C. 2002. How many *Wolbachia* supergroups exist? *Mol. Biol. Evol.*, 19, 341–346.

Lo, N., Paraskevopoulos, C., Bourtzis, K., O'Neill, S. L., Werren, J. H., Bordenstein, S. R. and Bandi, C. 2007. Taxonomic status of the intracellular bacterium *Wolbachia pipientis*. *Int J Syst Evol Microbiol.*, 57, 654–657.

Majerus, M. E. and Zakharov, I. A. 2000. Multiple causes of male-killing in a single sample of the two-spot ladybird, *Adalia bipunctata* (Coleoptera: Coccinellidae) from Moscow. *Heredity*, 84, 605–609.

Martin, G., Juchault, P., Sorokine, O., Van Dorsselaer, A. 1990. Purification and characterization of androgenic hormone from the terrestrial isopod *Armadillidium vulgare*. *Gen. Comp. Endocrinol.*, 80, 349–354.

Moreira, L. A., Iturbe-Ormaetxe, I., Jeffery, J. A., Lu, G., Pyke, A. T., Hedges, L. M., Rocha, B. C., Hall-Mendelin, S., Day, A., Riegler, M., Hugo, L. E., Johnson, K. N., Kay, B. H., McGraw, E. A., Van den Hurk, A. F., Ryan, P. A. and O'Neill, S. L. 2009. A Wolbachia symbiont in *Aedes aegypti* limits infection with dengue, Chikungunya, and Plasmodium. *Cell* 139:1268–1278.

Narita, S., Kageyama, D., Nomura, M. and Fukatsu, T. 2007. Unexpected mechanism of symbiont-induced reversal of insect sex: Feminizing *Wolbachia* continuously acts on the butterfly *Euremahecabe* during larval development. *Appl. Environ. Microbiol.*, 73, 4332–4341.

Negri, I. 2012. *Wolbachia* as an "infectious" extrinsic factor manipulating host signalling pathways. *Front. Endocrinol.*, 2, 115. doi:10.3389/fendo.2011.00115.

Negri, I., Franchini, A., Gonella, E., Daffonchio, D., Mazzoglio, P. J., Mandrioli, M. and Alma, A. 2009. Unravelling the Wolbachia evolutionary role: The reprogramming of the host genomic imprinting. *Proc. R. Soc. London B Biol. Sci.*, 276, 2485–2491.

O'Connor, L., Plichart, C., Sang, A. C., Brelsfoard, C. L., Bossin, H. C. and Dobson, S. L. 2012. Open release of male mosquitoes infected with a *Wolbachia* biopesticide: Field performance and infection containment. *PLoS Negl. Trop. Dis.*, 6(11), e1797. https://doi.org/10.1371/journal.pntd.0001797.

O'Neill, S. L., Giordano, R., Colbert, A. M., Karr, T. L. and Robertson, H. M. 1992. 16S rRNA phylogenetic analysis of the bacterial endosymbionts associated with cytoplasmic incompatibility in insects. *Proc. Natl. Acad. Sci.*, 89, 2699–2702.

O'Neill, S. L. and Karr, T. L. 1990. Bidirectional incompatibility between conspecific populations of *Drosophila simulans*. *Nature*, 348, 178–180

Pannebakker B. A., Pijnacker L. P., Zwaan B. J. and Beukeboom L. W. 2004. Cytology of *Wolbachia*-induced parthenogenesis in *Leptopilina clavipes* (Hymenoptera: Figitidae). *Genome*, 47, 299–303. 10.1139/g03-137.

Paraskevopoulos, C., Bordenstein, S. R., Wernegreen, J. J., Werren, J. H. and Bourtzis, K. 2006. Toward a *Wolbachia* multilocus sequence typing system: Discrimination of *Wolbachia* strains present in *Drosophila* species. *Curr. Microbiol.*, 53, 388–395.

Park, C. H., Lim, H. W., Kim, H. W., Lee, W. G., Roh, J. Y., Park, M. Y. and Shin, E. 2016. High prevalence of *Wolbachia* infection in Korean populations of *Aedes albopictus* (Diptera: Culicidae). *J. Asia-Pac. Entomol.*, 19(1), 191–194. doi:10.1016/j.aspen.2015.12.014.

Pinto, S. B., Stainton, K., Harris, S., Kambris, Z., Sutton, E. R., Bonsall, M. B., Parkhill, J. and Sinkins, S. P. 2013. Transcriptional regulation of *Culex pipiens* mosquitoes by *Wolbachia* influences cytoplasmic incompatibility. *PLoS Pathog.*, 9(10), e1003647. https://doi.org/10.1371/journal.ppat.1003647.

Plantard, O., Rasplus, J. Y., Mondor, G., Le Clainche, I. and Solignac, M. 1998. *Wolbachia* induced thelytoky in the rose gallwasp *Diplolepis spinosissimae* (Giraud) (Hymenoptera: Cynipidae) and its consequences on the genetic structure of its host. *Proc. R. Soc. Lond. B*, 265, 1075–1080.

Rasgon, J. L., Gamston, C. E. and Ren, X. 2006. Survival of *Wolbachia pipientis* in cell-free medium. *Appl. Environ. Microbiol.*, 72, 6934–6937.

Reed, K. M. and Werren, J. H. 1995. Induction of paternal genome loss by the paternal-sex-ratio chromosome and cytoplasmic incompatibility bacteria (*Wolbachia*): A comparative study of early embryonic events. *Mol. Reprod. Dev.*, 40, 408–418.

Rigaud, T. 1997. Inherited microorganisms and sex determination of arthropod hosts. In: *Influential passengers inherited microorganisms and arthropod reproduction*, S. L. O'Neill, A. A. Hoffman and J. H. Werren (eds.), pp. 81–101. New York: Oxford University Press.

Rigaud, T., Moreau, J. and Juchault, P., 1999. *Wolbachia* infection in the terrestrial isopod *Oniscus asellus*: Sex ratio distortion and effect on fecundity. *Heredity*, 83, 469–475.

Riparbelli, M. G., Giordana, R., Ueyama, M. and Callaini, G. 2012. *Wolbachia*-mediated male killing is associated with defective chromatin Remodelling. *PLos One*, 7(1), e30045. https://doi.org/10.1371/journal.pone.0030045.

Ros, V. I., Fleming, V. M., Feil, E. J. and Breeuwer, J. A. 2009. How diverse is the genus *Wolbachia*? Multiple-gene sequencing reveals a putatively new *Wolbachia* supergroup recovered from spider mites (Acari: Tetranychidae). *Appl. Environ. Microbiol.*, 75, 1036–1043.

Rousset, F., Vautrin, D. and Solignac, M. 1992. Molecular identification of *Wolbachia*, the agent of cytoplasmic incompatibility in *Drosophila simulans*, and variability in relation with host mitochondrial types. *Proc. R. Soc. London Ser. B*, 247, 163–168.

Saldana, M. A., Hegde, S. and Hughes, G. L. 2017. Microbial control of arthropod-borne disease. *Mem. Inst. Oswaldo Cruz*, 112(2), 81–93.

Salunke, B. K., Salunkhe, R. C., Patole, M. S. and Shouche, Y. S. 2010. *Wolbachia* and termite association: Present status and future implications. *J. Biosci.*, 35, 171–175.

Schmidt, T. L., Barton, N. H., Rasic, G., Turley, A. P., Montgomery, B. L., Inaki, I., Cook, P. E., Ryan, P. A., Ritchie, S. A., Hoffmann, A. A., O'Neill, S. L. and Turelli, M. 2017. Local introduction and heterogeneous spatial spread of dengue-suppressing *Wolbachia* through an urban population of *Aedes aegypti*. *PLos Biol.*, in press. https://doi.org/10.1371/journal. pbio.200189410.1371/journal.pntd.0005496.

Serbus, L. R., Casper-Lindley, C., Landmann, F. and Sullivan, W. 2008. The genetics and cell biology of Wolbachia-host interactions. *Ann. Rev. Genet.*, 42, 683–707.

Sinkins, S. P., Curtis, C. F. and O'Neill, S. L. 1997. The potential application of inherited symbiont systems to pest control. In: *Influential passengers: Inherited microorganisms and arthropod reproduction*, S. L. O'Neill, A. A. Hoffman and J. H. Werren (eds.), pp. 155–175. Oxford: Oxford University Press.

Stevens, L., Giordano, R. and Fialho, R. F. 2001. Male-killing, nematode infections, bacteriophage infection, and virulence of cytoplasmic bacteria in the genus *Wolbachia*. *Ann. Rev. Ecol. Syst.*, 32, 519–545.

Stouthamer, R. 1997. *Wolbachia* induced parthenogenesis. In: *Influential passengers: Inherited microorganisms and invertebrate reproduction*, S. L. O'Neill, A. A. Hoffmann and J. H. Werren (eds.), pp. 102–124. New York: Oxford University Press.

Stouthamer, R., Breeuwer, J. A. and Hurst, G. D. 1999. *Wolbachia pipientis*: Microbial manipulator of arthropod reproduction. *A. Rev. Microbiol.*, 53, 71–102.

Stouthamer, R. and Kazmer, D. J. 1994. Cytogenetics of microbe associated parthenogenesis and its consequences for gene flow in *Trichogramma* wasps. *Heredity*, 73, 317–327.

Suh, E., Mercer, D. R. and Dobson, S. L. 2017. Life-shortening *Wolbachia* infection reduces population growth of *Aedes aegypti*. *Acta Trop.*, 172, 232–239. doi:10.1016/j.actatropica.2017.05.015.

Sumida, Y., Katsuki, M., Okada, K., Okayama, K. and Leswis, Z. 2017. *Wolbachia* induces cost to life history and reproductive traits in the moth, *Ephestia kuehniella*. *J. Stored Prod. Res.*, 71, 93–98. http://dx.doi.org/10.1016/j.jspr.2017.02.003.

Tagami, Y., Toi, M., Sugiyama, K., Tataraand, A. and Saito, T., 2006. *Wolbachia*-induced cytoplasmic incompatibility in *Liriomyza tripoli*i and its possible use as a tool in insect pest control. *Biol. Control* 38, 205–209.

Tan, C. H., Wong, P. J., Li, M. I., Yang, H. T., Ng, L. C. and O'Neill, S. L. 2017. *w*Mel limits Zika and Chikungunya virus infection in a Singapore *Wolbachia* introgressed *Ae. aegypti* strain, *w*Mel-Sg. *PLoS Negl. Trop. Dis.*, 11(5), e0005496. https://doi.org/10.1371/journal.pntd.0005496.

Terradas, G. and McGraw, E. A. 2017. *Wolbachia*-mediated virus blocking in the mosquito vector *Aedes aegypti*. *Curr. Opin. Insect Sci.*, 22, 1–8. http://dx.doi.org/10.1016/j.cois.2017.05.005

Tram, U., Ferree, P. M. and Sullivan, W. 2003. Identification of *Wolbachia*–host interacting factors through cytological analysis. *Microbes Infect.*, 5, 999–1011.

Tram, U., Fredrick, K., Werren, J. H., Sullivan, W. 2006. Paternal chromosome segregation during the first mitotic division determines *Wolbachia*-induced cytoplasmic incompatibility phenotype. *J. Cell Sci.*, 119, 3655–3663.

Tram, U. and Sullivan, W. 2002. Role of delayed nuclear envelope breakdown and mitosis in *Wolbachia*-induced cytoplasmic incompatibility. *Science*, 296, 1124–1126.

Vandekerckhove, T. T., Watteyne, S., Willems, A., Swings, J. G., Mertens, J. and Gillis, M. 1999. Phylogenetic analysis of the 16S rDNA of the cytoplasmic bacterium *Wolbachia* from the novel host *Folsomia candida* (Hexapoda, Collembola) and its implications for *Wolbachial* taxonomy. *FEMS Microbiol. Lett.*, 180, 279–286.

Vavre, F., Dedeine, F., Quillon, M., Fouillet, P., Fleury, F. and Bouletreau, M. 2001. Within species diversity of *Wolbachia*-induced cytoplasmic incompatibility in haplodiploid insects. *Evolution*, 55, 1710–1714.

Vavre, F., Fleury, F., Lepetit, D., Fouillet, P. and Bouletreau, M. 1999. Phylogenetic evidence for horizontal transmission of *Wolbachia* in host-parasitoid associations. *Mol. Biol. Evol.*, 16, 1711–1723.

Vavre, F., Fleury, F., Lepetit, D., Fouillet, P. and Bouletreau, M. 2000. Evidence for female mortality in *Wolbachia*-mediated cytoplasmic incompatibility in haplodiploid insects, epidemiologic and evolutionary consequences. *Evolution*, 54, 191–200.

Veneti, Z., Bentley, J. K., Koana, T., Braig, H. R. and Hurst, G. D. D. 2005. A functional dosage compensation complex required for male killing in Drosophila. *Science*, 307, 1461–1463.

Walker, T., Johnson, P. H., Moreira, L. A., Iturbe-Ormaetxe, I., Frentiu, F. D., Mc Meniman, C. J., Leong, Y. S., Dong, Y., Axford, J., Kriesner, P., Lloyd, A. L., Ritchie, S. A., O'Neill, S. L. and Hoffmann, A. A. 2011. The *w*Mel *Wolbachia* strain blocks dengue and invades caged *Aedes aegypti* populations. *Nature*, 476, 450–453.

Weaver, S. C., Costa, F., Garcia-Blanco, M. A., Ko, A. I., Ribeiro, G. S., Saade G., Shi, P. Y. and Vasilakis, N. 2016. Zika virus: History, emergence, biology, and prospects for control. *Antiviral Research*, 130, 69–80.

Weeks, A. R. and Breeuwer, A. J. 2001. *Wolbachia*-induced parthenogenesis in a genus of phytophagous mice. *Proc. R. Soc. Lond B*, 268, 2245–2251.

Werren, J. H. 1997. Biology of *Wolbachia*. *Ann. Rev. Entomol.*, 42, 587–609.

Werren, J. H., Baldo, L. and Clark, M. E. 2008. *Wolbachia*: Master manipulators of invertebrate biology. *Nature Rev. Microbiol.*, 6, 741–751.

Werren, J. H., Guo, L. and Windsor, D. W. 1995. Distribution of *Wolbachia* in neotropical arthropods. *Proc. R. Soc. London Ser. B*, 262, 147–204.

Werren, J. H., Hurst, G. D. D., Zhang, W., Breeuwer, J. A. J., Stouthamer, R. and Majerus, M. E. N. 1994. Rickettsial relative associated with male-killing in the lady bird beetle (*Adalia bipunctata*) *J. Bacteriol.*, 176, 388–394.

Werren, J. H., Skinner, S. W. and Huger, A. M. 1986. Male-killing bacteria in a parasitic wasp. *Science*, 231, 990–992.

Werren, J. H. and Windsor, D. M. 2000. *Wolbachia* infection frequencies in insects: Evidence of a global equilibrium? *Proc. R. Soc. Lond. B*, 267, 1277–1285.

Werren, J. H., Zhang, W. and Guo, L. R. 1995. Evolution and phylogeny of *Wolbachia*: Reproductive parasites of arthropods. *Proc. R. Soc. London B Biol. Sci.*, 261, 55–63.

Wiwatanaratanabutr, I. 2013. Distribution, diversity and density of wolbachial infections in cladocerans and copepods from Thailand. *J. Invertebr. Pathol.*, 114, 341–345.

Wiwatanaratanabutr, I. and Grandjean, F. 2016. Impacts of temperature and crowding on sex ratio, fecundity and *Wolbachia* infection intensity in the copepod *Mesocyclops thermocyclopoides*. *J. Invertebr. Pathol.*, 141, 18–23.

Wiwatanaratanabutr, I. and Zhang, C. 2016. *Wolbachia* infections in mosquitoes and their predators inhabiting rice field communities in Thailand and China. *Acta Trop.*, http://dx.doi.org/doi:10.1016/j.actatropica.2016.03.026.

Wrensch, D. L. and Ebbert, M. A., eds. 1993. *Evolution and diversity of sex ratio in insects and mites*. New York: Chapman & Hall.

Ye, Y. H., Woolfit, M., Huttley, G. A., Rances, E., Caragata, E. P., Popovici, J., O'Neill, S. L. and McGraw, E. A. 2013. Infection with a virulent strain of wolbachia disrupts genome wide-patterns of cytosine methylation in the mosquito *Aedes aegypti*. *PloS One*, 8(6), e66482. https://doi.org/10.1371/journal.pone.0066482.

Yen, J. H. and Barr, A. R. 1971. New hypothesis of the cause of cytoplasmic incompatibility in *Culex pipiens*. *Nature*, 232, 675–678.

Zabalou, S., Riegler, M., Theodorakoppoulou, M., Stauffer, C., Savakis, C. and Bourtzis, K. 2004. *Wolbachia*-induced cytoplasmic incomptability as a means for insect pest control. *PNAS*, 101, doi:10.1073/pnas.0403853101.

Zchori-Fein, E., Roush, R. T., and Hunter, M. S. 1992. Male production by antibiotic treatment in *Encarsia formosa* (Hymenoptera: Aphelinidae) an asexual species. *Experentia*, 48, 102–105.

Zeh, D., Zeh, J. and Bonilla, M. 2005. *Wolbachia*, sex ratio bias and apparent male killing in the harlequin beetle riding pseudoscorpion. *Heredity*, 95, 41–49.

Zheng, Y., Ren, P. P., Wang, J. L. and Wang, Y. F. 2011. *Wolbachia*-induced cytoplasmic incompatibility is associated with decreased Hira expression in male *Drosophila*. *PLoS One* 6(4), e19512. https://doi.org/10.1371/journal.pone.0019512.

Zhou, W., Rousset, F. and O'Neill, S. 1998. Phylogeny and PCR–based classification of *Wolbachia* strains using wsp gene sequences. *Proc. R. Soc. London B Biol. Sci.*, 265, 509–515.

Zug, R., and Hammerstein, P. 2012. Still a host of hosts for *Wolbachia*: Analysis of recent data suggests that 40% of terrestrial arthropod species are infected. *PLoS One*, 7(6), e38544. https://doi.org/10.1371/journal.pone.0038544.

# 13 Microorganisms in the Management of Vectors

## *A Sustainable Ecological Approach*

*Sajal Bhattacharya and Probal Basu*

**CONTENTS**



## 13.1 INTRODUCTION

The phylum Arthropoda constitutes about 75% of all animal species described so far. The two most medically important arthropod groups are the insects and arachnids. Among the representatives of the class Insecta, mosquitoes, flies, and fleas play a central role in spreading different medically important pathogens. The mosquito spreads parasitic (malaria, filariasis) and arboviral diseases (Dengue, Japanese encephalitis, Zika virus, West Nile fever) (Becker, 2008; Tolle, 2009; Bhattacharya and Basu, 2014; Basu and Bhattacharya, 2016). According to the World Health Organization (WHO), over 17% of diseases among all fatal diseases are vector-borne diseases (WHO, 2016). Each year more than 1 billion people are infected and more than 1 million people die from the vector-borne diseases like dengue, malaria and chikungunya (WHO, 2014). Blackflies are responsible for the transmission of human and bovine onchocerciasis as well as *Leucocytozoon* spp. in fowl (Skovmand et al., 2007). Albeit chemical insecticidal control has been applied to reduce the vector burden, a strong deleterious environmental effect has been seen during the process (Kali, 2015). Thus, biological control, especially entomopathogenic microorganism-based control, of disease vectors is of poignant interest in modern-day advanced research. Different entomopathogens, including viruses, bacteria, protozoa, and fungi, result in severe decline in vector population (Skovmand et al., 2007). Keeping in view the above, an attempt has been made to explore the perspectives of disease-vector control using microorganisms.

The inadequacy of currently available control strategies, such as vaccines, drugs, and pesticides, has led to a dramatic increase in occurrence of vector borne diseases (Halstead, 2007; Kyle and Harris, 2008). Currently, two-fifths of the world's population is at risk from dengue fever, and 50 million to 100 million individuals are infected each year (Halstead, 2007).

## 13.2 MICROORGANISMS FOR VECTOR CONTROL

### 13.2.1 Bacteria and Virus

Bacteria are potent to cause infection in arthropod hosts. Entomopathogenic bacteria (Table 13.1) can significantly influence the host's reproduction. Actinobacteria have a major role in biological control of disease-causing insects like *Musca domestica* by the production of insecticidal active compounds (Gadelhak et al., 2005). It has been found that different strains of *Streptomyces* (Table 13.1) show strong larvicidal activity upon *Anopheles* larvae and *Culex quinquefasciatus* (Dhanasekaran and Thangaraj, 2014). Albeit the peritrophic membrane of first and second instar larva blocks the entry of virus, different viruses could show a diverse effect on vector insects like mosquitoes. Researchers have demonstrated that a marine bacteria, *Bacillus cereus*, isolated from the coastal area of Pondicherry, India, has shown significant toxicity against the filarial vector *Culex quinquefasciatus* (Poopathi et al., 2014). It has been reported that field populations of *Culex pipiens* complex have shown 150-fold resistances against *Bacillus sphaericus* in India. This resistance could be generated by several factors, including insect species with high reproductive rate, short generation time, and more genetic variations in the insect populations (Liu and Tabashnik, 1997). *Bacillus amyloliquefaciens*, a bacterium with mosquitocidal activity has been identified by the researchers from Andaman and Nicobar Islands, India (Geetha et al., 2011).

Viruses such as baculovirus, densovirus, iridovirus, and cytoplasmic polyhedrosis virus could significantly infect mosquitoes (Becnel and White, 2007; Kali, 2015). The occlusion bodies of the virus are ingested by the larvae of the vector insect (e.g., mosquitoes) and multiply in the gastric caeca and cause death of the larva (Becnel and White, 2007). The significance of the occlusion bodies of the virus as the potent microbial control agent relies on the fact that those bodies are thermostable, can be stored at 4°C, and can produce highly infectious progeny. Novel formulations like microcapsulation of viral occlusion bodies with magnesium have been considered as potential approaches to enhance the infection level (Becnel and White, 2007).

### 13.2.2 Fungus and Protozoa

Most of the fungi are aquatic and have been considered as facultative parasites of mosquito larvae (Scholte et al., 2004; Kali, 2015) (Table 13.1). The fungal spores adhere to the surface of the insect cuticle, and then germinate and differentiate in the

**TABLE 13.1**
**Major Microorganisms Used as Vector Control Agents**

| Major Groups | Examples | References |
|---|---|---|
| | **Bacteria** | |
| Actinobacteria | *Streptomyces avermectinus* | Goodfellow and O'Donell, 1989 |
| | *Streptomyces sp. VITDDK3* | Moncheva et al., 2002 |
| | *Streptomyces NRRL 30562* | Bèrdi, 1989 |
| Alpha proteobacteria | Wolbachia sp. | Kali, 2015 |
| Bacilli | *B. thuringiensis* subsp. *israelensis* (Bti) | Kali, 2015 |
| | *B. sphaericus* (Bs) | |
| | *B. thuringiensis* subsp. *morrisoni* | |
| | *B. thuringiensis* subsp. *jegathesan* | |
| | *B. thuringiensis* subsp. *medellin* | |
| Clostridia | *Clostridium bifermentans* | Kali, 2015 |
| | **Virus** | |
| Baculovirus | Nucleo polyhedro virus (NPV) | Andreadis et al., 2003 |
| Densovirus | *Anopheles gambiae* densovirus (AgDNV) | Ng et al., 2011 |
| Iridovirus | Invertebrate Iridovirus -6 (IIV-6) | Chinchar et al., 2005 |
| Cytoplasmic polyhedrosis virus | *Uranotaenia apphirina* cytoplasmic polyhedrosis virus (UsCPV) | Shapiro et al., 2004 |
| | **Fungi** | |
| Oomycota | *Lagenidium, Leptolegnia* | Kali, 2015; Scholte et al., 2004 |
| Chytridiomycota | *Coelomomyces* | |
| Zygomycota | *Entomophthora, Conidiobolus* | |
| Deuteromycota | *Culicinomyces, Beauveria* | |
| | **Protozoa** | |
| Microsporidia | *Nosema, Thelohania, Edhazardia* | Chapman et al., 1973; Becnel et al., 2005 |

form of appressorium, which facilitates the penetration of fungi into the vector insect cuticle. After the spores colonize within the hemocoel, they differentiate into conidia and conidiophores (Butt et al., 2013). These conidia cause the destruction of disease vectors (Riba et al., 1986; Lacey et al., 1988). Moreover, some studies show mosquitoes that have developed resistance against chemical insecticides are not only susceptible to fungus infection but also extend the life span of currently used insecticides making it a suitable complimentary control measure along with conventional chemical pesticides (Farenhorst et al., 2009). Research indicated that the effect of *Aspergillus flavus* and *Aspergillus parasiticus* is significantly toxic for the late third instar larva of *Culex quinquefasciatus* (Govindarajan et al., 2005). Different bacterial spore formers from a temporary flooded site in Orissa, India, have showed significant mosquitocidal activity against *Aedes aegypti* (Rout et al., 2011). Interestingly, a water mold, *Lagenidium giganteum*, is found to contain significant mosquitocidal activity (Vyas et al., 2007).

Few genera of microsporidian protozoa (viz., *Nosema, Thelohania*, and *Edhazardia*) are seen to infect mosquitoes (Chapman et al., 1973; Becnel et al., 2005). Earlier studies have revealed that mosquitoes get infected by microsporidian protozoa, *Nosema* and *Thelohania* (Table 13.1). *Nosema* spores can notably kill mosquito larva, pupa, and adults by attacking the epithelial cells of the midgut of the mosquitoes, whereas *Thelohania* commonly infect the adipose tissues of mosquitoes (National Academy of Sciences, 1973). *Edhazardia* is a microsporidian parasite of *Aedes aegypti* that produces spores that are ingested by mosquito larva from the environment, and the spore germinates and begins to multiply in the host tissue.

## 13.3 *BACILLUS THURINGIENSIS ISRAELENSIS*

*Bacillus thuringiensis israelensis* (Bti), is a gram-positive bacterium, has got significant attention in the field of microbial control of disease vectors. This bacterium produces insecticidal crystal protein toxins during its sporulation (Kali, 2015) (Figure 13.1). Bti produces two general groups of insecticidal crystal proteins: Cyt (cytolysins) and Cry (crystal delta-endotoxins). The *cry* genes have four classes, among which *cry II* and *cry IV* are active against diptera, primarily mosquito and blackflies (Poopathi, 2012). During sporulation, the bacterium produces proteins that then form crystals that are toxic to larva of vector mosquitoes like *Aedes*. The crystals

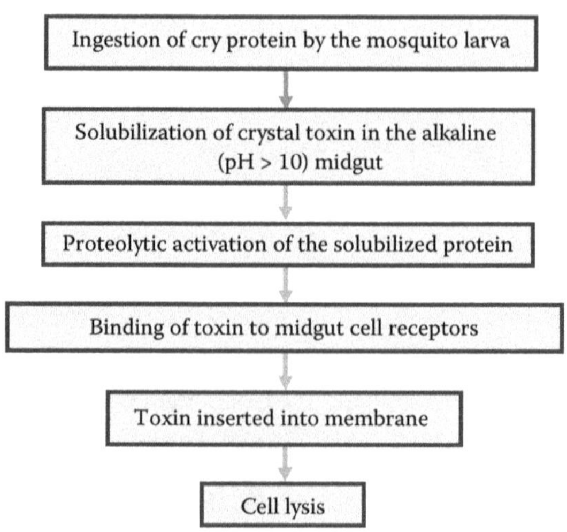


**FIGURE 13.1** Mechanism of *Bacillus thuringiensis israelensis* toxin action. (From Land, M. and Miljand, M., 2014, Biological control of mosquitoes using *Bacillus thuringiensis israelensis*: A pilot study of effects on target organisms, non-target organisms and humans, MistraEviEM Pilot Study PS4, www.eviem.se.)

contain four polypeptides: Cry IVA (125 kDa), Cry IVB (135 kDa), Cry IVD (68 kDa), and CytA (28 kDA). The gene encoding of these Cry toxins are situated in the 72 kDa plasmid (Poopathi, 2012).

A large-scale trial of bactoculicide, a formulation of Bti (strain 164) imported from Russia, has been conducted against *Anopheles*, *Culex*, and *Aedes* spp., the mosquito vectors of malaria, filariasis, Japanese encephalitis (JE), and dengue, in their breeding habitats in different regions of India (Mittal, 2003). It was observed that application of the formulation results in the reduction of *Ae. aegypti, Ae. albopictus, and Cx. quinquefasciatus* larva for few weeks. Studies showed that application of the formulation causes the 90% reduction for a week (Mittal, 2003).

## 13.4 *BACILLUS SPHAERICUS*

*Bacillus sphaericus* (Bs) has an excellent larvicidal activity against larva of certain mosquito species like *Culex* (Federici et al., 2003; Ricci et al., 2012). The bacterium enters the mosquito larva through the alimentary canal. The infected larva usually loses its normal turgor and its motility disappears after 3 days (Balaraman, 1981). All the bacteria are confined within the peritrophic membrane of the midgut region (Figure 13.2). The larvicide of this bacterium consists of 51 kD and 42 kD proteins, which are responsible for the toxicity of mosquito larva. The 42 kD protein is processed in the larval midgut and produces 39 kD protein, which acts as the major toxin against the larval tissue (Poopathi, 2012).

Additionally, the application of different formulations of *B. sphaericus* against *Culex* and *Anopheles* larvae has revealed its significant efficiency on them. It has been observed that formulation of this bacterium exhibited its effect on *Cx. quinquefasciatus* at polluted water habitats (Mittal, 2003).

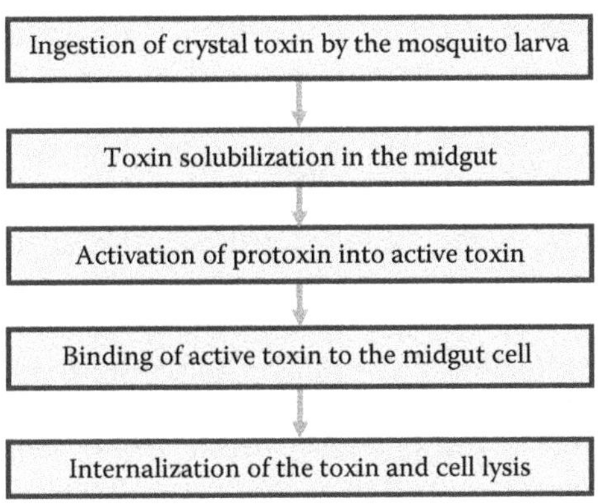


**FIGURE 13.2** Mechanism of action of *Bacillus sphaericus* toxin. (From Dhanasekaran, D. and Thangaraj, R., 2014, Microbial secondary metabolites are an alternative approaches against insect vector to prevent zoonotic diseases, *Asian Pac. J. Trop. Dis.*, **4**(4): 253–261, doi: 10.1016/S2222-1808(14)60569-7.)

## 13.5 *WOLBACHIA*

*Wolbachia* are maternally transmitted, symbiotic, gram-negative, intracellular, α-proteobacteria (Sanchez-Contreras and Vlisidou, 2008) that can spread within insect and nematodes populations because of their unique ability to manipulate host reproduction (Iturbe-Ormaetxe et al., 2011; Pan et al., 2012). *Wolbachia* can manipulate the host defense system to facilitate its own persistent infection, resulting in a compromise of the mosquito's ability to carry human pathogens (Pan et al., 2012). *Wolbachia* affects different mosquito genera that include *Aedes*, *Culex*, and *Mansonia* (Walker and Moreira, 2011). The bacterium causes reproductive disorders like cytoplasmic incompatibility (CI) and parthenogenesis (Walker and Moreira, 2011) within the insect populations. This outlines a new strategy for controlling mosquito-borne tropical diseases, such as dengue. The *Wolbachia* bacterium was first reported from the reproductive tissues of *Culex pipiens* mosquitoes and thus the species named as *Wolbachia pipientis* (Werren, 1997).

Research indicated that more than 20% of insects that are potential disease vectors harbors a *Wolbachia* strain (Sanchez-Contreras and Vlisidou, 2008). When a *Wolbachia*-infected male insect mates with a noninfected female, CI develops. It has been observed that sperm of a *Wolbachia*-infected male insect is incompatible with eggs from females that do not contain the bacteria. *Wolbachia*-mediated CI involves modification of sperm during spermatogenesis (Calvitti, 2011). Moreover, researchers report that the *Wolbachia* genome contains repetitive DNA elements, which is crucial for host–*Wolbachia* interaction (Wu et al., 2004; Foster et al., 2005). Furthermore, the bacteria contain a unique ankyrin (ANK) domain, which is typically a 33-residue L-shaped motif that mediates protein–protein interactions in diverse families of proteins, including cytoskeletal and membrane proteins, transcriptional and developmental regulators, toxins, and cyclin-dependent kinase (CDK) inhibitors (Sedgwick and Smerdon, 1999). Though the molecular mechanism underlying *Wolbachia*-mediated resistance to pathogens is not understood completely, it is associated with a number of immune genes, such as defensins, cecropins, and several Toll pathway genes, that were upregulated by *Wolbachia* in *Ae. aegypti* in order to boost mosquito innate immunity (Pan et al., 2011). In agreement with these observations, activation of the Toll pathway by means of RNAi depletion of cactus, an inhibitor of REL1, has been shown to block proliferation of dengue virus in *Ae. aegypti*. This bacterium elevates the expression of NADPH oxidase and reactive oxygen species (ROS). In turn, ROS upregulation results in activation of the Toll-mediated pathway, which mediates both the antioxidant and production of antimicrobial peptides (e.g., cecropins and defensins).

All known wild populations of *Ae. albopictus* are naturally infected with two strains of the *Wolbachia pipientis*, known as *w*AlbA and *w*AlbB. *Ae. aegypti* is naturally uninfected with the bacterium. Recent study has shown that when an over-replicating strain of *Wolbachia* from *Drosophila melanogaster*, *w*MelPop, was transferred into *Ae. aegypti*, the dissemination of dengue virus was strongly inhibited, as was the dissemination of chikungunya virus. When introduced into *Ae. aegypti* mosquitoes, the over-replicating *w*MelPop strain not only inhibits the replication of dengue virus strongly, but also stimulates chronic immune upregulation (Blagrove et al., 2011). In addition, transfer of the *w*AlbB strain from *Ae. albopictus* into *Ae. aegypti* had led to reduced susceptibility to dengue. Both *Wolbachia* strains also induced CI (sperm–egg cytoplasmic incompatibility) in *Ae. aegypti*, whereby uninfected females mated with infected males produce embryos that die shortly after fertilization. Thus, the combination of viral inhibition and a built-in self-spreading mechanism provides attractive prospects for the control of dengue transmission by *Ae. aegypti* (Marcus et al., 2011).

Their study provides an example of strong pathogen inhibition in a naturally *Wolbachia*-infected mosquito species, demonstrating that this inhibition is not limited to naturally naïve species, and suggests that the *Wolbachia* strain is more important than host background for viral inhibition. Complete bidirectional cytoplasmic incompatibility was observed with WT strains infected with the naturally occurring *Ae. albopictus Wolbachia*, and this provides a mechanism for introducing *w*Mel into natural populations of this species. Researchers have shown that *Wolbachia* can induce ROS-dependent activation of the Toll pathway to control dengue virus. The ability of *Wolbachia* to enhance immunity and block dengue virus proliferation in a newly acquired host, *Ae. aegypti*, makes it a potential "mosquito vaccine" that could be used efficiently to prevent pathogen transmission. More detailed understanding of molecular mechanisms of *Wolbachia* effects on dengue virus proliferation in *Ae. aegypti* is of paramount importance for refining the utilization of a *Wolbachia–Aedes* system approach for the future control of dengue virus transmission (Pan et al., 2011). Research by Blagrove et al. (2011) have shown that that *w*Mel infection can block dengue virus transmission in the increasingly widespread vector species *Ae. albopictus*. Their study has yielded an *Ae. albopictus* line that may provide the basis for a viable new option for dengue control in this species. Earlier studies reported the generation of a *w*Mel-transinfected *Ae. aegypti* line, dengue inhibition in this line (Walker et al., 2011), and successful field trials in Queensland, Australia (Hoffmann et al., 2011), demonstrated the feasibility of field implementation of a *Wolbachia* population replacement strategy. Blagrove et al. demonstrate that both of the two main vectors of dengue globally are amenable to such a strategy.

## 13.6 DISCUSSION

Although the use of entomopathogenic microorganisms in controlling vector populations confronts various pragmatic limitations like problems in large-scale production, storage, and lower infection rate (Burges, 1981; Usta, 2013; Kali, 2015), these control measures are considered to be more ecologically safe and sustainable (Usta, 2013). In this context, the changing environment should be taken into consideration during assessing the efficacy of entomopathogenic microorganisms-based control method. The interaction between microorganisms and its host insect in this changing climate might unveil certain interdependent survival strategy to coexist, which may contribute them to live, sustain, and propagate within the environment without significantly affecting each other. This notion can be strengthened by the fact that shows the presence of *Wolbachia* within the filarial vector mosquito, *Culex quinquefaciatus* in South India (Sunish et al., 2011). Thus, a sustained study regarding the interaction between the entomopathogenic microorganisms and disease vectors is required to ascertain the potential and prospect of such control agents against the disease vectors in the changing environment.

Combining the natural control agents with safe biopesticides will go a long way toward achieving satisfactory control of at least some species of mosquitoes in some habitats. Studies have shown that natural regulatory factors, such as oviposition cues and predators, play an important role in the natural decline of mosquito larvae (Mulla, 1990).

The warfare against mosquitoes involves few genera of microsporidian protozoa (viz., *Nosema, Thelohania*, and *Edhazardia*) that were seen to inhabit and develop pathogenicity in mosquitoes (Chapman et al., 1973; Becnel et al., 2005). Studies revealed that *Nosema* spores can notably kill mosquito larva, pupa, and adults. It generally attacks the epithelial cells of the midgut of the mosquitoes, whereas *Thelohania* commonly infects the adipose tissues of mosquitoes (National Academy of Sciences, 1973).

Hence, their potential role in controlling the vector mosquitoes can be evaluated further. Today, with the advent of the improvised tools, techniques in molecular biology, and substantial advancements in tissue culture techniques, it seems that parasitic or symbiotic microorganisms could be cultured in large scale for application.

However, it has been found that most of these agents are not sufficiently cost-effective to be used in such control programs. Despite the numerous potential control agents evaluated, the only cost-effective pathogen identified to date is the bacterium Bti. Its success is due primarily to its ability to be mass produced in large quantities on relatively inexpensive media using large-scale fermentation technology. Even so, this bacterium is only effective against larvae, and thus its use is limited to vector control programs that have traditionally used larviciding as a strategy. Though limited, the use of Bti has been quite valuable in the World Health Organization's Onchocerciasis Control Program in West Africa, and may ultimately prove useful in malaria, filariasis, and viral disease control programs that include larvicides. However, Bti is unlikely in the future to be of use against major malaria vectors such as *An. gambiae* and *An. stephensi*, where larvicide is not an important control strategy.

The lack of new agents and technologies to control vectors indicate that chemical insecticides will continue to be important, where resistance is not a major problem in vector control programs. New disease control technologies are being explored, such as disease-refractory transgenic mosquitoes (Collins and Besansky, 1994) and releasing transgenic insecticidal organisms used as larval food into breeding habitats. The former tactic has merit, but may ultimately lead to resistance in the population of the disease-causing organism, such as the malarial parasites.

The interaction of the microbes and its host insect in the changing climate might unveil a certain interdependent survival strategy to coexist, which may contribute them to live, sustain, and propagate within the environment without significantly affecting each other. This notion can be reinforced by the fact that shows the presence of *Wolbachia* within the filarial vector mosquito *Culex quinquefaciatus* in South India (Sunish et al., 2011).

## 13.7 CONCLUSION

To conclude, it can be stated that beside the efficiency, scientists are in agreement that the synthetic chemical insecticides are found to be ecologically unsafe and unsustainable. Thus, entomopathogenic microorganisms as biocontrol agents have become judicious alternatives to the scientific community. Over the years, a wealth of knowledge about the occurrence, basic biology, and control potential of numerous viruses, bacteria, fungi, and protozoa that infect the vectors has been generated. This knowledge is of value in its own right and serves as a foundation against which to measure the potential of upcoming control agents. However, the management of vector and vector-borne diseases using microorganisms is not without limitations. Its use is definitely an important strategy; though this control approach may not be applicable in every ecological and socioeconomic situation. However, serious scientific assessment of these microorganisms and its ecological impact are needed before their widespread application. This will help to exclude the adverse effects, if any on the nontarget organisms of the environment. Nevertheless, this environment-friendly management could be a valuable component of an integrated mosquito vector control program. Since, there is no effective candidate vaccines at our disposal, excepting yellow fever virus, the disease control through inclusive vector management would be the appropriate choice at the present juncture to contain the vector-borne diseases. Apart from potential entomopathogens such as Bti, other prospective microorganisms mentioned here should be taken up for in-depth study to control and combat mosquito populations in a sustainable manner.

## ACKNOWLEDGMENTS

The authors would like to extend their sincere gratitude to Dr. Dipak Kumar Kar, principal, Asutosh College, for providing necessary facilities for carrying out this work; and to Dr. Sriparna Datta Ray and Dr. Deep Chandan Chakraborty, assistant professors in zoology, Asutosh College, for their kind support.

## REFERENCES

Andreadis, T. G., Becnel, J. J. and White S. E. (2003). Infectivity and pathogenicity of a novel baculovirus, Cuni NPV from *Culex nigripalpus* (Diptera: Culicidae) for thirteen species and four genera of mosquitoes. *J. Med. Entomol.*, 40(4): 512–517.

Balaraman, K. (1981). Microbial control agent—Their effects on the biology of mosquitoes. *Proc. Indian Nat. Sci. Acad.*, B46(6): 868–871.

Basu, P. and Bhattacharya, S. (2016). A new dimension in the dengue epidemiology with special reference to genetic diversity of the virus: A review. *Int. J. Fauna Biol. Studies*, 3 (3): 29–41.

Becker, N. (2008). Influence of climate change on mosquito development and mosquito-borne diseases in Europe. *Parasitol. Res.*, 103(1): S19–28. doi: 10.1007/s00436-008-1210-2.

Becnel, J. J. and White, S. E. (2007). Mosquito pathogenic viruses-the last 20 years. *J. Am. Mosq. Control Assoc.*, 23(2 Suppl.): 36–49. doi: http://dx.doi.org/10.2987/8756-971X (2007)23[36:MPVLY]2.0.CO;2.

Becnel, J. J., White, S. E. and Shapiro A. M. (2005). Review of microsporidia-mosquito relationships: From the simple to the complex. *Folia Parasitol.*, 52: 41–50.

Bèrdi, J. (1989). The discovery of new bioactive microbial metabolites: Screening and identification. In: *Bioactive microbial metabolites, progress in industrial microbiology*, vol. 27, M. E. Bushell and U. Grafe, editors. Amsterdam, Netherlands: Elsevier, pp. 3–25.

Bhattacharya, S. and Basu, P. (2014). Japanese encephalitis virus infection in different vertebrates and its epidemiological significance: A review. *Int. J. Fauna Biol. Studies*, 1(6): 32–37.

Blagrove, M. S. C., Arias-Goeta, C., Failloux, A. B. and Sinkins, S. P. (2011). *Wolbachia* strain wMel induces cytoplasmic incompatibility and blocks dengue transmission in *Aedes albopictus*. *Proc. Nat. Acad. Sci. USA*, 109: 255–260. doi: 10.1073/pnas.1112021108.

Burges, H. D. (1981). Safety Testing and Quality Control of Microbial Pesticides. In: *Microbial control of pests and plant diseases*. London, Academic Press, pp. 738–768.

Butt, T. M., Greenfield, B. P. J., Greig, C., Maffeis, T. G. G., Taylor, J. W. D., Piasecka, J., Dudley, E., Abdulla, A., Dubovskiy, I. M., Garrido-Jurado, I., Quesada-Moraga, E., Penny M. W. and Eastwood, D. C. (2013). *Metarhizium anisopliae* pathogenesis of mosquito larvae: A verdict of accidental death. *PLoS One*, 8(12): e81686. doi: http://dx.doi.org/10.1371/journal.pone.0081686.

Calvitti, M. (2011). Bacteria endosymbionts: A source of innovation in biotechnology for the control of vector-borne diseases. *Energia, Ambiente e Innovazione*, 6: 49–57.

Chapman, H. C., Gorham J. R. and Fukuda, T. (1973). *Thelohania* (Nosematidae: Microsporida) in *Aedes* mosquitoes of Alaska. *Mosq. News*, 33: 465–466.

Collins, F. H. and Besansky, N. J. (1994). Vector biology and the control of malaria in Africa. *Science*, 264: 1874–1875. doi: 10.1126/science.8009215.

Dhanasekaran, D. and Thangaraj, R. (2014). Microbial secondary metabolites are an alternative approaches against insect vector to prevent zoonotic diseases. *Asian Pac. J. Trop. Dis.*, 4 (4): 253–261. doi: 10.1016/S2222-1808(14)60569-7.

Farenhorst, M., Mouatcho, J. C. Kikankie, C. K., Brooke, B. D., Hunt, R. H., Thomas, M. B., Koekemoer, L. L., Knols, B. G. and Coetzee, M. (2009). Fungal infection counters insecticide resistance in African malaria mosquitoes. *Proc. Natl. Acad. Sci.*, 106: 17443–17447.

Federici, B. A., Park, H. W., Bideshi, D. K., Wirth, M. C. and Johnson, J. J. (2003). Recombinant bacteria for mosquito control. *J. Exp. Biol.*, 206(21): 3877–3885. doi: 10.1242/jeb.00643.

Foster, J., Ganatra, M., Kamal, I., Ware, J., Makarova, K., Ivanova, N., Bhattacharyya, A., Kapatral, V., Kumar, S., Posfai, J. Vincze, T., Ingram, J., Moran, L., Lapidus, A., Omelchenko, M., Kyrpides, N., Ghedin, E., Wang, S., Goltsman, E., Joukov, V., Ostrovskaya, O., Tsukerman, K., Mazur, M., Comb D., Koonin, E. and Slatko, B. (2005). The *Wolbachia* genome of *Brugia malayi*: Endosymbiont evolution within a human pathogenic nematode. *PLoS Biol.*, 3: e121. doi: 10.1371/journal.pbio.0030121.

Gadelhak, G. G., El-Tarabily, K. A. and AL-Kaabi, F. K. (2005). Insect control using chitinolytic soil actinomycetes as biocontrol agents. *Int. J. Agric. Biol.*, 7(4): 627–633. doi://doi.org/10.22376/ijpbs.

Geetha, I., Manonmani, A. M. and Prabakaran, G. (2011). *Bacillus amyloliquefaciens*: A mosquitocidal bacterium from mangrove forests of Andaman & Nicobar Islands, India. *Acta Trop.*, 120: 155–159. doi: 10.1016/j.actatropica.2011.07.006.

Goodfellow, M. and O'Donell, A.G. (1989). Search and discovery of industrially significant actinomycetes. In: *Microbial products, new approaches*, S. Baumberg, I. Hunter, P. M. Rhodes, editors. Cambridge, UK: Cambridge University Press, pp. 343–383.

Govindarajan, M., Jebanesan, A. and Reetha, D. (2005). Larvicidal effect of extracellular secondary metabolites of different fungi against the mosquito, *Culex quinquefasciatus* Say. *Trop. Biomed.*, 22(1): 1–3.

Halstead, S. B. (2007). Dengue. *Lancet*, 370: 1644–1652. doi: 10.1016/S0140-6736(07)61687-0.

Hoffmann, A. A., Montgomery, B. L., Popovici, J., Iturbe-Ormaetxe, I., Johnson, P. H., Muzzi, F. Greenfield, M., Durkan, M., Leong, Y. S., Dong, Y., Cook, H., Axford, J., Callahan, A. G., Kenny, N., Omodei, C., McGraw, E. A., Ryan, P. A., Ritchie, S. A., Turelli, M. and O'Neill, S. L. (2011). Successful establishment of *Wolbachia* in *Aedes* populations to suppress dengue transmission. *Nature*, 476: 454–457. doi: 10.1038/nature10356.

Iturbe-Ormaetxe, I., Walker, T. and O'Neill, S. L. (2011). *Wolbachia* and the biological control of mosquito-borne disease. *EMBO Rep.*, 12: 508–518. [DOI: 10.1038/embor.2011.84]

Kali, A. (2015). Microbial entomopathogens in control of mosquito borne diseases. *Int. J. Pharm. Bio. Sci.*, 6(4): 232–237. doi://doi.org/10.22376/ijpbs.

Kyle, J. L. and Harris, E. (2008). Global spread and persistence of dengue. *Annu. Rev. Microbiol.*, 62: 71–92. doi: 10.1146/annurev.micro.62.081307.163005.

Lacey, L., Lacey, L. and Roberts, D. W. (1988). Route of invasion and histopathology of *Metarhizium anisopliae* in *Culex quinquefasciatus*. *J. Invertebr. Pathol.*, 52: 108–118. https://www.ncbi.nlm.nih.gov/pubmed/3418133.

Land, M. and Miljand, M. (2014). Biological control of mosquitoes using *Bacillus thuringiensis israelensis*: A pilot study of effects on target organisms, non-target organisms and humans. *MistraEviEM Pilot Study PS4* (www.eviem.se).

Liu, Y. B. and Tabashnik, B. E. (1997). Experimental evidence that refuges delay insect adaptation to *Bacillus thuringiensis*. *Proc. R. Soc. Lond. [Biol.]*, 264: 605–610. doi: 10.1098/rspb.1997.0086.

Mittal, P. K. (2003). Biolarvicides in vector control: Challenges and prospects. *J. Vect. Borne Dis.*, 40: 20–32.

Moncheva, P., Tishkov, S., Dimitrova, N., Chipeva, V., Antonova-Nikolova, S., and Bogatzevska, N. (2002). Characteristics of soil actinomycetes from Antarctica. *J. Culture Collection*, 3: 3–14.

Mulla, M. S. (1990). Activity, field efficacy, and use of *Bacillus thuringiensis israelensis* against mosquitoes. In: *Bacterial control of mosquitoes and black flies*, H. de Barjac and D. J. Sutherland, eds. New Brunswick, NJ: Rutgers University Press, pp. 134–160.

National Academy of Sciences (U.S.), Panel on Perspectives in Mosquito-Control Methods Suitable for Developing Countries. (1973). *Mosquito control: Some perspectives for developing countries: A report of an ad hoc panel of the Advisory Committee on Technological Innovation*. Washington, DC: National Academy of Sciences.

Ng, T. F., Willner, D. L., Lim, Y. W., Schmieder, R., Chau, B., Nilsson, C., Anthony, S., Ruan, Y., Rohwer, F. and Breitbart, M. (2011). Broad surveys of DNA viral diversity obtained through viral metagenomics of mosquitoes. *PLoS One*, 6: e584. doi: 10.1371/journal.pone.0020579.

Pan, X., Zhou, G., Wu, J., Bian, G., Lu, P., Raikhel, A. S. and Xi, Z. (2012). *Wolbachia* induces reactive oxygen species (ROS)-dependent activation of the Toll pathway to control dengue virus in the mosquito *Aedes aegypti*. *Proc. Nat. Sci. Acad. USA*, 109: E23–E31. doi: 10.1073/pnas.1116932108.

Poopathi, S. (2012). Current trends in the control of mosquito vectors by means of biological larvicides. *J. Biofertil. Biopestici.*, 3(125): 1–14. doi: 10.4172/2155-6202.1000125.

Poopathi, S., Mani, C., Thirugnanasambantham, K., Praba, V. L., Ahangar, N. Z. and Balagangadharan, K. (2014). Identification and characterization of a novel marine *Bacillus cereus* for mosquito control. *Parasitol. Res.*, 113: 323–332. doi: 1007/s00436-013-3658-y.

Riba, G., Keita, A., Soares, G. G., and Ferron, P. (1986). Comparative studies of *Metarhizium anisopliae* and *Tolypocladium cylindrosporum* as pathogens of mosquito larvae. *J. Am. Mosq. Control Assoc.*, 2(4): 469–473. http://www.biodiversitylibrary.org/content/part/JAMCA/JAMCA_V02_N4_P469-473.pdf.

Ricci, I., Valzano, M., Ulissi, U., Epis. S., Cappelli.A, and Favia, G. (2012). Symbiotic control of mosquito borne disease. *Pathog. Glob. Health*, 106(7): 380–385. doi: 10.1179/2047773212Y.0000000051.

Rout, R., Raina, V., Suar, M. and Luethy, P. (2011). High prevalence of bacterial spore-formers active against mosquito larvae in temporary monsoon flooded sites in Orissa, India. *J. Am. Mosq. Control. Assoc.*, 27(2): 159–161. doi: 10.2987/10-6067.1.

Sanchez-Contreras, M. and Vlisidou, I. (2008). The diversity of insect bacteria interactions and its applications for disease control. *Biotechnol. Genet. Eng. Rev.*, 25: 203–243. doi: 10.5661/bger-25-203.

Scholte, E. J., Knols, B. G. J., Samson, R. A. and Takken, W. (2004). Entomopathogenic fungi for mosquito control: A review. *J. Insect Sci.*, 4(19): 1–24. doi: http://dx.doi.org/10.1673/031.004.1901.

Sedgwick, S. G. and Smerdon, S. J. (1999). The ankyrin repeat: A diversity of interactions on a common structural framework. *Trends Biochem. Sci.*, 24: 311–316. doi: http://dx.doi.org/10.1016/S0968-0004 (99)01426-7.

Shapiro, A. M., Becnel, J. J. and White, S. E. (2004). A nucleopolyhedrovirus from Uranotaenia sapphirina (Diptera: Culicidae). *J. Invertebr. Pathol.*, 86: 96–103.

Skovmand, O., Kerwin, J. and Lacey, L. A. (2007). Microbial control of mosquitoes and black flies. In: *Field manual of techniques in invertebrate pathology*, L. A. Lacey and H. K. Kaya, editors. Dordrecht, The Netherlands: Springer, pp. 735–750.

Sunish, I. P., Rajendran, R., Paramasivan, R., Dhananjeyan, K. J. and Tyagi, B. K. (2011). *Wolbachia* endobacteria in a natural population of *Culex quinquefasciatus* from filariasis endemic villages of south India and its phylogenetic implication. *Trop. Biomed.*, 28(3): 569–576.

Tolle, M. A. (2009). Mosquito-borne diseases. *Curr. Probl. Pediatr. Adolesc. Health Care*, 39 (4): 97–140. doi: 10.1016/j.cppeds.2009.01.001.

Usta, C. (2013). Microorganisms in biological pest control—A review (bacterial toxin application and effect of environmental factors). In: *Current progress in biological research*, M. Silva-Opps, editor. InTech, pp. 287–317. doi: http://dx.doi.org/10.5772/55786.

Vyas, N., Dua, K. K. and Prakash, S. (2007). Efficacy of *Lagenidium giganteum* metabolites on mosquito larvae with reference to nontarget organisms. *Parasitol. Res.* 101: 385–390. doi: 10.1007/s00436-007-0496-9.

Walker, T., Johnson, P. H., Moreira, L. A., Iturbe-Ormaetxe, I., Frentiu, F. D., McMeniman, C. J., Leong, Y. S., Dong, Y., Axford, J., Kriesner, P., Lloyd, A. L., Ritchie, S. A., O'Neill, S. L. and Hoffmann, A. A. (2011). The wMel *Wolbachia* strain blocks dengue and invades caged *Aedes aegypti* populations. *Nature*, 476: 450–453. doi: 10.1038/nature10355.

Walker, T. and Moreira, L. A. (2011). Can *Wolbachia* be used to control malaria? *Mem. Inst. Oswaldo. Cruz.*, 106 (Suppl. I): 212–217. doi: 10.1590/S0074-02762011000900026.

Werren, J. H. (1997). Biology of *Wolbachia*. *Annu. Rev. Entomol.*, 42, 587–609.

World Health Organization (WHO). (2014). *India attains freedom from polio.* Geneva: World Health Organization. http://www.searo.who.int/india/en.

World Health Organization (WHO). (2016). *Vector-borne diseases.* Available from: http://www.who.int/mediacentre/factsheets/fs387/en/, accessed February 8, 2017.

Wu, M., Sun, L. V., Vamathevan, J., Riegler, M., Deboy, R., Brownlie, J. C., Mcgraw, E. A., Martin, W., Esser, C., Ahmadinejad, N., Wiegand, C., Madupu, R., Beanan, M. J., Brinkac, L. M., Daugherty, S. C., Durkin, A. S., Kolonay, J. F., Nelson, W. C., Mohamoud, Y., Lee, P., Berry, K., Young, M. B., Utterback, T., Weidman, J., Nierman, W. C., Paulsen, I. T., Nelson, K. E., Tettelin, H., O'Neill, S. L. and Eisen, J. A. (2004). Phylogenomics of the reproductive parasite *Wolbachia pipientis* wMel: A streamlined genome overrun by mobile genetic elements. *PloS Biol.*, 2: E69. doi: 10.1371/journal.pbio.0020069.

# 14 Midgut Bacterial Microbiota of Important Mosquito Disease Vectors and Its Role in Control of Parasite Transmission

*Kamlesh Kumar Yadav, Kshitij Chandel, Sameer S. Bhagyawant, and Vijay Veer*

## CONTENTS

## 14.1 INTRODUCTION

Disease vectors transmit various diseases among humans, animals, and plants. Disease vectors of humans and animals are blood feeders and along with the blood meal they take pathogens from an infected host and later transmit them to uninfected host(s) during their subsequent blood feedings either in the same stage (mechanical vector) or in a different stage (biological vector). Disease vectors mostly belong to the insects, ticks, and mite group of the phylum Arthropoda.

Vector-borne diseases (VBDs) are still regarded as a serious threat around the world as inflicting high human morbidity and mortality. VBDs are more common in tropical and subtropical countries, causing more than 1 billion cases and 1 million deaths annually and accounting for over 17% of all infectious diseases (World Health Organization [WHO] 2016). Important VBDs include malaria, dengue, chikungunya, human trypanosomiasis, leishmaniasis, Chagas disease, yellow fever, onchocerciasis, several different types of encephalitides including Japanese encephalitis, Kyasanur Forest disease (KFD), and scrub typhus.

During the last few years, the numbers of VBDs have increased due their emergence and reemergence. Several factors like unplanned urbanization, globalization of trade and travel, and climate change are mainly responsible for this shift in VBDs. VBDs numbers, prevalence, severity, and geographical distribution have increased. Diseases like chikungunya, dengue, and Japanese encephalitis in Asia and Oceania; West Nile virus in the Americas; and Rift Valley fever in Western and Eastern Africa are rapidly emerging (Gould and Solomon 2008), and more recently Zika virus is rapidly spreading in the countries of Central and Southern America, Africa, and Asia.

Mosquitoes are the best known disease vectors and medically most important group of insects, both in terms of the number of pathogens they transmit and also in the magnitude of health problems caused by the diseases worldwide (Service 1989). They transmit important diseases like malaria, dengue, chikungunya, Japanese encephalitis, filariasis, yellow fever, and Zika to man. Mosquito and mosquito-borne diseases affect half of the world's population, mostly in tropical and subtropical regions, by inflicting illness and death, and physical, psychological, and monetary losses to mankind. Mosquito-borne diseases are still a major public health problem and affecting at least one person out of every six throughout the world.

## 14.2 DISEASE MANAGEMENT STRATEGIES

Mosquito control is still the primary disease control intervention, as not many successes were achieved in the development of vaccines and drugs against diseases. Vector control programs are still primarily controlled by domiciliary insecticide applications as options to control them are limited. But prolong excessive and targeted use of insecticides leads to the development of insecticidal resistance in mosquitoes and deleterious effects on the environment. In India, mosquitoes such as *Anopheles culicifacies*, *Culex quinquefasciatus*, *Aedes albopictus*, and *Aedes aegypti* have already developed variable levels of resistance against commonly used insecticides like DDT, temephos, malathion, fenthion, and synthetic pyrethroids (Singh 2011; Tikar et al. 2011) out of 12 insecticides recommended by the WHO. These setbacks have underscored the need for the development of additional effective and ecofriendly methods for disease control so that the use of insecticides can be minimized. Moreover, resistance to the effective drug artemisinin in *Plasmodium falciparum* have already been reported from South East Asia and extending to Indian border (White 2004).

## 14.3 ALTERNATIVE METHODS OF VECTOR CONTROL

Several environment friendly methods are available with different efficacy, such as the use of predators (larvivorus fishes, Copepods), growth regulators autodissemination, biocides, lure and kill trapping systems using pheromones/kairomones, release of sterilized males (SIT), genetically modified vectors (transgenics), and paratransgenics vectors including *Wolbachia* endosymbiont-mediated transmission intervention.

A genetically modified (GM) mosquito is being developed with various degrees of success. Genetic modification of vector mosquitoes is one of such technologies that may be mainly used either to suppress or replace the wild populations of a vector so as to decrease vector populations or reduce vector ability for transmission. These applications include release of reared mosquitoes in the environment to introduce modified genetic traits in wild populations. Genetic modification encompasses multiple approaches, which are broadly categorized into two types. The first category includes the sterilized insect technique (SIT) for population suppression and the second category is gene drive systems for population replacement or manipulation. Gene drive systems include multiple approaches like the CRISPR gene-editing technique and homing endonuclease genes (HEG).

The sterilized insect technique includes the classical radiation-induced male sterility and dominant lethal gene systems. In classical radiation-induced male sterility (SIT), sterilized males are released into the wild for mating with wild females, which results in reduction of reproductive potential in the wild population. The SIT has successfully been used in eradication of the New World screwworm (*Cochliomyia hominivorax*) in Mexico and the United States, in controlling of the Mediterranean fruit fly population (*Ceratitis capitata*) in California and Florida, and the codling moth (*Cydia pomonella*) and pink bollworm moth (*Pectinophora gossypiella*) (Lindquist et al. 1992).

The release of insects carrying a dominant lethal gene (RIDL) technique is the genetic enhancement of the SIT. In the RIDL, a lethal gene has to be inserted in insects to produce the nontoxic, lethal protein (tTAV) that allows development of larvae but prevents the surviving of their offspring in the adulthood (Oxitec 2016).

A successful field trial of this techniques was done in the Cayman Islands to control the *Aedes aegypti* and about 80% of the wild population was suppressed

(Harris et al. 2011, 2012). Another evident of this technology has also been obtained from field trials in Brazil with the reduction of 81% and 95% of local *Ae. aegypti* populations (Carvalho et al. 2015).

## 14.4 MOSQUITO MIDGUT AND MICROBIOTA

The mosquito gut (Figure 14.1) presents a distinctive environment for microbial colonization, and bacteria in the gut potentially provide many beneficial services to their host such as nutrition, protection from parasites and pathogens, modulation of immune responses, and communication. The mosquito midgut is the second largest organ and is the site where the disease-causing parasite, harmful bacteria, viruses, and toxins as well as food and water comes in contact with the first. Each disease-causing pathogen has its distinct life cycle, but they face the same events after being ingested and exposed to the mosquito's midgut environment. Shortly after ingestion of blood meal in the midgut, temperature and pH change abruptly and the digestion process starts with proteolytic enzymes. These digestive enzymes can have a negative or positive impact on the pathogens' survival and therefore can influence vector competence. Pathogens must be passed over the peritrophic matrix during migration to their developmental site for their development (in the case of filarial worm) and multiplication (in the case of malaria and viral agents) (Beerntsen et al. 2000).

Only female mosquitoes are responsible for the transmission of disease, because only female mosquitoes need the blood meal as protein source for egg development in their gonotrophic cycle. Female mosquitoes picked up the parasites from the infected person or infected females bite the uninfected persons while taking their blood meal and transmit the disease parasites. Mosquitoes used wide range of hosts for blood-meal feeding, including mammals, birds, reptiles, amphibians, and fish. Mosquitoes generally prefer the single host individual for their blood meal, but due to the defensive response of the host to the painful bites their feeding is mainly interrupted. Some

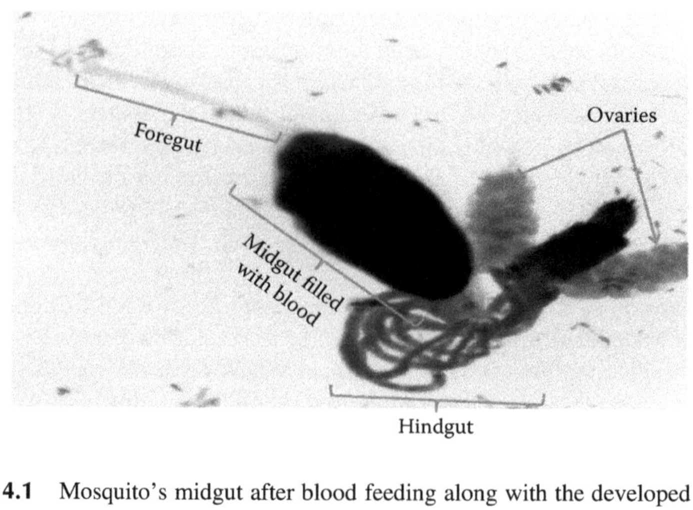


**FIGURE 14.1** Mosquito's midgut after blood feeding along with the developed ovaries.

mosquito species like *Aedes* feeds on multiple hosts, and hence show great epidemiological importance.

The midgut serves as the first contact point between parasites, midgut bacteria, and the epithelial surfaces, where significant parasite numbers are reduced (Azambuja et al. 2005). The defense mechanism in mosquitoes against invading pathogens is believed to be provided by the midgut microbiota, by means of raising the immunity or by impeding the parasite development in the mosquito midgut. In the recent years, it is reported that some midgut-inhabiting bacteria play an important role in vectorial capacity of mosquitoes through interaction with host and parasites (Gonzalez-Ceron et al. 2003; Dong et al. 2009; Kambris et al. 2010; Cirimotich et al. 2011a; Fang et al. 2011; Rasgon 2011).

### 14.4.1 Acquisition and Establishment of Microbiota Inside Mosquito Midgut

The mosquito midgut has its own normal microbiota, and it can be infected with certain microbes that are known to augment the immune response of the mosquito (Meister et al. 2005, 2009; Dong et al. 2006), and it believe that immunocompetent mosquitoes are thought to be less likely to transmit parasites such as malaria, dengue, and chikungunya (Abdul-Ghani et al. 2012).

The midgut microbial diversity of mosquitoes is known to vary according to the life stages of the mosquitoes (Minard et al. 2013). The larvae of mosquitoes is an aquatic stage, however the adult male and female is a terrestrial stage. Hence, midgut microbiota in mosquitoes are acquired from both the aquatic larval stage, and by nectar and blood feeding in the adult stage (Yadav et al. 2016a). In the aquatic larval stage, it might be inserted from breeding sites and passed on to the adults (Smith et al. 1998; Osei-Poku et al. 2012; Díaz-Nieto et al. 2016). Midgut microbes in the adult mosquito are thought to be acquired through vertical inheritance as well as from the surrounding environments (Minard et al. 2013; Moro et al. 2013; Buck et al. 2016). Recently, it has also been, proposed by Buck and colleagues (2016) that the midgut bacterial diversity of the mosquito is a reflection of microbiota acquired from various environments (Buck et al. 2016; Yadav et al. 2016). Midgut microbiotas of adults have also been determined by the native breeding sources of the larvae where they were grown and from where the bacteria have been acquired, which makes a specific bacterial profile for different environmental condition (Zouache et al. 2011; Boissiere et al. 2012; Minard et al. 2013; Tchioffo et al. 2015; Buck et al. 2016). This is the reason for variations in the midgut microbiota among adult mosquitoes, which breed at distinct sites. It has also been described that bacterial profiles associated with mosquitoes provided precise and predictive information about the spatial dynamics of the mosquito population (Buck et al. 2016).

Recently, Yadav et al. (2016a) identified a large number of bacterial genera, including *Microbacterium*, *Acinetobacter*, *Stenotrophomonas*, *Micrococcus*, *Pseudomonas*, *Klebsiella*, *Aeromonas*, *Bacillus*, *Clostridium*, and *Enterobacter*, that were already reported to be common in mosquitoes breeding sites and to be ingested by the larvae and passed on to the adults male and female (Smith et al. 1998; Osei-Poku et al. 2012; Díaz-Nieto et al. 2016). Moreover, it has also been reported that the additional bacteria could be inserted at adult stages through horizontal transfer from breeding environments (Coon et al. 2014; Díaz-Nieto et al. 2016).

The gut microbial diversity of mosquitoes varies with various developmental stages (Demaio et al. 1996; Wang et al. 2011; Minard et al. 2013; Ngwa et al. 2013). Ngwa et al. (2013) have also demonstrated that diversity of midgut microbiota decreases during the development of mosquitoes from egg to adult. They also identified a dominant gram-negative bacterium *Elizabethkingia meningoseptica* in the midgut of male and female adult *Anopheles stephensi* that can be transmitted from one generation to another. The variation of midgut bacterial communities also depends on the mosquito's sex (male and female) and ecological factors (Zouache et al. 2011; Minard et al. 2013).

### 14.4.2 Effect of Feeding Habits on Microbial Diversity

It has been described that bacterial communities vary according to different feeding regimes. Male adults solely depend upon plant sap, however, in addition to plant sugars, females also require a blood meal for their ovary development. This shifting leads to a significant shift from high carbohydrate levels to protein in mosquito midgut, resulting in increased levels of enteric bacteria, while reducing the overall microbiota diversity (Foster and Takken 2004; Manda et al. 2007; Rani et al. 2009; De Gaio et al. 2011; Wang et al. 2011; Chavshin et al. 2013; Yadav et al. 2016). In 1996, Pumpuni et al. described that the number of bacteria in *An. gambiae* have increased 11-fold 24 h after a blood meal while in *An. stephensi* it increased up to 40-fold. Increasing the bacterial count up to 2000-fold in *Culex* mosquito was also observed. In the presence of a blood meal, the bacterial population increases, but after 3 to 5 days it decreases to the pre-blood meal level (Pumpuni et al. 1996). After a blood meal, an interspecies competition between the bacterial isolates occurs in the female mosquito's midgut and the numbers of coexisting bacterial species become low. Terenius et al. (2012) have also suggested that midgut bacterial dynamics in the mosquito midgut is due to the possible existence of coadaptation between midgut bacteria and their host.

## 14.5 CHARACTERIZATION OF MIDGUT MICROBIOTA

The midgut of mosquitoes harbors a wide range of microbiota, as given in Table 14.1. Microbiota influence the fitness traits of their hosts including immunity, nutrition, and reproduction. Using the latest biotechnology techniques, midgut microbiota might be used for the development of an effective and efficient technique like paratransgenesis for the management of vector-borne disease. Prior to development of such techniques, it is essential to explore the midgut microbiota of mosquitoes and give attention to their important role. Mainly two techniques—culture dependent and culture independent—have been used for the characterization of the composition, dynamics, and diversity of mosquito midgut microbiota.

The first step of the characterization of midgut microbiota in both culture-dependent and culture-independent techniques is the sample collection and dissection of the mosquito population for midgut isolation.

**TABLE 14.1**
**Isolated Microbiota from the Different Mosquito Species Collected from Various Countries and Their Specific Activity**

| Mosquito Species | Country of Collection | Name of Microbiota | Specific Activities of Microbiota | Reference |
|---|---|---|---|---|
| *Culex tarsalis* (lab) | United States (California) | *Achromobacter*, *Aerobacter*, *Escherichia*, *Flavobacterium*, *Micrococcus*, *Proteus*, *Pseudomonas*, *Saccharomyces*, *Geotrichum* | Male mosquitoes showed fewer organisms than females | Chao and Wistreich (1959) |
| *Cx. quinquefasciatus* (lab) | United States (California) | *Achromobacter* sp., *Flavobacterium* sp., *Hafnia group*, *Klebsiella* sp., *Micrococcus caseolyticus*, *Micrococcus ureae*, *Pseudomonas* sp. | – | Chao and Wistreich (1960) |
| *Cx. fatigans* (lab) (=*Cx quinquefasciatus*) | United States (Texas) | *Alcaligenes* sp., *Lactobacillus* sp., *Pseudomonas* sp. | Study concluded that the removal of microorganisms either allows more nutriment to become available for the malaria parasites or deprives the insect of some metabolite necessary for defense against *Plasmodium relictum*. | Ferguson and Micks (1961) |
| Malaria mosquito | Democratic Republic of the Congo | *Pseudomonas* sp. | *Pseudomonas* play important role in sporogenesis of hematozoon of malaria in *Anopheles* mosquito. | Jadin et al. (1966) |
| *Aedes sollicitans* (field) | United States (New Jersey) | *Spiroplasma culicicola* | A novel species from the salt marsh mosquito *Aedes sollicitans* | Hung et al. (1987) |
| *Anopheles stephensi* (lab) | Germany | *Serratia marcescens* | Mortality rate in *Anopheles stephensi* increases after infection with *Plasmodium berghei* and *Serratia marcescens* | Seitz et al. (1987) |

*(Continued)*

**TABLE 14.1 (CONTINUED)**
**Isolated Microbiota from the Different Mosquito Species Collected from Various Countries and Their Specific Activity**

| Mosquito Species | Country of Collection | Name of Microbiota | Specific Activities of Microbiota | Reference |
|---|---|---|---|---|
| *Culex tritaeniorhynchus* (field) | Taiwan | *Spiroplasma taiwanense* | A novel species from the *Cx. tritaeniorhynchus* | Abalain-Colloc et al. (1988) |
| *Ae. aegypti* and *An. stephensi* | France | *Spiroplasma taiwanense* | Significantly reduces the survival of *Ae. aegypti* larvae (Abalain-Colloc et al. 1988) | Humphery Smith et al. (1991b) |
| *An. stephensi* | France | *Spiroplasma taiwanense* | *Spiroplasma* also significantly reduces the flight capacity of adult female *Ae. aegypti* (Abalain-Colloc et al. 1988) | Humphery Smith et al. (1991a) |
| *Cx. quinquefasciatus* | India | **Larvae**<br>*Bacillus*, *Pseudomonas*, *Shigella*, *Staphylococcus*, *Flavobacterium*, *Serratia*, *Enterobacter*, *Alcaligenes*, *Pectobacterium*, *Acinetobacter*, *Escherichia*, *Listeria*, *Proteus*, *Micrococcus* and *Salmonella* | During the early instar of larval development, *Bacillus*, *Pseudomonas*, *Shigella*, *and Staphylococcus* caused 100% mortality<br>Isolate *Escherichia* completely suppressed the adult emergence | Vasanthi and Hoti (1992) |
| *An. stephensi* | United States | *Ewingella americana*, *Serratia marcescens*, *Staphylococcus* sp. | Gram-negative bacterial strains partially or completely inhibit the oocyst formation | Pumpuni et al. (1993) |

(*Continued*)

**TABLE 14.1 (CONTINUED)**
**Isolated Microbiota from the Different Mosquito Species Collected from Various Countries and Their Specific Activity**

| Mosquito Species | Country of Collection | Name of Microbiota | Specific Activities of Microbiota | Reference |
|---|---|---|---|---|
| *An. albimanus*, *An. gambiae*, *An. stephensi* (lab) | United States (Baltimore, Maryland) | *Pseudomonas cepacia*, *Aeromonas hydrophila*, *Flavobacterium* spp., *Cedecea lapagei*, *Enterobacter agglomerans*, *Pseudomonas gladioli*, *Kluyvera cryocrescens*, *Serratia* spp., *Acinetobacter* spp. | In the presence of a blood meal, the bacterial population increases 11- to 40-fold, but it decreases to pre-blood meal levels after 3 to 5 days | Pumpuni et al. (1996) |
| *Cx. annulus* (field) | Taiwan | *Spiroplasma diminutum* | A novel bacterial species | Williamson et al. (1996) |
| *Cx. pipiens* (field) | United States | *Acinetobacter* spp., *Aeromonas* spp., *Comamonas* spp., *flavobacterium* spp., *Klebsiella* spp., *Pseudomonas* sp., *Sphingobacterium* spp. | During the development of mosquitoes, the number of bacterial count changes and the bacterial population decreases in between the emergence of larva and pupa (280–1100-fold), whereas increases after the blood feeding (70–16,000-fold) | Demaio et al. (1996) |
| *Ps. Columbiae* (field) | United States | *Acinetobacter* spp., *Cedecea* spp., *E. cloacae*, *Enterobacter* sp., *Flavibacterium* spp., *Ps. Aeruginosa*, *Pseudomonas* sp., *Sphingobacterium* sp., *Stenotrophomonas maltophilia* | | |
| *Ae. triseriatus* (field) | United States | *Acinetobacter lwoffii*, *Bacillus* spp., *Brevibacterium* spp., *Comamonas* spp., *E. Cloacae*, *Enterobacter* sp., *Enterococcus faecalis*, *Erwinia* spp., *Favovacterium* spp., *p. agglomerans*, *Providencia* spp., *Ps. aeruginosa*, *Serratia marcescens*, *Serratia* sp., *Xanthomonas maltiphilia* | | |

*(Continued)*

**TABLE 14.1 (CONTINUED)**
**Isolated Microbiota from the Different Mosquito Species Collected from Various Countries and Their Specific Activity**

| Mosquito Species | Country of Collection | Name of Microbiota | Specific Activities of Microbiota | Reference |
|---|---|---|---|---|
| *An. gambiae* (field) | Kenya and Mali | *Achromobacter xylosoiydans, Bacillus cereus, Bacillus coagulans, Bacillus mucoides, Bacillus thurigensis, Cedecea davisae, Escherichia coli, Hydrogenophaga pseudoflava, Klabsiella pneumonia, Morganella morgani, P. agglomerans, Pantoea ananas, Pseudomonas putida, Pseudomonas stutzeri, Salmonella choleraesuis, Salmonella enteritidis* | *An. funestus* females that harbored gram-positive bacteria were likely to be more infected with sporozoites compared with those with no cultivable bacteria or gram-negative bacteria in their midguts | Straif et al. (1998) |
| *An. funestus* (field) | | *Bacillus megaterium, Brevundiumonas diminuta, C. davisae, Comamonas testeroni, E. Coli, Erwinia chrysanthemum, Flavobacterium resinovorum, Gluconobacter cerinus, K. Pneumoniae, Klyvera cryosceens, P. agglomerans, P. ananas, Ps. Stutzeri, Pseudomonas mendocina, S. choleraesuis, St. maltophilia* | | |
| *Cx. pipiens* (lab) | Egypt | *Bacillus* sp., *Shigela* sp., *Staphylococcus* sp., *Streptococcus* sp. | *Bacillus* and *Staphylococcus* are essential for normal and high fecundity of *Cx. pipiens*<br>Gut bacteria are also essential for normal embryo development | Fouda et al. (2001) |

(*Continued*)

**TABLE 14.1 (CONTINUED)**
**Isolated Microbiota from the Different Mosquito Species Collected from Various Countries and Their Specific Activity**

| Mosquito Species | Country of Collection | Name of Microbiota | Specific Activities of Microbiota | Reference |
|---|---|---|---|---|
| *An. albimanus* (field) | Mexico | *E. cloacae*, *E. amnigenus*, *Enterobacter* sp., *S. marcescens*, *Serratia* sp. | Infection of *Plasmodium vivax* decreases in *Anopheles* mosquitoes when coinfected with *En. cloacae*, *En. Amnigenus* 2, and *S. marcensces* | Gonzalez-Ceron et al. (2003) |
| *Cx. quinquefasciatus* and *Ae. aegypti* | India | *Aeromonas culicicola* | A novel midgut bacterial species | Pidiyar et al. (2002) |
| *Cx. quinquefasciatus* (field) | India | *Acinetobacter calcoaceticus*, *Acinetobacter junii*, *Acinetobacter* spp., *P. aeruginosa*, *St. maltophilia*, *B. thuringensis*, *Enterococcus seriolicida*, *Microbacterium oxydans*, *Staphyloccus epidermidis*, *Lactococcus* spp., *unidentified* spp., *Aeromonas veroni*, *P. agglomerans*, *Aeromonas veroni*, *Staphyloccus epidermidis*, unidentified spp. | Midgut microbial diversity study | Pidiyar et al. (2004) |
| *An. arabiensis* (field) | Western Kenya | **Culture dependent**<br>*Bacillus simplex*, *Vibrio metschnikovii*, *Serratia odorifera*, *Nocardia corynebacterioides*, *Bacillus silvestris*, *Escherichia senegalensis*, *Janibacter limosus*<br>**Culture independent**<br>*Acidovorax temperans*, *Mycoplasma wenyonii*, *Stenotrophomonas maltophilia*, *Paenibacillus* sp., *Anaplasma ovis*, *Ehrlichia* sp. *strain Bom Pastor* | Several bacteria were also to be symbionts in other insects. *Nocardia corynebacterioides*, a symbiotic bacterium of Chagas disease vector, and has been proven as potential bacteria of paratransgenesis.<br>*Asaia*, a dominant mosquito-associated bacterial genus, was found in adult male and female midgut, and the reproductive system of male and larval gut | Lindh et al. (2005) |

*(Continued)*

**TABLE 14.1 (CONTINUED)**
**Isolated Microbiota from the Different Mosquito Species Collected from Various Countries and Their Specific Activity**

| Mosquito Species | Country of Collection | Name of Microbiota | Specific Activities of Microbiota | Reference |
|---|---|---|---|---|
| *An. gambiae* sensu stricto (field) | | **Culture dependent** *Pseudomonas putida* **Culture independent** *Stenotrophomonas maltophilia*, *Stenotrophomonas* sp., *Aeromonas hydrophila*, *Aeromonas* sp. | | |
| *An. funestus* (field) | Western Kenya | **Culture independent** *Spiroplasma* sp. | | |
| *Anopheles gambiae* (lab) | Sweden | *Pseudomonas otitidis*, *Delftia tsuruhatensis*, *Comamonas testosterone*, *Sphingobacterium multivorum*, *Chryseobacterium meningosepticum*, *Pantoea stewartii*, *Stenotrophomonas maltophilia* | *Pantoea* exhibited the transstadial and horizontal transmission properties. | Lindh et al. (2008) |
| *An. stephensi* | Italy (Camerino) | *Asaia* spp., *Gluconobacter asaii*, *Acetobacter aceti*, *Sphingomonas rhizogenes* | Among all isolates bacterial genera *Asaia* was dominant and reported from female and larval gut, salivary gland, and male reproductive system | Favia et al. (2007) |
| *An. maculipennis* | Italy (Orte) | *Serratia* spp., *Asaia* spp., *Staphylococcus* spp. | | |
| *An. gambiae* | Burkina Faso (West Africa) | *Sphingomonas* spp., *Phenilobacterium* spp., *Asaia* spp., *Burkolderia* spp., *Aquabacterium* sp., *Acinetobacter* spp., *Pseudomonas* spp. | | |
| *An. stephensi* (lab reared) | India | **Adult male culturable** *Chryseobacterium meninqosepticum*, *Agrobacterium* sp., *Pseudomonas mendocina*, *Serratia marcescens* | Midgut bacterial diversity study of field- and laboratory-reared *Anopheles* mosquitoes | Rani et al. (2009) |

*(Continued)*

**TABLE 14.1 (CONTINUED)**
**Isolated Microbiota from the Different Mosquito Species Collected from Various Countries and Their Specific Activity**

| Mosquito Species | Country of Collection | Name of Microbiota | Specific Activities of Microbiota | Reference |
|---|---|---|---|---|
| | | **Adult male unculturable**<br>*C. meninqosepticum, Elizabethkingia meninqosepticum, A. umefaciens, P. Tolaasii, Klebsiella* sp., *S. marcescens*<br>**Adult female culturable**<br>*C. meninqosepticum, Comamonas* sp., *P. Mendocina, S. marcescens*<br>**Adult female unculturable**<br>*C. meninqosepticum, E. Meninqosepticum, S. Marcescens, Serratia* sp. | | |
| *An. stephensi* (field collected) | India | **Adult male culturable**<br>*Micrococcus* sp., *Staphylococcus hominis, S. saprophyticus, Acinetobacter A. lwofii, A. radioresistens, A. johnsonii, Enterobacter cloacae, Escherichia hermani*<br>**Adult male unculturable**<br>*Bacillus* sp., *Paenibacillus alginolyticus, P. chondroitinus, Paenibacillaceae, Herbaspirillum sp., Photorhabdus luminescens*<br>**Adult female culturable**<br>*Chryseobacterium indologenes, Acinetobacter hemolyticus, A. radioresistens, Citrobacter freundii, Enterobacter cloacae, E. sakazaki, E. hermani* | | |

(*Continued*)

**TABLE 14.1 (CONTINUED)**
**Isolated Microbiota from the Different Mosquito Species Collected from Various Countries and Their Specific Activity**

| Mosquito Species | Country of Collection | Name of Microbiota | Specific Activities of Microbiota | Reference |
|---|---|---|---|---|
| | | **Adult female unculturable** | | |
| | | *Leuconostoc citreum, Achromobacter xylosoxidans, Acinetobacter hemolyticus, Acinetobacter*, sp., *Pseudomonas putida, Pseudomonas synxantha, Pseudomonas* sp., *S. marcescens, S. Nematodiphila, S. proteamaculans, Xenorhabdus nematodiphila, Leminorella grimontii* | | |
| | | **Larvae culturable** | | |
| | | *C. indologenes, Bacillus* sp., *B. Cereus, B. firmus, Exiguobacterium, Acinetobacter venetianus, Aeromonas sobria, A. Popoffii, P. anquilliseptica, Pseudoxanthomonas, Thorsellia anopheles, Vibrio chlorae, Deinococcus xinjiangensis* | | |
| | | **Larvae unculturable** | | |
| | | *Calothrix* sp., *Brevibacterium paucivorans, Dysqonomonas* sp., *Staphylococcus cohnii, S. suis, B. thermo amylovorans, Lactobacillus Azoarcus* sp., *Leptothrix* sp., *Hydroxenophaga, Ignatzschineria larvae* sp., *Enterobacter* sp., *Serratia*, sp., *Serratia* sp., *T. anopheles, S. Marcescens, S. Nematodiphila, D. xinjiangensis* | | |

*(Continued)*

**TABLE 14.1 (CONTINUED)**
**Isolated Microbiota from the Different Mosquito Species Collected from Various Countries and Their Specific Activity**

| Mosquito Species | Country of Collection | Name of Microbiota | Specific Activities of Microbiota | Reference |
|---|---|---|---|---|
| *An. gambiae* | United States | *Enterobacter asburiae*, *Microbacterium* sp., *Sphingomonas* sp., *Serratia* sp., *Chryseobacterium meningosepticum*, *Asaia bogorensis*, *Bacillus subtilis*, *Enterobacter aerogenes*, *Escherichia coli*, *Herbaspirillum* sp., *Pantoea agglomerans*, *Pseudomonas fluorescens*, *Pseudomonas straminea*, *Phytobacter diazotrophicus*, *Serratia marcescens* | Mosquito microbiota upregulated the immune genes including several other anti-*Plasmodium* factors<br>Susceptibility to *Plasmodium* infection in microbe-free mosquitoes increases compared to mosquito cofeeding with bacteria | Dong et al. (2009) |
| *An. stephensi*, *An. gambiae*, *Ae. aegypti*, *Ae. albopictus* (lab) | Sweden, Burkina Faso (West Africa), Switzerland, Italy | *Asaia* spp., *Burkholderia*, *Elizabethkingia*, *Bacillus spp.*, *Gluconacetobacter liquefaciens* | *Asaia* was dominant among all bacterial isolates and was identified from most of the mosquitoes.<br>The symbiotic bacteria *Asaia* play an important role in the normal development of *An. stephensi* larvae | Chouaia et al. (2010, 2012) |
| *Ae. aegypti* (lab) | Brazil | **Culturable**<br>*Klebsiella pneumonia*, *Asaia* sp., *Pichia ohmeri*, *Klebsiella pneumonia*, *Serratia marcescens*, *Enterobacter asburiae*, *Serratia* sp., *Bacillus subtilis*, *Asaia krungthepensis*, *Bacillus* sp., *Enterococcus caccae* | During blood digestion, number of bacteria were high in the midgut and reached its peak at 48 h after blood meal feeding and lumen was almost totally occupied by bacteria. | Gusmão et al. (2010) |

*(Continued)*

**TABLE 14.1 (CONTINUED)**
**Isolated Microbiota from the Different Mosquito Species Collected from Various Countries and Their Specific Activity**

| Mosquito Species | Country of Collection | Name of Microbiota | Specific Activities of Microbiota | Reference |
|---|---|---|---|---|
| | | **Unculturable** | | |
| | | **Bacterial clones** | | |
| | | *Serratia marcescens*, *Uncultured bacterium*, *Klebsiella pneumoniae*, *Enterobacter aerogenes*, *Pantoea agglomerans*, *Enterobacter* sp., *Bactéria endofítica*, *Citrobacter freundii*, *Cedecea davisae*, *Pantoea* sp., *Serratia* sp., *Kluyvera ascorbata* | | |
| | | **Yeast clones** | | |
| | | *Pichia guilliermondii*, *Candida nodaensis*, *Candida fermentati* | | |
| *Culex quinquefasciatus* (field) | India | *Chryseobacterium culicis* | A novel bacterial species | Kampfer et al. (2010) |
| *Ae. albopictus* (field) | Madagascar | *Acinetobacter genomo* sp., *Pseudomonas putida*, *Pseudomonas* sp., *Pantoea agglomerans*, *Stenotrophomonas maltophilia*, *Citrobacter* sp., *Yokenella regensburgei*, *Asaia* sp., *Delftia* sp., *Bradyrhizobium* sp., *Streptococcaceae*, *Shigella flexneri*, *Wolbachia pipientis*, *Rickettsia*, *Rhizobium* sp., *Herbaspirillum* sp., *Staphylococcus* sp., *Enterobacter* sp. | Bacterial composition and diversity of *Ae. albopictus* and *Ae. aegypti* were influenced by both the sex of individuals and by the environment inhabited by the mosquitoes<br>Members of alpha- and gamma proteobacteria phyla, as well as Bacteroidetes, responded to chikungunya infection | Zouache et al. (2011, 2012) |

(*Continued*)

**TABLE 14.1 (CONTINUED)**
**Isolated Microbiota from the Different Mosquito Species Collected from Various Countries and Their Specific Activity**

| Mosquito Species | Country of Collection | Name of Microbiota | Specific Activities of Microbiota | Reference |
|---|---|---|---|---|
| *Ae. aegypti* (field) | Madagascar | *Acinetobacter genomo* sp., *Staphylococcus saprophyticus*, *Asaia* sp., *Pseudomonas* sp., *Acinetobacter* sp., *Enterobacter* sp. | With chikungunya infection, the bacterial abundance from the Enterobacteriaceae family increased, whereas the abundance of known insect endosymbionts like *Wolbachia* and *Blattabacterium* decreased | |
| *An. gambiae*, *An. arabiensis* (field) | Zambia | *Acinetobacter* sp., *B. pumilus*, *Bacillus* sp., *Enterobacter* sp., *P. putida*, *B. cereus*, *Bacillus* sp. "Mali 51", *E. mexicanum*, *K. turfanensis*, *Pantoea* sp., *P. rhodesiae*, *Staphylococcus* sp., *Arthrobacter* sp., *Comamonas* sp., *Enterobacter* sp., *Knoellia* sp. | Bacteria-generated reactive oxygen species shows anti-*Plasmodium* effect | Cirimotich et al. (2011a) |
| *An. gambiae* (field) | Kenya | *Klebsiella*, *Raoultella*, *Serratia*, *Enterobacter*, *Aeromonas*, *Pseudomonas*, *Elizabethkingia*, *Acinetobacter*, *Comamonas*, *Propionibacterium*, *Stenotrophomonas*, *Bacillariophyta*, *Thorsellia*, *Finegoldia*, *Chlorophyta*, *Methylocystis*, *GpIIa*, *Roseomonas*, *Novosphingobium*, *Aerococcus*, *Corynebacterium*, *Lactobacillus*, *Cloacibacterium*, *Rhizobium*, *Porphyrobacter*, *Agromyces*, *GpV*, *Clostridium*, *Hydrogenophaga* *GpI*, *Methylophilus*, *Fusobacterium*, *Chryseobacterium*, *Pelagibacter*, *Sphingobium* | Gut bacterial composition varies at family level according to their different life stages<br>Diet (sugar meal and blood meal) significantly affects the midgut microbial structure, and blood meals favor the enteric bacterial communities and reduce the community diversity | Wang et al. (2011) |

*(Continued)*

**TABLE 14.1 (CONTINUED)**
**Isolated Microbiota from the Different Mosquito Species Collected from Various Countries and Their Specific Activity**

| Mosquito Species | Country of Collection | Name of Microbiota | Specific Activities of Microbiota | Reference |
|---|---|---|---|---|
| *Ae. albopictus* (field) | United States | *Erwinia quercina*, *Vagococcus salmoninarium*, *Kluyvera cryocrescens*, *Enterobacter ludwigii*, *Pseudomonas rhodesiae*, *Pantoea agglomerans*, *Bacillus megaterium*, *Chryseobacterium aquaticum*, *Erwinia quercina*, *Roseomonas cervicalis*, *Pedobacter agri*, *Curtobacterium flaccumfaciens*, *Leuconostoc mesenteroides*, *Curtobacterium flaccumfaciens*, *Paenibacillus borealis*, *Brenneria quercina*, *Leuconostoc mesenteroides*, *Vagococcus salmoninarium*, *Brenneria salicis*, *Erwinia persicinus* | La Crosse virus infectivity for Vero cells has been reduced by midgut bacteria and about 50% isolated bacteria played this significant role | Joyce et al. (2011) |
| *An. maculipennis* | Iran | *Pseudomonas mendocina*, *Aeromonas punctata*, *Aeromonas bivalvium*, *Lysinibacillus sphaericus* | Midgut bacteria may be used for development of paratransgenesis approach to control malaria transmission | Djadid et al. (2011) |
| *An. stephensi* (field) | Iran | *Bacillus pumilus*, *Brevundimonas aurantiaca*, *Lysinibacillus sphaericus*, *Sphingomonas paucimobilis*, *Rahnella aquatilis*, *Pantoea agglomerans*, *Pantoea stewartii* | | |
| *Ae. aegypti* (field) | Panama | *Micrococcus* sp., *Acetobacter ghanensis*, *Asaia Krungthepensis*, *Asaia bogorensis*, *Roseomonas* sp., *Bacillus subtilis*, *Staphylococcus capprae*, *Lactococcus lactis*, *Paenibacillus* sp., *Chromobacterium haemolyticum*, *Comamonas testosterone*, *Shinella* | A reciprocal interaction between the midgut microbiota and dengue virus and microbiota critically influences the virus transmission | Ramirez et al. (2012) |

*(Continued)*

**TABLE 14.1 (CONTINUED)**
**Isolated Microbiota from the Different Mosquito Species Collected from Various Countries and Their Specific Activity**

| Mosquito Species | Country of Collection | Name of Microbiota | Specific Activities of Microbiota | Reference |
|---|---|---|---|---|
| | | *kummerowiae*, *Elizabethkingia meningoseptica*, *Chryseobacterium* sp., *Acinetobacter* sp., *Aeromonas hydrophila*, *Aeromonas* sp., *Enterobacter hormaechei*, *Enterobacter hormaechei* subsp. *Steigerwaltii*, *Enterobacter ludwigii*, *Shigella* sp., *Pantoea dispersa*, *Pantoea agglomerans*, *Proteus mirabilis*, *Proteus penneri*, *Pseudomonas* sp., *Pseudomonas stutzeri*, *Serratia marcescens strain*, *Serratia marcescens* | | |
| *An. gambiae* (field) | Cameroon | *Asaia* sp., *Burkholderia* sp., *Acinetobacter* sp., *Ralstonia* sp., *Methylobacterium* sp., *Sphingomonas* sp., *Pseudomonas* sp., *Stenotrophomonas* sp., *Streptococcus* sp., *Comamonas* sp., *Sediminibacterium* sp., *Escherichia* sp., *Shigella* sp., *Bradyrhizobium* sp., *Staphylococcus* sp., *Fusobacterium* sp., *Elizabethkingia* sp., *Gluconacetobacter* sp., *Neisseria* sp., *Schlegelella* sp., *Prevotella* sp., *Cedecea* sp., *Serratia* sp. | Gut microbial communities are one of the major components of mosquito innate immune responses, which play an important role in controlling malaria infection | Boissiere et al. (2012) |
| *Ae. aegypti* (lab) | Sweden | *Elizabethkingia meningoseptica*, *Pantoea stewartii* ssp. *Stewartii*, *Sphingomonas* sp., *Burkholderiaceae* sp., | After blood meal feeding by female adults, the number of coexisting bacterial species becomes low | Terenius et al. (2012) |

*(Continued)*

**TABLE 14.1 (CONTINUED)**
**Isolated Microbiota from the Different Mosquito Species Collected from Various Countries and Their Specific Activity**

| Mosquito Species | Country of Collection | Name of Microbiota | Specific Activities of Microbiota | Reference |
|---|---|---|---|---|
| | | *Bacillus* sp., *Serratia marcescens*, *Klebsiella pneumonia*, *Enterococcus faecalis* | Bacterial dynamics among the mosquito midgut is due to the possible existence of coadaptation between midgut bacteria and their host | |
| *Ae. aegypti*, *Cx, quinquefasciatus*, *Ae. bromeliae*, *An. gambiae*, *An. funestus*, *An. coustani*, *Mansonia Africana*, *Ma. uniformis* | Kenya | *Aeromonas*, *Chryseobacterium*, Alphaproteobacteria *Asaia*, *Zymobacter*, *Acinetobacter*, *Bacilli*, *Staphylococcus*, *Actinobacteria*, *Propionibacterium*, *Sphingomonas*, *Pseudomonas*, *Gluconobacter*, *Streptococcus*, *Methylobacterium*, *Corynebacterium*, *Escherichia*, *Shigella*, *Enterococcus*, *Pantoea*, *Micrococcus*, *Providencia*, *Lactococcus*, *Acetobacter* | A simple microbiota has been found in the mosquito midgut, which is dominated by single bacterial taxon<br>The gut bacteria of different mosquitoes share a noteworthy similarity<br>An extreme variability is found in the individuals of a population and a little midgut bacteria can be overlap in their midgut, which cause the variability of disease rates within the mosquito population | Osei-Poku et al. (2012) |
| *An. stephensi* (field) | Iran | **Larva**<br>*Myroides*, *Chryseobacterium*, *Klebsiella*, *Aeromonas*, *Pseudomonas*, *Enterobacter*, *Shewanella Exiguobacterium*, *Kocuria*, *Enterococcus*, *Microbacterium*, *Rhodococcus*<br>**Adult**<br>*Pseudomonas*, *Alcaligenes*, *Bordetella*, *Myroides, Aeromonas* | *Pseudomonas* found dominantly in both larvae and adult stages of *Anopheles stephensi* and other malaria vectors makes it a potential paratransgenic candidate | Chavshin et al. (2011) |

*(Continued)*

**TABLE 14.1 (CONTINUED)**
**Isolated Microbiota from the Different Mosquito Species Collected from Various Countries and Their Specific Activity**

| Mosquito Species | Country of Collection | Name of Microbiota | Specific Activities of Microbiota | Reference |
|---|---|---|---|---|
| *Culex quinquefasciatus* (field) | India | *Halobacillus littoralis*, *Staphyloccocus cohnii* | These bacteria are halotolerant and can grow in hypersaline water having 18%–20% w/v NaCl concentration | Reegan et al. (2013) |
| *An. culicifacies* (field) | Iran | *Acinetobacter*, *Delftia*, *Aeromonas*, *Bacillus*, *Kurthia*, *Chryseobacterium*, *Exiguobacterium*, *Pseudomonas*, *Microbacterium*, *Thorsellia*, *Staphylococcus*, *Variovorax* | A variation of midgut microbiota occurs among the old age female and larvae and newly emerged female<br>Diversity of midgut microbiota varied between sampling sites and different stages of mosquito<br>*Pseudomonas* commonly retrieved from all sampling sites and from all larvae and adult samples, and also have transstadial passage property due to which it suggest a most adaptive bacterial genera for mosquitoes, hence may be a suitable for paratransgenesis approach | Chavshin et al. (2014) |
| *Ae. aegypti* (field) | India | *Aeromonas media*, *Aeromonas salmonicida smithia*, *Bacillus cereus*, *Brevibacillus agri*, *Edwardsiella tarda*, *Pantoea agglomerans*, *Pseudomonas alcaligenes*, *Burkholderia mallei*, *Pseudomonas putida*, *Serrattia odorifera*, *Xenorhybdus luminiscens* | *S. odorifera* blocked the midgut surface molecule prohibitin, and this is responsible for the enhancement of dengue virus type 2 susceptibility to *Ae. aegypti* | Apte-Deshpande et al. (2012, 2014) |

*(Continued)*

**TABLE 14.1 (CONTINUED)**
**Isolated Microbiota from the Different Mosquito Species Collected from Various Countries and Their Specific Activity**

| Mosquito Species | Country of Collection | Name of Microbiota | Specific Activities of Microbiota | Reference |
|---|---|---|---|---|
| *Ae. aegypti* (lab) | India | *Aeromonas hydrophila*, *Aeromonas media*, *Aeromonas salmonicida smithia*, *Pantoea agglomerans*, *Enterobacter cloacae*, *Microbacterium oxydans*, *Pseudomonas alcaligenes*, *Burkholderia mallei*, *Pseudomonas putida*, *Serrattia odorifera* | Susceptibility to chikungunya in adult *Ae. aegypti* increased when *Ae. aegypti* was coinfected with *S. odorifera* | |
| *Ae. albopictus* (field) | Madagascar | *Dermacoccus* sp., *Arsenicicoccus bolidensis*, *Terrabacter* sp., *Curtobacterium flaccumfaciens*, *Leucobacter* sp., *Microbacterium arborescens*, *Mi. esteraromaticum*, *Mi. flavescens*, *Arthrobacter albidus*, *Kocuria* sp., *Micrococcus pumilus*, *Micrococcus* sp., *Cellulosimicrobium cellulans*, *Streptomyces* sp., *Deinococcus* sp., *Bacillus isronensis*, *B. megaterium*, *B. pumilus*, *B. subtilis*, *Planococcus* sp., *Staphylococcus epidermidis*, *St. warneri*, *Haematobacter massiliensis*, *Skermanella aerolata*, *Sphingomonas yunnanensis*, *Neisseria* sp., *Asaia* sp., *Citrobacter freundii*, *Enterobacter* sp., *Klebsiella oxytoca*, *Pantoea* sp., *Acinetobacter baumannii*, *Ac. lwoffii*, *Pseudomonas oryzihabitans*, *Xanthomonas* sp. | Relative abundance and composition of midgut bacterial mosquitoes varied among the different collection sites of mosquitoes<br>Predominant genus *Pantoea* as a possible candidate for paratransgenesis, which might influence the vector competence of *Ae. albopictus* | Moro et al. (2013) |

(*Continued*)

**TABLE 14.1 (CONTINUED)**
**Isolated Microbiota from the Different Mosquito Species Collected from Various Countries and Their Specific Activity**

| Mosquito Species | Country of Collection | Name of Microbiota | Specific Activities of Microbiota | Reference |
|---|---|---|---|---|
| *An. gambiae* | Cameroon | **Male**<br>*Delftia* sp., *Acinetobacter septicus*, *Pseudomonas stutzeri*, *Pseudomonas stutzeri*, *Enterobacter* sp., *Klebsiella* sp., *Escherichia coli*, *Escherichia-Shigella*, *Shigella flexneri*, *Serratia marcescens*<br>**Female**<br>*Delftia* sp., *Pseudomonas stutzeri*, *Escherichia coli*, *Escherichia Shigella*, *Shigella flexneri*, *Serratia marcescens*, *Serratia* sp.<br>**Pupa**<br>*Comamonas* sp., *Acinetobacter* sp., *Acinetobacter septicus*, *Acinetobacter septicus*, *Aeromonas hydrophila*, *Pseudomonas stutzeri*, *Pseudomonas stutzeri*, *Pseudomonas mendocina*, *Escherichia coli*, *Escherichia-Shigella*, *Shigella flexneri*, *Serratia marcescens*<br>**Larva**<br>*Delftia* sp., *Acinetobacter septicus*, *Aeromonas hydrophila*, *Aeromonas veronii*, *Aeromonas caviae*, *Pseudomonas stutzeri*, *Pseudomonas stutzeri*, *Enterobacter* sp., *Escherichia coli*, *Escherichia Shigella*, *Shigella flexneri*, *Rahnella aquatilis*, *Serratia marcescens* | Intensity and prevalence of *Plasmodium falciparum* infection were significantly reduced by the natural bacterial isolates<br>*Serratia* are significantly correlated with *P. falciparum* infection and found abundantly<br>Some bacteria acquired to mosquito tissue from the surrounding environment | Tchioffo et al. (2013, 2105) |

(*Continued*)

**TABLE 14.1 (CONTINUED)**
**Isolated Microbiota from the Different Mosquito Species Collected from Various Countries and Their Specific Activity**

| Mosquito Species | Country of Collection | Name of Microbiota | Specific Activities of Microbiota | Reference |
|---|---|---|---|---|
| *An. stephensi* (lab) | Germany | **Eggs**<br>*Elizabethkingia meningoseptica*, *Flectobacillus major*, *Pelomonas saccharophila*, *Sediminibacterium salmoneum*, *Undibacterium pigrum*, *Brevibacillus formosus*<br>**Larvae**<br>*Elizabethkingia meningoseptica*, *Flectobacillus major*, *Microbacterium dextranolyticum*, *Bradyrhizobium japonicum*, *Nubsella zeaxanthinifaciens*, *Lishizhenia caseinilytica*, *Rhizobium daejeonense*, *Rhizobium huautlense*, *Mesorhizobium loti*<br>**Pupae**<br>*Elizabethkingia meningoseptica*, *Rhizobium daejeonense*, *Microbacterium dextranolyticum*, *Rhizobium daejeonense*, *Aquabacterium parvum*<br>**Adult**<br>*Elizabethkingia meningoseptica*, *Sphingobacterium siyangense*, *Asaia siamensis* | Diversity of midgut microbiota decreases during the development of mosquitoes from egg to adult<br>*E. meningoseptica* was dominant species and transmitted from one generation to another and it dominancy in the midgut was not altered by any diet (sugar and blood) | Ngwa et al. (2013) |
| *Culex quinquefasciatus* (field) | India | *Acinetobacter* sp., *Aerococcus* sp., *Aeromonas* sp., *Arthrobacter creatinolyticus*, *Bacillus* sp., *Citrobacter braakii*, *Delftia lacustris*, *Enterobacter* sp., *Enterococcus* sp., *Escherichia* | The mosquito harbors diverse midgut microbiota, which might be one of the factors for variation in disease | Chandel et al. (2013) |

*(Continued)*

**TABLE 14.1 (CONTINUED)**
**Isolated Microbiota from the Different Mosquito Species Collected from Various Countries and Their Specific Activity**

| Mosquito Species | Country of Collection | Name of Microbiota | Specific Activities of Microbiota | Reference |
|---|---|---|---|---|
| | | *coli*, *Es. hermannii*, *Exiguobacterium* sp., *Janibacter melonis*, *Klebsiella* sp., *Kocuria* sp., *Kytococcus schroeteri*, *Lactococcus lactis*, *Leucobacter tardus*, *Lysinibacillus macroides*, *Microbacterium* sp., *Micrococcus* sp., *Morganella morganii*, *Pantoea anthophila*, *Pantoea dispersa*, *Proteus vulgaris*, *Providencia alcalifaciens*, *Providencia rettgeri*, *Pseudomonas* sp., *Serratia marcescens*, *Shigella flexneri*, *Sporosarcina luteola*, *Staphylococcus* sp., *Stenotrophomonas maltophilia*, *Vagococcus fluvialis* | transmission rates or vector competence | |
| *A. arabiensis* (field) | Zambia | *Comamonas* sp., *Acinetobacter* sp., *Pseudomonas putida*, *Pantoea* sp., *Pseudomonas rhodesiae*, *Serratia marcescens*, *Elizabethkingia anophelis* | Some midgut microbiotas have the capabilities to activation of mosquito's immune system and blocking the development of *Plasmodium. Serratia marcescens* inhibits sexual and asexual stages of *Plasmodium*<br>Bacterial isolates also have the ability to affect the mosquito life span | Bahia et al. (2014) |
| *Ae. aegypti*, *An. gambiae*, *Georgecraigius atropalpus* (lab) | United States | Armatimonadaceae, *Paenibacillus*, Clostridiaceae, Corynebacteriaceae, Microbacteriaceae, Propionibacteriaceae, | During the development of *Ae. aegypti*, composition of bacterial community changes | Coon et al. (2014) |

(*Continued*)

**TABLE 14.1 (CONTINUED)**
**Isolated Microbiota from the Different Mosquito Species Collected from Various Countries and Their Specific Activity**

| Mosquito Species | Country of Collection | Name of Microbiota | Specific Activities of Microbiota | Reference |
|---|---|---|---|---|
| | | Rhodocyclaceae, Oxalobacteraceae, Neisseriaceae, Procabacteriaceae, Comamonadaceae, Aeromonadaceae, Xanthomonadaceae, Moraxellaceae, Enterobacteriaceae, Pseudomonadaceae, Sphingobacteriaceae, Chitinophagaceae, Flavobacteriaceae | Development of axenic mosquito larva does not occurs<br>Isolated bacterial species like *Aeromonas*, *Acinetobacter*, *Chryseobacterium*, *Aquitalea*, *Microbacterium*, and *Paenibacillus* rescue the development<br>Noncommunity bacterial species like DH5α strain of *E. coli* can also have the same developmental effect | |
| *Cx. quinquefasciatus* | India | *Vagococcus carniphilus*, *Vagococcus fluvialis* | *V. fluvialis* first reported from midgut of *Cx. quinquefasciatus* mosquito collected from Arabian Sea coastline of India | Chandel et al. (2015) |
| *Anopheles barbumbrosus*, *An. crawfordi*, *An. dirus*, *An. maculatus*, *An. gigas* | Vietnam | Acetobacteraceae, *Acinetobacter junii*, *Acinetobacter* sp., *Asaia spathodeae*, *Bartonella* sp., *Diplorickettsia massiliensis*, *Cellvibrio ostraviensis*, *Enhydrobacter aerosaccus*, *Enhydrobacter* sp., *Enterobacter aerogenes*, *Hafnia paralvei*, *Klebsiella pneumonia*, *Serratia* sp., *Tatumella* sp., *Thorsellia anopheles*, *Yersinia* sp. *Moraxella osloensis*, *Novosphingobium* sp., *Pseudomonas aeruginosa*, *Pseudomonas* sp., | Microbial diversity of *Anopheles* will lead to development of new approach to control *Anopheles* vector and transmission of malaria parasites | Ngo et al. (2015) |

(*Continued*)

**TABLE 14.1 (CONTINUED)**
**Isolated Microbiota from the Different Mosquito Species Collected from Various Countries and Their Specific Activity**

| Mosquito Species | Country of Collection | Name of Microbiota | Specific Activities of Microbiota | Reference |
|---|---|---|---|---|
| | | Sphingomonadaceae, Sphingomonadales, *Sphingobium* sp., *Sphingomonas* sp., *Stenotrophomonas maltophilia*, *Stenotrophomonas* sp., *Xanthomonas* sp., *Staphylococcus pasteuri*, *S. sciuri*, *S. warneri*, *Bacillus* sp., *Enterococcus faecium*, *Brachybacterium* sp., *Brevibacterium* sp., *Janibacter* sp., *Leucobacter chromiiresistens*, *Microbacterium* sp., *M. radiodurans*, *M. testaceum*, Micrococcaceae, *Chryseobacterium* sp., Flavobacteriaceae, *Knoellia* sp. | | |
| *Aedes albopictus* (field) | India | *Dietzia maris* | *Dietzia maris* might be a suitable candidate for paratransgenesis | Yadav et al. (2015b) |
| *Ae. albopictus* (field) | India | *Enterobacter cloaca*, *Bacillus aryabhattai*, *Staphylococcus hominis*, *Acinetobacter pittii*, *Pseudomonas monteilii*, *Klebsiella michiganensis*, *Pseudomonas geniculata*, *Elizabethkingia anophelis*, *Klebsiella oxytoca*, *Pseudomonas aeruginosa*, *Enterobacter hormaechei*, *Lysinibacillus fusiformis*, *Alcaligenes faecalis*, *Enterobacter asburiae*, *Bacillus tequilensis*, *Bacillus cereus* | Variation in the bacterial species among the interspecies of mosquitoes, for example, *Ae. albopictus* and *Ae. aegypti* collected from same habitats. *Enterobacter* was the dominant genus in both *Ae. albopictus* and *Ae. aegypti* and involved in various important functions such as: | Yadav et al. (2015a) |

(*Continued*)

**TABLE 14.1 (CONTINUED)**
**Isolated Microbiota from the Different Mosquito Species Collected from Various Countries and Their Specific Activity**

| Mosquito Species | Country of Collection | Name of Microbiota | Specific Activities of Microbiota | Reference |
|---|---|---|---|---|
| *Ae. aegypti* (field) | India | *Enterobacter cloacae*, *Stenotrophomonas maltophilia*, *En. xiangfangensis*, *Pseudomonas monteilii*, *Lysinibacillus fusiformis*, *Staphylococcus hominis*, *Bacillus aerophilus*, *Pantoea dispersa*, *Micrococcus yunnanensis*, *Klebsiella michiganensis*, *Aeromonas veronii*, *Ps. mosselii*, *Ba. aryabhattai*, *Kl. pneumonia* | Blocked *Plasmodium falciparum* development in *An. gambiae* and sporogonic development of *P. vivax* in *An. albimanus*<br>Induced the expression of mosquito-immune components in midgut of *An. stephensi* have already been described<br>A potential candidate of paratransgenic approach | |
| *Aedes albopictus* (field) | India | *Staphylococcus saprophyticus*, *Brevibacillus agri* | Midgut bacteria interacted with each other and showed swarming motility | Yadav et al. (2016b) |
| *Ae. albopictus* (field) | India | **Sugar fed adult**<br>*Acinetobacter pittii*, *Enterobacter asburiae*, *Micrococcus endophyticus*, *Aerococcus viridians*, *Staphylococcus cohnii*, *Pseudomonas monteilii*, *Delftia lacustris*, *Kocuria palustris*, *Micrococcus yunnanensis*, *Bacillus subtilis*, *Staphylococcus haemolyticus*, *Klebsiella michiganensis* and *Pseudomonas geniculata* | Midgut bacteria are generally acquired through vertical inheritance as well as from surrounding environments<br>Diversity of midgut microbiota is also related to the gender of the mosquito<br>Diversity of midgut microbiota is also depends on the food (sugar and blood) | Yadav et al. (2016a) |

(*Continued*)

**TABLE 14.1 (CONTINUED)**
**Isolated Microbiota from the Different Mosquito Species Collected from Various Countries and Their Specific Activity**

| Mosquito Species | Country of Collection | Name of Microbiota | Specific Activities of Microbiota | Reference |
|---|---|---|---|---|
| | | **Bloode fed adult** | | |
| | | *Pseudomonas monteilii*, *Staphylococcus cohnii*, *Staphylococcus pasteuri*, *Stenotrophomonas maltophilia*, *Chryseobacterium rhizoplanae*, *Delftia lacustris*, *Staphylococcus saprophyticus*, *Pseudomonas mosselii*, *Micrococcus yunnanensis* and *Pseudomonas geniculata* | | |
| | | **Male adult** | | |
| | | *Pantoea dispersa*, *Enterobacter xiangfangensis*, *Staphylococcus warneri*, *Staphylococcus hominis*, *Pseudomonas aeruginosa*, *Acinetobacter pittii*, *Micrococcus yunnanensis*, *Kocuria palustris*, *Bacillus subtilis*, *Staphylococcus arlettae*, *Bacillus aerophilus* and *Enterobacter asburiae* | | |
| | | **Larvae** | | |
| | | *Acinetobacter pittii*, *Janibacter hoylei*, *Clostridium sporogenes*, *Pseudomonas geniculata*, *Klebsiella pneumoniae*, *Microbacterium paraoxydans*, *Bacillus subtilis* subsp. *subtilis*, *Bacillus cereus*, *Staphylococcus cohnii*, *Pseudomonas monteilii*, *Aeromonas veronii*, *Leucobacter kyeonggiensis*, *Kocuria palustris*, *Bacillus aryabhattai*, *Bacillus subtilis* subsp. *inaquosorum* and *Staphylococcus pasteuri* | | |

*(Continued)*

**TABLE 14.1 (CONTINUED)**
**Isolated Microbiota from the Different Mosquito Species Collected from Various Countries and Their Specific Activity**

| Mosquito Species | Country of Collection | Name of Microbiota | Specific Activities of Microbiota | Reference |
|---|---|---|---|---|
| *An. gambiae* | Burkina Faso (West Africa) | Caulobacterales, Rhizobiales, Rhodobacterales, Rickettsiales, Sphingomonadales, Alteromonadales, Enterobacteriales, Gammaproteobacteria, Oceanospirillales, Pseudomonadales, Cytophagales, Flavobacteriales, Sphingobacteriales, Bifidobacteriales, Corynebacteriales, Frankiales, Micrococcales, Propionibacteriales, Clostridiales, Lactobacillales, Comamonadaceae, Oxalobacteraceae, Burkholderiaceae, Staphylococcaceae, Bacillaceae, Planococcaceae | Mosquito-associated microbiota depends upon the environmental condition from where they have been acquired | Buck et al. (2016) |
| *Culex pipiens* | Argentina | *Klebsiella* sp. and *Aeromonas* sp. | Native bacteria like *Klebsiella* sp. and *Aeromonas* sp., and exogenous yeast *Saccharomyces cerevisiae* greatly influences the development of *Culex pipiens*<br>*Klebsiella* sp. and *Aeromonas* sp. serve as attractants for the selection oviposit ion habitat<br>It also nourishes the most fragile larval stage, L1, to more resilient stage second instar, L2 | Díaz-Nieto et al. (2016) |

(*Continued*)

**TABLE 14.1 (CONTINUED)**
**Isolated Microbiota from the Different Mosquito Species Collected from Various Countries and Their Specific Activity**

| Mosquito Species | Country of Collection | Name of Microbiota | Specific Activities of Microbiota | Reference |
|---|---|---|---|---|
| *Aedes vexans complex*, *Ochlerotatus canadensis*, *Oc. japonicas*, *Oc. stimulans*, *Oc. triseriatus*, *Oc. trivitatus*, *Coquilletidia perturbans*, *Culex pipiens/restuans*, *Anopheles punctipennis*, *An. quadrimaculatus*, *Cu. salinarius*, *Cu. tarsalis* | Canada | *Flaviflexus*, *Curtobacterium*, *Rothia*, *Kocuria*, *Gardnerella*, *Bacteroides*, *Staphylococcus*, *Oscillibacter*, *Wolbachia*, *Novosphingobium*, *Sphingomonas*, *Comamonas*, *Algimonas*, *Aggregatibacter*, *Bacillus cereus*, *Enterococcus*, *Methylobacterium*, *Asaia*, *Pantoea*, *Acinetobacter johnsonii*, *Pseudomonas*, *Mycoplasma* | A significant phylosymbiosis occurs between the host and microbiota<br>Impact of region on microbiota occurs only a little bit<br>Seasonal shifts of microbiota consistently repeated<br>Microbiota may affect the variation in vector competence of mosquitoes and was found that *Wolbachia* influences the viral pathogens carried by the mosquitoes | Novakova et al. (2017) |

*(Continued)*

**TABLE 14.1 (CONTINUED)**
**Isolated Microbiota from the Different Mosquito Species Collected from Various Countries and Their Specific Activity**

| Mosquito Species | Country of Collection | Name of Microbiota | Specific Activities of Microbiota | Reference |
|---|---|---|---|---|
| *Aedes albopictus*, *Ae. japonicas*, *Ae. triseriatus*, *Ae. trivittatus*, *Ae. vexans*, *Anopheles crucians*, *An. punctipennis*, *An. quadrimaculatus*, *Culex pipiens*, *Cu. restuans*, *Culiseta inornata*, *Psorophora ferox* | United States | All isolates belongs to 11 phyla: Proteobacteria, Actinobacteria, Firmicutes, Bacteroidetes, Acidobacteria, Cyanobacteria, Tenericutes, Spirochaetes, Planctomycetes, Parcubacteria, Fusobacteria | A similar gut microbiota has been shared by the species of different mosquito individuals<br>The individuals of same mosquito species commonly harbored a different gut microbiota collected from the same study site and collection date | Muturi et al. (2017) |

### 14.5.1 Sample Collection

Samples may be either adult mosquito (male or female) or the larvae that has been collected generally from the field in a sterilized net cage or sterilized plastic or glass bottle. The collected samples are immediately brought to the laboratory for further study.

### 14.5.2 Dissection of Mosquito Midgut

The collected samples are immediately brought to the laboratory for dissection under a stereomicroscope in a sterilized environmental condition using sterilized apparatus. Prior to dissection, the mosquito samples should be surface sterilized with 75% ethanol for 5 min, followed by washing twice with phosphate buffered saline (PBS) (Chandel et al. 2013; Yadav et al. 2015a). The dissected midgut is homogenized in the PBS solution. For the culture-dependent identification techniques, it is directly pour plated on the agar plate, while for the culture-independent approach it stored at −20°C for further analysis.

#### 14.5.2.1 Culture-Dependent-Based Identification

Culture-dependent is a conventional technique in which the microbiota have been isolated from the midgut of the mosquitoes and cultured in laboratory conditions followed by polymerase chain reaction (PCR) amplification of a specific gene like 16S rRNA gene from their whole genome, sequencing, sequencing analysis, and comparison to the GenBank database. In the culture-dependent approach, the synthetic media supplemented with various nutrients are used for growing the microbial population under controlled laboratory conditions (Nold et al. 1996; Burns et al. 2004; Simbahan et al. 2005). This technique may be used for the identification of only those bacterial isolates that are able to grow in laboratory conditions, whereas those that are unable to grow in laboratory conditions may not be identified.

##### *14.5.2.1.1 Isolation and Purification of Midgut Bacteria*

For the isolation of midgut bacteria, the homogenized solution has to be serial diluted and transferred on the solid agar media, then incubated for 12 to 24 hr at 37°C. The bacterial colonies have to be purified using streaking techniques until the purified colonies have been obtained (Yadav et al. 2016a).

##### *14.5.2.1.2 Characterization of Bacterial Isolates*

The isolated purified bacterial isolates have been characterized by mainly two techniques: biochemical and molecular.

*14.5.2.1.2.1 Biochemical Characterization* Biochemical testing is the conventional method for the identification of bacteria based on the different biochemical activities of different bacterial species. Each bacterial species has a specific physiological characteristic and performs a specific carbohydrate, protein, and fat metabolism. They also have the ability to utilize specific compounds as their food sources. These entire characteristic is specific for each bacterial species and used to identify them. Biochemical testing is the first line of characterization of bacterial isolates and commonly used, however, it may not be used for high accurately identification at species level.

*14.5.2.1.2.2 Molecular Characterization* Molecular characterization based techniques is used to identification of bacteria more precisely. Molecular characterization techniques may be further characterized by either 16S rRNA sequence techniques or matrix-assisted laser desorption ionization–time-of-flight mass spectrometry (MALDI-TOF MS).

In the 16S rRNA sequence-based identification techniques, about 1.5 kb long specific genes are amplified from the whole genome of bacterial isolates and sequenced. The obtained sequences are further manually checked, analyzed, aligned, and compared to the GenBank database using the BLAST algorithm (http://www.ncbi.nlm.nih.gov/BLAST) and the EzTaxon server (http://www.ezbiocloud.net/eztaxon) to search the homologous sequences for the identification of bacterial species.

The MALDI-TOF MS detection technique is used for the differentiation of microorganisms based on their organism-specific biomarker molecules. The biomarker molecules may be the intact proteins, their proteolytic peptides, or nonribosomal peptides. Tandem mass spectra of intact proteins or proteolytic peptides generate the sequence-specific fragments for the biomarkers obtained after microwave-assisted acid hydrolysis. In addition to searching with proteome database, the individual biomarker proteins or proteolytic peptides are unambiguously identified from their tandem mass spectra; furthermore the organism has to be identified (Demirev and Fenselau 2008).

#### 14.5.2.2 Culture-Independent-Based Identification

The noncultivable predominant bacterial isolates of mosquito midgut do not explicate by the culture-dependent-based technique (Amann et al. 1995). Cultured fraction of microbiota poorly represents the microbial diversity and only less than 1% of the whole bacterial diversity is explored by the culture (Riesenfeld et al. 2004). Recently, the culture-independent method has been increasingly used to rapidly describe the microbial diversity of any ecosystem (Xu et al. 2012). The culture-independent technique is used for the study of midgut microbes without culturing in laboratory conditions. In this technique the microbial diversity has to be determined by the accession genetic diversity and the metagenomic approach focuses on one common gene like 16S rRNA that has been shared by all microbial community members (Nossa et al. 2010).

In 1993, Muyzer and coworkers developed a culture-independent technique, denaturing gradient gel electrophoresis (DGGE), that has been used to quickly determine microbial diversity. In the DGGE technique the microbial community can be directly measured by the separation of the PCR products, which are amplified by using the universal primers. The technique has been used for the characterization of microbial communities on the basis of the culture-independent approach from mosquito midgut (Lindh et al. 2005; Rani et al. 2009; Gusmao et al. 2010; Buck et al. 2016). Apart from mosquito's midgut microbial communities, it also has been used for the characterization the microbiota from several other habitats like soil (Stephen et al. 1998), polluted river (Cébron et al. 2004), marine (Díez et al. 2001), and gas industry pipelines (Zhu et al. 2003).

In the culture-independent approach the biodiversity studies of microbial communities have been carried out by the amplification of 16S rRNA gene followed by the separation or identification either by cloning and Sanger sequencing or by direct pyro-sequencing (Nossa et al. 2010). After bulk PCR amplification, several tools have recently been used for the sequence-specific separation such as DGGE (Muyzer et al. 1993), temperature gradient gel electrophoresis (TGGE) (Zoetendal et al. 1998), terminal-restriction fragment length polymorphism (T-RFLP) (Kitts et al. 2001), denaturing high-performance liquid chromatography (DHPLC) (Penny et al. 2010), single-stranded conformation polymorphism (SSCP) (Ege et al. 2011), constant denaturing capillary electrophoresis (CDCE) (Thompson et al. 2004), and temporal temperature gradient gel electrophoresis (TTGE) (Ogier et al. 2002). Denaturing electrophoresis has been used for the separation which allows the comparison of diversity of microbiota at a low or medium level (Roudiere et al. 2007). All these methods provide a fingerprint of the microbial communities, and therefore these techniques might be used for the study of microbial diversity samples at a large scale.

Therefore, in the recent years the use of both culture-dependent and culture-independent approaches for the analysis of microbial community has increased significantly (Orphan et al. 2000; Ellis et al. 2003; Kisand and Wikner 2003). However, both techniques have some limitations like in culture-dependent the bacteria that are unable to grow in laboratory conditions may not be identified, while in the culture-independent techniques all the microbiota inhabiting mosquito midgut can be identified but they may not be used for the development of any defense strategies like the paratransgenic approach against invading pathogens. Only those microbes that have been cultured in the laboratory can be used for the genetic manipulation for the expression of effectors molecules against any pathogen in the paratransgenic approach.

### 14.5.3 Fluorescent In Situ Hybridization (FISH)

Apart from the aforementioned techniques, fluorescent in situ hybridization (FISH) is another technique that is commonly used for the study of microbial communities in insect gut. In this technique the sequencing of 16S rRNA gene is not needed. This technique is based on the visualization of symbiotic bacteria in the insect gut by targeting the 16S rRNA gene sequences of bacterial species or genus-specific fluorescently labeled probes (Cheung et al. 1977; Aminov et al. 200). During the 1990s, this technique was commonly used for the study (visualization, detection, and characterization) of intracellular symbiotic bacteria of several insects like termites (Berchtold et al. 1999), crickets (Santo Domingo et al. 1998), aphids (Fukatsu et al. 1998), and others insects. In 1999, Berchtold et al. used group-specific, rRNA targeting fluorescently labeled oligonucleotide probes for the study of bacterial communities in the hindguts of the Australian lower termite *Mastotermes darwiniensis*. This technique is highly useful for the study of bacterial dynamics of uncultivable microbiota of the insects (Shi et al. 2010).

To visualize the bacterial communities inside insect gut, two other techniques—DAPI (4′,6-diamidino-2-phenylindole) and green fluorescent protein (GFP)—may also be used along with FISH techniques. DAPI may be used for the staining of bacterial cells in different insect species reared in the same condition to highlight significant differences in their numbers (Cazemier et al. 1997a, 1997b). GFP is mainly used for tracking microbial cells inside the host body. In a study Hurst and Jackson (2002) used it to track the colonization of *Serratia entomophila* in the *Costelytra zealandica* (grass grub) gut. Chavshin et al. (2013) transformed the *Escherichia coli* for the expression of GFP (*E.coli*-GFP) and tracked the *E. coli*-GFP in larvae, pupae, and adults of *Anopheles* mosquitoes, and found that *E. coli*-GFP persisted during molting and metamorphosis events, and confirmed the transstadial transmission property. Recently, Chavshin and coworkers (2015) successfully transformed the *Pseudomonas* isolated from *An. stephensi* for the expression of GFP and inserted it into larvae and adults. *Pseudomonas*-GFP was tracked in the gut of larvae, pupae, and adults, and through fluorescent microscopy and PCR assays it was conformed that the *Pseudomonas* have the transstadial transmission property and transmitted from larvae to pupae and subsequently to adults. After the conformation of the transstadial transmission property of the *Pseudomonas and E. coli*, there is a possibility of the genetic modification of these bacterial for the development of paratransgenic approaches for the management of vector-borne disease. Chen et al. (2015) also used the genetic modified midgut bacteria *Elizabethkingia anophelis* for the expression of GFP and showed that this bacterial species is highly associated with *Anopheles* mosquitoes and might be used for the development of paratransgenic applications due to having the transstadial transmission property.

## 14.6 IMPORTANT ROLES OF MOSQUITO MIDGUT MICROBIOTA

### 14.6.1 Midgut Microbiota in Defense against Parasites/Pathogens

The importance of midgut bacteria in relation to their various functional roles is still unexplored. It has been reported that some midgut bacteria are involved in disease prevention through various ways. Pumpuni et al. (1996) suggested that introduced or indigenous bacteria could modify vectorial competence of *Anopheles* through hindering pathogen development and weakening the ability of *Plasmodium* to establish infection (Dong et al. 2009; Cirimotich et al. 2011a).

It has been reported that gram-negative bacteria is involved in inhibition of sporogonic development in the midgut of mosquito (Pumpuni et al. 1993; Straif et al. 1998; Dong et al. 2006). Gonzalez-Ceron (2003) observed that midgut microbiota influence the parasite transmission through blocking the *Plasmodium vivax* sporogonic development in *An. albimanus*. The bacterial species *Enterobacter* generated the reactive oxygen species (ROS) in *An. gambiae* and blocked the development of *P. falciparum* (Cirimotich et al. 2011a).

There is a possibility that the midgut microbiota could modulate the vectorial capacity of *An. gambiae* through inhibition of *Plasmodium* and other pathogen development. However, the mechanism is still not understood (Dong et al. 2009). The diverse microbiota affects the development of *Plasmodium*, and intensity and prevalence of *Plasmodium* infection were significantly reduced by the natural bacterial isolates (Tchioffo et al. 2013; Ngo et al. 2015).

Pumpuni et al. (1993) reported that gram-negative bacterial strains partially or completely inhibit the oocyst formation, whereas gram-positive bacteria do not show any inhibitory effect. *An. funestus* females that harbor gram-positive bacteria were likely to be more infected with sporozoites compared with those with no cultivable bacteria or gram-negative bacteria in their midguts (Straif et al. 1998). In 2003, Gonzalez and colleagues isolated midgut bacteria from field-collected *An. albimanus* Weidemann and observed that infection of *Plasmodium vivax* decreases in *Anopheles* mosquitoes when it was coinfected with midgut bacteria *Enterobacter cloacae*, *En. Amnigenus* 2, and *S. marcensces* (Gonzalez et al. 2003). Cirimotich et al. (2011a) have also proved that *Enterobacter* interfere with *P. falciparum* development through generation of ROS, which shows anti-*Plasmodium* activity.

In 2012, Boissiere et al. demonstrated that gut microbial communities are one of the major components of mosquito innate immune responses and significantly influences the *Plasmodium* infection in *An. gambiae*. Some gut bacterial isolates are able to activate the mosquito's immune system in *Anopheles* mosquito.

Moreover, it has also been reported that the midgut bacteria also influences the susceptibility to some viral diseases. In the presence of midgut bacteria *Serratia odorifera*, immune response of *Ae. aegypti* has been suppressed, due to which susceptibility of *Ae. aegypti* to dengue and chikungunya virus increases (Apte-Deshpande et al. 2012, 2014). Susceptibility of *Ae. aegypti* to dengue virus type 2 also increases when fed with *Aeromonas* sp. and *E. coli* (Rani et al. 2009).

### 14.6.2 Midgut Microbiota in Food Digestion and Assimilation

Midgut bacteria can produce some compounds that can be directly assimilated by the host. Bacteria can improve the digestion by producing some degrading enzymes that facilitate the assimilation of complex molecules. In phytophagous insects, microbiotas generally provide vitamins, amino acids, and sterol that complement limited plant diets. The best known example is the involvement of the bacterium *Buchnera* in providing essential amino acids to aphids (Douglas 1998). Another interesting example is the bacteria that provide vitamin B that is not present in vertebrate blood, the sole nutrient source of *Glossina* tsetse flies (Aksoy 2000).

It is well known that the midgut microbiota plays an important role in the digestion of food (Douglas et al. 1998; Fouda et al. 2001; De Gaio et al. 2011). A male adult mosquito never takes a blood meal and solely depends on the plants sap (Rani et al. 2009). However, in addition to plant sap, the female takes a blood meal. Hence, in the male adults, the bacterial species helps in the digestion of plant sugar (high carbohydrate) while in the females it also helps in the digestion of blood meal (high protein) (Yadav et al. 2016). This causes the variation of midgut microbiota in male and female mosquito adults. It has been demonstrated that *Serratia* and *Enterobacter* contain hemolytic enzymes and play an important role in blood digestion in mosquito (Campbell 2004; Gusmao 2010; De Gaio et al. 2011). Minard et al. (2013) described that bacteria *Acinetobacter baumannii* and *A. johnsonii* could be involved in both blood digestion and nectar assimilation in *Ae. albopictus*. *Acinetobacter* strains of mosquito midgut were able to metabolize the amino acids α-ketovaleric acid and glycine, which are blood components, as well as 4-hydroxybenzoic acid and xylose, which are common constituents of plant sap. The bacterial species *Asaia bogorensis* isolated from *An. stephensi* was shown to be phototrophic with respect to vitamins, suggesting it may provide vitamins to the mosquito (Crotti et al. 2010).

It has been reported that bacteria release some compounds that are essential for mosquito larval development. For instance, it has been demonstrated that a high level of *Pseudomonas aeruginosa* improved larval growth of *Culex quinquefasciatus* in a phosphorus-rich medium, whereas that of *Cx. tarsalis* was slowed (Peck and Walton 2006). The level of phosphorus in breeding sites could be a factor explaining how mosquitoes can adapt to a specific condition according to their bacterial load, possibly with a trade-off between the nutritional and toxic roles of bacteria. Differential tolerance of larvae to putative toxins present in *P. aeruginosa* could explain why the two mosquito species are not found in the same aquatic habitat (Minard et al. 2013).

It has been proven that microorganisms are an important food source for larval development, and in absence of microorganisms, the larval growth be hampered. *Ae. aegypti* larvae cannot be grown on sterile media (Rozeboom 1935). It also affects the normal larval growth, and in absence of bacteria the size was smaller than the larvae with water containing bacteria (Wotton et al. 1997). In contrast, one report showed the successful rearing of *Cx. quinquefasciatus* larvae in water with tetracycline (Mourya et al. 2002). These studies indicate that bacteria are very important, if not essential for larval development of mosquito. Some studies have demonstrated a link between the presence of bacteria in insect hosts and their ability to degrade some insecticide molecules ingested inside of mosquito midgut. The acquisition of these bacteria by each generation could be an easy way for an insect to detox itself from the insecticide without any genetic cost (Kikuchi et al. 2011).

### 14.6.3 Midgut Microbiota in Fertility and Fecundity

Fouda et al. (2001) demonstrated that symbiotic bacteria influence the potential of reproduction (fertility and fecundity), preoviposition, and blood meal digestion in *Cx. pipiens* mosquitoes. Previously, it had been shown that reproductive organs of insects had been colonized by bacteria that could manipulate the host reproduction. One important symbiotic bacterial genus—*Wolbachia*—has the ability to involve itself in the cytoplasmic incompatibility mechanism and control mosquito mating in which the production of viable progeny have been prevented when an infected male mates with an uninfected female or a female infected with an incompatible *Wolbachia* strain. For the production of viable offspring, some mosquito species like *Aedes* and *Culex* are dependent on *Wolbachia* strains (Minard et al. 2013). Apart from *Wolbachia strains, some other bacterial genera like Bacillus* and *Staphylococcus* are essential for normal and high fecundity of *Cx. pipiens*. It has also been proven that gut bacteria are essential for the normal development of the embryo (Fouda et al. 2001).

The bacterial isolates also affect the life span of the mosquitoes. It has also been observed that *Serratia marcescens* is efficiently colonized in mosquito gut and comprises the survival and inhibition of sexual and asexual stages of *Plasmodium* through secreting some effector molecules (Bahia et al. 2014; Tchioffo et al. 2015).

### 14.6.4 Midgut Microbiota as Biolarvicides

Midgut bacteria may be used as biolarvicides. Recently, the highly potent larvicidal activities have been reported from one anaerobic gram-positive bacterium *Clostridium bifermentans* subsp. *malaysia.* The toxicity of bacteria is very high to *Anopheles* and comparatively low to *Culex* and *Aedes* larvae (Lee and Seleena 1990; Thiery et al. 1992). *Lysinibacillus sphaericus* contains two potent toxins, namely, Mtx and Bin, that have larvicidal properties (Berry 2012). These toxins paralyze the digestive system and disrupt the nervous system of insects (Majambere et al. 2007; Nartey et al. 2013). Some other bacterial species, including *Bacillus thuringiensis israelensis* (Bti) and *Bacillus sphaericus* (Bs), are highly effective against mosquito's larvae. Their protoxins in parasporal crystals and the spore coat are soluble in the alkaline pH medium in larval midgut and functions as stomach poison (Porter 1996; Mittal 2003; Raghavendra et al. 2011).

### 14.6.5 Gut Microbiota and Insecticide Degradation in Insecticide-Resistant Insects

Insects harbor a wide range of microbiota inside their gut that influence the fitness traits of hosts including immunity, nutrition, and reproduction. The involvement of microbiota in degradation and detoxification of xenobiotics through catabolization of organic molecules also have been reported (Boush and Matsumura 1967; Dowd and Shen 1990). The degradation of xenobiotics, like synthetic organic compounds used in pest control, leads to the development of insecticide-resistant insects. Recently, Almeida and coworkers (2017) hypothesized that insects carry a bacteria that degrades a particular pesticide to which insects are resistant. They have isolated different bacterial species from the gut of fifth instars of *Spodoptera frugiperda* strains, which were resistant to various insecticides, namely, deltamethrin, lambda-cyhalothrin, spinosad, chlorpyrifos ethyl, and lufenuron. The bacterial isolates *Enterococcus casseliflavus* and *Enterococcus mundtii* were commonly isolated gut microbiota of larval strain resistant to lambda-cyhalothrin, chlorpyrifos ethyl, and spinosad. However, *E. casseliflavus* were isolated from deltamethrin-resistant larval strain, while *E. mundtii* from lufenuron resistant. The bacterial species *Delftia lacustris*, *Microbacterium paraoxydans*, and *Leclercia adecarboxylata* were identified only from the chlorpyrifos ethyl-resistant larval gut, *Pseudomonas psychrotolerans* from spinosad, *Arthrobacter nicotinovorans* from deltamethrin, *Pseudomonas stutzeri* from lambda-cyhalothrin, and *Staphylococcus sciuri* and *Microbacterium arborescens* from lufenuron (Almeida et al. 2017).

Apart from *Spodoptera frugiperda* strain, the insecticide degrading bacteria also have been reported from some other insects. The bacterial species of genus *Burkholderia* found in stinkbug and provide the resistant against fenithrotion (Kikuchi et al. 2011). *Citrobacter* sp. (CF-BD) a gut symbiont of *Bactrocera dorsalis* (Hendel) provides the resistance against the trichlorphon through biodegradation (Cheng et al. 2017).

Resistant development in mosquitoes against insecticides is a serious problem in vector-borne disease management. Mosquitoes harbor a wide range of gut microbiota and it might be possible to degrade insecticides in the mosquito midgut and make them resistant, but it is still unexplored. However, the presence of some bacterial species in mosquito midgut influencing insecticide-resistance development has also been reported. The presence of *Wolbachia* is positively correlated with insecticide resistance in *Cx. pipiens mosquito* (Berticat et al. 2002).

Some bacterial species of insecticide-resistance insects have been reported from the mosquito midgut, including the bacterial species *Delftia lacustris* isolated from chlorpyrifos ethyl-resistant *S. frugiperda* larval strains were also identified by Chandel et al. (2013) and Yadav et al. (2016a) from *Culex* and *Aedes* mosquitoes, respectively. Yadav and coworkers (2016a) also identified the *Microbacterium paraoxydans* from midgut of *Ae. albopictus* larvae in India. *Microbacterium arborescens* isolated from the lufenuron-resistant *S. frugiperda* strains were also reported from *Ae. albopictus* mosquitoes collected from Madagascar (Moro et al. 2013). Development of insecticidal resistance in pests may be the result of combined effect of midgut bacteria and genetic mutation.

## 14.7 PARATRANSGENESIS

The paratransgenesis technique might be a suitable, highly effective, and environment-friendly solution for vector-borne disease management. In paratransgenesis, bacteria are used as vehicles to express molecules in the gut, which negatively or positively affect the health of the host or suppress parasite colonization (Ren et al. 2008; Coutinho-Abreu et al. 2010; Rasgon 2011) (Figure 14.2). Symbiotic bacteria associated with vectors are isolated and genetically modified for expression of effector

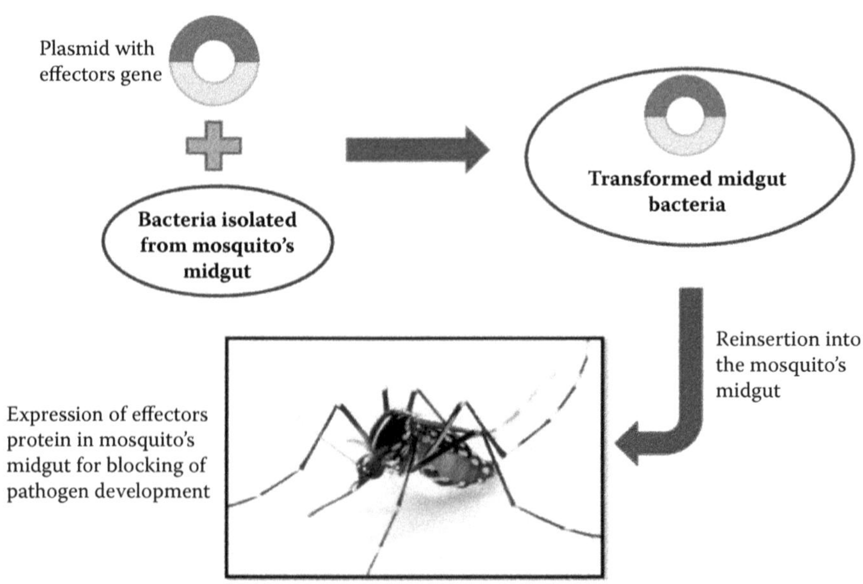


**FIGURE 14.2** Process of paratransgenesis techniques.

molecules and reintroduced through food to the mosquito midgut. Intense research on gut microbiota and their role in humans and other invertebrates are being carried out, and our knowledge of gut-inhabiting microbes has grown exponentially during the last decade. It has been observed that microbiota play important roles in the growth, development, pathogenesis, and environmental adaptation of host insects. The molecular and systems-level analysis of insect gut symbiotic microbial community will allow us to discover novel biocatalysts for biomass deconstruction and to develop innovative strategies for pest management. Thus the vector in the paratransgenic technique is indirectly transformed through its symbionts. This technology has been developed as a potential control strategy for malaria (Riehle et al. 2007).

The technique was first developed and successfully used to control Chagas disease (Durvasula et al. 1997). A genetic manipulated symbiotic bacteria, *Rhodococcus rhodnni*, from the midgut of kissing bug (reduviid) *Rhodnius prolixus*, expressed an effector molecule Cecropin A, which reduces the intensity of *Trypanosoma cruzi* infection. Genetically transformed *Sodalis*, a symbiont of tsetse flies, the vector of sleeping sickness appears a promising strategy for reduction of African *trypanosomes* transmission (Cheng and Aksoy 1999).

Suitable midgut bacteria should be identified for the development of paratransgenic techniques and might be established inside the mosquito's midgut and transmitted from one generation to the next (Pidiyar et al. 2004). Several studies have been carried out on midgut microbiota of different mosquito species including *Aedes*, *Anopheles*, and *Culex*, and it was found that the mosquito midgut harbors a diverse group of bacterial genera, such as *Acinetobacter*, *Aeromonas*, *Pantoea*, *Wolbachia*, *Bacillus*, *Enterobacter*, *Enterococcus*, *Klebsiella*, *Pseudomonas*, *Serratia*, *Staphylococcus*, *Stenotrophomonas*, *Flavobacterium*, *Microbacterium*, *Nocardia*, *Janibacter*, *Stenotrophomonas*, *Ehrlichia*, *Delftia*, *Chryseobacterium*, *Asaia*, *Escherichia*, *Micrococcus*, *Paenibacillus*, *Citrobacter*, *Thorsellia*, *Brevibacterium*, *Microbacterium*, *Sphingomonas*, *Corynebacterium*, *Shewanella*, *Alcaligenes*, *Brevibacillus*, *Neisseria*, *Xanthomonas*, *Shigella*, *Elizabethkingia*, *Providencia*, *Janibacter*, *Lysinibacillus*, *Exiguobacterium*, *Leucobacter*, and *Kocuria*. Among them *Wolbachia*, *Pantoea*, *Asaia*, *Escherichia*, *Pseudomonas* and *Thorsellia* have transstadial transmission properties and can be easily transferred from one generation to the next, making them potential candidates for genetic manipulation and paratransgenic techniques. However, potential application remains to be done (Pumpuni et al. 1996; Alphey et al. 2002; Gonzalez-Ceron et al. 2003; Pidiyar et al. 2004; Lindh et al. 2005; Favia et al. 2007; Damiani et al. 2008; Terenius et al. 2008; Rani et al. 2009, Zouache et al. 2011; Osei-Poku et al. 2012; Chandel et al. 2013; Yadav et al. 2015a, 2016a). The midgut bacteria used for a paratransgenic system should be colonized in a wide variety of mosquito populations in which they can be easily established and produce sufficient numbers of recombinant bacteria for the expression of sufficient amount of effectors molecules (Sayler and Ripp 2000; Riehle et al. 2007).

Recently, Yadav and coworkers (2015b) isolated and identified a bacterial species, *Dietzia maris*, from the midgut of female *Ae. albopictus*. *Dietzia maris* have the capability to survive in a wide range of temperatures and pH meaning they could establish themselves in a wide variety of mosquito population. This bacterial species is highly similar to the genus *Rhodococcus*, and *Rhodococcus rhodnni* have already been used as paratransgenesis for prevention of Chagas disease. Hence, the bacterial species *Dietzia maris* might be a suitable candidate of paratransgenic system.

In addition, Yadav et al. (2015a) have also identified the midgut bacteria *Enterobacter cloacae* in great abundance from wild *Ae. albopictus* and *Ae. aegypti* mosquitoes. It has been suggested that transformed *Enterobacter cloacae* bacteria secreted the antileishmanial peptides within the gut of sand fly *Phlebotomus papatasi* and might be a potential candidate of paratransgenic approach to reduce the Leishmania transmission (Maleki-Ravasan et al. 2015).

The isolated bacteria *Asaia* sp. from the midgut of *Ae. aegypti* were successfully transformed and the adult mosquitoes reinfected through sugar or blood meals representing a candidate for expressing antipathogen molecules within the mosquito. It has been demonstrated that the bacteria *Asaia* can be vertically transmitted to the progeny by maternal, paternal, and transstadial routes and horizontally transmitted among individuals with mating and cofeeding (Favia et al. 2007, 2008; Damiani et al. 2010). This fact may make it possible to successfully introduce modified bacteria *Asaia* sp. into wild mosquito populations in the field (Ricci et al. 2012). *Asaia* was the most frequently identified mosquito-associated bacterial genus among the *An. stephensi*, *An. maculipennis*, and *An. gambiae* (Favia et al. 2007). Apart from midgut of adult males and females, the genus *Asaia* was also colonized in the reproductive system of male and larval gut. Chouaia and coworkers also reported the *Asaia* as dominant bacterial genera among the other four different mosquito species, namely. *An. gambiae*, *An. stephensi*, *Ae. aegypti*, and *Ae. albopictus*, and play a beneficial role in the normal development of *An. stephensi* larvae (Chouaia et al. 2010, 2012).

In 2007, *Escherichia coli* was genetically engineered to display anti-*Plasmodium* effector molecules for inhibition of *P. berghei* development in *An. stephensi*, though its survivorship in mosquitoes is very poor (Riehle et al. 2007). Bisi and Lampe (2011) proved that symbiotic bacteria *Pantoea agglomerans* express and secrete the anti-*Plasmodium* effector proteins (SM1, anti-Pbs21, and PLA2).

## 14.8 *WOLBACHIA* ENDOSYMBIOTIC BACTERIA

The transinfection of mosquitoes with the maternally inherited, endosymbiotic bacteria *Wolbachia* is a promising new biocontrol approach. Fruit fly *Wolbachia* strains can invade and sustain themselves in mosquito populations, reduce adult life span, affect mosquito reproduction, and interfere with pathogen replication. *Wolbachia*-infected *Aedes aegypti* mosquitoes have been released in areas of Australia where outbreaks of dengue fever occur, as a prelude to the application of this technology in dengue-endemic areas of Southeast Asia.

*Wolbachia* is gram-negative, obligate intracellular bacteria and shows the maternal inheritance properties. It can infect mosquito gonads and therefore in the paratransgenic systems it may act as a potential target for the delivering of effector molecules (Yoshida et al. 2001). The *Wolbachia* can naturally infect *Ae. albopictus* and *Cx. pipiens* and might also infected many other wild mosquito species in which its prevalence is high.

*Wolbachia* have some important features such as direct insertion of transgene into the *Wolbachia* genome and cytoplasmic incompatibility mechanisms used to suppression of the target mosquito's population; fixation of transgene in other cytoplasmic elements that are coinherited in the mosquito with *Wolbachia* bacteria; and may be used for the transformation of the target mosquito's genome. These important characteristics make the *Wolbachia* ideal to use in the control the vector-borne disease (Townson et al., 2002).

In the paratransgenic approach, such bacteria should be selected to spread in a large populations of hosts, and the transovarian transmission property of *Wolbachia* makes it a highly potential candidate for this system (Weiss et al. 2006; Walker et al. 2011). There is a great possibility to use *Wolbachia* to infect many mosquito species and produce different outcomes (Zimmer 2001).

A *Wolbachia* strain plays an important role in infected *Ae. aegypti* mosquitoes, including shortening the life span of adult mosquitoes, reducing pathogen transmission, decreasing the susceptibility to certain pathogens like dengue, chikungunya virus, and *Plasmodium* parasites (Baldini et al., 2014); and inducing cytoplasmic incompatibility without any significant fitness cost (Weiss et al., 2006).

Hughes et al. (2014) described that the wild population of *An. gambiae* mosquitoes from the Burkina Faso, Africa, have the capability to be naturally infected with *Wolbachia*. However, it has also been proven that the midgut microbiota of some adult mosquitoes species like *An. gambiae* and *An. stephensi* prevent the infection of *Wolbachia*, but when the midgut microbiota of these mosquito species was removed, the *Wolbachia* infected the mosquito and was successfully transmitted to the progeny (Bian et al. 2013).

### 14.8.1 *Wolbachia*-Mediated Cytoplasmic Incompatibility

During the last decade, one of the efficient strategies to reduce crop insect pests was the introduction of sterile males in a population. This technique succeeded in limiting the expansion of the fruit fly *Ceratitis capitata* (Robinson 2002). The incompatible insect technique (IIT) was developed based on *Wolbachia*-mediated cytoplasmic incompatibility (Werren et al. 2008). This process prevents infected males from producing viable progeny when mating with an uninfected female or a female infected with an incompatible *Wolbachia* strain (Figure 14.3).

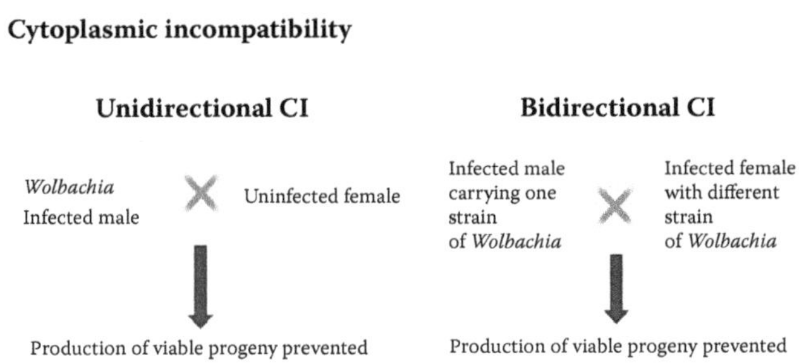


**FIGURE 14.3** Process of cytoplasmic incompatibility.

As early as 1967, *Wolbachia*-induced cytoplasmic incompatibility (CI) was proposed as a tool for the control of *Culex* mosquitoes (Laven 1967). In the 1970s this strategy was also tried in India to eradicate the mosquito population (Curtis and Adak 1974). There has been some field testing, but it has never been operationally implemented.

### 14.8.2 *Wolbachia*-Mediated Life Shortening and Vector Incompetency

The life-shortening effect of *Wolbachia* strain *w*MelPop was first time discovered in *Drosophila melanogaster* (Min and Benzer 1997). Later, it was reported in *Ae. aegypti* mosquitoes where the life span became half in *Wolbachia* strain *w*MelPop-CLA recipient mosquitoes (Ricci et al. 2012). A female mosquito must survive an extrinsic incubation period to transmit the dengue virus or other pathogens. *Wolbachia* induced the shifting of the population age toward younger females that were incapable to pathogen transmission (Brownstin et al. 2003; Cook et al. 2008). This *Wolbachia*-based, biocontrol approach is a potential strategy to prevention of disease transmission and vector-borne disease management.

*Wolbachia* strains *w*Ri and *w*Pip Istanbul significantly reduce the hatching rate in *Ae. albopictus* mosquitoes (Fu et al. 2010; Atyame et al. 2011). Interestingly, no impact was observed on the fitness of *w*Pip Istanbul transinfected *Cx. pipiens*, making this a more promising approach than SIT (Alphey 2002; Benedict and Robinson 2003; Atyame et al. 2011). Besides *Wolbachia*, other bacterial genera such as *Bacillus* and *Staphylococcus* importantly influence the fertility of the *Cx. pipiens*, although the mechanisms are not yet determined (Fouda et al. 2001). *Wolbachia* directly provides the fitness benefit to their host through involvement in nutritional and development, fecundity or oogenesis, and resistance to pathogens (Dedeine et al. 2001; Hedges et al. 2008; Teixeira et al. 2008; Brownlie et al. 2009; Bian et al. 2010; Glaser and Meola 2010; Hosokawa et al. 2010; Kambris et al. 2010). Some *Wolbachia* strains can be involved in blood-feeding alteration; interference to *Plasmodium*, dengue virus, chikungunya, and bendy proboscis; increasing metabolism; and increasing the inhibition activity of filarial nematodes (Turley et al. 2009; Moreira et al. 2009a, 2009b; Evans et al. 2009; Kambris et al. 2009; Iturbe-Ormaetxe et al. 2011).

## 14.9 CONCLUSION

Midgut microbiota of mosquitoes play various important roles in immunity, food digestion, fertility, and fecundity. Hence, the interest in microbiota associated with mosquitoes has been rapidly increasing. Due to the important role of midgut microbiota, it might be potentially used for the development of an ecofriendly, effective, and cheap techniques like paratransgenesis without fear from the development of resistance in mosquito populations.

## REFERENCES

Abalain-Colloc, M. L., Rosen, L., Tully, J. G., Bove, J. M., Chastel, C., and Williamson, D. L. 1988. *Spiroplasma taiwanense* sp. nov. from *Culex tritaeniorhynchus* mosquitos collected in Taiwan. *Int. J. Syst. Bacteriol.* 38:103–107.

Abdul-Ghani, R., Al-Mekhlafi, A. M., and Alabsid, M. S. 2012. Microbial control of malaria: Biological warfare against the parasite and its vector. *Acta Trop.* 121:71–84.

Aksoy, S. 2000. Tsetse—A haven for microorganisms. *Parasitol. Today* 16:114–118.

Almeida, L. G., Moraes, L. A., Trigo, J. R., Omoto, C., CoÃnsoli, F. L. 2017. The gut microbiota of insecticide-resistant insects houses insecticide-degrading bacteria: A potential source for biotechnological exploitation. *PLoS One* 12(3):e0174754.

Alphey, L. 2002. Re-engineering the sterile insect technique. *Insect Biochem. Mol. Biol.* 32:1243–1247.

Amann, R. I., Ludwig, W., and Schleifer, K. H. 1995. Phylogenetic identification and in situ detection of individual microbial cells without cultivation. *Microbiol. Rev.* 59:143–169.

Aminov, R. I., Walker, A. W., Duncan S. H., Harmsen H. J. M., Welling G. W., and Flint H. J. 2006. Molecular diversity cultivation and improved detection by fluorescent in situ hybridization of a dominant group of human gut bacteria related to *Roseburia spp* or *Eubacterium rectal. Appl. Environ. Microbiol.* 72(9):6371–6376.

Apte-Deshpande, A., Paingankar, M., Gokhale, M. D., and Deobagkar, D. N. 2012. *Serratia odorifera* a midgut inhabitant of *Aedes aegypti* mosquito enhances its susceptibility to dengue-2 Virus. *PLoS One* 7:e40401.

Apte-Deshpande, A. D., Paingankar, M. S., Gokhale, M. D., and Deobagkar, D. N. 2014. *Serratia odorifera* mediated enhancement in susceptibility of *Aedes aegypti* for chikungunya virus. *Ind. J. Med. Res.* 139:762–768.

Atyame, C. M., Duron, O., Tortosa, P., Pasteur, N., Fort, P., and Weill, M. 2011. Multiple *Wolbachia* determinants control the evolution of cytoplasmic incompatibilities in *Culex pipiens* mosquito populations. *Mol. Ecol.* 20:286–298.

Azambuja, P., Garcia, E. S., and Ratcliffe, N. A. 2005. Gut microbiota and parasite transmission by insect vectors. *Trends Parasitol.* 21(12):568–572.

Bahia, A. C., Dong, Y., Blumberg, B. J., Mlambo, G., Tripathi, A., BenMarzouk-Hidalgo, O. J., Chandra, R., and Dimopoulus, G. 2014. Exploring *Anopheles* gut bacteria for *Plasmodium* blocking activity. *Environ. Microbiol.* 16(9):2980–2994.

Baldini, F., Segata, N., Pompon, J., Marcenac, P., Robert Shaw, W., Dabire, R K., Diabate, A., Levashina, E. A., and Catteruccia, F. 2014. Evidence of natural *Wolbachia* infections in field populations of *Anopheles gambiae*. *Nat. Commun.* 5:3985.

Beerntsen, B. T., James, A. A., and Christensen, B. M. 2000. Genetics of mosquito vector competence. *Microbiol. Mol. Biol. Rev.* 64(1):115–137.

Benedict, M. Q., and Robinson, A. S. 2003. The first releases of transgenic mosquitoes: An argument for the sterile insect technique. *Trends Parasitol.* 19:349–355.

Berchtold, M., Chatzinotas, A., Schönhuber, W., Brune, A., Amann, R., Hahn, D., and König, H. 1999. Differential enumeration and in situ localization of microorganisms in the hindgut of the lower termite *Mastotermes darwiniensis* by hybridization with rRNA-targeted probes. *Arch. Microbiol.* 172(6):407–416.

Berry, C. 2012. The bacterium, *Lysinibacillus sphaericus*, as an insect pathogen. *J. Invertebr. Pathol.* 109:1–10.

Berticat, C., Rousset, F., Raymond, M., Berthomieu, A., and Weill, M. 2002. High *Wolbachia* density in insecticide-resistant mosquitoes. *Proc. Biol. Sci.* 269:1413–1416.

Bian, G., Joshi, D., Dong, Y., Lu, P., Zhou, G., Pan, X., Xu, Y., Dimopoulus, G., and Xi., Z. 2013. *Wolbachia* invades *Anopheles stephensi* populations and induces refractoriness to *Plasmodium* infection. *Science* 340:748–751.

Bian, G., Xu, Y., Lu, P., Xie, Y., and Xi, Z. 2010. The endosymbiotic bacterium *Wolbachia* induces resistance to dengue virus in *Aedes aegypti*. *PLoS Pathog.* 6:e1000833.

Bisi, D. C., and Lampe, D. J. 2011. Secretion of anti-*Plasmodium* effector proteins from a natural *Pantoea agglomerans* isolate by using PelB and HlyA secretion signals. *Appl. Environ. Microbiol.* 77:4669–4675.

Boissiere, A., Tchioffo, M. T., Bachar, D., Abate, L., Marie, A., Nsango, S. E., Shahbazkia, H. R., et al. 2012. Midgut microbiota of the malaria mosquito vector *Anopheles gambiae* and interactions with *Plasmodium falciparum* infection. *PLoS Pathog.* 8:e1002742.

Boush, G. M., and Matsumura, F. 1967. Insecticidal degradation by *Pseudomonas melophthora*, the bacterial symbiote of the apple maggot. *J. Econ. Entomol.* 60:918–920.

Brownlie, J. C., Cass, B. N., Riegler, M., Witsenburg, J. J., Iturbe-Ormaetxe, I., McGraw, E. A., and O'Neill, S. L. 2009. Evidence for metabolic provisioning by a common invertebrate endosymbiont, *Wolbachia pipientis*, during periods of nutritional stress. *PLoS Pathog.* 5:e1000368.

Brownstin, J. S., Hett, E., and O'Neill, S. L. 2003. The potential of virulent *Wolbachia* to modulate disease transmission by insects. *J. Invertebr. Pathol.* 84:24–29.

Buck, M., Nilsson, L. K. J., Brunius, C., Dabiré, R. K., Hopkins, R., and Terenius, O. 2016. Bacterial associations reveal spatial population dynamics in *Anopheles gambiae* mosquitoes. *Nat. Sci. Rep.* 6:22806.

Burns, D. G., Camakaris, H. M., Janssen, P. H., and Dyall-Smith, M. L. 2004. Combined use of cultivation-dependent and cultivation-independent methods indicates that members of most haloarchaeal groups in an Australian crystallizer pond are cultivable. *Appl. Environ. Microbiol.* 70(9):5258–5265.

Campbell, C. L., Mummey, D. L., Schmidtmann, E. T., and Wilson, W. C. 2004. Culture independent analysis of midgut microbiota in the arbovirus vector *Culicoides sonorensis* (Diptera: Ceratopogonidae). *J. Med. Entomol.* 41:340–348.

Carvalho, D. O., McKemey, A. R., Garziera, L., Lacroix, R., Donnelly, C. A., Alphey, L., Malavasi, A., and Capurro, M. L. 2015. Suppression of a field population of *Aedes aegypti* in Brazil by sustained release of transgenic male mosquitoes. *PLoS Negl. Trop. Dis.* 9(7):e0003864.

Cazemier, A. E., Hackstein, J. H. P., Op den Camp, H. J. M., Rosenberg, J., and van der Drift, C. 1997a. Bacteria in the intestinal tract of different species of arthropods. *Microb. Ecol.* 33:189–197.

Cazemiera, A. E., Op den Camp, H. J. M., Hacksteina, J. H. P., and Vogelsa, G. D. 1997b. Fibre digestion in arthropods. *Comp. Biochem. Physiol. Part A Physiol.* 118:101–109.

Cébron A., Coci M., Garnier, J., and Laanbroek, H. J. 2004. Denaturing gradient gel electrophoretic analysis of ammonia-oxidizing bacterial community structure in the lower Seine River: Impact of Paris wastewater effluents. *Appl. Environ. Microbiol.* 70 (11):6726–6737.

Chandel, K., Mendki, M. J., Parikh, R. Y., Kulkarni G., Tikar, S.N., Sukumaran D., Shri Prakash, Parashar, B. D., Shouche Y. S., and Veer, V. 2013. Midgut microbial community of *Culex quinquefasciatus* mosquito populations from India. *PLoS One* 8(11):e80453.

Chandel, K., Parikh, R. Y., Mendki, M. J., Shouche, Y. S., and Veer, V. 2015. Isolation and characterization of *Vagococcus* sp. from midgut of *Culex quinquefasciatus* (Say) mosquito. *J. Vector Borne Dis.* 52:52–57.

Chao, J., and Wistreich, G. 1959. Microbial isolation from the midgut of *Culex tarsalis* Conquillet. *J. Ins. Path.* 1:311–318.

Chao, J., and Wistreich, G. 1960. Microbial isolation from the midgut of larval and adult *Culex quinquefasciatus* Say. *J. Ins. Path.* 2:220–224.

Chavshin, A. R., Oshaghi, M. A., Vatandoost, H., Pourmand, P. R., Raeisi, A., Enayati, A. A., Mardai, N., and Ghoorchian, S. 2011. Identification of bacterial microflora in the midgut of the larvae and adult of wild caught *Anopheles stephensi*: A step toward finding suitable paratransgenesis candidates. *Acta Trop.* 121:129–34.

Chavshin, A. R., Oshaghi, M. A., Vatandoost, H., Pourmand, M. R., Raeisi, A., and Terenius, O. 2014. Isolation and identification of culturable bacteria from wild *Anopheles culicifacies*, a first step in a paratransgenesis approach. *Parasit. Vectors* 4(7):419.

Chavshin, A. R., Oshaghi, M. A., Vatandoost, H., Yakhchali, B., Raeisi, A., and Zarenejad, F. 2013. *Escherichia coli* expressing a green fluorescent protein (GFP) in *Anopheles stephensi*: A preliminary model for paratransgenesis. *Symbiosis* 60:17–24.

Chavshin, A. R., Oshaghi, M. A., Vatandoost, H., Yakhchali, B., Zarenejad, F., and Terenius, O. 2015. Malpighian tubules are important determinants of *Pseudomonas* transstadial transmission and longtime persistence in Anopheles stephensi. *Parasit. Vectors* 8:36.

Chen, S., Bagdasarian, M., and Walker, E. D. 2015. Elizabethkingia anophelis: Molecular manipulation and interactions with mosquito hosts. *Appl. Environ. Microbiol.* 81 (6):2233–2243.

Cheng, D., Guo, Z., Riegler, M., Xi, Z., Liang, G., and Xu, Y. 2017. Gut symbiont enhances insecticide resistance in a significant pest, the oriental fruit fly *Bactrocera dorsalis* (Hendel). *Microbiome* 5(1):13.

Cheng, Q., and Aksoy, S. 1999. Tissue tropism, transmission and expression of foreign genes in vivo in midgut symbionts of tsetse flies. *Insect Mol. Biol.* 8(1):125–32.

Cheung, S. W., Tishler, P. V., Atkins, L., Sengupta, S. K., Modest, E. J., and Forget, B. G. 1977. Gene mapping by fluorescent in situ hybridization. *Cell Biol. Int. Rep.* 1(3):255–262.

Chouaia, B., Rossi, P., Epis, S., Mosca, M., Ricci, I., Damiani, C., Ulissi, U. et al. 2012. Delayed larval development in *Anopheles* mosquitoes deprived of *Asaia* bacterial symbionts. *BMC Microbiol.* 12(Suppl 1):S2. doi: 10.1186/1471-2180-12-S1-S2.

Chouaia, B., Rossi, P., Montagna, M. et al. 2010. Molecular evidence for multiple infections as revealed by typing of *Asaia* bacterial symbionts of four mosquito species. *Appl. Environ. Microbiol.* 76(22):7444–7450.

Cirimotich, C. M., Dong, Y., Clayton, A. M., Sandiford, S. L., Souza-Neto, J. A., Mulenga, M., and Dimopoulos, G. 2011a. Natural microbe-mediated refractoriness to *Plasmodium* infection in *Anopheles gambiae*. *Science* 332:855–859.

Cirimotich, C. M., Ramirez, J. L., Dimopoulos, G. 2011b. Native microbiota shape insect vector competence for human pathogens. *Cell Host Microbe.* 10:307–310.

Cook, P. E., McMeniman, C. J., and O'Neill, S. L. 2008. Modifying insect population age structure to control vector-borne disease. *Adv. Exp. Med. Biol.* 627:126–140.

Coon, K. L., Vogel, K. J., Brown, M. R., and Strand, M. R. 2014. Mosquitoes rely on their gut microbiota for development. *Mol. Ecol.* 23:2727–2739.

Coutinho-Abreu, I. V., Zhu, K. Y., and Ramalho-Ortigao, M. 2010. Transgenesis and paratransgenesis to control insect-borne diseases: Current status and future challenges. *Parasitol. Int.* 59:1–8.

Crotti, E., Rizzi, A., Chouaia, B., Ricci, I., Favia, G., Alma, A., Sacchi, L. et al. 2010. Acetic acid bacteria, newly emerging symbionts of insects. *Appl. Environ. Microbiol.* 76:6963–6970.

Curtis, C. F., and Adak, T. 1974. Population replacement in *Culex fatigans* by means of cytoplasmic incompatibility: Laboratory experiments with non-overlapping generations. *Bull. World Health Organ.* 51:249–255.

Damiani, C., Ricci, I., Crotti, E., Rossi, P., Rizzi, A., Scuppa, P., Capone, A. et al. 2010. Mosquito-bacteria symbiosis: The case of *Anopheles gambiae* and *Asaia*. *Microb. Ecol.* 60:644–654.

Damiani, C., Ricci, I., Crotti, E., Rossi, P., Rizzi, A., Scuppa, P., Esposito, F. et al. 2008. Paternal transmission if symbiotic bacteria in malaria vectors. *Curr Biol.* 18:1087–1088.

De Gaio, A. O., Gusmão, D. S., Santos, A. V., Berbert-Molina, M. A., Pimenta, P. F., and Lemos, F. J. 2011. Contribution of midgut bacteria to blood digestion and egg production in *Aedes aegypti* (diptera: culicidae) (L.). *Parasit. Vectors* 4:105.

Dedeine, F., Vavre, F., Fleury, F., Loppin, B., Hochberg, M. E., and Bouletreau, M. 2001. Removing symbiotic *Wolbachia* bacteria specifically inhibits oogenesis in a parasitic wasp. *Proc. Natl. Acad. Sci. USA.* 98:6247–6252.

Demaio, J., Pumpuni, C. B., Kent, M., and Beier, J. C. 1996. The midgut bacterial flora of wild *Ae. triseriatus*, *Culex pipiens* and *Phosphora columbiae* mosquito. *Am. J. Trop. Med. Hyg.* 54:219–223.

Demirev, P. A., and Fenselau, C. 2008. Mass spectrometry for rapid characterization of microorganisms. *Annu. Rev. Anal. Chem. (Palo Alto Calif.).*1:71–93.

Díez, B., Pedros-Alio, C., Marsh, T. L., and Massana, R. 2001. Application of denaturing gradient gel electrophoresis (dgge) to study the diversity of marine picoeukaryotic assemblages and comparison of dgge with other molecular techniques. *Appl. Environ. Microbiol.* 67:2942–2951.

Díaz-Nieto, L. M., D'Alessio, C., Perotti, M. A., and Berón, C. M. 2016. *Culex pipiens* development is greatly influenced by native bacteria and exogenous yeast. *PLoS ONE* 11 (4):e0153133.

Djadid, N. D., Jazayeri, H., Raz, A., Favia, G., Ricci, I., and Zakeri, S. 2011. Identification of the midgut microbiota of *An. stephensi* and *An. maculipennis* for their application as a paratransgenic tool against malaria. *PLoS One* 6(12):e28484.

Dong, Y., Aguilar, R., Xi, Z., Warr, E., Mongin, E., and Dimopoulos, G. 2006. *Anopheles gambiae* immune responses to human and rodent *Plasmodium* parasite species. *PLoS Pathog.* 2:e52.

Dong, Y., Manfredini, F., and Dimopoulos, G. 2009. Implication of the mosquito midgut microbiota in the defense against malaria parasites. *PLoS Pathog.* 5(5):e1000423.

Douglas, A. E. 1998. Nutritional interactions in insect-microbial symbioses: Aphids and their symbiotic bacteria Buchnera. *Annu. Rev. Entomol.* 43:17–37.

Dowd, P. F., and Shen, S. K. 1990. The contribution of symbiotic yeast to toxin resistance of the cigarette beetle (Lasioderma serricorne). *Entomol. Exp. Appl.* 56:241–248.

Durvasula, R. V., Gumbs, A., Panackal, A., Kruglov, O., Aksoy, S., Merrifield, R. B., Richards, F. F., and Beard, C. B. 1997. Prevention of insect-borne disease: An approach using transgenic symbiotic bacteria. *Proc. Natl. Acad. Sci. USA.* 94(7):3274–3278.

Ege, M. J., Mayer, M., Normand, A. C., Genuneit, J., Cookson, W., Braun-Fahrlander, Heederik, D., Piarroux, R., and von Mutius, E. 2011. Exposure to environmental microorganisms and childhood asthma. *N. Engl. J. Med.* 364:701–709.

Ellis, R. J., Morgan, P., Weightman, A. J., and Fry, J. C. 2003. Cultivation-dependent and -independent approaches for determining bacterial diversity in heavy-metal-contaminated soil. *Appl. Environ. Microbiol.* 69:3223–3230.

Evans, O., Caragata, E. P., McMeniman, C. J., Woolfit, M., Green, D. C., Williams, C. R., Franklin, C. E., O'Neill, S. L., and McGraw, E. A. 2009. Increased locomotor activity and metabolism of *Aedes aegypti* infected with a life-shortening strain of *Wolbachia pipientis*. *J. Exp. Biol.* 212:1436–1441.

Fang, W., Vega-Rodriguez, J., Ghosh, A. K., Jacobs-Lorena, M., Kang, A., and St. Leger, R. J. 2011. Development of transgenic fungi that kill human malaria parasites in mosquitoes. *Science* 331:1074–1077.

Favia, G., Ricci, I., Damiani, C., Raddadi, N., Crotti, E., Marzorati, M., Rizzi, A. et al. 2007. Bacteria of the genus *Asaia* stably associate with *Anopheles stephensi*, an Asian malarial mosquito vector. *Proc. Natl. Acad. Sci. USA.* 104:9047–9051.

Favia, G., Ricci, I., Marzorati, M., Negri, I., Alma, A., Sacchi, L., Bamdi, C., and Daffonchio, D. 2008. Bacteria of the genus Asaia: A potential paratransgenic weapon against malaria. *Adv. Exp. Med. Biol.* 627:49–59.

Ferguson, M. J., and Micks, D. W. 1961. Microorganisms associated with mosquitoes. I. Bacteria isolated from the adult *Culex fatigans* Wiedmann. *J. Insect. Pathol.* 3:112–119.

Foster, W. A., and Takken, W. 2004. Nectar-related vs. human-related volatiles: Behavioural response and choice by female and male *Anopheles gambiae* (Diptera: Culicidae) between emergence and first feeding. *Bull. Entomol. Res.* 94:145–157.

Fouda, M. A., Hassan, M. I., Al-Daly, A. G., and Hammad, K. M. 2001. Effect of midgut bacteria of *Culex pipiens* L. on digestion and reproduction. *J. Egypt. Soc. Parasitol.* 31:767–780.

Fu, Y., Gavotte, L., Mercer, D. R., and Dobson, S. L. 2010. Artificial triple *Wolbachia* infection in *Aedes albopictus* yields a new pattern of unidirectional cytoplasmic incompatibility. *Appl. Environ. Microbiol.* 76:5887–5891.

Fukatsu, T., Watanable, K., and Sekiguchi, Y. 1998. Specific detection of intracellular symbiotic bacteria of aphids by oligonucleotide-probed in situ hybridization. *App. Entomol. Zool.* 33:461–472.

Glaser, R. L., and Meola, M. A. 2010. The native *Wolbachia* endosymbionts of *Drosophila melanogaster* and *Culex quinquefasciatus* increase host resistance to West Nile virus infection. *PLoS ONE* 5:e11977.

Gonzalez-Ceron, L., Santillan, F., Rodriguez, M. H., Mendez, D., and Hernandez-Avila, J. E. 2003. Bacteria in midguts of field-collected *Anopheles albimanus* block *plasmodium vivax* sporogonic development. *J. Med. Entomol.* 40:371–374.

Gould, E. A., and Solomon, T. 2008. Pathogenic flaviviruses. *Lancet* 371(9611):500–509.

Gusmao, D. S., Santos, A. V., Marini, D. C., Bacci, Jr. M., Berbert-Molina, M. A., and Lemos, F. J. 2010. Culture-dependent and culture-independent characterization of microorganisms associated with *Aedes aegypti* (Diptera: Culicidae) (L.) and dynamics of bacterial colonization in the midgut. *Acta Trop.* 115:275–281.

Harris, A. F., McKemey, A. R., Nimmo, D., Curtis, Z., Black, I., Morgan, S. A., Oviedo, M. N. et al. 2012. Successful suppression of a field mosquito population by sustained release of engineered male mosquitoes. *Nature Biotechnol.* 30(9):828–830.

Harris, A. F., Nimmo, D., McKemey, A. R., Kelly, N., Scaife, S., Donnelly, C. A., Beech, C., Petrie, W. D., and Alphey, L. 2011. Field performance of engineered male mosquitoes. *Nature Biotechnol.* 29(11):1034–1037.

Hedges, L. M., Brownlie, J. C., O'Neill, S. L., and Johnson, K. N. 2008. *Wolbachia* and virus protection in insects. *Science* 322:702.

Hosokawa, T., Koga, R., Kikuchi, Y., Meng, X. Y., and Fukatsu, T. 2010. *Wolbachia* as a bacteriocyte-associated nutritional mutualist. *Proc. Natl. Acad. Sci. USA.* 107:769–774.

Hughes, G. L., Dodson, B. L., Johnson, R. M., Murdock, C. C., Tsujimoto, H., Suzuki, Y., Patt, A. A. et al. 2014. Native microbiome impedes vertical transmission of Wolbachia in Anopheles mosquitoes. *Proc. Natl. Acad. Sci. USA.* 111:12498–12503.

Humphery-Smith, I., Grulet, O., and Chastel, C. 1991a. Pathogenicity of *Spiroplasma taiwanense* for larval *Aedes aegypti* mosquitoes. *Med. Vet. Entomol.* 5:229–232.

Humphery-Smith, I., Grulet, O., Le Goff, F., and Chastel, C. 1991b. *Spiroplasma* (Mollicutes: Spiroplasmataceae) pathogenic for *Aedes aegypti* and *Anopheles stephensi* (Diptera: Culicidae). *J. Med. Entomol.* 28:219–222.

Hung, S. H. Y., Chen, T. A., Whitcomb, R. F., Tully, J. G., and Chen, Y. X. 1987. *Spiroplasma culicicola* sp. nov. from the salt marsh mosquito *Aedes sollicitans*. *Int. J. Syst. Bacteriol.* 37:365–370.

Hurst, M. R., and Jackson, T. A. 2002. Use of the green fluorescent protein to monitor the fate of *Serratia entomophila* causing amber disease in the New Zealand grass grub, Costelytra zealandica. *J. Microbiol. Methods* 50(1):1–8.

Iturbe-Ormaetxe, I., Walker, T., and O'Neill, S. L. 2011. *Wolbachia* and the biological control of mosquito-borne disease. *EMBO Rep.* 12(6):508–518.

Jadin, J., Vincke, I. H., Dunjic, A., Delville, J. P., Wery, M., Bafort, J., and Scheepers-Biva, M. 1966. Role of *Pseudomonas* in the sporogenesis of the hematozoon of malaria in the mosquito. *Bull. Soc. Pathol. Exot. Filiales.* 59:514–525.

Joyce, J. D., Nogueira, J. R., Bales, A. A., Pittman, K. E., and Aanderson, J. R. 2011. Interactions between La Crosse virus and bacteria isolated from the digestive tract of *Aedes albopictus* (Diptera: Culicidae). *J. Med. Entomol.* 48(2):389–394.

Kambris, Z., Blagborough, A. M., Pinto, S. B., Blagrove, M. S, Godfray, H. C., Sinden, R. E., and Sinkins, S. P. 2010. *Wolbachia* stimulates immune gene expression and inhibits *plasmodium* development in *Anopheles gambiae*. *PLoS Pathog.* 6:e1001143.

Kambris, Z., Cook, P. E., Phuc, H. K., and Sinkins, S. P. 2009. Immune activation by life-shortening *Wolbachia* and reduced filarial competence in mosquitoes. *Science* 326: 134–136.

Kampfer, P., Chandel, K., Prasad, G. B., Shouche, Y. S., and Vijay Veer. 2010. *Chryseobacterium culicis* sp. nov., isolated from the midgut of the mosquito *Culex quinquefasciatus*. *Int. J. Syst. Evol. Microbiol.* 60:2387–2391.

Kikuchi, Y., Hosokawa, T., and Fukatsu, T. 2011. An ancient but promiscuous host symbiont association between *Burkholderia* gut symbionts and their heteropteran hosts. *ISME J.* 5:446–460.

Kisand, V., and Wikner, J. 2003. Combining culture-dependent and -independent methodologies for estimation of richness of estuarine bacterioplankton consuming riverine dissolved organic matter. *Appl. Environ. Microbiol.* 69:3607–3616.

Kitts, C. L. 2001. Terminal restriction fragment patterns: A tool for comparing microbial communities and assessing community dynamics. *Curr. Issues Intest. Microbiol.* 2:17–25.

Laven, H. 1967. Eradication of *Culex pipiens* fatigans through cytoplasmic incompatibility. *Nature* 216:383–384.

Lee, H. L., and Seleena, P. 1990. Isolation and evaluation of larvicidal *Clostridium bifermentans* against mosquitoes of public health importance. *Trop. Biomed.* 7:103–106.

Lindh, J. M., Terenius, O., and Faye, I. 2005. 16S rRNA gene-based identification of midgut bacteria from field-caught *Anopheles gambiae* sensu lato and *An. funestus* mosquitoes reveals new species related to known insect symbionts. *Appl. Environ. Microbiol.* 71:7217–7223.

Lindh, J. M., Borg-Karlson, A. K., and Faye, I. 2008. Transstadial and horizontal transfer of bacteria within a colony of *Anopheles gambiae* (Diptera: Culicidae) and oviposition response to bacteria-containing water. *Acta Tropica.* 107:242–250.

Lindquist, D., Abusowa, M., and Hall, M. 1992. The New World screwworm fly in Libya: A review of its introduction and eradication. *Med. Vet. Entomol.* 6:2–8.

Majambere, S., Lindsay, S. W., Green, C., Kandeh, B., and Fillinger, U. 2007. Microbial larvicides for malaria control in the Gambia. *Malar. J.* 6:76.

Maleki-Ravasan, N., Oshaghi, M. A., Afshar, D., Arandian, M. H., Hajikhani, A. A., Yakhchali, B., Shiraz, M. H., Rassi, Y. et al. 2015. Aerobic bacterial flora of biotic and abiotic compartments of a hyperendemic Zoonotic Cutaneous Leishmaniasis (ZCL) focus. *Parasit. Vectors* 8:63.

Manda, H., Gouagna, L. C., Nyandat, E., Kabiru, E. W., Jackson, R. R., Foster, W. A., Githure, J. I., Beier, J. C., and Hassanali, A. 2007. Discriminative feeding behaviour of *Anopheles gambiae* s.s. on endemic plants in western Kenya. *Med. Vet. Entomol.* 21:103–111.

Meister, S., Agianian, B., Turlure, F., Relógio, A., Morlais, I., Kafatos, F. C., Christophides, G. K. 2009. *Anopheles gambiae* PGRPLC-mediated defense against bacteria modulates infections with malaria parasites. *PLoS Pathog.* 5(8):e1000542.

Meister, S., Kanzok, S. M., Zheng, X. L. et al. 2005. Immune signaling pathways regulating bacterial and malaria parasite infection of the mosquito *Anopheles gambiae*. *Proc. Natl. Acad. Sci. USA*. 102:11420–11425.

Min, K. T., and Benzer, S. 1997. *Wolbachia*, normally a symbiont of Drosophila, can be virulent, causing degeneration and early death. *Proc. Natl. Acad. Sci. USA*. 94(20): 10792–10796.

Minard, G., Mavingui, P., and Moro, C. V. 2013. Diversity and function of bacterial microbiota in the mosquito holobiont. *Parasit Vectors* 6:146.

Mittal, P. K. 2003. Biolarvicides in vector control: Challenges and prospects. *J. Vector Borne Dis.* 40:20–32.

Moreira, L. A., Iturbe-Ormaetxe, I., Jeffery, J. A., Lu, G., Pyke, A. T., Hedges, L. M., Rocha, B. C. et al. 2009a. A *Wolbachia* symbiont in *Aedes aegypti* limits infection with dengue, chikungunya, and *Plasmodium*. *Cell* 139:1268–1278.

Moreira, L. A., Saig, E., Turley, A. P., Ribeiro, J. M., O'Neill, S. L., and McGraw, E. A. 2009b. Human probing behavior of *Aedes aegypti* when infected with a life-shortening strain of *Wolbachia*. *PLoS. Negl. Trop. Dis.* 3:e568.

Moro, C. V., Tran, F. H., Raharimalala, F. N., Ravelonandro, P., and Mavingui, P. 2013. Diversity of culturable bacteria including *Pantoea* in wild mosquito *Aedes albopictus*. *BMC Microbiol.* 13:70.

Mourya, D. T., Gokhale, M. D., Pidiyar, V., Barde, P. V., Patole, M., Mishra, A. C., and Shouche, Y. 2002. Study of the effect of the midgut bacterial flora of *Culex quinquefasciatus* on the susceptibility of mosquitoes to Japenese encephalitis virus. *Acta Virologica.* 46(4):257–260.

Muturi, E. J., Ramirez, J. L., Rooney, A. P., and Kim, C.-H. 2017. Comparative analysis of gut microbiota of mosquito communities in central Illinois. *PLoS. Negl. Trop. Dis.* 11(2): e0005377.

Muyzer, G., De Waal, E. C., and Uitterlinden, A. G. 1993. Profiling of complex microbial populations by denaturing gradient gel electrophoresis analysis of polymerase chain reaction-amplified genes coding for 16S rRNA. *Appl. Environ. Microbiol.* 59:695–700.

Nartey, R., Owusu-Dabo, E., Kruppa, T., Baffour-Awuah, S., Annan, A. Oppong, S., Becker, N., and Obiri-Danso, K. 2013. Use of *Bacillus thuringiensis* var israelensis as a viable option in an integrated malaria vector control programme in the Kumasi Metropolis, Ghana. *Parasit. Vectors* 6:116.

Ngo, C. T., Aujoulat, F., Veas, F., Jumas-Bilak, E., and Manguin, S. 2015. Bacterial diversity associated with wild caught *Anopheles* mosquitoes from Dak Nong Province, Vietnam using culture and DNA fingerprint. *PLoS One* 10(3):e0118634.

Ngwa, C. J., Ckner, V. G., Abdelmohsen, U. R., Scheuermayer, M., Fischer, R., Hentschel, U., and Pradel, G. 2013. 16S rRNA gene-based identification of *Elizabethkingia meningo-septica* (Flavobacteriales: Flavobacteriaceae) as a dominant midgut bacterium of the Asian malaria vector *Anopheles stephensi* (Dipteria: Culicidae) with antimicrobial activities. *J. Med. Entomol.* 50(2):404–414.

Nold, S. C., Kopczynski, E. D., and Ward, D. M. 1996. Cultivation of aerobic chemo-organotrophic proteobacteria and gram-positive bacteria from a hot spring microbial mat. *Appl. Environ. Microbiol.* 62:3917–3921.

Nossa, C. W., Oberdorf, W. E., Yang, L., Aas, J. A., Paster, B. J., Desantis, T. Z., Brodie, E. L. Malamud, D., Poles, M. A., and Pei, Z. 2010. Design of 16S rRNA gene primers for 454 pyrosequencing of the human foregut microbiome. *World J. Gastroenterol.* 16: 4135–4144.

Novakova, E., Woodhams, D. C., Rodríguez-Ruano, S. M., Brucker, R. M., Leff, J. W., Maharaj, A., Amir, A., Knight, R., and Scott, J. 2017. Mosquito microbiome dynamics, a background for prevalence and seasonality of West Nile virus. *Front. Microbiol.* 8:526.

Ogier, J. C., Son, O., Gruss, A., Tailliez, P., and Delacroix-Buchet, A. 2002. Identification of the bacterial microflora in dairy products by temporal temperature gradient gel electrophoresis. *Appl. Environ. Microbiol.* 68:3691–3701.

Orphan, V. J., Taylor, L. T., Hafenbradl, D., and Delong, E. F. 2000. Culture-dependent and culture-independent characterization of microbial assemblages associated with high-temperature petroleum reservoirs. *Appl. Environ. Microbiol.* 66:700–711.

Osei-Poku, J., Mbogo, C. M., Palmer, W. J., and Jiggins, F. M. 2012. Deep sequencing reveals extensive variation in the gut microbiota of wild mosquitoes from Kenya. *Mol. Ecol.* 21:5138–5150.

Oxitec's Eco-Friendly Solution Addresses the Challenges of the *Aedes aegypti* Mosquito. 2016. Oxitec Ltd. http://www.oxitec.com/oxitecs-eco-friendly-solution-addresses-challenges-aedes-aegypti-mosquito/ (Accessed Sep, 2017).

Peck, G. W., and Walton, W. E. 2006. Effect of bacterial quality and density on growth and whole body stoichiometry of *Culex quinquefasciatus* and *Culex tarsalis* (Diptera: Culicidae). *J. Med. Entomol.* 43:25–33.

Penny, C., Nadalig, T., Alioua, M., Gruffaz, C., Vuilleumier, S., and Bringel, F. 2010. Coupling of denaturing high-performance liquid chromatography and terminal restriction fragment length polymorphism with precise fragment sizing for microbial community profiling and characterization. *Appl. Environ. Microbiol.* 76:648–3651.

Pidiyar, V. J., Jangid, K., Patole, M. S., and Shouche, Y. S. 2004. Studies on cultured and uncultured microbiota of wild *Culex quinquefasciatus* mosquito midgut based on 16S ribosomal RNA gene analysis. *Am. J. Trop. Med. Hyg.* 70:597–603.

Pidiyar, V., Kaznowski, A., Narayan, N. B., Patole, M., and Shouche, Y. S. 2002. *Aeromonas culicicola* sp. nov., from the midgut of *Culex quinquefasciatus*. *Int. J. Syst. Evol. Microbiol.* 52(5):1723–1728.

Porter, A. G. 1996. Mosquitocidal toxins, genes and bacteria: The hit squad. *Parasitol. Today* 12:175.

Pumpuni, C. B., Beier, M. S., Nataro, J. P., Guers, L. D., and Davis, J. R. 1993. *Plasmodium falciparum*: Inhibition of sporogonic development in *Anopheles stephensi* by gram-negative bacteria. *Exp. Parasitol.* 77:195–199.

Pumpuni, C. B., DeMaio, J., Kent, M., Davis, J. R., and Beier, J. C. 1996. Bacterial population dynamics in three *Anopheline* species: The impact on *Plasmodium* sporogonic development. *Am. J. Trop. Med. Hyg.* 54:214–218.

Raghavendra, K., Barik, T. K., Reddy, B. P., Sharma, P., and Dash, A. P. 2011. Malaria vector control: From past to future. *Parasitol. Res.* 108:757–779.

Ramirez, J. L., Souza-Neto, J., Torres Cosme R, Rovira, J., Ortiz, A., Pascale, J. M., and Dimopoulos, G. 2012. Reciprocal tripartite interactions between the *Aedes aegypti* midgut microbiota, innate immune system and dengue virus influences vector competence. *PLoS Negl. Trop. Dis.* 6(3):e1561.

Rani, A., Sharma, A., Rajagopal, R., Adak, T., and Bhatnagar, R. K. 2009. Bacterial diversity analysis of larvae and adult midgut microflora using culture-dependent and culture-independent methods in lab-reared and field-collected *Anopheles stephensi*—An Asian malarial vector. *BMC Microbiol.* 9:96.

Rasgon, J. L. 2011. Using infections to fight infections: Paratransgenic fungi can block malaria transmission in mosquitoes. *Future Microbiol.* 6:851–853.

Reegan A. D., Paulraj, M. G., and Ignacimuthu S. 2013. Isolation and characterization of Halotolerent bacteria associated with the midgut of *Culex quinquefasciatus* Say (Diptera: Culicidae). *Pakistan J. Biol. Sci.* doi: 10.3923/pjbs.2013.

Ren, X., Hoiczyk, E., and Rasgon, J. L. 2008. Viral paratransgenesis in the malaria vector *Anopheles gambiae*. *PLoS Pathog.* 4:e1000135.

Ricci, I., Valzano, M., Ulissi, U., Epis, S., Cappelli, A., and Favia, G. 2012. Symbiotic control of mosquito borne disease. *Pathog. Glob. Health* 106:380–385.

Riehle, M. A., Moreira, C. K., Lampe, D., Lauzon, C., and Jacobs-Lorena, M. 2007. Using bacteria to express and display anti-*Plasmodium* molecules in the mosquito midgut. *Int. J. Parasitol.* 37(6):595–603.

Riesenfeld, C. S., Schloss, P. D., and Handelsman, J. 2004. Metagenomics: Genomic analysis of microbial communities. *Annu. Rev. Genet.* 38:525–552.

Robinson, A. S. 2002. Genetic sexing strains in medfly, *Ceratitis capitata*, sterile insect technique programmes. *Genetica.* 116:5–13.

Roudiere, L., Lorto, S., Tallagrand, E., Marchandin, H., Jeannot, J. L., and Jumas-Bilak, E. 2007. Molecular fingerprint of bacterial communities and 16S rDNA intra-species heterogeneity: A pitfall that should be considered. *Pathol. Biol.* 55:434–440.

Rozeboom, L. E. 1935. The relation of bacteria and bacterial filtrates to the development of mosquito larvae. *Am. Hyg.* 1:167–179.

Santo Domingo, J. W., Kaufman, M. G., Klug, M. J., and Tiedje, J. M. 1998b. Characterization of the cricket hindgut microbiota with fluorescently labeled rrna-targeted oligonucleotide probes. *Appl. Environ. Microbiol.* 64(2):752–755.

Sayler, G. S., and Ripp S. 2000. Field applications of genetically engineered microorganisms for bioremediation processes. *Curr. Opin. Biotechnol.* 11:286–289.

Seitz, H. M., Maier, W. A., Rottok, M., and Feldmamm, H. B. 1987. Concominant infection of *Anopheles stephensii* with *Plasmodium berghei* and *Serratia marcescens*, additive detrimental effects. *Zentralbl. Bakteriol. Mikrobiol. Hyg. A.* 266:155–166.

Service, M. W. 1989. *Demography and vector borne disease.* CRC Press, Boca Raton, FL.

Shi, W., Syrenne, R., Sun, J., and Yuan, J. S. 2010. Molecular Approaches to Study the Insect Gut Symbiotic Microbiota at the "Omics" Age. *Insect Sci.* 17:199–219.

Simbahan, J., Kurth, E., Schelert, J., Dillman, A., Moriyama, E., Jovanovich, S., and Blum, P. 2005. Community analysis of a mercury hot spring supports occurrence of domain-specific forms of mercuric reductase. *Appl. Environ. Microbiol.* 71: 8836–8845.

Singh, R. K., Dhiman, R. C., Mittal, P. K., and Dua, V. K. 2011. Susceptibility status of dengue vectors against various insecticides in Koderma (Jharkhand), India. *J. Vector Borne Dis.* 48:116–118.

Smith, T. W., Walker, E. D., and Kaufman, M. G. 1998. Bacterial density and survey of cultivable heterotrophs in the surface water of a freshwater marsh habitat of *Anopheles quadrimaculatus* larvae (Diptera: Culicidae). *J. Am. Mosq. Control. Assoc.*14: 72–77.

Stephen, J. R., Kowalchuk, G. A., Bruns, M. V., McCaig, A. E., Phillips, C. J., Embley, T. M., and Prosser, J. I. 1998. Analysis of b-subgroup proteobacterial ammonia oxidizer populations in soil by denaturing gradient gel electrophoresis analysis and hierarchical phylogenetic probing. *Appl. Environ. Microbiol.* 64:2958–2965.

Straif, S. C., Mbogo, C. N., Toure, A. M., Walker, E. D., Kaufman, M., Toure, Y. T., and Beier, J. C. 1998. Midgut bacteria in *Anopheles gambiae* and *An. funestus* (Diptera: Culicidae) from Kenya and Mali. *J. Med. Entomol.* 35:222–226.

Tchioffo, M. T., Boissière, A., Abate, L., Nsango, S. E., Bayibeki, A. N., Awono-Ambene, P. H., Christen, R., Gimonneau, G., and Morlais, I. 2015. Dynamics of bacterial community composition in the malaria mosquito's epithelia. *Front. Microbiol.* 6:1500.

Tchioffo, M. T., Boissiere, A., Churcher, T. S., Abate, L., Gimonneau, G., Nsango, S. E., Awono-Ambene, P. H., Christen, R., Berry, A., and Morlais, I. 2013. Modulation of malaria infection in *Anopheles gambiae* mosquitoes exposed to natural midgut bacteria. *PLoS One* 8(12):e81663.

Teixeira, L., Ferreira, A., and Ashburner, M. 2008. The bacterial symbiont *Wolbachia* induces resistance to RNA viral infections in *Drosophila melanogaster*. *PLoS Biol* 6: e2.

Terenius, O., de Oliveira C. D., Pinheiro, W. D., Tadei, W. P., James, A. A., and Marinotti, O. 2008. 16S rRNA gene sequences fro bacteria associated with adult *Anopheles darlingi* (Diptera: Culicidae) mosquitoes. *J. Med. Entomol.* 45:172–175.

Terenius, O., Lindh, J. M., Eriksson-Gonzales, K., Bussiere, L., Laugen, A. T., Bergquist, H., Titanji, K., and Faye, I. 2012. Midgut bacterial dynamics in *Aedes aegypti*. *FEMS Microbiol. Ecol.* 80:556–565.

Thiery, I., Hamon, S., Gaven, B., and De Barjac, H. 1992. Host range of *Clostridium bifermentans* serovar Malaysia, a mosquitocidal anaerobic bacterium. *J. Am. Mosq. Control. Assoc.* 8:272–277.

Thompson, J. R., Randa, M. A., Marcelino, L. A., Tomita-Mitchell, A., Lim, E., Polz, M. F. 2004. Diversity and dynamics of a North Atlantic coastal Vibrio community. *Appl. Environ. Microbiol.* 70:4103–4110.

Tikar, S. N., Mendki, M. J., Sharma A. K., Sukumaran D., Veer, V., Prakash, S., and Parashar, B. D. 2011. Resistance status of the malaria vector mosquitoes, *Anopheles stephensi* and *Anopheles subpictus* towards adulticides and larvicides in arid and semi-arid areas of India. *J. Insect Sci.* 11:85.

Townson, H. 2002. *Wolbachia* as a potentialtool for suppressing filarial transmission. *Ann. Trop. Med. Parasitol.* 96:117–127.

Turley, A. P., Moreira, L. A., O'Neill, S. L., and McGraw, E. A. 2009. *Wolbachia* infection reduces blood-feeding success in the dengue fever mosquito, *Aedes aegypti*. *PLoS Negl. Trop. Dis.* 3:e516.

Vasanthi, V. and Hoti, S.L. 1992. Microbial flora in gut of *Culex quinquefasciatus* breeding in cess pits. *Southeast Asian J. Trop. Med. Public Health.* 23:312–317.

Walker, T., Johnson, P. H., Moreira, L. A. et al. 2011, The *wMel Wolbachia* strain blocks dengue and invades caged *Aedes aegypti* populations. *Nature* 476:450–3.

Wang, Y., Gilbreath, T. M., Kukutla, P., Yan, G., and Xu, J. 2011. Dynamic gut microbiome across life history of the malaria mosquito *Anopheles gambiae* in Kenya. *PLoS One* 6: e24767.

Weiss, B. L., Mouchotte, R., Rio, R. V., Wu, Z., Heddi, A., and Aksoy, S. 2006. Interspecific transfer of bacterial endosymbionts between tsetse fly species: Infection establishment and effect on host fitness. *Appl. Environ. Microbiol.*72:7013–7021.

Werren, J. H., Baldo, L., and Clark, M. E. 2008. *Wolbachia*: Master manipulators of invertebrate biology. *Nat. Rev. Microbiol.* 6:741–751.

White, N. J. 2004. Antimalarial drug resistance. *J. Clin. Invest.* 113(8):1084–1092.

Williamson, D. L., Tully, J. G., Rosen, L., Rose, D. L., Whitcomb, R. F., Abalain-Colloc, M. L., Carle, P., Bove, J. M., and Smyth, J. 1996. *Spiroplasma diminutum* sp. nov., from *Culex annulus* mosquitoes collected in Taiwan. *Int. J. Syst. Bacteriol.* 46:229–233.

World Health Organization (WHO). 2016. *Vector-borne diseases*. http://www.who.int /mediacentre/factsheets/fs387/en/ (accessed February 2017).

Wotton, R. S., Chaloner, D. T., Yardley, C. A., and Merritt, R. W. 1997. Growth of *Anopheles* mosquito larvae on dietary micro-biota in aquatic surface microlayers. *Med. Vet. Entomol.* 11:65–70.

Xu, Y., Moser, C., Al-Soud W. A., Sorensen, S., Hoiby, N., Nielsen P. H., and Thomsen, T. R. 2012. Culture-dependent and -independent investigations of microbial diversity on urinary catheters. *J. Clin. Microbiol.* 50(12):3901–3908.

Yadav, K. K., Bora, A., Datta, S., Chandel, K., Gogoi, H. K., Prasad, G. B., and Vijay Veer. 2015a. Molecular characterization of midgut microbiota of *Aedes albopictus* and *Aedes aegypti* from Arunachal Pradesh, India. *Parasit. Vectors* 18(8):641.

Yadav, K. K., Chandel, K., Bora, A., and Veer, V. 2015b. Isolation and characterization of *Dietzia maris* from midgut of *Aedes albopictus*: A suitable candidate for paratransgenesis. *Int. J. Mos. Res.* 7(24):7–12.

Yadav, K. K., Datta, S., Naglot, A., Bora, A., Vanlal, H., Bhagyawant, S., Veer, V., and Raju, P. S. 2016a. Diversity of cultivable midgut microbiota at different stages of the Asian tiger mosquito, *Aedes albopictus* from Tezpur, India. *PLoS One* 11(12):e0167409.

Yadav, K. K., Naglot, A., Chandel, K., Bhagyawant, S., Bora, A., Gogoi, H. K., Vijay Veer, and Raju, P. S. 2016b. Molecular characterization of midgut bacteria of *Aedes albopictus* exhibiting swarming motility property. *Afr. J. Microbiol. Res.* 10(25):949–959.

Yoshida, S., Ioka, D., Matsuoka, H., Endo, H., and Ishii A. 2001. Bacteria expressing single-chain immunotoxin inhibit malaria parasite development in mosquitoes. *Mol. Biochem. Parasitol.* 113:89–96.

Zhu, X. Y., Lubeck, J., and Kilbane J. J. 2003. Characterization of microbial communities in gas industry pipelines. *Appl. Environ. Microbiol.* 69:5354–5363.

Zimmer, C. 2001. *Wolbachia*, a tale of sex and survival. *Science* 292:1093–1095.

Zoetendal, E. G., Akkermans, A. D., and De Vos, W. M. 1998. Temperature gradient gel electrophoresis analysis of 16S rRNA from human fecal samples reveals stable and host-specific communities of active bacteria. *Applied Environ. Microbiol.* 64:3854–3859.

Zouache, K., Michelland, R. J., Failloux, A.-B., Grundmann, G. L., and Mavingui, P. 2012. Chikungunya virus impacts the diversity of symbiotic bacteria in mosquito vector. *Mol. Ecol.* 21:2297–2309.

Zouache, K., Raharimalala, F. N., Raquin, V., Tran-Van, V., Raveloson, L. H., Ravelanandro, P., and Mavingui, P. 2011. Bacterial diversity of field-caught mosquitoes, *Aedes albopictus* and *Aedes aegypti*, from different geographic regions of Madagascar. *FEMS Microbiol. Ecol.* 75:377–389.

# 15 A Review of Control of Blood Sucking Parasites by Marine Actinomycetes-Derived Compounds

*Krishnan Kannabiran*

**CONTENTS**



## 15.1 INTRODUCTION

Mosquitoes are the most prevalent vectors of several vertebrate blood parasites. Mosquito transmits diseases such as malaria, dengue fever, yellow fever, filariasis, chikungunya, and viral encephalitis all contribute to much of the health problems worldwide (Matasyoh et al. 2011). A high proportion of the population is at risk of developing mosquito-borne diseases (Ottesen et al. 2008; World Health Organization [WHO] 2008a). Mosquito-borne diseases currently pose a major challenge to both tropical and subtropical countries. The diseases like malaria, dengue and filariasis are the most significant cause of morbidity and mortality every year (WHO 2009). In India, these diseases are still causing high morbidity and mortality and two to three million people are affected every year (Sharma et al. 2009).

Malaria and lymphatic filariasis (LF) are the most common mosquito-borne parasitic diseases worldwide (Manguin et al. 2010). The overall prevalence and health impact of malaria and LF have made them important diseases for control and global elimination (Kyelem et al. 2008; WHO 2008b). Nearly 3.3 billion people are living in malaria risk areas around the world and about 250 million people are infected annually. Malaria is believed to be responsible for approximately 1 million deaths per year, particularly among young children (<5 years) and pregnant women. Africa is endemic and the highest malaria burden of 212 million cases (86% globally) is distributed across 45 countries (Manguin et al. 2010).

Malaria is caused by *Plasmodium* parasites and transmitted by the female *Anopheles* mosquito acting as a host and a vector of the parasite. Malaria is highly endemic in Assam, India, and causes high morbidity and mortality (Baruah et al. 2007; Pal et al. 2009). The emergence of drug-resistant *Plasmodium falciparum* strains has been found in many endemic areas of Assam (Dua et al. 2003). In India during the years 2000 to 2007, malaria-associated mortality was reported between 43 and 300, with 300 being the highest in the year 2006 (Sharma and Dev 2006).

*Anopheles subpictus* is a vector of malaria and some helminth and arbovirus species (Chandra et al. 2010). *A. subpictus* breeds in accumulated rainwater; rice fields (Dhanda and Kaul 1980); and wastewater disposal and irrigated sites (Mukhtar et al. 2003). Approximately 48.11 million are infected and 420 million are residing in endemic areas. *Culex tritaeniorhynchus*, a mosquito vector, breeds in stagnant waters of paddy fields and drainage waste waters (Innis 1995). The transmission cycle of *C. tritaeniorhynchus* is confined to rice field areas (Sunish and Reuben 2001) and pigs and birds as amplifier hosts.

*Culex quinquefasciatus* Say (Diptera: Culicidae) is a mosquito vector species responsible for transmission of LF in many countries including India (Raghavendra et al. 2011). *C. quinquefasciatus* breeds in waters mixed with organic debris containing rooting vegetation, household refuse, and excreta. Larvae of this vector are found in partially blocked wastewater drains, pits, septic tanks, and in abandoned wells containing water unfit for drinking (Kovendan et al. 2011). Bancroftian filariasis is caused by *Wuchereria bancrofti* and transmitted by *C. quinquefasciatus*. It accounts for 95% of the total LF cases, and in India LF is the second important mosquito-borne diseases next to malaria (Ramaiah 2000). *C. quinquefasciatus* transmits viruses, which include West Nile virus (WNV) (Kent et al. 2010). Among the various *Aedes* species present in India, the dengue vector *A. aegypti* was the most common species (86.89%) occurring throughout the year, whereas *A. albopictus* and *A. vitattus* were found during monsoon periods both in the tropical and subtropical zones. Out of 53 *Anopheles* species present in India, only nine species are considered to be major vectors of malarial disease.

Tick infection affects almost all dairy and meat animals in India, which leads to huge economic loss (Ghosh et al. 2006). Ticks and tick-borne diseases affect cattle and cause severe losses to farmers due to high cattle mortality, loss of body weight, and milk loss, and it also costs more for the effective control of infection (Homewood et al. 2006; Kivaria 2006). *Haemaphysalis bispinosa* infection was found in about 40% of cattle, 55.4% of goats, and 13.2% of pigs (Ghosh et al. 2007). *Rhipicephalus (Boophilus) microplus* is responsible for cattle deaths, resulting in loss of milk, meat, and leather productions. Ghosh et al. (2007) reported the disease transmission potential for *Haemaphysalis* (theileriosis and babesiosis in sheep and goats) and *Rhipicephalus* (bovine and canine babesiosis, ehrlichiosis, and equine babesiosis). In the recent past, changes in public health policy in several countries, social factors, and development of drug-resistance mosquitoes and the pathogens transmitted by mosquito vectors resulted in a resurgence in the incidence of mosquito-borne diseases (Geetha et al. 2007).

## 15.2 VECTOR CONTROL AND DRUG RESISTANCE

At present, vector control is one of the most serious problems in developing countries like India, due to the lack of general awareness of the public and the poor socio-economic conditions (Gueye et al. 2016). Larval source management (LSM), including habitat modification, manipulation, and larviciding, has historically been used for mosquito control in many parts of the world (Afrane et al. 2016). A more common method for the control of mosquito-borne diseases is the use of insecticides, and several synthetic agents developed in the past have been used successfully. Continuous use of insecticides provoked undesirable effects, including toxicity to nontarget organisms, and fostered environmental and human health concerns (Lee et al. 2001). Residual toxicity of synthetic insecticides such as dichlorodiphenyltrichloroethane (DDT) and benzene hexachloride (BHC) leads to a serious problem due to the slow decomposition of these chemicals (Prashith Kekuda et al. 2010), and these chemicals remain in nature for a long time. Easily decomposable pesticides with low toxicity would be very advantageous. There is an ever-increasing demand for the development of new types of antibiotics to control mosquitoes.

Even though vector control methods are being improved in many aspects, there are still significant drawbacks and challenges, which include resistance developed to the routinely used synthetic insecticides and phytochemicals. Synthetic insecticides produce several ecological problems, such as the development of resistant strains, ecological imbalance, and impact on mammals. Mosquito vectors had developed a resistance to several pesticides in the recent past due to which treating mosquito-borne diseases is becoming increasingly difficult in underdeveloped and developing countries.

Tick control, in general, is based on the use of chemical acaricides. Unfortunately, use of chemical acaricides has produced many negative side effects, including the development of chemical resistance in some tick populations, and food and environmental contamination (Rosado-Aguilar et al. 2010). Continuous and indiscriminate use of acaricides leads to the development of chemical-resistant ticks along with contamination of the environment and animal products (Graf et al. 2004). A recent survey on acaricide resistance conducted through a questionnaire reported large-scale acaricide resistance in India (Food and Agriculture Organization [FAO] 2004). Impact on the environment and safety of the chemical insecticides used are needed to avoid the possible consequences on humans.

Lack of an alternative, cost-effective, and safer insecticide led to tremendous increase in mosquito populations. Increasing insecticide resistance stressed the need for the development of new methods for effective mosquito control and elimination of vector-borne diseases. Hence, there is an urgent need for developing biologically active materials and isolating microbial secondary metabolites for effective insect and mosquito control, which are expected to reduce the hazards to humans and other organisms by minimizing the accumulation of harmful chemical residues in the environment. Hence, an alternative approach for mosquito control is the application of natural products of microbial origin. Natural bioactive compounds are considered as promising alternative for mosquito and tick control. They might offer additional advantages such as low toxicity to mammals and more environmental friendliness (Fernandes and Freitas 2007; Ribeiro et al. 2007).

## 15.3 MICROBIAL VECTOR CONTROL

Microbial control of insect vectors was found to be highly effective and generally has several advantages over chemical control methods because many are host specific and safe for nontarget organisms. The inherent dangers associated with the synthetic insecticides can be replaced by the process for the production of microbial products which are safe and less hazardous (Misato 1983). *Bacillus thuringiensis* subsp. *israelensis* and *Bacillus sphaericus* 2362 were used as microbial insecticides for control of mosquito larvae (Federici 1995). Marine microbes and their metabolites represent a potential source for effective vector control.

## 15.4 MARINE/TERRESTRIAL ACTINOMYCETES-MEDIATED VECTOR CONTROL

Marine microorganisms serve as a valuable and inexhaustible source for several novel products such as antibiotics, antitumor agents, and other novel therapeutic substances (Subramani and Aalbersberg 2012). Most of the antibiotics developed for various diseases to date are from natural products derived from marine microorganisms. Some of the antibiotics include erythromycin, vancomycin, and streptomycin, and these are used for treating bacterial infections (Kelecom 2002). Marine microbial natural products are very important not only for their potent therapeutic properties but also for the pharmacokinetic properties useful for clinical drug development. Marine microbes have not been exploited extensively for novel chemical compounds and new chemical entities, as their terrestrial counterparts. Most of the marine bioactive compounds that have been extracted and identified successfully are from actinomycetes. It has been shown that marine actinomycetes produce bioactive substances that are different from known compounds from terrestrial counterparts (Fenical 1993). Marine actinomycetes are not been completely explored for novel bioactive compounds, and there are still underexplored and unexplored regions of salt pans, deep sea trenches, and hydrothermal vents that need to be studied.

Actinobacteria have gained much importance being the potent source for novel antibiotics and other bioactive secondary metabolites (Newman et al. 2000). *Streptomyces* genera are important, and they are widely distributed in marine environment and terrestrial habitats (Suthindhiran and Kannabiran 2010). *Streptomyces* species are of commercial importance due to their unique capacity to produce several novel secondary metabolites and new chemical entities. Most of the studies on marine *Streptomyces* have focused on antibiotic production for the application of human (Dharmaraj and Alagarsamy 2009) and fish pathogens (Patil et al. 2001; Suthindhiran and Kannabiran 2011).

Marine actinomycetes extracts have been reported to be very effective larvicidal agents against blood-sucking parasites. Larvicidal activity of actinomycete isolate (S-70) against *Toxocara canis* larvae was reported (Al-Doori et al. 1991). *Streptomyces sioyaensis* SA-1758 isolated from Gamo, Miyagi Prefecture, Japan, has been shown to possess acaricidal activity (Takahashi et al. 1989). The alarming death rate caused by parasites need to be controlled by using a new and effective larvicidal agent (Pimentel-Elardo et al. 2010). Insecticidal activity of *Streptomyces* species isolated from soil samples collected from the Western Ghats of Agumbe, Karnataka, India, has been reported against *A. aegypti* (Prashith Kekuda et al. 2010). Extracellular secondary metabolites of *Streptomyces microflavus* showed significant larvicidal activity against *C. pipiens* and the $LC_{50}$ value was calculated to be 77.3 ppm with 95% fiducial limits (El-Bendary et al. 2010). Larvicidal activity of *Streptomyces* sp. isolated from Muthupet mangrove, Tamil Nadu, India, against *Anopheles* mosquitoes has been reported (Vijayakumar et al. 2010). Larvicidal, mosquito repellent, and ovicidal activity of marine-derived actinobacteria, *Saccharomonospora* sp. (LK-1), *Streptomyces roseiscleroticus* (LK-2), and *Streptomyces gedanensis* (LK-3) against *C. tritaeniorhynchus* and *C. gelidus* have been reported (Karthik et al. 2011). Marine *streptomyces* sp. VITDDK3 isolate showed acaricidal and larvicidal activity against blood-sucking parasites (Deepika et al. 2012). The antiparasitic, larvicidal, and insecticidal activity of actinomycetes extracts reported from 2010 onward is given in Table 15.1.

Marine *streptomyces* VITSVK5 sp. has been reported to possess larvicidal activity against *R.* (*Boophilus*) *microplus*, *A. stephensi*, and *C. tritaeniorhynchus* (Saurav et al. 2013). The culture filtrates of *Streptomyces citreofluorescens* exhibited larvicidal activity against *A. stephensi*, *C. quinquefasciatus*, and *A. aegypti* vectors for malaria, filarial, and dengue (Singh and Prakash 2012). The ethyl acetate (EA) extract of *Streptomyces* VITSTK7 sp. has been shown to exhibit acaricidal and insecticidal activity against the larvae of cattle ticks *H. bispinosa* and *R.* (*Boophilus*) *microplus* (Acari: Ixodidae), fourth instar larvae of malaria vector *A. subpictus*, and filarial vector *C. quinquefasciatus* (Diptera: Culicidae) (Thenmozhi, 2013). The crude extract from *Saccharomonas* spp. was shown to be effective against the larvae of *A. stephensi*, *C. tritaeniorhynchus*, and adults of *H. bispinosa*, *R. (B.) microplus* and *Hippobosca maculate* (Karthik et al. 2013). The EA extract of *Streptomyces* sp. VITPK9 showed larvicidal activity against *A. subpictus*, *C. tritaeniorhynchus*, *C. gelidus* and *R. (B.) microplus* (Sanjenbam and Kannabiran 2013). Actinomycetes isolated from salt range region of Pakistan showed significant larvicidal activity against the third instar larvae of mosquitoes *C. quinquefasciatus* and *Tribolium castaneum* (Anwar et al. 2014). Larvicidal and mosquito repellent activities of marine *Streptomyces* sp. VITJS4 isolated from marine soil samples of Puducherry coast, Tamil Nadu, India, against *A. stephensi*, *A. aegypti*, and *C. quinquefasciatus* have been reported (Naine and Devi 2014). Endophytic actinomycetes *Streptomyces albovinaceus* and *Streptomyces badius* showed larvicidal activity against fourth instar stages of *C. quinquefasciatus* Say mosquito larvae (Tanvir et al. 2014). EA extract from actinomycetes isolated from soil samples of the southeast coast of India are shown to be effective against *C. quinquefasciatus*, *A. aegypti*, and *A. stephensi* (Vinothkumar et al. 2015). Actinomycetes isolated from terrestrial soil samples

**TABLE 15.1**
**Antiparasitic, Larvicidal, and Insecticidal Activity of Actinomycetes**

| Actinomycetes Species | Extract | Activity | References |
|---|---|---|---|
| *Streptomyces cacaoi* subsp. *cacaoi*-M20 | Culture filtrate | *A. aegypti* and *C. quinquefasciatus* | Janaki (2016a, 2016b) |
| *Streptomyces* sp. | Culture filtrate | *A. aegypti* and *A. stephensi* | Balkrishnan et al. (2016) |
| Actinomycetes sp. | Culture filtrate | *C. quinquefasciatus* | Rajesh et al. (2015) |
| Actinomycetes sp. | Ethyl acetate | *A. aegypti*, *A. stephensi*, and *C. uinquefasciatus* | Vinothkumar et al. (2015) |
| *Streptomyces albovinaceus* and *Streptomyces badius* | Culture filtrate | *C. uinquefasciatus* | Tanvir et al. (2014) |
| Actinomycetes sp. | Culture filtrate | *C. uinquefasciatus* and *Tribolium castaneum* | Anwar et al. (2014) |
| *Streptomyces* sp. VITPK9 | Ethyl acetate | *A. subpictus*, *C. tritaeniorhynchus*, *C. gelidus*, and *R. microplus* | Sanjenbam and Kannabiran (2013) |
| *Saccharomonas* spp. | Crude extract | *A. stephensi*, *C. tritaeniorhynchus*, *H. bispinosa*, and *R. microplus* | Karthik et al. (2013) |
| *Streptomyces citreofluorescens* | Culture filtrate | *A. stephensi*, *C. quinquefasciatus*, and *A. aegypti* | Singh and Prakash (2012) |
| *Saccharomonospora* sp., *Streptomyces roseiscleroticus*, and *Streptomyces gedanensis* | Culture filtrate | *C. tritaeniorhynchus* and *C. gelidus* | Karthik et al. (2011) |
| *Streptomyces microflavus* | Extracellular crude extract | *C. pipiens* | El-Bendary et al. (2010) |
| *Streptomyces* sp. | Culture filtrate | *A. aegypti* | Prashith Kekuda et al. (2010) |
| *Streptomyces* sp. | *Streptomyces* sp. | *Anopheles* mosquitoes | Vijayakumar et al. (2010) |

exhibited larvicidal activity against third instar *C. quinquefasciatus* (Rajesh et al. 2015). *Streptomyces* species isolated from the Parangipettai mangrove ecosystem showed significant larvicidal activity against *A. aegypti* and *A. stephensi* mosquitoes (Balkrishnan et al. 2016). Larvicidal activity of *Streptomyces cacaoi* subsp. *cacaoi*-M20 isolated from backwater soil of Ariyankuppam, Puducherry, India, against *A. aegypti* (Janaki 2016a) and *C. quinquefasciatus* has been reported (Janaki 2016b).

## 15.5 MARINE ACTINOMYCETES-DERIVED NANOPARTICLES AS VECTOR CONTROL AGENTS

Silver nanoparticles (AgNPs) synthesized using microbial source have been identified as a potential replacement of synthetic chemical insecticides. Hence the need to use green synthesized AgNPs for the control of mosquitoes causing many diseases. AgNPs synthesized using the extracellular metabolites of *Streptomyces* sp. RD (JX512257) showed larvicidal activity against *C. quinquefasciatus* and *A. aegypti* (Rajesh et al. 2013). AgNPs synthesized from terrestrial actinomycetes species *Saccharopolyspora erythraea* showed larvicidal activity against *A. aegypti* and *A. stephensi* (Prabha et al. 2014). Recently the larvicidal potential of *Streptomyces* sp. M25-mediated synthesized AgNPs has been reported (Shanmugasundaram and Balagurunathan 2015). The biosynthesized AgNPs (10–35 nm) exhibited significant larvicidal activity against malarial vector *A. subpictus* ($LC_{50}$ 51.34 mg/L and $\chi^2$ value of 8.228), filarial vector *C. quinquefasciatus* ($LC_{50}$ 48.98 mg/L and $\chi^2$ value of 14.307), and dengue vector *A. aegypti* ($LC_{50}$ 60.23 mg/L and $\chi^2$ value of 4.042), respectively.

## 15.6 ANTIPARASITIC COMPOUNDS FROM ACTINOMYCETES

Screening of actinomycetes secondary metabolites has led to the discovery of several antibiotics and biologically active compounds with desired pharmacological activities. Few secondary metabolites isolated from marine actinomycetes have been shown to exhibit strong larvicidal activity against blood-sucking parasites. Larvicidal and insecticidal compounds reported from actinomycetes isolates are given in Table 15.2.

The marine actinomycetes-derived compound spinosad is a novel insecticide produced by the fermentation of *Saccharopolyspora spinosa* (Snyder et al. 2007). Depsipeptides produced by actinomycetes species were used as a potential antimalarial compound (Fotie and Morgan 2008). The salinosporamide A, produced by the marine actinomycete *Salinisporatropica*, exhibited potent inhibitory activity against the *Plasmodium* during the erythrocytic stages of growth (Prudhomme et al. 2008). The antihelminthic activity of abamectin and ivermectin derived from *Streptomyces avermectinius* has been reported (Molinari et al. 2010). Bioactive compounds extracted from *Streptomyces* sp. isolated from Mediterranean sponges, cyclic depsipeptide valinomycin, indolocarbazole, alkaloid staurosporine, and butanolide were found to be active against *Leishmania* and *Trypanosoma brucei* (Pimental-Elardo et al. 2010). The compound (2S,5R,6R)-2-hydroxy-3,5,6-trimethyloctan-4-one extracted from *Streptomyces* sp. VITDDK3 showed strong acaricidal and larvicidal activity against

**TABLE 15.2**
**Larvicidal and Insecticidal Compounds Reported from Actinomycetes Isolates**

| Active Compound | Organism | Activity | Reference |
|---|---|---|---|
| Cyclopentanepropanoic acid, 3,5-bis(acetyloxy)-2-[3-(methoxyimino)octyl], methyl ester; 5-azidomethyl-3-(2-ethoxycarbonyl-ethyl)-4-ethoxycarbonylmethyl-1H-pyrrole-2-carboxylic acid, ethyl ester; akuammilan-16-carboxylic acid,17-(acetyloxy)-10-methoxy, methyl ester (16R) | *Streptomyces* sp. VITSTK7 | *H. bispinosa*, *R. (B.) microplus*, *A. subpictus*, and *C. quinquefasciatus* | Thenmozhi and Kannabiran (2013) |
| 5-(2,4-dimethylbenzyl) pyrrolidin-2-one | *Streptomyces* VITSVK5 spp. | *R. (B.) microplus*, *A. stephensi*, and *C. quinquefasciatus* | Saurav et al. (2013) |
| 2S,5R,6R)-2-hydroxy- 3,5,6- trimethyloctan-4-one (HTMO) | *Streptomyces* spp. VITDDK3 | *R. (B.) microplus*, *A. subpictus*, and *C. quinquefasciatus* | Deepika and Kannabiran (2012) |
| Cyclic depsipeptide valinomycin, indolocarbazole, alkaloid staurosporine, and butanolide | *Streptomyces* sp. | *Leishmania* and *Trypanosoma brucei* | Pimental-Elardo et al. (2010) |
| Abamectin and ivermectin | *Streptomyces avermectinius* | Antihelminthic | Molinari et al. (2010) |
| Salinosporamide A | *Salinisporatropica* | *Plasmodium* species | Prudhomme et al. (2008) |
| Depsipeptides | Actinomycetes | Antimalarial | Fotie and Morgan (2008) |
| Spinosad | *Saccharopolyspora spinosa* | Insecticide | Snyder et al. (2007) |

blood-sucking parasites (Deepika et al. 2012). The larvicidal activity of the compound 5-(2,4-dimethylbenzyl) pyrrolidin-2-one extracted from marine *Streptomyces* VITSVK5 sp. against *R.* (*Boophilus*) *microplus*, *A. stephensi*, and *C. tritaeniorhynchus* was already been reported (Saurav et al. 2013). Three major compounds cyclopentanepropanoic acid, 3,5-bis(acetyloxy)-2-[3-(methoxyimino)octyl], methyl ester; 5-azidomethyl-3-(2-ethoxy carbonyl-ethyl)-4-ethoxycarbonylmethyl-1H-pyrrole-2-carboxylic acid, ethyl ester; and akuammilan-16-carboxylic acid, 17-(acetyloxy)-10-methoxy, methyl ester (16R) extracted from *Streptomyces* sp. VITSTK7 have been shown to exhibit larvicidal activity against the larvae of cattle ticks *H. bispinosa* and *R.* (*B.*) *microplus*; fourth-instar larvae of malaria vector *A. subpictus*; and filarial vector *C. quinquefasciatus* (Thenmozhi et al. 2013).

## 15.7 CONCLUSION

The marine and costal area of the Indian Peninsula has not been fully explored for novel actinobacteria and their secondary metabolites. An extensive survey of the available literature clearly indicates that marine actinomycetes are not being fully explored for mosquito larvicidal/insecticidal compounds. Even though few studies report the larvicidal activity of marine actinomycetes extracts and nanoparticles, there are few reports on extraction and identification of mosquitocidal/antiparasitic compounds from marine/terrestrial actinomycetes. Secondary metabolites from marine actinomycetes are an alternative source for developing new, effective, and eco-friendly larvicidal/insecticidal compounds. Hence further studies on marine natural products are very much needed to alleviate the challenges posed by mosquito vectors.

## ACKNOWLEDGMENTS

The author thanks VIT University for all the support and encouragement provided to write this chapter.

## REFERENCES

Afrane, Y. A., N. G. Mweresa, C. L. Wanjala, T. M. Gilbreath, G. Zhou, M. Lee, A. K. Githeko, and G. Yan. 2016. Evaluation of long-lasting microbiallarvicide for malaria vector control in Kenya. *Malar J* 15:577.

Al-Doori, M., A. A. al-Tae, S. Jalil, and S. A. Hassa. 1991. Larvicidal activity of actinomycete isolate against *Toxocara canis*. *Folia Parasitologica* 38:379–382.

Anwar, S., B. Ali, F. Qamar, and I. Sajid. 2014. Insecticidal activity of actinomycetes isolated from salt range, Pakistan against mosquitoes and red flour beetle. *Pakistan J Zool* 46:83–92.

Balakrishnan, S., P. Santhanam, and M. Srinivasan. 2016. Larvicidal potency of marine actinobacteria isolated from mangrove environment against *Aedes aegypti* and *Anopheles stephensi*. *J Parasit Dis*. DOI 10.1007/s12639-016-0812-3.

Baruah, I., N. G. Das, and J. Kalita. 2007. Seasonal prevalence of malaria vectors in Sonitpur district of Assam, India. *J Vec Borne Dis* 44:149–153.

Chandra, G., I. Bhattacharjee, and S. Chatterjee. 2010. A review on *Anopheles subpictus* Grassi—A biological vector. *Acta Trop* 115:142–154.

Deepika, T. L., K. Kannabiran, V. Gopiesh Khanna, G. Rajakumar, C. Jayaseelan, T. Santhoshkumar, and A. A. Rahuman. 2012. Isolation and characterization of acaricidal and larvicidal novel compound (2S,5R,6R)-2-hydroxy-3,5,6-trimethyloctan-4-one from *Streptomyces* sp. against blood-sucking parasites. *Parasitol Res* 111:1153–1163.

Dhanda, V., and H. N. Kaul. 1980. Mosquito vectors of Japanese encephalitis virus and their bionomics in India. *Proc Indian Natl Sci Acad* 46B:759.

Dharmaraj, S., and S. Alagarsamy. 2009. Bioactive potential of *Streptomyces* isolated from marine sponges. *World J Microbiol Biotechnol* 25:1971–1979.

Dua, V. K., V. Dev, S. Phookam, N. C. Gupta, V. P. Sharma, and S. K. Subbarao. 2003. Multidrug resistant *Plasmodium falciparum* malaria in Assam, India: Timing of recurrence and anti-malarial drug concentrations in whole blood. *Am J Trop Medi Hyg* 69: 555–557.

El-Bendary, M. A., H. M. Rifaat, and A. A. Keera. 2010. Larvicidal activity of extracellular secondary metabolites of *Streptomyces microflavus* against *Culex pipiens*. *Can J Pure Appl Sci* 4:1021–1026.

Food and Agriculture Organization of the United Nations (FAO). 2004. Ticks: Acaricide resistance: Diagnosis management and prevention. In: *Guidelines resistance management and integrated parasite control in ruminants*. Rome: FAO Animal Production and Health Division.

Federici, B. A. 1995. The future of microbial insecticides as vector control agents. *J Am Mosquito Control Ass* 11(2):260–268.

Fernandes, F. F., and E. P. S. Freitas. 2007. Acaricidal activity of an oleoresinous extract from *Copaifera reticulata* (Leguminosae: Caesalpinioideae) against larvae of the southern cattle tick, *Rhipicephalus (Boophilus) microplus* (Acari: Ixodidae). *Vet Parasitol* 147:150–154.

Fenical, W. 1993. Chemical studies of marine bacteria: Developing a new resource. *Chem Rev* 93:1673–1683.

Fotie, J., and R. E. Morgan. 2008. Depsipeptides from microorganisms: A new class of antimalarials. *Mini Rev Med Chem* 8:1088–1094.

Geetha, I., G. Prabakaran, K. P. Paily, A. M. Manonmani, and K. Balaraman. 2007. Characterisation of three mosquitocidal *Bacillus* strains isolated from mangrove forest. *Biol Control* 42:34–40.

Ghosh, S., P. Azhahianambi, and J. de la Fluente. 2006. Control of ticks of ruminants with special emphasis on livestock farming system in India-present and future possibilities for integrated control: A review. *Exp Appl Acarol* 40:49–66.

Ghosh, S., G. C. Bansal, S. C. Gupta, D. Ray, M. Q. Khan, H. Irshad, M. Shahiduzzaman, U. Seitzer, and J. S. Ahmed. 2007. Status of tick distribution in Bangladesh, India and Pakistan. *Parasitol Res* 101:S207–S216.

Graf, J. F., R. Gogolewski, and B. N. Leach. 2004. Tick control: An industry point of view. *Parasitol* 129:S247–S442.

Gueye, C. S., G. Newby, R. D. Gosling, M. A. Whittaker, D. Chandramohan, L. Slutsker, and M. Tanner. 2016. Strategies and approaches to vector control in nine malaria-eliminating countries: A cross-case study analysis. *Malar J* 15:2.

Homewood, K., P. Trench, L. S. Randal, G. Lynen, and B. Bishop. 2006. Livestock health and socio-economic impacts of veterinary interventions in Masailand: Infection and treatment vaccine against East Coast fever. *Agric Syst* 89:248–271.

Innis, B. L. 1995. Japanese encephalitis. In: *Exotic viral infections*, J. S. Potterfield (ed.), 147–174. London: Chapman & Hall.

Janaki, T. 2016a. Larvicidal activity of *Streptomyces* cacaoi subsp. cacaoi-M20 against *Aedes aegypti*. *Int J Bot Studies* 1:47–49.

Janaki, T. 2016b. Larvicidal activity of Streptomyces cacaoi subsp. cacaoi-M20 against *Culex quinquefasciatus* (III Instar). *Int J Mosquito Res* 3:47–51.

Karthik, L., K. Gaurav, K. V. Rao, G. Rajakumar, and A. A. Rahuman. 2011. Larvicidal, repellent, and ovicidal activity of marine actinobacteria extracts against *Culex tritaeniorhynchus* and *Culex gelidus*. *Parasitol Res* 108:1447–1455.

Karthik, L., G. Kumar, A. V. Kirthi, K. V. Rao, and A. A. Rahuman. 2013. Entomopathogenic marine actinomycetes as potential and low-cost biocontrol agents against bloodsucking arthropods. *Parasitol Res* 112:3951–3959.

Kelecom, A. 2002. Secondary metabolites from marine microorganisms. *An Acad Bras Ciênc* 74:151–170.

Kent, R. J., M. B. Crabtree, and B. R. Miller. 2010. Transmission of West Nile Virus by *Culex quinquefasciatus* Say infected with Culex Flavivirus Izabal. *PLoS Negl Trop Dis* 4: e671.

Kivaria, F. M. 2006. Estimated direct economic costs associated with tick-borne diseases on cattle in Tanzania. *Trop Anim Health Prod* 38:291–299.

Kovendan, K., K. Murugan, S. Vincent, and S. Kamalakannan. 2011. Larvicidal efficacy of *Jatropha curcas* and bacterial insecticide, *Bacillus thuringiensis*, against lymphatic filarial vector, *Culex quinquefasciatus* Say (Diptera: Culicidae). *Parasitol Res* 109: 1251–1257.

Kyelem, D., G. Biswas, M. J. Bockarie, M. H. Bradley, M. El-Setouhy, P. U. Fischer, R. H. Henderson, J. W. Kazura, P. J. Lammie, S. M. Njenga, E. A. Ottesen, K. D. Ramaiah, F. O. Richards, G. J. Weil, and S. A. Williams. 2008. Determinants of success in national programs to eliminate lymphatic filariasis: A perspective identifying essential elements and research needs. *Am J Trop Med Hyg* 79:480–484.

Lee, S. E., J. E. Kim, and H. S. Lee. 2001. Insecticide resistance in increasing interest. *Agric Chem Biotechnol* 44:105–112.

Manguin, S., M. J. Bangs, J. Pothikasikorn, and T. Chareonviriyaphap. 2010. Review on global co-transmission of human *Plasmodium* species and *Wuchereria bancrofti* by *Anopheles* mosquitoes. *Infect Genet Evol* 10:159–177.

Matasyoh, J. C., B. Dittrich, A. Schueffler, and H. Laatsch. 2011. Larvicidal activity of metabolites from the endophytic *Podospora* sp. against the malaria vector Anopheles gambiae. *Parasitol Res* 108:561–566.

Misato, T. 1983. Recent status and future aspects of agricultural antibiotics. In: *Pesticides chemistry: Human welfare and environment, vol. 2: Natural products*, 241–246. Oxford: Pergamon Press.

Molinari, G., S. Soloneski, and M. L. Larramendy. 2010. New Ventures in the genotoxic and cytotoxic effects of macrocyclic lactones, abamectin and ivermectin. *Cyto Geno Res* 128:37–45.

Mukhtar, M., N. Herrel, F. P. Amerasinghe, J. Ensink, W. van der Hoek, and F. Konradsen. 2003. Role of wastewater irrigation in mosquito breeding in south Punjab, Pakistan. *Southeast Asian J Trop Med Public Health* 34:72–80.

Naine, S. J., and C. S. Devi. 2014. Larvicidal and repellent properties of Streptomyces sp. VITJS4 crude extract against *Anopheles stephensi*, *Aedes aegypti* and *Culex quinquefasciatus* (Diptera: Culicidae). *Pol J Microbiol* 63:341–348.

Newman, D. J., G. M. Cragg, and K. M. Snader. 2000. The influence of natural products upon drugs discovery. *Nat Prod Rep* 17:215–234.

Ottesen, E. A., P. J. Hooper, M. Bradley, and G. Biswas. 2008. The global programme to eliminate lymphatic filariasis: Health impact after 8 years. *PLoS Negl Trop Dis* 2:317.

Pal, P. S. M., R. Bhalwar, V. K. Mehta, A. Mahendraker, and A. K. Mehta. 2009. Malaria in Assam: A challenge. *Indian J Comm Med* 34:94–96.

Patil, R., G. Jeyasekaran, S. A. Shanmugam, and R. Jeya Shakila. 2001. Control of bacterial pathogens, associated with fish diseases, by antagonistic marine actinomycetes isolated from marine sediments. *Indian J Mar Sci* 30:264–267.

Pimentel-Elardo, S. M., S. Kozytska, T. S. Bugni, C. M. Ireland, H. Moll, and U. Hentschel. 2010. Anti-parasitic compounds from Streptomyces sp. strains isolated from Mediterranean sponges. *Mar Drugs* 8:373–380.

Prabha, S. B., A. Vignesh, D. Elumalai P. K. Kaleena, and K. Murugeason. 2014. Antimicrobial activity of extracellularly synthesized silver naoparticles using *Saccharopolyspora erythraea*, a potent actinomycete and its efficacy againts mosquito larvae. *Int J Nanotechnol Appl* 4:9–18

Prashith Kekuda, T. R., K. S. Shobha, and R. Onkarappa. 2010. Potent insecticidal activity of two *Streptomyces* species isolated from the soils of the Western ghats of Agumbe, Karnataka. *J Nat Pharm* 1:30–32.

Prudhomme, J., E. McDaniel, N. Ponts, S. Bertani, W. Fenical, P. Jensen, and L. Roch. 2008. Marine actinomycetes: A new source of compounds against the human malaria parasite. *PLoS One* 3:e2335.

Raghavendra, K., T. K. Barik, R. M. Bhatt, H. C. Srivastava, U. Sreehari, and A. P. Dash. 2011. Evaluation of the pyrrole insecticide chlorfenapyr for the control of *Culex quinquefasciatus* Say. *Acta Trop* 118:50–55.

Rajesh, K., D. Dhanasekaran, and B. K. Tyagi. 2015. Mosquito survey and larvicidal activity of actinobacterial isolates against *Culex* larvae (Diptera: Culicidae). *J Saudi Soc Agric Sci* 14:116–122.

Rajesh, K., K. C. Padmavathi, A. Ranjani, P. M. Golpinath, D. Dhanasekaran, and G. Archunan. 2013. Green synthesis, characterisation and larvicidal activity of AgNPs against *Culex quinquefasciatus* and *Aedes aegypti* larvae. *Am J Drug Discov Dev* 3: 245–253.

Ramaiah, K. D., P. K. Das, E. Michael, and H. Guyatt. 2000. The economic burden of lymphatic filariasis in India. *Parasitol Today* 16:251–253.

Ribeiro, V. L. S., E. Toigo, S. A. L. Bordignon, K. Goncalves, and G. von Poser. 2007. Acaricidal properties of extracts from the aerial parts of *Hipericum polyanthemum* on the cattle tick *Boophilus microplus*. *Vet Parasitol* 147:199–203.

Rosado-Aguilar, J. A., A. J. Aguilar-Caballero, R. I. Rodríguez-Vivas, R. Borges-Argaez, Z. García-Vázquez, and M. Méndez-González. 2010. Acaricidal activity of extracts from *Petiveria alliacea* (Phytolaccaceae) against the cattle tick, *Rhipicephalus (Boophilus) microplus* (Acari: Ixodidae). *Vet Parasitol* 168:299–303.

Sanjenbam, P., and K. Kannabiran. 2013. Antimicrobial and larvicidal activity of *Streptomyces* sp. VITPK9 isolated from a brine spring habitat of Manipur, India. *Der Pharmacia Lett* 5:65–70.

Saurav, K., G. Rajakumar, K. Kannabiran, A. A. Rahuman, K. Velayutham, G. Elango, C. Kamaraj, and A. A. Zahir. 2013. Larvicidal activity of isolated compound 5-(2,4-dimethylbenzyl) pyrrolidin-2-one from marine *Streptomyces* VITSVK5 sp. against *R. microplus*, *Anopheles stephensi*, and *Culex tritaeniorhynchus*. *Parasitol Res* 112: 215–226.

Shanmugasundaram, T., and R. Balagurunathan. 2015. Mosquito larvicidal activity of silver nanoparticles synthesised using actinobacterium, *Streptomyces* sp. M25 against *Anopheles subpictus*, *Culex quinquefasciatus* and *Aedes aegypti*. *J Parasit Dis* 39(4):677–684.

Sharma, P., L. Mohan, and C. N. Srivastava. 2009. Amaranthus oleracea and Euphorbia hirta: Natural potential larvicidal agents against the urban Indian malaria vector, *Anopheles stephensi* Liston (Diptera: Culicidae). *Parasitol Res* 106:171–176.

Sharma, V. P., and V. Dev. 2006. Prospects of malaria control in northeastern India with particular reference to Assam. In: *Proceedings of National Symposium on Tribal Health*, 21–30.

Singh, G., and S. Prakash. 2012. Lethal effect of *Streptomyces citreofluorescens* against larvae of malaria, filaria and dengue vectors. *Asian Pac J Trop Med* 5(8):594–597.

Snyder, D. E., J. Meyer, A. G. Zimmermann, M. Qiao, S. J. Gissendanner, L. R. Cruthers, R. L. Slone, and D. R. Young. 2007. Preliminary studies on the effectiveness of the novel pulicide, spinosad, for the treatment and control of fleas on dogs. *Vet Parasitol Res* 150:345–351.

Subramani, R., and W. Aalbersberg. 2012. Marine actinomycetes: An ongoing source of novel bioactive metabolites. *Microbiol Res* 167:571–580.

Sunish, I. P., and R. Reuben. 2001. Factors influencing the abundance of Japanese encephalitis vectors in rice fields in India-I. *Abiotic Med Vet Entomol* 15:381–392.

Suthindhiran, K., and K. Kannabiran. 2010. Diversity and exploration of bioactive marine actinomycetes in the Bay of Bengal of the Puducherry coast of India. *Indian J Microbiol* 50:76–82.

Suthindhiran, K., and K. Kannabiran. 2011. Anti-fish nodaviral activity of furan-2-yl acetate extracted from marine *Streptomyces* spp. *Nat Prod Res* 25:834–843.

Takahashi, A., S. Kurasawa, D. Ikeda, Y. Okami, and T. Takeuchi. 1989. Altemicidin, a new acaricidal and antitumor substance. I. Taxonomy, fermentation, isolation and physico-chemical and biological properties. *J Antibiotics (Tokyo)* 42:1556–1561.

Tanvir, R., I. Sajid, and S. Hasnain. 2014. Biotechnological potential of endophytic actino-mycetes associated with Asteraceae plants: Isolation, biodiversity and bioactivities. *Biotechnol Lett* 36:767–773.

Thenmozhi, M., J. Vinay Gopal, K. Kannabiran, G. Rajakumar, K. Velayutham, and A. A. Rahuman. 2013. Eco-friendly approach using marine actinobacteria and its compounds to control ticks and mosquitoes. *Parasitol Res* 112:719–729.

Vijayakumar, R., S. Murugesan, A. Cholarajan, and V. Sakthi. 2010. Larvicidal potentiality of marine actinomycetes isolated from Muthupet Mangrove, Tamilnadu, India. *Int J Microbiol Res* 1:179–183.

Vinodhkumar, M., G. Subhapriya, J. Ramanathan, I. Suresh, and Kalpana. 2015. Screening of bioactive potential for larvicidal activity of marine actinomycetes. *Int J Curr Microbiol App Sci* 4:972–980.

World Health Organization (WHO). 2008a. Global programme to eliminate lymphatic filari-asis. *Weekly Epidemiol Rec* 83:333–341.

World Health Organization (WHO). 2008b. *Global programme to eliminate lymphatic fila-riasis (GPELF)*. http://www.who.int/lymphatic_filariasis/disease/en/.

World Health Organization (WHO). 2009. *World malaria report 2009*. Geneva, Switzerland: WHO.

# 16 Cyanobacterial Toxins as Biolarvicides for Blood-Sucking Vectors

*Reehana Nazar, Mohamed Imran Mohamed Yousuff, Thajuddin Nooruddin, and Dhanasekaran Dharumadurai*

**CONTENTS**

## 16.1 INTRODUCTION

Mosquitoes play a predominant role in the transmission of dengue, malaria, yellow fever, filariasis, and several other diseases that today are the greatest health problems in the world (Hafeez et al., 2011). The present resurgence of these diseases is due to a higher number of breeding places in today's throwaway society. Further, the indiscriminate use of synthetic insecticides is creating multifarious problems such as environmental pollution, insecticide resistance, and toxins hazardous to human beings. Synthetic insecticides such as organochlorine, organophosphorous, carbamates, pyrethrins, and pyrethroids are commonly used for controlling the ever-increasing population of vectors. The overuse of these chemical insecticides can harm the environment and nontarget organisms, and has also resulted in resistant development in vectors (Bilal et al., 2012; Shaalan, 2012). Hence, alternative approaches free from such problems are needed; newly developed approaches should be environmentally safe, biodegradable, and cost-effective indigenous methods consisting of legal, scientific, and technological weapons with more powerful combatable properties against such vectors. Natural products are generally preferred because of their less harmful nature to target organisms and their innate biodegradability.

Cyanobacteria produce many secondary metabolites that have been found to be toxic or otherwise bioactive. These compounds potentially help outcompete other photosynthetic organisms and/or act as deterrents to grazing invertebrates (Kearns and Hunter, 2001; Lürling, 2003). Considering the sympatry of cyanobacteria with mosquito larvae (Pope et al., 2005), cyanobacteria are a potential source for compounds with mosquito larvicidal activity (Harada et al., 2000). With this sympatric association and the benefits of larviciding, cyanobacteria were evaluated for the potential of containing mosquito larvicidal compounds.

## 16.2 MOSQUITOS AS VECTORS FOR DISEASE

About 3000 species of mosquito have been identified (the Culicidae family), but only several of them are of concern as disease vectors (Gillot et al., 2005). Due to the nature of mosquitoes they obtain blood meals from multiple hosts they are able to acquire pathogens and parasites from one vertebrate host and pass them to another. An efficient vector species requires that an appropriately long life stage takes place in proximity to disease hosts and that their longevity is sufficient to enable the pathogens/parasites to proliferate and/or develop to the infective stage within the vector before being transmitted. Other factors are also important when investigating the competence of a mosquito disease vector (Johnson et al., 2002; Jupp and Kemp, 2002; Becker et al., 2010; World Health Organization [WHO], 2013b):

- The time of day when a mosquito takes a blood meal
- The habitat where mosquitoes breed, develop, and live (how close these habitats might be to populations of concern, i.e., humans)
- The mobility of a species and willingness to immigrate to new areas
- The tenacity by which mosquitoes pursue a blood meal

- The diet of the mosquito (some mosquitoes preferentially dislike feeding on humans)
- The immune system of the mosquito (whether a pathogen is still viable after transportation in a blood meal)
- The longevity of the mosquito species as a biting adult
- The fecundity of the mosquito species and the relative number of individuals in the environment

### 16.2.1 Diseases Vectored by Mosquitoes

A number of microorganisms are vectored by mosquitoes worldwide. Tables 16.1 to 16.3 summarize the diseases malaria, dengue, and yellow fever and West Nile fever, respectively, which that have the greatest effect on human health.

## 16.3 CYANOBACTERIA AS A POTENTIAL SOURCE FOR MOSQUITO LARVICIDAL COMPOUNDS

### 16.3.1 Cyanobacteria and Mosquito Sympatry

Cyanobacteria can associate with eukaryotic algae, bacteria, and fungi forming benthic biofilms or mats called periphyton. Periphyton has been recognized as a

**TABLE 16.1**
**Malaria**

| Disease | Etiology and Transmission | Pathogenesis and Symptoms | Source of Vector | Effect on Human Health |
|---|---|---|---|---|
| **Malaria** | A mother infected with the *Plasmodium* parasite can also pass the disease to her baby at birth called congenital malaria.<br>Five species of *Plasmodium*: *P. knowlesi*, *P. falciparum*, *P. vivax*, *P. ovale*, and *P. malariae*. | Symptoms develop within 10 days to 4 weeks following infection.<br>Common symptoms of malaria include: shaking chills, high fever, profuse sweating, headache, nausea, vomiting diarrhea, anemia, muscle pain, convulsions, coma, bloody stools. | Transmitted by mosquitoes of the *Anopheles* genus.<br>Transmission from mosquito to human only occurs between the temperatures of 16°C and 33°C (Becker et al., 2010). | There were 219 million cases of malaria in 2010, with 660,000 deaths (WHO, 2012).<br>The estimated disability adjusted life years (DALYs) (a measure of morbidity and mortality lost) in 2010 was 82,685,000 (Murray et al., 2012).<br>Elimination of malaria is a UN millennium development goal (WHO, 2012b). |

**TABLE 16.2**
**Dengue Fever Virus**

| Disease | Etiology and Transmission | Pathogenesis and Symptoms | Source of Vector | Effect on Human Health |
|---|---|---|---|---|
| **Dengue fever virus** | Four closely related viruses cause dengue fever. The viruses are transmitted from *Aedes aegypti* and *Aedes albopictus* mosquitoes to humans in a viral life cycle that requires both humans and mosquitoes. There is no human-to-human dengue fever transmission. The viruses belong to the *Flaviviridae* family. There are four types of dengue virus (DENV 1 to 4) and all four serotypes are closely related. | Dengue is prevalent throughout the tropics and subtropics. Dengue hemorrhagic fever (DHF) is a specific syndrome that tends to affect children under 10 years of age. This complication of dengue causes abdominal pain, hemorrhage (bleeding), and circulatory collapse (shock). Symptoms of dengue fever include severe joint and muscle pain, swollen lymph nodes, headache, fever, exhaustion, and rash. The presence of fever, rash, and headache (the "dengue triad") is characteristic of dengue fever. | *Aedes aegypti* mosquitos are the principle (and most efficient) vector. *A. albopictus*, *A. polynesiensis*, and *A. scutellaris* can also vector the disease, though with less efficiency (Becker et al., 2010). | 284 million to 528 million infections in 2010. Responsible for approximately 16,000 deaths (WHO, 2013b). Loss of 825,000 DALYs globally in 2010 (Murray et al., 2012). |

significant source of primary productivity and habitat for invertebrates (McCormick et al., 1998). In combination with detritus, periphyton acts as the base for many food webs (Browder et al., 1994; Rader and Richardson, 1994). Cyanobacteria are present in many mosquito-breeding areas (Laird, 1988), as mosquitoes will lay their eggs within cyanobacterial mats (Rejmankova et al., 1996). Mats act as shelter, sources of food, and substrate for egg attachment during mosquito oviposition (Pope et al., 2005). However, some debate the potential for cyanobacteria to be a potential source of food for aquatic invertebrates such as mosquito larvae.

### 16.3.2 Grazer Choice of Food within Periphyton

Grazers within periphyton have been found to lack the discriminatory ability to reject certain sources of food (Resh and Houp, 1986). These grazers were found to lack the

## TABLE 16.3
## Yellow Fever and West Nile Fever

| Disease | Etiology and Transmission | Disease Pathogenesis and Symptoms | Source of Vector | Effect on Human Health |
|---|---|---|---|---|
| **Yellow fever** | Viral disease of typically short duration. Transmitted usually through the bite of the yellow fever mosquito *Aedes aegypti*, but other *Aedes* mosquitoes, such as the tiger mosquito *Aedes albopictus*, as well. | Yellow fever virus replicates initially in macrophages before traveling through the lymph system to become a systemic infection. Symptoms include fever, chills, loss of appetite, nausea, muscle pains (particularly in the back), and headaches. Liver damage begins causing yellow skin. | Spread primarily by *Aedes aegypti*, a type of mosquito found throughout the tropics and subtropics (Becker et al., 2010). | About 200,000 people became infected with yellow fever worldwide in 2010 resulting in about 30,000 deaths (WHO, 2013a). |
| **West Nile Fever** | Arbovirus of the *Flavivirus* kind in the family *Flaviviridae*. Spread by various species of mosquitoes, with birds being the most commonly infected animal and serving as the prime reservoir host. | West Nile virus enables the cell-free virus to enter the central nervous system without disturbing the barrier's integrity. Increased mRNA and protein expression of the tight-junction protein claudin-1 and of two cell adhesion molecules were seen 2 to 3 days after cellular infection, the same time at which West Nile virus replication had peaked. Symptoms include fever, headaches, feeling tired, muscle pain or aches, nausea, loss of appetite, vomiting, and rash. Less than 1% of the cases is severe and result in neurological disease when the central nervous system is affected. | Transmitted by *Culex* mosquitoes especially *Culex pipiens* and *Culex restuans* (Becker et al., 2010). | In 2011, infections included 220 people in Canada, 5387 people in the United States, and 1200 people in Europe (killing 0, 2243, and 104 people, respectively) (CDC, 2012; ECDC, 2012; PHAC, 2012). |

sensory capability for food selection. For these particular grazers, food "selection" is likely determined by the morphology of their feeding parts (Karouna and Fuller, 1992), which influences the efficiency of feeding on particular algal species (Steinman, 1996).

The appearance of selectivity is likely caused by random feeding leading to the discovery of compatible food items, that is, food morphology versus feeding morphology (Steinman, 1996). Cyanobacteria do not normally offer much nutritional values for grazers (Schmidt and Jonasdottir, 1997). Much research has shown that the production of secondary metabolites common to cyanobacteria would hinder their palatability (Porter, 1977; Gregory, 1983). However, some grazers do prefer cyanobacteria such as the chironomid midge *Cricotopus nostocicola*, which forms endosymbiotic associations with the cyanobacterium *Nostoc* in which the midge feeds on exclusively. *Nostoc* benefits from this association by somehow having a higher rate of photosynthesis when compared to midgeless strains.

### 16.3.3 Mosquito Larvae Feeding on Periphyton

Mosquito larvae are often abundant in cyanobacterial mats (Lange and Lopez, 1996). Despite this knowledge, there is no conclusive information on whether mosquito larvae actually target specific food sources, like cyanobacteria, in these environments. Researchers have found cyanobacteria in the guts of mosquito larvae (Howland, 1930; Jones, 1960). Whether the mosquito larvae were targeting the cyanobacteria or ingesting it because of the association with other sources of food remains unclear. Other research involves rearing mosquito larvae with specific strains of cyanobacteria to observe ingestion. Results indicate that mosquito larvae can ingest cyanobacteria and that non-filamentous forms are preferred (Thiery et al., 1991). However, whether the cyanobacteria are filamentous or not may be of less importance within cyanobacterial mats. Large grazers can dislodge food particles allowing for previously inedible algal strains to be accessible to small grazers (Dudley, 1992). The determination of larval gut contents has been problematic in that partially digested food items are difficult to identify from larvae collected in the field. More recent techniques such as DNA barcoding and the use of stable isotopes offer tools that circumvent the problems in visual identification of partially digested food (Garros et al., 2008; Belicka et al., 2012). However, the technique is still plagued by the same shortcomings previously mentioned. The presence of particular food items, confirmed with DNA or stable isotope analyses, does not confirm that mosquito larvae were targeting these food items specifically.

### 16.3.4 Cyanobacteria: Novel Insecticides and Candidates for Biological Control Agents

Cyanobacteria produce a wide array of toxins that have been found to reduce the grazing rates of primary consumers. The true nature of these toxic compounds is unknown but has been linked to allelopathic associations (Berry et al., 2008). Toxins produced by cyanobacteria are much more likely to be detoxified by microorganisms than synthetic insecticides because they already occur naturally in the environment in which they are to be employed (Lange and Lopez, 1996). Given the variability of their secondary metabolites, cyanobacteria are recognized as an enormous potential for future insecticides (Namikoshi and Rinehart, 1996). The use of cyanobacterial toxins as mosquito larvicides is an emerging field. Several advancements have taken place during the last two decades.

Nassar et al. (1999) discovered mosquito larvicidal activity against *Culex pipiens* using crude extracts in microgram per milliliter (μg/mL) concentrations derived from unspecified cyanobacteria. These extracts did not exhibit any mammalian toxicity at the same concentrations using mouse serum in an acetylcholinesterase assay. Rao et al. (1999) discovered serendipitously that cyanobacterial strains of the genus *Westiellopsis* inhibited mosquito larvae growth when investigating the biofertilizer properties of the cyanobacterium. Subsequent analysis revealed that the methanolic extracts contained the activity and showed to be larvicidal at microgram per milliliter concentrations against various disease vectors including strains of *Aedes aegypti* (dengue fever and yellow fever), *Anopheles stephensi* (malaria), and *Culex quinquefaciatus* (encephalitis).

Kiviranta et al. (1993) screened 76 cyanobacterial isolates of which several were found to contain compounds with mosquito larvicidal activity against the larvae of *Aedes aegypti*. The most pronounced activity was associated with isolates containing the neurotoxic anatoxin-a and the hepatotoxic microcystins. These findings were not of much use for commercial application, considering the potential nontarget toxicity, but demonstrate that purified compounds from cyanobacteria can act as mosquito larvicides (Table 16.4).

The following year, several compounds were purified from *Oscillatoria agardhii* with mosquito larvicidal properties other than the neurotoxic and hepatotoxic compounds common to this species (Kiviranta and Abdel-Hameed, 1994). One of the more notable discoveries of mosquito larvicidal compounds derived from cyanobacteria was the appearance of unsaturated fatty acids derived from *Oscillatoria agardhii* strain 27, which were found to be toxic to the larvae of *Aedes albopictus*. Compounds were characterized, and the common structure is relatively safe for mammals and the environment. Some of these fatty acids were found to have antimicrobial and enzyme-inhibiting properties. These compounds were characterized based on the bioassay-guided fractionation of cyanobacteria using a mosquito larvicidal assay. Although they were not purified as a result of bioassay-guided fractionation using a mosquito larvicidal assay, they still provide useful information. Only extracts have been mentioned so far and thus mosquito larvicidal activity could have resulted from synergism of several components. It is important to demonstrate that purified compounds can produce the activity as well (Harada et al., 2000).

Beyond the use of cyanobacterial compounds for commercial mosquito larvicides, there is also considerable interest in allowing the cyanobacteria to propagate and produce desired compounds in mosquito breeding sites. Genes involved in the production of insecticidal proteins derived from *Bacillus thuringiensis* have been incorporated into four strains of the ubiquitous *Synechococcus* (Sangthongpitag et al., 1996). The authors chose *Synechococcus* because they collected several strains of it from mosquito breeding sites. Also, *Synechococcus* was found to be particularly appropriate as a toxin gene carrier because it is found worldwide and tolerant to various salinities, temperatures, and insecticides of biological and chemical origin. The unicellular organization of *Synechococcus* would facilitate its digestion by mosquito larvae.

**TABLE 16.4**
**Larvicidal Potential of Cyanotoxins against Blood-Sucking Vectors**

| Biological Source | Cyanotoxins | Chemical Nature of Cyanotoxins | Larvicidal Activity | Reference |
|---|---|---|---|---|
| Cyanobacteria | Microcystin-LR (Hepatotoxic) | Protein phosphatase inhibitors | *Aedes aegypti* | Codd and Bell, 1996 |
| Cyanobacteria | Anatoxin-a | Acetylcholine | *Aedes aegypti* | Kiviranta et al., 1991 |
| Cyanobacterial isolates | Neurotoxic anatoxin-a and hepatotoxic microcystins | – | *Aedes aegypti* | Kiviranta et al., 1993 |
| Unspecified cyanobacteria | Crude extracts | – | *Culex pipiens* | Nassar et al., 1999 |
| *Westiellopsis* | Methanolic extracts | – | *Aedes aegypti* (dengue fever and yellow fever), *Anopheles stephensi* (malaria), and *Culex quinquefaciatus* (encephalitis) | Rao et al., 1999 |
| *Oscillatoria agardhii* strain 27 | Unsaturated fatty acids | – | *Aedes albopictus* | Harada et al., 2000 |
| *Oscillatoria agardhii* | Purified compounds other than neurotoxic and hepatotoxic compounds | – | – | Kiviranta and Abdel-Hameed, 1994 |

Transgenics may be a premature approach in that many fear the release of such organisms into the environment could have unforeseen consequences. With more research on cyanobacteria-derived mosquito larvicides, transgenics may not be necessary, that is, inoculating mosquito breeding sites with cyanobacteria capable of producing mosquito larvicides.

### 16.3.5 Effects of Anatoxin-a and Microcystin-LR on Mosquito Larvae

No products derived from cyanobacteria are currently in use commercially as mosquito larvicides. However, several compounds have been explored, including the well-documented cyanobacterial toxins anatoxin-a and the microcystin-LR. Although these are unlikely candidates for mosquito control considering their toxicities to nontarget organisms, it is important to illustrate that purified compounds (versus extracts) from cyanobacteria can house mosquito larvicidal properties. Anatoxin-a has a half-life of 14 days under normal photoperiods and its degradation is accelerated by direct sun exposure (Chorus and Bartram, 1999). The $LC_{50}$ range of purified anatoxin-a against *Ae. aegypti* was calculated to be 17 to 37 μg/ml. Its mechanism of action is assumed to be the same as with other insects, a neurotoxin. Anatoxin-a is an acetylcholine analogue that is not degraded like acetylcholine causing muscle cells to be stimulated continuously (Kiviranta et al., 1991) (Figure 16.1).

The half-life of microcystin-LR is roughly 7 days (Codd and Bell, 1996). The $LC_{50}$ range of "purified" microcystin against *Ae. aegypti* was calculated to be 2.5 to 6.1 mg/ml (Kiviranta and Abdel-Hameed, 1993). These large values do seem to indicate a crude extract although the authors claim to have gone through several stages of purification.

Microcystins are protein phosphatase inhibitors and are classified as hepatotoxins in regard to the location where they accumulate in vertebrates. However, classification as a hepatotoxin is not useful when describing its effects on organisms that do not have hepatic tissue. Utilizing histopathology microscopy, it was discovered that microcystins lysed midgut epithelial cells in the mosquito larvae of *Ae. aegypti.* The activity of both anatoxin-a and microcystin-LR on mosquito larvae demonstrate that purified compounds from cyanobacteria can house mosquito larvicidal activity. These results help to justify the rationale behind searching for mosquito larvicides within cyanobacteria (Saario et al., 1994).

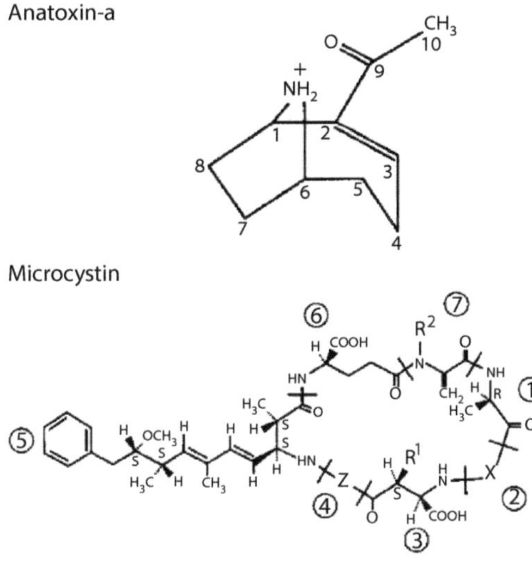


**FIGURE 16.1** Structure of anatoxin-a and microcystin.

## 16.4 ISOLATION AND CHARACTERIZATION OF MOSQUITO LARVICIDAL COMPOUNDS FROM CYANOBACTERIA

Cyanobacteria (blue-green algae) produce a diverse array of toxic or otherwise bioactive metabolites. These allelochemicals may also play a role in defense against potential predators and grazers, particularly aquatic invertebrates and their larvae, including mosquitoes. Compounds derived from cyanobacteria were collected and investigated as insecticides against the mosquito. Screening of cyanobacterial biomass revealed several strains that exhibited mosquito larvicidal activity. Scientists have been purifying compounds from natural sources since ancient times. These efforts date back to 2600 B.C. and have been recorded on clay tablets found in Mesopotamia. Scientists attempted to pull everything they can out of the organism being studied and test individual components separately with several experiments, such as bioassays, in order to deduce the nature and utility of the derived compound(s) (Cragg and Newman, 2005).

### 16.4.1 Bioassays

Well plates are often used in bioassay-guided fractionation experiments. The well size must be considered on the basis of size of a single organism, including their potential for growth, as well as the number of organisms to be placed into each well (Robertson et al., 2007). The latter should be tested, as adverse effects may be discovered as a result of group density issues. Mosquito larvae in particular are vulnerable to density-related phenomena, as delayed development is directly correlated to density (Clements, 1992).

### 16.4.2 Fractionation

The protocol commences with the identification of the organism to be analyzed for potential natural products. In the case of deriving mosquito larvicidal compounds from cyanobacteria, a small aliquot of cyanobacterial biomass is placed with the larvae. This exposure repeated with as many strains of cyanobacteria as possible to help ensure the most useful (in housing desired compounds) will be discovered in the screening process. Once the most effective cyanobacteria have been identified, in regard to mosquito larvicidal activity, active components are purified. The biomass is subjected to polar and nonpolar solvent extractions, which are tested individually for activity. Active extracts are separated into numerous fractions via column chromatography, which in turn are tested individually for activity. The solvent system is determined by those giving the best resolution with thin-layer chromatography (TLC). If the activity is lost at this step or seems to not be in any particular location, the protocol should be revised. Fractionation can be done in small scale at first with the use of solid phase extraction cartridges and scaled up for a large glass column once the active fractions have been identified. Active fractions are separated by another column and then tested once again, hence the term "bioassay-guided fractionation." Column chromatography is performed until samples are pure enough for high-pressure liquid chromatography (HPLC). To consider fractions to be HPLC ready, they are required to have baseline separation between peaks, which will allow for their individual collection. Standard acetonitrile–water or methanol–water sloped gradients are first used and are then fine-tuned to achieve satisfactory baseline separation. Finally, individual peaks are collected and tested for activity. During the HPLC phase, all wavelengths should be considered when dealing with unknown compounds. It is possible for a compound to have several chromophores, that is, be visible at more than one wavelength. If multiple chromophores are suggested, one must be sure the peaks at the multiple wavelengths have exactly the same retention time and are similarly symmetrical, otherwise several compounds may still be present at that specific retention time. Since elution via HPLC likely means the sample contains water, a rotary evaporator should be used to remove the more volatile solvents such as methanol and acetonitrile. Once the sample can freeze, the remainder of the water can be removed with a freeze dryer (Berry, 2014).

### 16.4.3 Characterization

The compound purity is verified by a single mass present via mass spectrometry that helps to confirm that it is indeed pure. High-resolution mass spectrometry can give masses up to four decimal places. With this information, molecular formulas can be generated. However, the amount of formulas generated can be quite numerous if the atomic composition is not known. Discovering the functional groups with infrared spectroscopy (IR) adds another layer to characterization. In combination with proton and carbon nuclear magnetic resonance (NMR) techniques, the remainder of the structure can be deduced. This section is an overly simplistic view of characterization, because NMR spectra do not look as clean as shown in textbooks, and many complications can arise when working with unknowns, as the properties of prospective compounds are obtained through trial and error. Important discoveries can result from mistakes, particularly if meticulous observations are constantly being made.

### 16.4.4 Toxicological Analyses

Toxicological analyses were performed using mosquito larvae, zebrafish embryos, brine shrimp, and bacteria. The analysis of the mosquito larvae was the most comprehensive as it has higher relevance and thus deserving of a large portion of the purified sample. Zebrafish (*Danio rerio*) are a widely used vertebrate model encompassing such subjects as embryonic development, genetics, natural products, and toxicology (Hill et al., 2005; Crawford et al., 2008). Many cyanobacteria-derived metabolites have been evaluated using zebrafish assays (Wright et al., 2006; Berry et al., 2008, 2009). The assay has proven to be a useful tool in the purification of bioactive metabolites via bioassay guided fractionation. The zebrafish model can potentially detect the bioactivity of many types of molecules because embryos will not develop properly if any of the many processes involved in embryogenesis are affected. Other benefits of the model include the small embryo size, which allows for their evaluation in 24–96 well plates, transparency of the embryos allowing for microscopic analyses, and quick embryogenesis (3–5 days) allowing for more evaluations hastening the bioassay guided-fractionation process (Berry et al., 2008).

Prior research suggests a link between the toxicity of fatty acids to mosquito larvae and brine shrimp. Studies have shown the ability of unsaturated fatty acids to reduce

$Na^+/K^+$-ATPase activity in *Artemia salina* (Morohashi et al. 1991). Harada et al. (2000) correlated this activity with that observed on mosquito larvae. Unsaturated fatty acids were found to have toxicity comparable to microcystin on *Daphnia magna* (Reinikainen et al., 2001). Cumulatively, these studies suggest the potential for arthrocidal effects of unsaturated fatty acids.

The antibacterial activity of unsaturated fatty acids has been well documented (McGaw et al., 2002; Mundt et al., 2003; Zheng et al., 2005). Confirmation of antibacterial activity would strengthen the proposed structures as unsaturated fatty acids.

## 16.5 SIGNIFICANCE

### 16.5.1 Novelty

The current study encompassed a search for mosquito larvicides derived from cyanobacteria. The idea emerged as an enticing prospect, particularly with the occurrences of cyanobacteria in mosquito breeding sites. Photosynthetic organisms such as cyanobacteria produce chemical defenses as inhibitors to competitors and potentially as deterrents to grazing pressure. An experiment emerged from this concept with prospects in discovering novel control agents for disease vectors that could be used in rotation with contemporary strategies.

Considering the vast amount of compounds produced by cyanobacteria, the exact nature of the discovered mosquito larvicide was highly speculated. The purification yields another peptide-like molecule such as a microcystin or a small neurotoxic alkaloid such as anatoxin-a. As in the case of Harada et al. (2000), when all cyanobacterial components are screened based on bioassay-guided fractionation using a mosquito larvicidal assay, the results indicate that the most toxic components are in fact unsaturated fatty acids. Harada et al. also demonstrated that unsaturated fatty acids (oleic acid and linoleic acid) were toxic to the larvae of *Aedes albopictus*, whereas saturated fatty acids (palmitic acid and stearic acid) were nontoxic. This trend in unsaturation causing higher levels of activity seems to hold for other types of lipids including the sulfated glycolipids (SGL-3). As a result of Harada et al., the first cyanobacteria derived compounds were discovered using a mosquito larvicidal assay and bioassay-guided fractionation. The characterization of these fatty acids is not entirely conclusive. A mass spectrum is provided and some $^1$HNMR spectra comparisons are made between a 90% methanol fraction and that of $\gamma$- linolenic acid. The NMR spectrum of the fraction does show appropriate chemical shifts but fails to integrate as the spectra of $\gamma$-linolenic acid. Linoleic acid has been documented to be mosquito larvicidal when used in combination with oleic acid and/or as part of a crude extract (Harada et al., 2000; Rao, 2008).

### 16.5.2 Evaluation of the Potential Environmental Effects of Using Unsaturated Fatty Acids to Control Mosquito Populations

In natural conditions, fatty acids would float and thus not interact with any organism located below the microfilm layer, such as fish embryos. There have been no documented cases of mammalian toxicity with these types of fatty acids. In fact, they are linked to decreased blood glucose levels in diabetes research (Heydarizadeh et al., 2013). These results include the fatty acid most closely related to palmitoleic acid (Yang et al., 2011). Other beneficial properties of unsaturated fatty acids to mammals include but are not limited to cardioprotection (Lee et al. 2010), antiobesity (Micallef et al., 2009), and anticholesterol (Valdivielso et al., 2009). Matsunaga et al. (1995) showed the potential for cyanobacteria being a great source of palmitoleic acid for pharmaceutical use. With all these studies considered, a prospective mosquito larvicide with medicinal uses to mammals would seem like a strong candidate to combat disease vectors with little impact on the environment. Research exists as far back as 1970 in regard to the application of fatty acids as mosquito repellents (Skinner et al., 1970) and ovipositional repellents (Hwang et al., 1984). It is somewhat surprising that these efforts in mosquito control using fatty acids have not been more fruitful. However, because of the advent of synthetic antibiotics and insecticides, many of the plant- or algae-derived compounds were not fully explored (Cowan, 1999). With increased demand for novel antibiotics and insecticides to prevent future bacterial and disease vector resistance, these venues should be considered more carefully in large part because the prospective compounds have beneficial properties to mammals and already exist in the environments that they are to be used.

### 16.5.3 Palmitoleic Acid as a Mosquito Control Agent

Before making conclusive remarks in regard to the use of unsaturated fatty acids as mosquito larvicides, it is important to compare them to mosquito larvicides used in the field today. The highly insect-specific Cry toxins from *Bacillus thurengiensis* are some of the most prospective mosquito larvicides to date. $LC_{50}$ values were calculated to be 3.8 to 7.54 ng/mL (Sun et al., 1996). They have no known toxicity to vertebrates or plants, and they are naturally degraded (Bravo et al., 2005). Cry toxins consist of multidomain proteins. The different domains contribute to a stepwise mechanism responsible for its activity. However, any one of these components that are required for its performance can be a source of tolerance and even potential future resistance (Bravo et al., 2005). The first of these components is the requirement for proteolytic activation by proteases derived from the insect's midgut. Resistance of this kind has been documented as early as the late 1990s and corresponded to reduced production of particular proteinases (Oppert et al., 1997). The second component involves the interaction with receptors in the midgut.

Mutations in these receptors have corresponded to the lack of toxin binding (Ferre and Van Rie, 2002). The third component involves carbohydrate-like binding domains found on the Cry toxins. These are integral in the binding of the toxin to target receptors and can be impaired by products derived from oligosaccharide synthesis (Griffits et al., 2001). These were, however, mostly speculated means of resistance using organisms in optimal conditions to promote selection, and there are no examples of resistance in the field thus far. The only shortcoming of these proteins that exists today is that they are susceptible to sedimentation and have to be periodically reapplied. Sedimentation is likely negligible in smaller bodies of water considering the extremely low $LC_{50}$ values and the fact that commercial formulations (mosquito dunks) are composed of bacterial spores and not just the toxic proteins. The juvenile hormone analogue methoprene is used in larger bodies of water where the Cry proteins would sediment out. However, methoprene is more toxic to nontarget organisms, as it inhibits emergence from the pupa stage in all insects.

In conclusion, current mosquito control strategies targeting the larval stage are not without shortcomings. With the advent of range expansion of disease vectors, it is hardly the time and place to be content with Cry proteins and methoprene. These mosquito larvicides need to immediately be rotated with other insecticides to help prevent resistance and negative impacts on the environment. Unsaturated fatty acids could be a good candidate considering they lack the ability to sediment, do not bioaccumulate in active form, lack toxicity toward nontarget organisms, already exist in the environment that they are to be used, and could be produced continuously by allowing particular strains of cyanobacteria to propagate in mosquito breeding sites. Resistance is inevitable and is merely delayed by control strategies. The delay may only be long enough to discover the next generation of insecticides. Considering the encroachment of disease vectors aided by changes in global climate, the issue should be addressed with great urgency.

## 16.6 CONCLUSION

Vector-borne diseases like filariasis, malaria, dengue, and Japanese encephalitis remain a serious public health problem, all over the world. The control of these important mosquito borne diseases relies heavily on the extensive use of chemical insecticides, although, they are very expensive and toxic to nontarget organisms. Owing to these constraints, biological control agents have been found to apply, widely and effectively, in mosquito control programs. But, for operational purposes, there is an urgent need to produce these bacteria, utilizing cheap and commonly available biological waste materials, through simple fermentation biotechnology.

Nature represents a formidable pool of bioactive compounds and is a strategic source for new and successful pesticidal products. Recent advances made in genomics, proteomics, and combinatorial chemistry show that newer strategies for vector control can be successfully devised. In view of the foregoing review on mosquito control strategies, it may be concluded that vector control operations for the prevention and control of vector borne-diseases must be carried out at a cost not exceeding what the communities concerned can afford to allocate for such a purpose. Vector control operations must also be based on the ecological and population dynamics characteristics of vectors concerned. An integrated vector management (IVM) strategy is defined as a vector management practice that utilizes all suitable techniques and methods in an orderly and compatible manner and that aims at maintaining the vector population at a level below which disease transmission ceases to occur. The prime requisites for designing an IVM strategy are careful appraisal of the ecological situation, application of an appropriate method at the appropriate time with primary emphasis on environmental management utilizing highly skilled and disciplined man power, continuous monitoring and feedback, enlisting community participation, and coordination of the functioning of different departments involved directly or indirectly in preventing mosquitogenic conditions. If designed and implemented properly, an IVM strategy is cost-effective in the long run. More emphasis must be placed on field experiments with proper designs. A newer and novel bacterial toxin with different structures and modes of action has to be identified so as to minimize the risk of developing insect resistance. Studies should be pursued toward developing effective and environment friendly green technologies.

## ACKNOWLEDGMENTS

Author N. Reehana is grateful to the Secretary and Correspondent, Principal of Jamal Mohamed College, Tiruchirappalli, Tamilnadu, India, and the other authors are grateful to the Department of Biotechnology (DBT), Government of India, New Delhi toward the Indo–UK project (Ref.: BT/IN/Indo-UK/SuBB/23/NT/2013 Dt: 30.08 .2013).

## REFERENCES

Becker, N., Petric, D., Zgomba, M., Boase, C., Madon, M., Dahl, C., and Kaiser, A. 2010. *Mosquitoes and Their Control*, 2nd ed. Springer: New York.

Belicka, L. L., Sokol, E. R., Hoch, J. M., Jaffé, R., and Trexler, J. C. 2012. A molecular and stable isotopic approach to investigate algal and detrital energy pathways in a freshwater marsh. *Wetlands*, 32(3): 531–542.

Berry, G. A. 2014. Mosquito larvicides from cyanobacteria. PhD dissertation, Florida International University, Miami.

Berry, J., Gibbs, P., Schmale, M., and Saker, M. 2009. Toxicity of cylindrospermopsin, and other apparent metabolites from *Cylindrospermopsis raciborskii* and *Aphanizomenon ovalisporum*, to the zebrafish (*Danio rerio*) embryo. *Toxicon*, 53: 289–299.

Berry, J. P., Gantar, M., Perez, M. H., Berry, G., and Noriega, F. G. 2008. Cyanobacteria toxins as allelochemicals with potential applications as algaecides, herbicides and insecticides. *Mar. Drugs*, 6: 117–146.

Bilal, H., Akram, W., Din, S., Khan, I. A., Hassan, S. A., and Arshad, M. 2012. Larvicidal acitivity of selected plant extracts against *Aedes albopictus* Skuse (Diptera: Culicidae). *Afr. Entomol.*, 20(1): 8–12.

Bravo, A., Soberon, M., and Gill, S. S. 2005. *Bacillus thuringiensis*, mechanism and use. *Comp. Mol. Insec. Sci.*, 6: 175–197.

Browder, J. A., Gleason, P. J., and Swift, D. R. 1994. Periphyton in the Everglades: Spatial variation, environmental correlates, and ecological implications. In *Everglades: The Ecosystem and Its Restorations*, 379–418. Delray Beach, FL: St. Lucie Press.

Centers for Disease Control and Prevention (CDC). 2012. West Nile virus activity and statistics. http://www.cdc.gov/ncidod/dvbid/westnile/surv&controlCaseCount12_detailed .htm (accessed January 12, 2013).

Chorus, I., and Bartram, J. (eds.). 1999. *Toxic Cyanobacteria in Water: A Guide to Their Public Health Consequences, Monitoring and Management*. London: E & FN Spon, World Health Organization.

Clements, A. N. 1992. *The Biology of Mosquitoes. Vol. 1: Development, Nutrition and Reproduction*. New York: Chapman & Hall.

Codd, G. A., and Bell, S. G. 1996. The occurrence and fate of blue-green algal toxins in freshwaters. London: Her Majesty's Stationery Office (National Rivers Authority R and D Report 29).

Cowan, M. M. 1999. Plant products as antimicrobial agents. *Clin. Microbio. Rev.*, 12(4): 564–582.

Cragg, G. M., and Newman, D. J. 2005. Biodiversity: A continuing source of novel drug leads. *Pure Appl. Chem.*, 77: 7–24.

Crawford, A. D., Esguerra, C. V., and de Witte, P. A. 2008. Fishing for drugs from nature: Zebrafish as a technology platform for natural product discovery. *Planta Medica*, 74: 624–632.

Dudley, T. L. 1992. Beneficial effects of herbivores on stream macroalgae via epiphyte removal. *Oikos*, 65: 121–127.

European Centre for Disease Prevention and Control (ECDC). 2012. West Nile fever maps and statistics. http://ecdc.europa.eu/en/healthtopics/west_nile_fever/West-Nile-fevermaps /Pages/index.aspx (accessed January 14, 2013).

Ferre, J., and Van Rie, J. 2002. Biochemistry and genetics of insect resistance to *Bacillus thuringiensis*. *Annu. Rev. Entomol.*, 47: 501–533.

Garros, C., Ngugi, N., Githeko, A. E., Tuno, N., and Yan, G. 2008. Gut content identification of larvae of the *Anopheles gambiae* complex in western Kenya using a barcoding approach. *Mol. Ecol. Res.*, 8: 512–518.

Gillott, C. 2005. *Entomology*. 3rd ed. New York: Springer.

Gregory, S. V. 1983. Plant-herbivore interactions in stream systems. *Stream Ecol.*, 157–189.

Griffits, J. S., Whitacre, J. L., Stevens, D. E., and Aroian, R. V. 2001. *Bacillus thuringiensis* toxin resistance from loss of a putative carbohydrate-modifying enzyme. *Science*, 293: 860–864.

Hafeez, F., Akram, W., Abdul, E., and Shaalan, S. 2011. Mosquito larvicidal activity of citrus limonoids against *Aedes albopictus*. *Parasitol Res.*, 109: 221–229.

Harada, K., Suomalainen, M., Uchida, H., Masui, H., Ohmura, H. K., Kiviranta, J., Niku-Paavola, M. L., and Ikemoto, T. 2000. Insecticidal compound against mosquito larvae form *Oscillatoria agardhii* strain 27. *Environ. Toxicol.*, 15(2): 114–119.

Heydarizadeh, P., Poirier, I., Loizeau, D., Ulmann, L., Mimouni, V., Schoefs, B., and Bertrand, M. 2013. Plastids of marine phytoplankton produce bioactive pigments and lipids. *Mar. Drugs*, 11: 3425–3471.

Hill, A. J., Teraoka, H., Heideman, W., and Peterson, R. E. 2005. Zebrafish as a model vertebrate for investigating chemical toxicity. *Toxicol. Sci.*, 8(1): 6–19.

Howland, L. J. 1930. The nutrition of mosquito larvae, with special reference to their algal food. *Bull. Entomol. Res.*, 21: 431–439.

Hwang, Y. S., Schultz, G. S., and Mulla, M. S. 1984. Structure-activity relationship of unsaturated fatty acids as mosquito ovipositional repellent. *J. Chem. Ecol.*, 10(1): 145–151.

Johnson, B. W., Chambers, T. V., Crabtree, M. B., Filippis, A. M., Vilarinhos, P. T., Resende, M. C., Macoris, M. L., and Miller, B. R. 2002. Vector competence of Brazilian *Aedes aegypti* and *Aedes albopictus* for a Brazilian yellow fever virus isolate. *Trans. R. Soc. Trop. Med. Hyg.*, 96(6): 611–613.

Jones, J. C. 1960. The anatomy and rythmical activities of the alimentary canal of *Anopheles* larvae. *Ann. Entomol. Soc. Am.*, 53: 459–474.

Jupp, P. G., and Kemp, A. 2002. Laboratory vector competence experiments with yellow fever virus and five South African mosquito species including *Aedes aegypti*. *Trans. R. Soc. Trop. Med. Hyg.*, 96(6): 493–498.

Karouna, N. K., and Fuller, R. L. 1992. Influence of four grazers on the periphyton community associated with clay tiles and leaves. *Hydrobiologia*, 245: 53–62.

Kearns, K. D., and Hunter, M. D. 2001. Toxin-producing *Anabaena flos-aquae* induces settling of *Chlamydomonas reinhardtii*, a competing motile alga. *Microb. Ecol.*, 42: 80–86.

Kiviranta, J., and Abdel-Hameed, A. 1994. Toxicity of the blue green algae *Oscillatoria agardhii* to the mosquito *Aedes aegypti* and the shrimp *Artemia salina. World J. Microbiol. Biotechnol.*, 10: 517–520.

Kiviranta, J., Abdel-Hameed, A., Sivonen, K., Niemeläl, S. I., and Carlberg, G. 1993. Toxicity of cyanobacteria to mosquito larvae—Screening of active compounds. *Environ. Toxicol. Water Qual.*, 8: 63–71.

Kiviranta, J., Sivonen, K., Niemeläl, S. I., and Huovinen, K. 1991. Detection of toxicity of cyanobacteria by *Artemia salina* bioassay. *Environ. Toxicol. Water Qual.*, 6(4): 423–436.

Laird, M. 1988. *The Natural History of Larval Mosquito Habitats*. London: Academic Press.

Lange, L., and Lopez, C. S. 1996. *Protection Agents from Nature: Natural Products and Analogues, Chapter 1*. Cambridge, UK: Royal Society of Chemistry.

Lee, S. H., Kang, H. J., Lee, H. J., Kang, M. H., and Park, Y. K. 2010. Six-week supplementation with *Chlorella* has favorable impact on antioxidant status in Korean male smokers. *Nutrition*, 26: 175–183.

Lürling, M. 2003. Effects of Microcystin-Free and Microcystin-Containing Strains of the Cyanobacterium *Microcystis aeruginosa* on Growth of the Grazer *Daphnia magna. Environ. Toxicol.*, 18: 202–210.

Matsunaga, T., Takeyama, H., Miura, Y., Yamazaki, T., Furuya, H., and Sode, K. 1995. Screening of marine cyanobacteria for high palmitoleic acid production. *FEMS Microbiol. Lett.*, 133(1–2): 137–141.

McCormick, P. V., Shuford III, R.B.E., Backus, J. G., and Kennedy, W. C. 1998. Spatial and seasonal patterns of periphyton biomass and productivity in the northern Everglades, Florida, USA. *Hydrobiologia*, 362: 185.

McGaw, L. J., Jager, A. K., and Van Staden, J. 2002. Isolation of antibacterial fatty acids from *Schotia brachypetala. Fitoterapia*, 73: 431–433.

Micallef, M., Munro, I., Phang, M., and Garg, M. 2009. Plasma n-3 polyunsaturated fatty acids are negatively associated with obesity. *Br. J. Nutr.*, 102: 1370–1374.

Morohashi, M., Tsuchiya, K., Mita, T., and Kawamura, M. 1991. Identification of (Na,K) ATPase inhibitor in brine shrimp, *Artemia salina*, as long-chain fatty acids. *J. Comp. Physiol. B*, 161: 69–72.

Murray, C., Vos, T., and Lozano, R. 2012. Disability-adjusted life years (DALYs) for 291 diseases and injuries in 21 regions, 1990–2010: A systematic analysis for the Global Burden of Disease. *The Lancet*, 380(9859): 2197–2223.

Mundt, S., Kreitlow, S., and Jansen, R. 2003. Fatty acids with antibacterial activity from the cyanobacterium *Oscillatoria redekei* HUB 051. *J. Appl. Phycol.*, 15: 263–267.

Namikoshi, M., and Rinehart, K. L. 1996. Bioactive compounds produced by cyanobacteria. *J. Ind. Microbiol. Biotechnol.*, 17(5): 373–384.

Nassar, M. M., Hafez, S. T., Nagaty, I. M., and Khalaf, S. A. 1999. The insecticidal activity of Cyanobacteria against four insects, two of medical importance and two agricultural pests with reference to the action on albino mice. *J. Egypt Soc. Parasitol.*, 29(3): 939–949.

Oppert, B., Kramer, K.J., Beeman, R. W., Johnson, D., and McGaughey, W. H. 1997. Proteinase-mediated insect resistance to *Bacillus thuringiensis* toxins. *J. Biol. Chem.*, 272: 23473–23476.

Pope, K., Masuoka, P., Rejmankova, E., Grieco, J., Johnson, S., and Roberts, D. 2005. Mosquito habitats, land use, and malaria risk in Belize from satellite imagery. *Eco. Soc. Amer.*, 15(5): 1223–1232.

Porter, K. G. 1977. The plant-animal interface in freshwater ecosystems. *Am. J. Sci.*, 65: 159–170.

Public Health Agency of Canada (PHAC). 2012. West Nile virus map and statistics monitor 2012. http://www.phac-aspc.gc.ca/wnv-vwn/index-eng.php (accessed January 14, 2013).

Rader, R. B., and Richardson, C. J. 1994. Response of macroinvertebrates and small fish to nutrient enrichment in the northern Everglades. *Wetlands*, 14:134.

Rao, D. R., Thangavel, C., Kabilan, L., Suguna, S., Mani, T. R., and Rahuman, S. A. A., Venkatesan, P., and Gopalakrishnan, G. 2008. Mosquito larvicidal activity of oleic and linoleic acids isolated from *Citrullus colocynthis* (Linn.) Schrad. *Parasitol. Res.*, 103: 1383–1390.

Rao, D. R., Thangavel, C., Kabilan, L., Suguna, S., Mani, T. R., and Shanmugasundaram, S. 1999. Larvicidal properties of the cyanobacterium *Westiellopsis* sp. (blue-green algae) against mosquito vectors. *Trans. R. Soc. Trop. Med. Hyg.*, 93(3): 232.

Reinikainen, M., Meriluoto, J. A., Spoof, L., and Harada, K. 2001. The toxicities of a polyunsaturated fatty acid and a microcystin to *Daphnia magna*. *Environ. Toxicol.*, 16(5): 444–448.

Rejmankova, E., Roberts, D. R., Manguin, S., Pope, K. O., Komarek, J., and Post, R. A. 1996. *Anopheles albimanus* (Diptera: Culicidae) and cyanobacteria: An example of larval habitat selection. *Environ. Entomol.*, 25(5): 1058–1067.

Resh, V. H., and Houp, R. E. 1986. Life history of the caddisfly *Dibusa angata* and its association with the red alga *Lemanea australis*. *J. N. Am. Benthol. Soc.*, 5: 28–40.

Robertson, J. L., Russell, R. M., Preisler, H. K., and Savin, N. E. 2007. *Bioassays with Arthropods*. 2nd ed. Boca Raton, FL: CRC Press.

Saario, E., Abdel-Hameed, A., and Kiviranta, J. 1994. Larvicidal microcystin toxins of cyanobacteria affect midgut epithelial cells of *Aedes aegypti* mosquitoes. *Med. Vet. Entomol*, 8: 398–400.

Sangthongpitag, K., Delaney, S. F., and Rogers, P. L. 1996. Evaluation of four fresh-water unicellular cyanobacteria as potential hosts for mosquitocidal toxins. *Biotechnol. Lett.*, 18(2): 175–180.

Schmidt, K., and Jonasdottir, S. H. 1997. Nutritional quality of two cyanobacteria: How rich is "poor" food? *Mar. Ecol. Prog. Ser.*, 151: 1–10.

Shaalan, E. A. 2012. Predation capacity of *Culiseta longiareolata* mosquito against some mosquitoes species larvae. *J. Entomol*, 9(3): 183–186.

Skinner, W. A., Tong, H. C., Maibach, H. I., and Skidmore, D. 1970. Human skin-surface lipid fatty acids—Mosquito repellents. *Experientia*, 26(7): 728–730.

Steinman, A. D. 1996. Effects of grazers on freshwater benthic algae. In *Algal Ecology*, 1st ed., 341–373. San Diego, CA: Academic Press.

Sun, M., Luo, X., Dai, J., Qu, K., Liu, Z., Yu, L., Chen, Y., and Yu, Z. 1996. Evaluation of *Bacillus thuringiensis* and *Bacillus sphaericus* strains from Chinese soils toxic to mosquito larvae. *J. Invertebr. Pathol.*, 68(1): 74–77.

Thiery, I., Nicolas, L., Rippka, R., and Tandeau de Marsac, N. 1991. Selection of cyanobacteria isolated from mosquito breeding sites as a potential food source for mosquito larvae. *Appl. Environ. Microbiol.*, 57(5): 1354–1359.

Valdivielso, P., Rioja, J., Garcia-Aras, C., Sanchez-Chaparro, M. A., and Gonzalez-Santos, P. 2009. Omega 3 fatty acids induce a marked reduction of apolipoprotein B48 when added to fluvastatin in patients with type 2 diabetes and mixed hyperlipidemia: A preliminary report. *Cardiovasc. Diabeto*, 8: 1–7.

World Health Organization (WHO). 2012. World Malaria Report 2012: Fact sheet. http://www.who.int/malaria/publications/world_malaria_report_2012/wmr2012_factsheet.pdf (accessed February 13, 2013).

World Health Organization (WHO). 2013a. Yellow fever: Fact sheet. http://www.who.int/mediacentre/factsheets/fs100/en/ (accessed May 3, 2013).

World Health Organization (WHO). 2013b. World top ten causes of death: Fact sheet. http://www.who.int/mediacentre/factsheets/fs310/en/ (accessed February 18, 2013).

Wright, A. D., Papendorf, O., König, G. M., and Oberemm, A. 2006. Effects of cyanobacterium *Fischerella ambigua* isolates and cell free culture media on zebra fish (*Danio rerio*) embryo development. *Chemosphere*, 65: 604–608.

Yang, Z. H., Miyahara, H., and Hatanaka, A. 2011. Chronic administration of palmitoleic acid reduces insulin resistance and hepatic lipid accumulation in KK-Ay Mice with genetic type 2 diabetes. *Lipids Health Dis.*, 10: 120.

Zheng, C. J., Yoo, J. S., Lee, T. G., Cho, H. Y., Kim, Y. H., and Kim, W. G. 2005. Fatty acid synthesis is a target for antibacterial activity of unsaturated fatty acids. *FEBS Lett.*, 579: 5157–5162.

# 17 The Genetically Altered Microbes and Viruses in Control of Mosquito-Borne Diseases

*Amrita Kumari*

## CONTENTS



## 17.1 INTRODUCTION

Mosquito borne diseases continue to remain a major public health problem in tropical and subtropical regions of the world despite decades of national and international efforts to control the spread of the vectors and disease. The magnitude of mosquito-borne diseases depends on the interaction of various biological, ecological, socio-demographic, and economic factors. The increase in the geographic distribution of mosquitoes is followed by the emergence of viruses and diseases in new areas (Wilke and Marrellii, 2015). The various diseases transmitted by mosquitoes are given in Table 17.1.

Human activities, such as urbanization, new irrigation projects, mass migration of workmen, and the production of a large amount of environmental debris that holds water pools (including disposable bottles and cans and discarded tires) and storage of water on or around living premises when reliable piped home water supplies are unavailable or unreliable, may markedly increase available mosquito breeding sites that have given mosquitoes abundant opportunities to breed, proliferate, and help in the spread of mosquito-borne diseases.

Due to increased resistance of malaria vector to insecticides, *Plasmodium* resistance to antimalarial drugs and nonavailability of an effective vaccine, the number of

**TABLE 17.1**
**Diseases Transmitted by Mosquitoes**

| Mosquito (Culicidae) | Diseases |
|---|---|
| *Anopheles* | Malaria, lymphatic filariasis |
| *Culex* | Lymphatic filariasis, Japanese encephalitis, other viral diseases |
| *Aedes* | Yellow fever, dengue, dengue hemorrhagic fever, other viral diseases, lymphatic filariasis |
| *Mansonia* | Lymphatic filariasis |

*Source:* Adapted from Rozendaal, J. A., 1997, *Vector control: Methods for use by individuals and communities*. Geneva: World Health Organization.

malaria cases has increased over the world as a result of deteriorating health systems. According to the World Health Organization (WHO), it has been observed that malaria is the world's most epidemiologically important vector-borne disease. Globally, over 2400 million people are at risk due to malaria with around 250 million cases and at least 0.6 million deaths per year are solely attributable to malaria alone, of which many are children below the age of 5 years (WHO, 2012a).

Dengue, including dengue hemorrhagic fever and dengue shock syndrome (DSS), is an acute febrile, fast-emerging viral disease transmitted by the infective bite of the *Aedes* mosquito (Nalongsack et al., 2009). The incidence of this disease has increased 30-fold over the last 50 years. Dengue virus comprises four serotypes, DENV 1–4 (Izabela et al., 2010), that infect about 50 million to 100 million people annually across the world and 2.5 billion people living in about 100 endemic countries are at risk of acquiring dengue infection. The mosquito *Aedes aegypti* is the primary vector of dengue in urban and some rural areas of tropical and subtropical regions, and it frequently bites multiple people during each feeding period, whereas *Aedes albopictus* is a secondary dengue vector in semiurban and rural areas (WHO, 2012b). Bed nets are ineffective, since *Ae. aegypti* has its peak activity during the day, making breeding-site management and the use of insecticides the main forms of control (Forattini, 2002).

Chikungunya is specifically a tropical disease that was first reported from Tanzania in 1952–1953. Subsequently, the virus spreads throughout Sub-Saharan Africa, India, and other countries of Southeast Asia affected millions of people leading to great epidemics (Edelman et al., 2000; Sudeep and Parashar, 2008). Chikungunya virus is transmitted by the bite of mosquitoes of genus *Aedes* such as *Ae. aegypti* and *Ae. albopictus* in Asia (Zeller, 1998; Jeandel et al., 2004).

Japanese encephalitis is a mosquito-borne viral disease of children below 10 years and adolescents that causes significant epidemics in South Asia, Southeast Asia, East Asia, and the Pacific (Saxena, 2008; Erlanger et al., 2009). Japanese encephalitis is an RNA virus that belongs to family Flaviviridae and genus *Flavivirus*, and is transmitted by mosquito *Culex*. Around 3 billion globally are living at risk in the epidemic countries with global incidence of 30,000 to 50,000 cases annually. In 2002, annual numbers of human deaths were estimated to be 10,000 to 15,000 and the global impact was 709,000 disability-adjusted life years (DALYs) (Erlanger et al., 2009). Japanese encephalitis causes an estimated 50,000 cases and 10,000 deaths every year, mostly of children aged less than 5 (WHO, World Health Day 2014).

Although great efforts have been made on national and international levels all around the world, there are currently no effective vaccines against dengue, filariasis, or malaria, and specific treatments are only available for malaria and filariasis. Therefore, the most effective way to prevent vector-borne diseases is to control mosquito populations. This chapter reviews the current state of knowledge about genetically altered microbes and viruses in control of mosquito-borne diseases, which was motivated by the aforementioned situation. It is a much more flexible and efficient approach than the use of genetically altered (transgenic) mosquitoes because effector molecules and symbiotic bacteria can be replaced if they do not achieve the desired result. This technique may therefore become an important integrated pest management tool for mosquito control.

## 17.2 MOSQUITO CONTROL STRATEGIES

Vector control is a central and critical component among all the key strategies that is widely promoted by the WHO for the prevention and control of mosquito-borne diseases, as no vaccines or specific therapeutics are effective against these diseases (Mutero et al., 2012). Traditional vector control strategies in the last century were based on use of chemical agents such as dichlorodiphenyltrichloroethane (DDT). Although insecticides have been successfully used to control mosquitoes of the genera *Aedes* and *Anopheles*, current ecological and environmental protection standards do not allow such approaches because of the adverse effects of many insecticides on nontarget species, including humans, their environmental impact, the contamination of soil and water, and the development of selective processes and subsequent mosquito resistance to insecticides (Wilke and Marrellii, 2015). New strategies therefore had to be created to replace the use of insecticides. For the sustainable control of vectors, the WHO recommended a pragmatic approach, integrated vector management (IVM), that can be applied in various combinations to suit different ecological and socioeconomic settings. It includes five key elements (Beier et al., 2008; Mutero et al., 2012):

- Advocacy, social mobilization, and legislation
- Collaboration within the health sector and with other sectors
- Integrated approach
- Evidence-based decision making
- Capacity building

High reproductive capacity and genomic flexibility of mosquitoes is demonstrated by the resistance of mosquito populations to chemical and biological insecticides as well as by their ability to adapt to different environmental conditions. This makes management of these insects very difficult (Besansky and Collins, 1992; Bracco et al., 1997; Nicholson, 2007; Pocquet et al., 2013; Govindarajan and Sivakumar, 2014).

Therefore, effective alternative forms of control are urgently needed that can be used on a large scale and are environmentally friendly and should be included in IVM (Taipe-Lagos and Natal, 2003; Natal et al., 2004; Cardoso et al., 2005).

Recently, quite a lot of attention has been addressed to the associations between symbionts (bacteria and viruses) and insects, which open interesting perspectives for control of mosquito borne diseases. A good deal of recent evidence clearly indicates that microbial symbionts, particularly bacteria, may influence specific traits of insect–host biology such as mating behavior, insecticide resistance, fitness, nutrition, defense, and reproduction in a significant number of insect species (Miller et al., 2010; Ferrari and Vavre, 2011; Simon et al., 2011; Kikuchi et al., 2012). In general symbiotic microorganisms offer some possibilities to control the mosquitoes, and two approaches are particular attractive: The first one implies the disruption of microbial symbionts required by insect pests. The second approach is the manipulation of microorganisms with major impacts on insect traits contributing to their "pest status" (Favia et al., 2012). The second approach—genetically altered bacteria and viruses (paratransgenesis)—would appear to be a more promising strategy to control mosquitoes.

## 17.3 USE OF GENETICALLY ALTERED MICROBES

### 17.3.1 Genetically Altered Bacteria

Large communities of diverse microorganisms reside in insects in different tissues within the organism, and at varying pH conditions. Their presence is tolerated by the immune system, as microbial presence confers some barrier to pathogens for the host. Furthermore, microbiota have been known to produce regulatory molecules that also act as a defense mechanism against pathogens that infect the host organism. These symbiotic microbes tend to out compete pathogenic microbes as a defense mechanism for the host, creating a bottleneck at crucial regions of pathogen development. In the mosquito vector, the midgut microbiota also play a fundamental, beneficial role as an innate barrier to pathogens. They allow the mosquito vector to adapt to diverse environments by providing them with nutrients that they may not be able to obtain themselves, as well as facilitate species-specific metabolism.

Studies have shown that mosquito vector microbiota may play a very crucial role in controlling the spread of mosquito vector diseases such as malaria and African sleeping sickness (Abreu et al., 2010). The use of microbiota as an innate system to manage the development of *Plasmodium falciparum* in mosquito vectors is currently being developed to enhance disease control. Paratransgenesis (the genetic modification of symbiotic organisms, such as bacteria and viruses) would appear to be the most suitable solution to control mosquito-borne diseases. This approach includes the manipulation of symbionts that can express antipathogen molecules within the host. The symbiotic bacteria or viruses are genetically modified (GM) to express effector molecules and then reintroduced into the mosquito where they produce the desired effect (Yoshida et al., 2001; Favia et al., 2007; Aksoy et al., 2008; Abreu et al., 2010). Thus, to develop an efficient paratransgenic approach to control mosquito-borne diseases, suitable symbiotic microorganisms with specific and well-defined features must be identified.

For a bacteria to be used in paratransgenesis three key components are required: (1) an effector molecule that achieves the desired effect, (2) a mechanism to display or excrete the effector molecule on the surface of the bacteria, and (3) bacteria that can survive in the mosquito long enough to produce the expected amount of effector molecules and thereby achieve the desired effect in the mosquito (Riehle and Jacobs-Lorena, 2005; Riehle et al., 2007). The chosen bacteria should be capable of colonizing a wide variety of mosquito species so that they can be deployed in different species and isolated strains. Producing recombinant bacteria in sufficient numbers is much simpler than creating transgenic mosquitoes (Sayler and Ripp, 2000; Riehle et al., 2007). Furthermore, the number of bacteria increases dramatically (100 to 1000 of times) after ingestion of blood (Briones et. al., 2008), resulting in a proportional increase in the amount of effector molecules expressed and secreted by GM bacteria (Wang et al., 2012), leading to various possible outcomes: obstructing pathogen transmission, reducing the mosquito's vector capacity, preventing fertilization of eggs, interfering with embryogenesis, and causing the death of the mosquito (Favia et al., 2007; Aksoy et al., 2008; Abreu et al., 2010; Chavshin et al., 2012). To date, quite a few bacteria have already been engineered to express antipathogen effector molecules.

Pidiyar et al. (2004) found a wide range of symbiotic bacteria in various mosquito species: *Bacillus* spp., *Staphylococcus* spp., *Pseudomonas* spp., and *Aeromonas culicicola* in *Culex quinquefasciatus*; *Serratia marcescens* and *Klebsiella ozonae*; *Pseudomonas aeruginosa* and *Enterobacter agglomerans* in *Culex pipiens*; and two strains of *A. culicicola* in *Ae. aegypti*. Dinparast et al. (2011) reported thirteen species of bacteria in *Anopheles maculipennis* and *Anopheles stephensi.* Weiss and Aksoy reported *Asaia, Enterobacter, Mycobacterium, Sphingomonas, Serratia*, and *Chryseobacterium* in midgut of *Anopheles gambiae* (Weiss and Aksoy, 2011). In southern Mexico, Gonzalez-Ceron et al. (2003) found *Enterobacter amnigenus*, *Enterobacter cloacae*, *Enterobacter* sp., *S. marcescens*, and *Serratia* sp. in *Anopheles albimanus.*

In *Anopheles*, paratransgenesis was first described by Yoshida who showed that a genetically transformed *Escherichia coli* expressing a single-chain immunotoxin could block *Plasmodium berghei* oocyst development in *An. stephensi* by 58% and significantly reduce the oocyst number from a mean of 45.4 to 8.6 oocysts per gut (Yoshida et al., 2001). Riehle et al. (2007) also genetically engineered *E. coli* to express molecules (SM1 and PLA2) in *An. stephensi* that inhibited *P. berghei*, leading to a reduction in oocyst numbers by 41% and 23%, respectively (Riehle et al., 2007). Wang et al. (2011) used *P. agglomerans* to deliver different kinds of effector molecules in *An. stephensi* and *An. gambiae* midguts and looked at the effect on *P. berghei* and *P. falciparum.* This paratransgenesis inhibited parasite development by up to 98% and the prevalence decreased by up to 84% for two of the effector molecules expressed, scorpine and Enolase Plasminogen Interaction Peptide (EPIP) (Wang et al., 2011).

Another bacterium with potential for use in paratransgenesis is *Thorsellia.* These bacteria are so far only found in vector mosquitoes and their breeding waters, which is a positive characteristic for paratransgenesis. It was first isolated in *An. arabiensis* from Kenya (Lindh et al., 2005). It was shown that it can utilize blood to enhance growth and it can tolerate alkaline environments (pH 9.1) and displays α-hemolysis (Briones et al., 2008). Beside this, another promising strategy used for the development of novel methods for pathogen transmission is the infection of *Wolbachia*-negative mosquitoes (like *Anopheles* spp. and *Ae. Aegypti*) with specific *Wolbachia* strains. *Wolbachia* are endosymbiotic, intracellular, gram-negative bacteria that can be found in the cytoplasmic vacuoles of insects, isopods, mites, and nematodes (Laven, 1967). They were first described within the reproductive tissues of the mosquito *Culex pipiens* in 1924 (Walker and Moreira, 2011). The bacteria are vertically and maternally transmitted through the egg cytoplasm and manipulate the host reproduction in order to enhance their spread by, for example, cytoplasmic incompatibility (CI), feminization, killing of male embryos, and parthenogenesis (meaning asexual reproduction when embryos develop from unfertilized eggs) (Walker and Moreira, 2011). For example, infecting *Ae. aegypti* with *Wolbachia* may cause different effects in the host. In particular, three outcomes open interesting perspectives in diseases control:

- Shortened life span, reducing their potential to transmit pathogens that have to infect adult mosquitoes to complete their development cycle (Weiss et al., 2006)
- Limited susceptibility to infection with the dengue or chikungunya virus or the *Plasmodium* parasite (Baldini et al., 2014)
- Depending on the strain of *Wolbachia*, induction of cytoplasmic incompatibility, with apparently no significant fitness cost and high horizontal transmission (Weiss et al., 2006)

Furthermore, alphaproteobacterium *Asaia* is also considered a potential candidate for use in mosquito paratransgenesis, which has attracted considerable interest due to its peculiar relationship with mosquitoes. This peculiarity relates to several elements: First, *Asaia* localizes in the gut, in the salivary glands, and in the reproductive organs of mosquitoes of both sexes (Damiani et al., 2008, 2010; Ricci et al., 2012). For example, transformed *Asaia* strains produce fluorescent proteins that showed a remarkable capacity to colonize the midgut, reproductive organs, and salivary glands of recipient mosquitoes (Favia et al., 2007; Damiani et al., 2008, 2010). These studies underline the possibility of using this bacterium to express antiparasite molecules within the mosquito body to inhibit the transmission of the parasite. At present, attempts are underway to modify *Asaia* to produce strains able to express and secrete anti-*Plasmodium* effector molecules to be used in "malaria transmission blocking" experiments (Ricci et al., 2012).

Wang et al. (2012) describe a substantially improved strategy to deliver effector molecules by engineering a natural symbiotic bacterium *Pantoea agglomerans* (previously known as *Enterobacter agglomerans*) to secrete antimalaria proteins in the mosquito midgut. They found that the development of the human parasite *Plasmodium falciparum* and the rodent parasite *P. berghei* in the mosquito was efficiently suppressed by up to 98%.

For bacteria to be used in paratransgenesis programs in the field they must not only be adapted to the digestive tract of mosquitoes, but also be able to compete with other bacteria in the digestive tract as well as have a lifetime compatible with the adult mosquito vector so they can readily colonize the midgut (Riehle et al., 2007).

### 17.3.2 Genetically Altered Fungi

Fungi are eukaryotic organisms found in nearly every ecosystem across the world having distinctive cell walls composed of chitin. Many species are only observed in their asexual form producing nonmotile spores called conidia (Osherov, 2001). Common fungi include molds and yeasts, such as the medically important genus *Aspergillus* and the commonly used baker's yeast *Saccharomyces* (Jayaram et al., 2014; Virginio et al., 2014). Fungi play diverse roles in the ecosystem spanning from commensal to mutualist to pathogenic organisms (George et al., 2013; Underhill and Iliev, 2014). Recently, there has been renewed interest in the role of fungi in the mycobiota of humans and other organisms such as mosquitoes (Ricci et al., 2011a; Cappelli et al., 2014; Underhill and Iliev, 2014).

Entomopathogenic fungi are fungal pathogens of insects. They are ubiquitous in the environment and can often be found in a free-living state. When spores of these fungi come into contact with an appropriate insect host, they attach to the surface of the insect's exoskeleton and germinate, producing fungal hyphae that secrete chitinase enzymes, which allow them to penetrate the insect cuticle and enter the hemocoel. Once inside the insect, they propagate in the insect body cavity, ultimately killing the insect. Fungal spores are released from the dead insect, completing the life cycle (Kanzok and Jacobs-Lorena, 2006).

A number of true entomopathogenic fungi parasitize mosquitoes, including fungi that belong to the genera *Beauveria* and *Metarhizium* (Scholte et al., 2004). These pathogens are capable of reducing mosquito life span as well as vector competence for pathogens (Dong et al., 2012; Garza-Hernández et al., 2013). In addition to filamentous fungi, many yeasts have been isolated from mosquitoes in both the lab and field (Frants and Mertvetsova, 1986; Gusmaõ et al., 2010; Ricci et al., 2011a, 2011b; Cappelli et al., 2014). It is clear that fungi are naturally associated with mosquitoes and this has warranted investigation into their potential for paratransgenic applications.

*Beauveria bassiana* and *Metarhizium anisopliae* are naturally occurring pathogens of mosquitoes, and they have historically been used to control insect pests like caterpillars (Lynch et al., 2012; Liu et al., 2013). The conidia or spores of these fungi germinate on the surface of their insect host, penetrate through the cuticle, spread systemically in the hemolymph, and then candidate resulting in the death of the host. Fungal infection causes a reduced propensity for blood feeding and a reduction in pathogen numbers that lowers an infected mosquito's ability to transmit disease (Blanford et al., 2005, 2012; Garza-Hernández et al., 2013). *B. bassiana* and *M. anisopliae* fungi tend to kill the mosquito slowly. This seems to be advantageous because it causes only moderate reductions in reproductive fitness, resulting in low selective pressure for resistance (Thomas and Read, 2007; Fang et al., 2011). Slow killing can still effectively prevent pathogen transmission, because the time from when a mosquito takes an infectious blood meal until it can transmit the pathogen through an infectious bite (the extrinsic incubation period, or EIP) can be days or even weeks (Chan and Johansson, 2012). The average EIP for *Plasmodium* is 12 to 14 days, but in nature, this process can take as long as 30 days (Fang et al., 2011; Thomas and Read, 2007). *B. bassiana* infection results in high mortality of *An. stephensi* at 14 days after a blood meal (Thomas and Read, 2007). Studies shows that most mosquitoes infected with *B. bassiana* shortly after a blood meal would be dead prior to being able to transmit the parasite (Blanford et al., 2005; Lynch et al., 2012). However, this strategy assumes that mosquitoes become infected with fungi immediately following a blood meal (Thomas and Read, 2007; Fang et al., 2011).

Genetic manipulation of *M. anisopliae* (entomopathogenic fungi) can improve their efficacy in reducing vector competence (Fang et al., 2011). Fang et al. (2011) genetically transformed *M. anisopliae* to express the effector molecule SM1 and the antimicrobial toxin scorpine, which interfere with *Plasmodium falciparum* development, reducing mosquito infectivity. Infection with these transgenic fungi dramatically reduces the number of *Plasmodium* sporozoites in the salivary gland, effectively reducing the likelihood of transmission (Fang et al., 2011). The advantage of using transgenic fungi over wild-type strains is that it avoids the requirement of the mosquito becoming infected immediately following a blood meal (Fang et al., 2011). Instead, the capacity for transmission is reduced even if the mosquito becomes infected with the fungus at a late stage of *Plasmodium* infection (Fang et al., 2011). Multiple lines of transgenic *M. anisopliae* can be created to produce different effector molecules to reduce the evolution of resistance (Fang et al., 2011). Furthermore, *M. anisopliae* could be engineered to produce antiviral effector molecules to interrupt dengue virus transmission. In summary, genetic manipulation of entomopathogenic fungi is a highly feasible strategy to decrease mosquito vector competence for pathogens.

### 17.3.3 Genetically Altered Yeasts

Yeasts are found in many insect species and benefit insects by accounting for nutritional dietary deficiency (Suh et al., 2004; Benda, 2008; Cappelli et al., 2014). Yeasts have been identified in *Aedes* and *Anopheles* mosquitoes in both the lab and field, which belong to the genera *Candida, Pichia*, and *Wickerhamomyces* (Frants and Mertvetsova, 1986; Gusmaõ et al., 2010; Ricci et al., 2011a). In nature, yeasts are typically found as plant nectars, which may be the source of acquisition for mosquitoes (Ricci et al., 2011a, 2011b). These yeasts can be found in the mosquito midgut, which places them in close proximity to invading pathogens. Yeasts do not appear to be pathogenic to mosquitoes, but their role in modulating pathogen susceptibility is unknown. In other biological systems, *Wickerhamomyces anomalus* is used as a biocontrol agent to prevent fungal contamination of grains and produce (Lahlali et al., 2004; Olstorpe and Passoth, 2011). It has been observed that some strains of *W. anomalus* produce a substance known as killer toxin (KT), which can compete with other microbes (Wang et al., 2007; Guo et al., 2013). KTs are known to be active against many microorganisms including *Leishmania* parasites, and future studies on a KT-producing strain of *W. anomalus* should determine if KT is active against *Plasmodium* (Savoia et al., 2006; Cappelli et al., 2014).

Besides midgut, *W. anomalus* has also been isolated from the gonads of mosquitoes (Ricci et al., 2011b). It has been discovered that yeasts are also present outside of the midgut, which may indicate that yeasts are able to escape the mosquito immune response through an unknown mechanism (Ricci et al., 2011b).

Finally, it will be important to study the potential for vertical transmission of paratransgenic *W. anomalus*, as related *Saccharomyces* yeasts can survive passage through the *Drosophila* midgut (Coluccio et al., 2008). Furthermore, genetically modified *W. anomalus* could produce antipathogen effector molecules. Related *Pichia* yeasts are highly responsive to genetic manipulation and are frequently used for protein expression (Mattanovich et al., 2012). Nevertheless, further research exploring the genetic tractability of *W. anomalus* is needed for vector management strategy.

## 17.4 USE OF GENETICALLY ALTERED VIRUSES

Many viruses are capable of infecting mosquitoes and such mosquito-associated viruses include baculoviruses, bunyaviruses, flaviviruses, mesoviruses, parvoviruses, and rhabdiviruses (Bolling et al., 2015). Mosquitoes serve as a host for many viruses that do not act as human pathogens and have been used as an efficient tool for elucidating molecular processes and explored for their applicability in mosquito control. For example, densoviruses, which are nonenveloped single-stranded DNA icosahedral viruses in the family *Parvoviridae* (subfamily *Densovirinae*) that infect arthropods such as mosquitoes and are potential candidates for paratransgenesis because of their tractability and exclusive infective of mosquitoes (Carlson et al., 2006). Mosquito densoviruses (MDVs) have narrow host ranges and are maintained in natural populations by a cycle that includes both horizontal and vertical transmission from infected adults to larvae. They are capable of infecting and disseminating in *Anopheles gambiae* and have been recently proposed as a paratransgenic tool for malaria control strategies (Ren et al., 2008).

Genetic manipulation of MDVs is an attractive option for controlling mosquito populations and in turn mosquito transmitted diseases. Genetically modified MDVs with different transgenes reduce the life span of mosquitoes (Gu et al., 2010; Suzuki et al., 2014). For instance, the insertion of transgene expressing toxins could increase killing efficiency and reduce selective pressure for resistance to a single mechanism.

*Ae. aegypti* mosquitoes, which transmit dengue and yellow fever, are highly susceptible to infection by the *Ae. aegypti* densovirus (AeDNV) (Buchatsky, 1989). It infects other mosquitoes of the genera *Aedes, Culex*, and *Culiseta*, and thus has the potential for use as a biological control agent (Buchatsky et al., 1987).

Ren and collaborators (2008) discovered, cloned, and characterized the first known DNV capable of infection and dissemination in *An. gambiae*: AgDNV. By the development of an AgDNV-based expression vector to express gene(s) of interest in *An. gambiae*, an exogenous gene (enhanced green fluorescent protein, EGFP) was expressed in *An. gambiae* mosquitoes. Wild-type and EGFP-transducing AgDNV virions were able to consistently infect *An. gambiae* larvae. Expressed EGFP was detected in adult tissues such as the midgut, fat body, and ovaries, and were transmitted to subsequent mosquito generations. This work produced clear evidence that AgDNV could be used as part of a paratransgenic malaria control strategy by transduction of anti-*Plasmodium* effectors or insect-specific toxins in *Anopheles* mosquitoes (Paula et al., 2011).

## 17.5 CONCLUSION

Despite various other methods of mosquito control, use of genetically modified bacteria and viruses offers several favorable features compared to other control methods. First, genetic manipulation of bacteria and viruses is much simpler and faster than genetic manipulation of mosquitoes, and are much easier to introduce into mosquito populations than transgenes. Second, they bypass the genetic barriers of reproductively isolated mosquito populations that often occur in areas of high malaria transmission, and thus which present a great obstacle to the spread of mosquito transgenes. Third, genetically modified bacteria and viruses can be produced on a large scale at low cost, even in some developing countries, and possible inactivation of bacterial transgenes in the field is not a major concern because of the easier logistics of introducing newly transformed bacteria or viruses. Finally, concerning regulatory and ethical requirements, already-existing regulations on the release of genetically modified bacteria and viruses into the environment are a sufficient starting point for formulating more specific guidelines for field applications of mosquito symbionts (Ricci et al., 2012).

Further, the same GM bacteria is used for different mosquito species and even different genera, the only requirement being that the bacteria used be able to survive and colonize the mosquito host in sufficient numbers so that effector molecule expression reaches the desired level, thus overcoming the problem highlighted earlier and increasing the range of mosquitoes that can be controlled (Sayler and Ripp, 2000; Pidiyar et al., 2004; Riehle et al., 2007). The successful transformation of *P. agglomerans* has substantially improved the expression of effector molecules and proteins in the mosquito midgut because of its capacity to rapidly replicate after a blood meal (Pidiyar et al., 2004). If the GM bacteria successfully express the effector molecule but fail to excrete it or display it on their surface, an apoptotic mechanism must be used, leading to extra fitness costs. However, if the effector molecules are excreted by the bacteria, they can be produced continually, significantly increasing the effectiveness of the paratransgenic system. Fortunately, there are mechanisms such as the outer membrane protein A (OmpA) that allow bacteria to export molecules to the environment (Riehle et al., 2007; Paula et al., 2011). Paratransgenic strategies may prove very useful for controlling mosquito species such as *Culex* that are, for as yet unknown reasons, inherently difficult to transform and have been genetically modified only once by the embryo microinjection technique (Miller et al., 1987; Allen and Christense, 2004). In contrast, many other mosquito species have been successfully transformed more than once (*Ae. aegypti*, *Ae. albopictus*, *Ae. fluviatilis*, *An. gambiae*, and *An. stephensi*) (Jasinskiene et al., 1998; Catteruccia et al., 2000; Rodrigues et al., 2006; Labbé et al., 2010). Microorganisms associated with vectors can exert pathogenic effects on the host, interfere with its reproduction, and reduce its vector competence (Favia et al., 2007; Pocquet et al., 2013).

Although our knowledge of the mosquito mycobiota is quite limited compared to the bacterial microbiota, the prospect of fungi for the control of vector-borne disease is high. In view of this, the path forward to field application is much more straightforward for entomopathogenic fungi than for other potential paratransgenic agents such as mosquito midgut bacteria. Fungi are easy to genetically modify, can be cultured and mass produced in the endemic environment, can be deployed in the field, and have already been used to control other arthropod pest species.

Finally, paratransgenesis may prove to be a highly valuable tool for use in IPM and is potentially of particular value in areas where there are species complexes or more than one species whose population needs to be reduced or made refractory to a given pathogen, as well as in the control of mosquito species that are not easily transformed.

## REFERENCES

Abreu, I. V. C., Zhu, K. Y., and Ortigao, M. R. 2010. Transgenesis and paratransgenesis to control insect-borne diseases: Current status and future challenges. *Parasitol. Int.* 59 (1):1–8.

Aksoy, S., Weiss, B., and Attardo, G. 2008. Paratransgenesis applied for control of tsetse transmitted sleeping sickness. *Adv. Exp. Med. Biol.* 627:35–48.

Allen, M. L., and Christense, B. M. 2004. Flight muscle-specific expression of act88F: GFP in transgenic *Culex quinquefasciatus* Say (Diptera: culicidae). *Parasitol. Int.* 53:307–314.

Baldini, F., Segata, N., Pompon, J., Marcenac, P., Shaw, W. Robert, Dabiré, R. K. et al. 2014. Evidence of natural *Wolbachia* infections in field populations of *Anopheles gambiae. Nat. Commun.* 5:3985.

Beier, J. C., Keating, J., Githure, J. I., Macdonald, M. B., Impoinvil, D. E., and Novak, R. J. 2008. Integrated vector management for malaria control. *Malar. J.* 7(1):S4. doi: 10.1186/1475-2875-7-S1-S4.

Benda, N. D. 2008. Detection and characterization of *Kodamaea ohmeri* associated with small hive beetle *Aethina tumida* infesting honey bee hives. *J. Apic. Res.* 47:194–201.

Blanford, S., Chan, B. H. K., Jenkins, N., Sim, D., Turner, R. J., Read, A. F. et al. 2005. Fungal pathogen reduces potential for malaria transmission. *Science* 308:1638–41.

Blanford, S., Jenkins, N. E., Read, A. F., and Thomas, M. B. 2012. Evaluating the lethal and pre-lethal effects of a range of fungi against adult *Anopheles stephensi* mosquitoes. *Malar. J.* 11:365.

Besansky, N. J., and Collins, F. H. 1992. The mosquito genome: Organization, evolution and manipulation. *Parasitol. Today* 8:186–192.

Bolling, B. G., Weaver, S. C., Tesh, R. B., and Vasilakis, N. 2015. Insect-specific virus discovery: Significance for the arbovirus community. *Viruses* 7: 4911–4928.

Bracco, J. E., Dalbon, M., Marinotti, O., and Barata, J. M. 1997. Resistance to organophosphorous and carbamates insecticides in a population of *Culex quinquefasciatus. Rev. Saude. Publica.* 31:182–183.

Briones, A. M., Shililu, J., Githure, J., Novak, R., and Raskin, L. 2008. *Thorsellia anophelis* is the dominant bacterium in a Kenyan population of adult *Anopheles gambiae* mosquitoes. *ISME J.* 2:74–82.

Buchatsky, L. P. 1989. Densonucleosis of blood sucking mosquitoes. *Dis. Aquat. Org.* 11: 145–150.

Buchatsky, L. P., Kuznetsova, M. A., Lebedinets, N. N., and Kononko, G. A. 1987. Development and basic properties of a virus preparation viroden. *Vopr. Virusol.* 6:729–733.

Cappelli, A., Ulissi, U., Valzano, M., Damiani, C., Epis, S., Gabrielli, M. G. et al. 2014. A *Wickerhamomyces anomalus* killer strain in the malaria vector *Anopheles stephensi. PLoS One.* 9:e95988.

Cardoso, J. C., Corseuil, E., and Barata, J. M. S. 2005. Culicinae (Diptera, Culicidae) ocorrentesno Estado do Rio Grande do Sul, Brasil. *Rev. Bras. Entomol.* 49:275–287.

Carlson, J., Suchman, E., and Buchatsky, L. 2006. Densoviruses for control and genetic manipulation of mosquitoes. *Adv. Virus. Res.* 68:361–392.

Catteruccia, F., Nolan, T., Loukeris, T. G., Blass, C., Savakis, C., Kafatos, F. C. et al. 2000. Stable germline transformation of the malaria mosquito *Anopheles stephensi*. *Nature* 405:959–962.

Chan, M., and Johansson, M. A. 2012. The incubation periods of dengue viruses. *PLoS One*. 7: e50972.

Chavshin, A. R., Oshaghi, M. A., Vatandoost, H., Pourmand, M. R., Raeisi, A., Enayati, A. A. et al. 2012. Identification of bacterial microflora in the midgut of the larvae and adult of wild caught *Anopheles stephensi*: A step toward finding suitable paratransgenesis candidates. *Acta. Trop.* 121:129–134.

Coluccio, A. E., Rodriguez, R. K., Kernan, M. J., and Neiman, A. M. 2008. The yeast spore wall enables spores to survive passage through the digestive tract of *Drosophila*. *PLoS One* 3: e2873.

Damiani, C., Ricci, I., Crotti, E., Rossi, P., Rizzi, A., Scuppa, P. et al. 2008. Paternal transmission of symbiotic bacteria in malaria vectors. *Curr. Biol.* 18:R1087–R1088.

Damiani, C., Ricci, I., Crotti, E., Rossi, P., Rizzi, A., Scuppa, P. et al. 2010. Mosquito-bacteria symbiosis: The case of *Anopheles gambiae* and *Asaia*. *Microb. Ecol.* 60:644–654.

Dinparast, D. N., Jazayeri, H., Raz, A., Favia, G., Ricci, I., and Zakeri, S. 2011. Identification of the midgut microbiota of *An. stephensi* and *An. maculipennis* for their application as a paratransgenic tool against malaria. *PLoS One*. 6:e28484.

Dong, Y., Morton, J. C., Ramirez, J. L., Souza-Neto, J. A., and Dimopoulos, G. 2012. The entomopathogenic fungus *Beauveria bassiana* activate toll and JAK-STAT pathway-controlled effector genes and anti-dengue activity in *Aedes aegypti*. *Insect Biochem. Mol. Biol.* 42:126–132.

Edelman, R., Tacket, C. O., Wasserman, S. S., Bodison, S. A., Perry, J. G., and Mangiafi, J. A. 2000. Phase II safety and immunogenicity study of live chikungunya virus vaccine tsi-gsd-218. *Am. J. Trop. Med. Hyg.* 62:681–685.

Erlanger, T. E., Weiss, S., Keiser, J., Utzinger, J., and Wiedenmayer, K. 2009. Past, present, and future of Japanese encephalitis. *Emer. Inf. Dis.* 15(1):1–7.

Fang, W., Vega-Rodríguez, J., Ghosh, A. K., Jacobs-Lorena, M., Kang, A., and St. Leger, R. J. 2011. Development of transgenic fungi that kill human malaria parasites in mosquitoes. *Science* 331:1074–1077.

Favia, G., Ricci, I., Damiani, C., Raddadi, N., Crotti, E., Marzorati, M. et al. 2007. Bacteria of the genus Asaia stably associate with *Anopheles stephensi*, an Asian malarial mosquito vector. *Proc. Natl. Acad. Sci. USA* 104:9047–9051.

Favia, G., Ricci, I., Scuppa, P., Damiani, C., Capone, P. R. A., Freece, C. D. et al. 2012. Facing malaria parasite with mosquito symbionts, malaria parasites. In *Malaria parasites*, Omolade Okwa (ed.), chap. 4. InTech.

Ferrari, J., and Vavre, F. 2011. Bacterial symbionts in insects or the story of communities affecting communities. *Philos. Trans. R. Soc. Lond. B. Biol. Sci.* 366 (1569):1389–1400.

Forattini, O. P. 2002. *Culicidologia médica*. São Paulo: EDUSP.

Frants, T. G., and Mertvetsova, O. A. 1986. Yeast associations with mosquitoes of the genus *Aedes* Mg. (Diptera, Culicidae) in the Tom-Ob river region. *Nauchnye. Doki. Vyss. Shkoly. Biol. Nauki.*, 94–98.

Garza-Hernández, J. A., Rodríguez-Pérez, M. A., Salazar, M. I. et al. 2013. Vectorial capacity of *Aedes aegypti* for dengue virus type 2 is reduced with co-infection of *Metarhizium anisopliae*. *PLoS Negl. Trop. Dis.* 7:e2013.

George, J., Jenkins, N. E., Blanford, S., Thomas, M. B., and Baker, T. C. 2013. Malaria mosquitoes attracted by fatal fungus. *PLoS One* 8:e62632.

Gonzalez-Ceron, L., Santillan, F., Rodriguez, M. H., Mendez, D., and Hernandez-Avila, J. E. 2003. Bacteria in midguts of field-collected *Anopheles albimanus* block *Plasmodium vivax* sporogonic development. *J. Med. Entomol.* 40:371–374.

Govindarajan, M., and Sivakumar, R. 2014. Larvicidal, ovicidal, and adulticidal efficacy of *Erythrina indica* (Lam.) (Family: Fabaceae) against *Anopheles stephensi, Aedes aegypti*, and *Culex quinquefasciatus* (Diptera: Culicidae). *Parasitol. Res.* 113:777–791.

Gu, J. B., Dong, Y. Q., Peng, H. J., and Chen, X. G. 2010. A recombinant AeDNA containing the insect specific toxin, BmK IT1, displayed an increasing pathogenicity on *Aedes albopictus*. *Am. J. Trop. Med. Hyg.* 83:614–623.

Guo, F.-J., Ma, Y., Xu, H.-M., Wang, X.-H., and Chi, Z.-M. 2013. A novel killer toxin produced by the marine-derived yeast *Wickerhamomyces anomalus* YF07b. *Antonie. Van. Leeuwenhoek* 103:737–746.

Gusmaõ, D. S., Santos, A. V., Marini, D. C., Bacci, M., Berbert-Molina, M. A., and Lemos, F. J. A. 2010. Culture dependent and culture-independent characterization of microorganisms associated with *Aedes aegypti* (Diptera: Culicidae) (L.) and dynamics of bacterial colonization in the midgut. *Acta Trop.* 115:275–281.

Izabela, A., Zybert, R., Wilschut, J., and Smit, J. M. 2010. Dengue virus life cycle: Viral and host factors modulating Infectivity. *Cell Mol. Life Sci.* 67:2773–2786.

Jasinskiene, N., Coates, C. J., Benedict, M. Q., Cornel, A. J., Rafferty, C. S., James, A. A. et al. 1998. Stable transformation of the yellow fever mosquito, *Aedes aegypti*, with the Hermes element from the housefly. *Proc. Natl. Acad. Sci.* 95:3743–3747.

Jayaram, V. B., Cuyvers, S., Verstrepen, K. J., Delcour, J. A., and Courtin, C. M. 2014. Succinic acid in levels produced by yeast (*Saccharomyces cerevisiae*) during fermentation strongly impacts wheat bread dough properties. *Food Chem.* 151:421–428.

Jeandel, P., Josse, R., and Durand, J. P. 2004. Exotic viral arthritis: Role of alphavirus. *Med. Trop.* 64: 81–88.

Kanzok, S. M., and Jacobs-Lorena, M. 2006. Entomopathogenic fungi as biological insecticides to control malaria. *Trends Parasitol.* 22(2):49–51.

Kikuchi, Y., Hayatsu, M., Hosokawa, T., Nagayama, A., Tago, K., and Fukatsu, T. 2012. Symbiont-mediated insecticide resistance. *Proc. Natl. Acad. Sci.* 109(22): 8618–8622.

Labbé, G. M., Nimmo, D. D., and Alphey, L. 2010. Piggybac and PhiC31 mediated genetic transformation of the Asian tiger mosquito, *Aedes albopictus* (Skuse). *PLoS. Negl. Trop. Dis.* 17:4–8.

Lahlali, R., Serrhini, M. N., and Jijakli, M. H. 2004. Efficacy assessment of *Candida oleophila* (strain O) and *Pichia anomala* (strain K) against major postharvest diseases of citrus fruits in Morocco. *Commun. Agric. Appl. Biol. Sci.* 69:601–609.

Laven, H. 1967. Eradication of *Culex pipiens* fatigans through cytoplasmic incompatibility. *Nature* 216:383–384.

Lindh, J. M., Terenius, O., and Faye, I. 2005. 16S rRNA gene-based identification of midgut bacteria from field-caught *Anopheles gambiae* sensu lato and *A. funestus* mosquitoes reveals new species related to known insect symbionts. *App. Env. Microbiol.* 71 (11):7217–7223.

Liu, Y.-J., Liu, J., Ying, S.-H., Liu, S.-S., and Feng, M.-G. 2013. A fungal insecticide engineered for fast per os killing of caterpillars has high field efficacy and safety in full-season control of cabbage insect pests. *Appl. Environ. Microbiol.* 79:64528.

Lynch, P. A., Grimm, U., Thomas, M. B., and Read, A. F. 2012. Prospective malaria control using entomopathogenic fungi: Comparative evaluation of impact on transmission and selection for resistance. *Malar. J.* 11:383.

Mattanovich, D., Branduardi, P., Dato, L., Gasser, B., Sauer, M., and Porro, D. 2012. Recombinant protein production in yeasts. *Methods Mol. Biol.* 824:329–358.

Miller, L. H., Sakai, R. K., Romans, P., Gwadz, R. W., Kantoff, P., and Coon, H. G. 1987. Stable integration and expression of a bacterial gene in the mosquito *Anopheles gambiae*. *Science* 237:779–781.

Miller, W. J., Ehrman, L., and Schneider, D. 2010. Infectious speciation revisited: Impact of symbiont-depletion on female fitness and mating behavior of *Drosophila* paulistorum. *PLoS Pathog.* 6:e1001214.

Mutero, C. M., Schlodder, D., Kabatereine, N., and Kramer R. 2012. Intergrated vector management for malaria control in Uganda: Knowledge, perceptions and policy development. *Malar. J.* 11:21.

Nalongsack, S., Yoshida, Y., Morita, Y., Sosouphanh, K., and Sakamoto, J. 2009. Knowledge, attitude and practice regarding dengue among people in Pakse, Laos. *Nagoya J. Med. Sci.* 71:29–37.

Natal, D., Araújo, F. A. A., Vianna, R. S. T., Pereira, L. E., and Ueno, H. M. O. 2004. Mosquito das águas poluídas. *São Paulo. Saneas* 26–31.

Nicholson, G. M. 2007. Fighting the global pest problem: Preface to the special toxicon issue on insecticidal toxins and their potential for insect pest control. *Toxicon* 49:413–422.

Olstorpe, M., and Passoth, V. 2011. *Pichia anomala* in grain biopreservation. *Antonie Van. Leeuwenhoek* 99:57–62.

Osherov, N. 2001. The molecular mechanisms of conidial germination. *FEMS Microbiol. Lett.* 199:153–160.

Paula, A. R., Carolino, A. T., Silva, C. P., and Samuels, R. I. 2011. Susceptibility of adult females *Aedes aegypti* (Diptera: Culicidae) to the entomopathogenic fungus *Metarhizium anisopliae* is modified following blood feeding. *Parasit. Vectors* 4:91.

Pidiyar, V. J., Jangid, K., Patole, M. S., and Shouche, Y. S. 2004. Studies on cultured and uncultured microbiota of wild *Culex quinquefasciatus* mosquito midgut based on 16s ribosomal RNA gene analysis. *Am. J. Trop. Med. Hyg.* 70:597–603.

Pocquet, N., Milesi, P., Makoundou, P., Unal, S., Zumbo, B., Atyame, C. et al. 2013. Multiple insecticide resistances in the disease vector *Culex p. quinquefasciatus from Western Indian Ocean. PLoS One.* 21:e77855.

Ren, X., Hoiczyk, E., and Rasgon, J. L. 2008. Viral paratransgenesis in the malaria vector *Anopheles gambiae. PLoS. Pathog.* 4(8):e1000135.

Ricci, I., Damiani, C., Scuppa, P., Mosca, M., Crotti, E., Rossi, P. et al. 2011a. The yeast *Wickerhamomyces anomalus* (Pichia anomala) inhabits the midgut and reproductive system of the Asian malaria vector *Anopheles stephensi. Environ. Microbiol.* 13:911–921.

Ricci, I., Mosca, M., Valzano, M., Damiani, C., Scuppa, P., Rossi, P., et al. 2011b. Different mosquito species host *Wickerhamomyces anomalus* (Pichia anomala): Perspectives on vector-borne diseases symbiotic control. *Antonie Van Leeuwenhoek* 99:43–50.

Ricci, I., Valzano, M., Ulissi, U., Epis, S., Cappelli, A., and Favia, G. 2012. Symbiotic control of mosquito borne disease. *Pathog. Glob. Health* 106(7):380–385.

Riehle, M. A., and Jacobs-Lorena, M. 2005. Using bacteria to express and display anti-parasite molecules in mosquitoes: Current and future strategies. *Insect Biochem. Mol. Biol.* 35:699–707.

Riehle, M. A., Moreira, C. K., Lampe, D., Lauzon, C., and Jacobs-Lorena, M. 2007. Using bacteria to express and display anti-Plasmodium molecules in the mosquito midgut. *Int. J. Parasitol.* 37:595–603.

Rodrigues, F. G., Oliveira, S. B., Rocha, B. C., and Moreira, L. A. 2006. Germline transformation of *Aedes fluviatilis* (Diptera: Culicidae) with the piggyBac transposable element. *Mem. Inst. Oswaldo. Cruz.* 101:755–757.

Rozendaal, J. A. 1997. *Vector control: Methods for use by individuals and communities.* Geneva: World Health Organization.

Savoia, D., Scutera, S., Raimondo, S., Conti, S., Magliani, W., and Polonelli, L. 2006. Activity of an engineered synthetic killer peptide on *Leishmania major* and *Leishmania infantum* promastigotes. *Exp. Parasitol.* 113:186–192.

Saxena, S. K. 2008. Japanese encephalitis: Perspectives and new developments. *Future Neurol.* 3(5):515–521.

Sayler, G. S., and Ripp, S. 2000. Field applications of genetically engineered microorganisms for bioremediation processes. *Curr. Opin. Biotechnol.* 11:286–289.
Scholte, E. J., Knols, B. G. J., Samson, R. A., and Takken, W. 2004. Entomopathogenic fungi for mosquito control: A review. *J. Insect Sci.* 4:19.
Simon, J. C., Boutin, S., Tsuchida, T., Koga, R., Le Gallic, J. F., Frantz, A. et al. 2011. Facultative symbiont infections affect aphid reproduction. *PLoS ONE.* 6 (7):e21831.
Suh, S.-O., Gibson, C. M., and Blackwell, M. 2004. *Metschnikowia chrysoperlae* sp. nov., *Candida picachoensis* sp. nov. and *Candida pimensis* sp. nov., isolated from the green lacewings *Chrysoperla comanche* and *Chrysoperla carnea* (Neuroptera: Chrysopidae). *Int. J. Syst. Evol. Microbiol.* 54:1883–1890.
Sudeep, A. B., and Parashar, D. 2008. Chikungunya: An overview. *J. Biosci.* 33(4):443–449.
Suzuki, Y., Niu, G., Hughes, G. L., and Rasgon, J. L. 2014. A viral over-expression system for the major malaria mosquito *Anopheles gambiae*. *Sci. Rep.* 4:5127.
Taipe-Lagos, C. B., and Natal, D. 2003. Abundância de culicídeos em área metropolitan Preservada e suas Implicações Epidemiológicas. *Rev. Saude. Publica.* 37:275–279.
Thomas, M. B., and Read, A. F. 2007. Can fungal biopesticides control malaria? *Nat. Rev. Microbiol.* 5:377–383.
Underhill, D. M., and Iliev, I. D. 2014. The mycobiota: Interactions between commensal fungi and the host immune system. *Nat. Rev. Immunol.* 14:405–416.
Virginio, E. D., Kubitschek-Barreira, P. H., Batista, M. V., Schirmer, M. R., Abdelhay, E., Shikanai-Yasuda, M. A. et al. 2014. Immunoproteome of *Aspergillus fumigatus* using sera of patients with invasive aspergillosis. *Int. J. Mol. Sci.* 15:14505–14530.
Wang, X., Chi, Z., Yue, L., Li, J., Li, M., and Wu, L. 2007. A marine killer yeast against the pathogenic yeast strain in crab (*Portunus trituberculatus*) and an optimization of the toxin production. *Microbiol. Res.* 162:77–85.
Wang, Y., Gilbreath, T. M., Kukutla, P., Yan, G., and Xu, J. 2011. Dynamic gut microbiome across life history of the malaria mosquito *Anopheles gambiae* in Kenya. *PLoS One* 6(9): e24767.
Wang, S., Ghosh, A. K., Bongio, N., Stebbings, K. A., Lampe, D. J., and Lorena, M. J. 2012. Fighting malaria with engineered symbiotic bacteria from vector mosquitoes. *PNAS* 109 (31):12734–12739.
Walker, T., and Moreira, L. A. 2011. Can *Wolbachia* be used to control malaria? *Mem. Inst. Oswaldo. Cruz.* 106(1): 212–217.
Weiss, B. L., Mouchotte, R., Rio, R. V., Wu, Y. N., Wu, Z., Heddi, A. et al. 2006. Interspecific transfer of bacterial endosymbionts between tsetse fly species: Infection establishment and effect on host fitness. *Appl. Environ. Microbiol.* 72:7013–7021.
Weiss, B., and Aksoy, S. 2011. Microbiome influences on insect host vector competence. *Trends Parasitol.* 27(11):514–522.
Wilke, B. B., and Marrellii, M. T. 2015. Paratransgenesis: A promising new strategy for mosquito vector control. *Parasite. Vector.* 8:342.
Word Health Organization (WHO). 2012a. New report signals slowdown in the fight against malaria. Press release, December 17.
Word Health Organization (WHO). 2012b. World dengue report.
Word Health Organization (WHO). 2014. Vector borne disease. Available from http://www.who.int/campaigns/world-health-day/2014/aboutdiseases/en/insect.html.
Yoshida, S., Ioka, D., Matsuoka, H., Endo, H., and Ishii, A. 2001. Bacteria expressing single chain immunotoxin inhibit malaria parasite development in mosquitoes. *Mol. Biochem. Parasitol.* 113:89–96.
Zeller, H. G. 1998. Dengue, arbovirus and migrations in the Indian Ocean. *Bull. Soc. Pathol. Exot.* 91:56–60.

# *Section II*

## *Microbial Control of Leishmania Vector*

# 18 Antileishmanial Compounds from Microbial Sources for Leishmaniasis Prevention

*Natarajan Amaresan, Hardik Naik Jinal, and Alwarappan Sankaranarayanan*

## CONTENTS



## 18.1 INTRODUCTION

Leishmaniasis is a protozoan disease caused by 20 different species of intracellular protozoan parasites belonging to the genus *Leishmania.* It was identified by William Leishman and Charles Donovan in 1903. *Leishmania* is a trypanosomatid protozoan that is transmitted between mammalian hosts by phlebotomies sand flies and causes leishmaniasis (Neva and Brown, 1994). Generally, leishmaniases are present as one of three forms: cutaneous (skin), mucocutaneous (mucous membrane), and visceral (internal organ). Cutaneous leishmaniasis is a chronic, slow-to-heal disease that causes skin ulcerations, which appear like volcanoes having raised borders with central craters. The ulcerations may appear for months to years and typically result in severe scarring (Herwaldt and Magill, 2010). *Leishmania* spp., such as *L. major*, *L. mexicana*, *L. amazonensis*, and *L. braziliensis* primarily cause cutaneous lesions. Some species of *Leishmania* belonging to Central and South America cause damage that may affect initial cutaneous sores on the face and the mucous membrane of nose and mouth. Visceral leishmaniasis affects the internal organs such as the spleen, liver, and bone marrow. Apart from the aforementioned infections, some unidentified species cause fever, weight loss, hepatosplenomegaly, and pancytopenia.

Of the three forms of leishmaniases, visceral leishmaniasis (VL) is the most severe. In India VL is known as Kala azar. The disease is caused by the parasite *L. donovani* and transmitted by the sandfly *Phlebotomus argentipes*. The infection particularly occurs in the liver and other organs. If left untreated, it become more serious. Many rural regions in the Indian state of Bihar are endemic for VL, with conservative estimates of nearly 100,000 deaths per year attributable to this disease. In that case, few possibilities and facilities are available for curing the disease (Hurwitz et al., 2011). The World Health Organization (WHO) considers leishmaniasis one of the most serious parasitic diseases in terms of prevalence and geographical distribution. Worldwide, more than 20 *Leishmania* species are known to infect humans across 98 countries or territories (Alvar et al., 2012).

Primary treatment for this disease is prevention of sandfly bites. Requena et al. (2004) declared that there is no immunoprotection available for leishmaniasis. Herwaldt and Magill (2010) reported that there are no preventive medications or vaccinations available either. Antimicrobial compounds such as liposomal amphotericin B and sodium stibogluconate (Pentostam) are used for VL and cutaneous leishmaniasis, respectively (Herwaldt and Magill, 2010). The lack of a vaccine or chemoprophylaxis limits the options for prevention of leishmaniasis. As leishmaniasis is a promising disease with an ability to change, interests have been raised about identifying biocontrol agents from biological sources. Varieties of antibiotic drugs are available for treatment of VL. The main drugs for the treatment include antimony, amphotericin, paromomycin, and miltefosine (Moore and Lockwood, 2010).

Biocontrol agents are natural compounds having the potential to antagonize disease-causing pathogens. These natural compounds are mostly isolated from biological sources such as medicinal plants, soil, water, and various microorganisms. The identification of antibiotics from the natural biological sources are less toxic to the human body as compared to synthetic drugs, which produce numerous side effects. It has been estimated that 1.3 million new cases occur annually, adding to the 10 million to 12 million cases found globally (Noazin et al., 2008). The estimates from the Global Burden of Disease Study (Institute for Health Metrics and Evaluation, 2010) indicate that more than 50,000 annual deaths result from leishmaniasis. This result urges the identification of safe and potential drug against leishmaniasis.

Common and easy prevention against leishmaniasis is to apply insecticide on the internal walls of houses, also known as indoor residual spraying (IRS) because it has been reported that phlebotomine sandflies are highly sensitive to insecticides (González et al., 2015). IRS is the most widely used intervention for controlling endophagic phlebotomine sandflies, but needs to be repeated regularly, which decreases its long-term sustainability (Davies, 2003). Alternatively, the number of phlebotomine sandflies could be reduced by removing breeding sites from the environment through activities such as replastering of cracks in walls with mud or lime (Kishore et al., 2006). Another method to control disease is by controlling the host of zoonotic leishmaniasis, depending on which animals act as reservoirs. Dogs play an important role as leishmaniasis reservoirs in some areas, and development of appropriate control measures is necessary (Dogan et al., 2006; Courtenay et al., 2009; Quinnell and Courtenay, 2009). Other animal reservoirs, such as rodents, have been targeted through poisonous baits (Roberts, 2006). In addition, diseases control planning focuses on the reducing sandfly–human contact or sandfly populations (Alvar et al., 2006).

## 18.2 VECTOR CONTROL USING BIOINSECTICIDES

Several methods were attempted to control phlebotomines, like usage of pesticides, insecticidal plants, and bioinsecticides. Among these, bioinsecticides are usually target specific, showing less toxicity on nontarget living organisms, and environmental friendly (Hynes and Boyetchko, 2006). However, the work on control of phlebotomines using entomopathogens is still scanty and application on control of leishmaniasis is limited (Kishore et al., 2006; Amóra et al., 2009) due to its difficulty to employ them in the field and diversity of habitat in which the vector can reproduce. The efficiency of *Bacillus thuringiensis* as an entomopathogen against phlebotomines was first demonstrated by de Barjac et al. (1981) against *Phlebotomus papatasi* and *Lutzomyia longipalpis* and by *B. sphaericus* against *P. martini* (Robert et al., 1997). Wahba et al. (1999) recommended *B. thuringiensis* var. *israelensis* to control *P. papatasi* larvae and pupae. The work on larval mortality and reduction in fertility of females was greatly shown with *B. sphaericus* (Wahba, 2000). However, it is still questionable to what extent this entomopathogen can be used for the biological control of phlebotomines (Lainson and Rangel, 2005).

Among the biological control agents, entomopathogenic fungi in general present rates that are more efficient and exhibit more specificity on the target insect than chemical insecticides. Fungal species that attack several hosts should be favored in control programs because they can be used in several areas, such as *Beauveria bassiana* (Barreto et al., 2004). Warburg (1991) demonstrated that filter paper impregnated with *B. bassiana* conidia increased *P. papatasi* and *Lu. longipalpis* adult mortality. Studies on biological control of phlebotomines are still scarce and its practical application seems to be limited to adult VL vector control (Kishore et al., 2006), since the employment of biolarvicides in the field is difficult due to the diversity of habitats in which this vector can reproduce.

## 18.3 PERSPECTIVE ON ANTILEISHMANIAL DRUGS

Treatment of leishmaniasis is usually done with nonleishmanial drugs (Singh et al., 2012). The drugs that are presently available to treat leishmaniasis are highly toxic. The pentavalent antimonials sodium stibogluconate and meglumine antimoniate have been used as first-line drugs for more than 70 years in most countries, despite severe side effects such as cardiac arrhythmia and acute pancreatitis (Mishra et al., 2007; Frézard et al., 2009). Due to increased parasitic resistance, these drugs are nearly outdated in North Bihar (India) (Sundar, 2001) but can still be used in other countries and other Indian states. The antifungal macrolide amphotericin B, which was first isolated from *Streptomyces nodosus*, is one of the most effective antileishmanial drugs that induces high cure rates. However, the disadvantage of amphotericin B includes adverse effects such as fever, chills, thrombophlebitis, myocarditis, severe hypokalemia, renal dysfunction, and even death. The need to develop a less toxic, more effective formulation of amphotericin B has led to new clinical formulations of amphotericin B in which deoxycholate has been replaced by other lipids.

Pentamidine is used as a second-line treatment for leishmaniasis; however, it can cause irreversible insulin-dependent diabetes mellitus and death. In addition to its severe toxic effects, pentamidine has poor efficacy, resulting in the abandonment of this treatment in India (Sundar and Chatterjee, 2006). Miltefosine, an alkyl phospholipids developed as an antitumor agent, has an excellent antileishmanial activity. Paromomycin is an aminoglycoside antibiotic with unique antileishmanial activity. Sitamaquine, a primaquine analogue (8-aminoquinolene), is another orally administrable compound. To date, little is known about its efficacy and toxicity. However, there is a need to go for combinatorial chemotherapies, which are under process and provide better results.

## 18.4 MICROORGANISMS INVOLVED IN MANAGEMENT OF LEISHMANIASIS

Management of diseases with the help of associated microorganisms is known as biological controls (Table 18.1). Scant information is available on the biological control of sandfly. Due to an increase in disease frequency, it is necessary to evolve new approaches to control leishmaniasis. Most sandflies have association with *Bacillus* species such as *Bacillus substilis* and *Bacillus megaterium*, which can form antileishmaniasis proteins (Hurwitz et al., 2011).

### 18.4.1 Bacteria

The violacein {3-[1,2-dihydro-5- (5-hydroxy-1H-indol-3-yl)-2-oxo-3H-pyrrol-3-ylidene]1,3-dihyro-2H-indol-2-1} extracted from *Chromabacterium violaceum* showed a $EC_{50}$/24 h value of 4.3 ± 1.15 μmol/L compared with pentamidine ($EC_{50}$/24 h = 0.46 μmol/L) against *L. amazonensis* (Leon et al., 2001). The antagonistic effect of *Pantoea ananatis*, and *Enterobacter cloacae* isolated from the digestive tract of *Lutzomyia evansi* showed inhibition of exponential *L. infantum* promastigotes by 77.29% ± 0.6%, and 70.17% ± 1.1% respectively. The result suggests that the presence of these bacteria in the vector intestine may affect the parasite development to metacyclic stages, infective to human hosts (Gómez et al., 2016). The 43 kDa toxin produced by *B. thuringiensis* serovar thuringiensis (H14) was tested against *L. major* promastigotes. The toxin showed $LC_{50}$ at 4.95 μg/ml against *L. major* promastigotes indicating a potential application as a therapeutic agent against leishmaniasis (El-Sadawy et al., 2008). A prodigiosin pigment producing *Serratia marcescens* SM365 lysed the promastigotes of *L. chagasi*. Further, the researchers suggested that bacterial mannose-sensitive fimbriae are important to *S. marcescens* SM 365 in the lysis of *L. chagasi* (Moraes et al., 2008).

**TABLE 18.1**
**Different Microbial Extracts Showing Antileishmanial Activity**

| Organism Name | Source | Preparation | Compound | Species of *Leishmania* | Reference |
|---|---|---|---|---|---|
| **Bacteria** | | | | | |
| *Chromobacterium violaceum* | | Ethanol | Violacein | *L. amazonensis* | Leon et al., 2001 |
| *Enterobacter cloacae* | *Lutzomyia evansi* | Methanol | – | *L. infantum* | Vivero Gómez et al., 2016 |
| *Pantoea ananantis* | *L. evansi* | Methanol | – | *L. infantum* | Vivero Gómez et al., 2016 |
| *Ochrobactrum* sp. | *L. evansi* | Methanol | – | *L. infantum* | Vivero Gómez et al., 2016 |
| *Bacillus thuringiensis* | – | Acetone | 43 kDa crystal protein | *L. major* | El-Sadawy and Abou El-Hag, 2008 |
| *Serratia marcescens* | – | – | Prodigiosin | *L. chagasi* | Moraes et al., 2008 |
| *Asaia* sp. | *L. longipalpis* | Spent media | – | *L. mexicana* | Sant'Anna et al., 2014 |
| *Ochrobactrum intermedium* | *L. longipalpis* | Spent media | – | *L. mexicana* | Sant'Anna et al., 2014 |
| *S. marcescens* | *L. longipalpis* | Spent media | – | *L. mexicana* | Sant'Anna et al., 2014 |
| **Fungi** | | | | | |
| *Diaporthe phaseolorum* | *Viguiera arenaria* | Ethyl acetate | APRT activity | *L. tarentolae* | Guimaraes et al., 2008 |
| *Phyllosticta* sp. or *Guignardia mangiferae* | *V. arenaria* | Ethyl acetate | APRT activity | *L. tarentolae* | Guimaraes et al., 2008 |
| *Phomopsis* sp. | *V. arenaria* | Ethyl acetate | APRT activity | *L. tarentolae* | Guimaraes et al., 2008 |

(*Continued*)

**TABLE 18.1 (CONTINUED)**
**Different Microbial Extracts Showing Antileishmanial Activity**

| Organism Name | Source | Preparation | Compound | Species of *Leishmania* | Reference |
|---|---|---|---|---|---|
| *Cochliobolus sativus* | *Vernonia polyanthes* | Ethanol | Cochlioquinone A, Isocochlioquinone A, Anhydrocochlioquinone A | *L. amazonensis* | do Nascimento et al., 2015 |
| *Purpureocillium lilacinum* | *Rauvolfia macrophylla* | Ethyl acetate | Purpureone | *L. donovani* | Lenta et al., 2016 |
| *Trichosporum* sp. | *Trigonella foenum-graecum* | Ethyl acetate | Diketopiperazine alkaloid isomers | *L. donovani* | Metwaly et al., 2015 |
| *Edenia* sp. | *Petrea volúbilis* | Hexanes<br>Chloroform<br>Ethyl acetate | Preussomerin EG1<br>Palmarumycin $CP_2$<br>Palmarumycin $CP_{17}$<br>Palmarumycin $CP_{18}$<br>CJ-12,371 | *L. donovani* | Martínez-Luis et al., 2008 |
| *Aspergillus terreus* | *Carthamus lanatus* | Ethyl acetate | (22*E*,24*R*)-stigmasta-5,7,22-trien-3-β-ol<br>Terrenolide S<br>Stigmast-4-ene-3-one | *L. donovani* | Elkhayat et al., 2015 |
| *Drechslera rostrata* | – | Methanol | di-2- ethylhexyl phthalate | *L. major* | Awaad et al., 2014 |
| *Eurotium tonpholium* | – | Methanol | 1,8-dihydroxy-3-methoxy-6-methyl-anthraquinone | *L. major* | Awaad et al., 2014 |
| *Edenia* sp. | *Petrea volubilis* | Ethyl acetate | 1,8-dihydroxynaphthalene | *L. donovani* | Martínez-Luis et al., 2011 |
| *Penicillium* sp. | *Philodendron tripartitum* | Ethyl acetate | – | *L. donovani* | Martínez-Luis et al., 2011 |

(*Continued*)

**TABLE 18.1 (CONTINUED)**
**Different Microbial Extracts Showing Antileishmanial Activity**

| Organism Name | Source | Preparation | Compound | Species of *Leishmania* | Reference |
|---|---|---|---|---|---|
| *Diaporthe* sp. | *Unidentified* (Cyperaceae) | Ethyl acetate | – | *L. donovani* | Martínez-Luis et al., 2011 |
| *Nectria* sp. | *Macuna mutisiana* | Ethyl acetate | – | *L. donovani* | Martínez-Luis et al., 2011 |
| *Mycosphaerella* sp. | *Psychotria* sp. | Ethyl acetate | Cercosporin | *L. donovani* | Martínez-Luis et al., 2011 |
| *Hypocrea* sp. | *Drymonia serrulata* | Ethyl acetate | – | *L. donovani* | Martínez-Luis et al., 2011 |
| *Fusarium* sp. | – | Ethyl acetate | Unidentified* | *L. mexicana* | Gamboa-Angulo et al., 2013 |
| *Verticillium* sp. | – | Ethyl acetate | Unidentified* | *L. mexicana* | Gamboa-Angulo et al., 2013 |
| *Phaeosphaeria herpotrichoides* | *Adenocystis utricularis* | Ethanol | – | *L. amazonensis* | Santiago et al., 2012 |
| Unidentified *Basidiomycetes* | Rainforest Brazil | Ethyl acetate | – | *L. amazonensis* | Rosa et al., 2009 |
| *Gymnopilus. cf. areolatus* | Rainforest | Ethyl acetate | – | *L. amazonensis* | Rosa et al., 2009 |
| *Irpex lacteus* | Brazil | Ethyl acetate | – | *L. amazonensis* | Rosa et al., 2009 |
| *Lentinus cf. strigosus* | Rainforest | Ethyl acetate | – | *L. amazonensis* | Rosa et al., 2009 |
| *Nothopanus hygrophanus* | Brazil | Ethyl acetate | – | *L. amazonensis* | Rosa et al., 2009 |
| *Pleurotus flabellatus* | Rainforest Brazil | Ethyl acetate | – | *L. amazonensis* | Rosa et al., 2009 |

(*Continued*)

**TABLE 18.1 (CONTINUED)**
**Different Microbial Extracts Showing Antileishmanial Activity**

| Organism Name | Source | Preparation | Compound | Species of *Leishmania* | Reference |
|---|---|---|---|---|---|
| *Cochliobolus* sp. | *Piptadenia adiantoides* | Ethyl acetate | Cochlioquinone A<br>Isocochlioquinone A | *L. amazonensis* | Campos et al., 2008 |
| **Yeast like fungi** | | | | | |
| *Pseudozyma* sp. | *L. longipalpis* | Spent media | – | *L. mexicana* | Sant'Anna et al., 2014 |
| **Cyanobacteria** | | | | | |
| *Lyngbya majuscula* | Shallow waters, Panama | $CH_2Cl_2$/MeOH | Dragonamides A and E and herbamide B | *L. donovani* | Balunas et al., 2010 |
| *Lyngbya majuscula* | Mangrove roots | $CH_2Cl_2$/MeOH | Almiramides A–C | *L. donovani* | Sanchez et al., 2010 |
| *Symploca* sp. | Marine water | $CH_2Cl_2$/MeOH | Symplocamide A | *L. donovani* | Linington et al., 2008 |
| *Nostoc* sp. | – | – | Nostocarboline | *L. donovani* | Barbaras et al., 2008 |
| *Oscillatoria nigroviridis* | Marine water | – | Viridamides A and B | *L. mexicana* | Simmons et al., 2008 |
| *Oscillatoria* sp. | Marine wate | $CH_2Cl_2$/MeOH | Coibacins A–D | *L. donovani* | Balunas et al., 2012 |
| **Algae** | | | | | |
| *Nannochloris* sp. | Red Sea | Methanol | Phenolic compounds | *L. infantum* | Pereira et al., 2015 |
| *Laurencia microcladia* | Marine water | Dichloromethane : methanol (7:3) | – | *L. mexicana* | Pelegrin et al., 2008 |
| *Dictyota caribaea* | Marine water | Dichloromethane : methanol (7:3) | – | *L. mexicana* | Pelegrin et al., 2008 |
| *Turbinaria turbinata* | Marine water | Dichloromethane : methanol (7:3) | – | *L. mexicana* | Pelegrin et al., 2008 |

*(Continued)*

**TABLE 18.1 (CONTINUED)**
**Different Microbial Extracts Showing Antileishmanial Activity**

| Organism Name | Source | Preparation | Compound | Species of *Leishmania* | Reference |
|---|---|---|---|---|---|
| *Lobophora variegata* | Marine water | Dichloromethane : methanol (7:3) | – | *L. mexicana* | Pelegrin et al., 2008 |
| *Laurencia dendroidea* | Marine wate | *n*-hexanes and EtOAc | (-)-elatol and obtusol | *L. amazonensis* | da S. Machado et al., 2014 |
| *Cystoseira baccata* | – | Hexane | Meroditerpenoids | *L. infantum* | de-Sousa et al., 2017 |
| *Dictyota pfaffii* | – | – | Dolabelladienetriol | *L. amazonensis* | Soares et al., 2012 |
| *Laurencia dendroidea* | Marine water | $CH_2Cl_2$-MeOH | Sesquiterpenes | *L. amazonensis* | da Silva Machado et al., 2011 |
| *Dictyota menstrualis* | Coastline of Brazil | Dichloromethane | Pachydictyol A and Isopachydictyol A | *L. amazonensis* | Lira et al., 2015 |
| *Canistrocarpus cervicornis* | Coast of Paraíso | Dichloromethane | 4-Acetoxydolastane Diterpene | *L. amazonensis* | dos Santos et al., 2011 |
| *Asparagopsis taxiformis* | Straits of Messina | Ethanol | – | *L. infantum* | Vitale et al., 2015 |
| Macroalgae | Lagoon | Ethanol | Phycocolloids | *L. infantum* | Minicante et al., 2016 |
| *Asparagopsis taxiformis* | | Hexane and dichloromethane | – | *L. donovani* | Genovese et al., 2009 |

*Do not match with compound registered in the MS database.

### 18.4.2 Fungi

The endophytic fungi isolated from *Viguiera arenaria* showed the inhibition of adenine phosphoribosyl transferase (APRT) enzyme from *L. tarentolae*. The authors found that three endophytic fungi—*Diaporthe phaseolorum* (60.7%), *Phyllosticta* sp. or *Guignardia mangiferae* (57.4%), and *Phomopsis* sp. (61.4%)—were able to decrease the APRT catalytic activity (inhibitory activity ≥50%). APRT has been proposed as a target for the rational search of new leishmanicidal drugs (Guimaraes et al., 2008). In another study, endophytic fungi *Cochliobolus sativus* isolated from *Vernonia polyanthes* showed inhibition of *L. amazonensis*. The extraction of compounds from the fungi resulted in three compounds: cochlioquinone A, isocochlioquinone A, and anhydrocochlioquinone A. The mixture of cochlioquinone A and isocochlioquinone A exhibited good antileishmanial activity, with an $IC_{50}$ value of 10.16 μg/mL, whereas anhydrocochlioquinone A showed 50.53 μg/mL (do Nascimento et al., 2015).

The endophytic fungi such as *Purpureocillium lilacinum*, *Trichosporum* sp., *Edenia* sp., *Aspergillus terreus*, *Penicillium* sp., *Diaporthe* sp., *Nectria* sp., *Mycosphaerella* sp., and *Hypocrea* sp. isolated from different plants showed strong inhibitory activity against *L. donovani*. The compounds responsible for inhibitory activity are very diverse. The purpureone compound isolated from *P. lilacinum* (Lenta et al., 2016) showed potent antileishmanial activity against *L. donovani* ($IC_{50}$ value of 0.174 μg $mL^{-1}$). Diketopiperazine alkaloid isomers—(6-S)-3-(1,3-dihydroxypropyl)-6-(2-methylpropyl)piperazine-2,5-dione and (6-R)-3-(1,3-dihydroxypropyl)-6-(2-methylpropyl)piperazine-2,5-dione—isolated from *Trichosporum* sp. (Metwaly et al., 2015) showed inhibition of *L. donovani* activities with $IC_{50}$ values of 96.3 and 82.5 μg/ml, respectively.

The endophytic fungus *Edenia* sp. isolated from leaves of *Petrea volubilis* L., collected in Coiba National Park, Veraguas, Panama, yielded five antileishmanial compounds, namely, preussomerin EG1, palmarumycin $CP_2$, palmarumycin $CP_{17}$, palmarumycin $CP_{18}$, and CJ-12,371. All the compounds showed significant inhibition of the amastigote form of *L. donovani*, with $IC_{50}$ values of 0.12, 3.93, 1.34, 0.62, and 8.40 μM, respectively (Martínez-Luis et al., 2008). In another study with *Edenia* sp. and *Mycosphaerella* sp., Martínez-Luis et al. (2011) showed marked inhibition of amastigotes of *L. donovani*. The compound responsible for inhibition of *L. donovani* from *Edenia* sp. was identified as nonsymmetrical dimers of 1,8-dihydroxynaphthalene (six compounds) with $IC_{50}$ values of 3.93, 1.34, 0.62, 11.6, 8.40, and 0.12 μM. The compound identified from *Mycosphaerella* sp. was cercosporin with $IC_{50}$ 0.46 and 0.64 μM, respectively. However, compounds of other endophytic fungi reported to have antileishmanial activity, including *Penicillium* sp., *Diaporthe* sp., *Nectria* sp., and *Hypocrea* sp., were unidentified (Martínez-Luis et al., 2011).

Terrenolide S, a new butenolide derivative compound extracted from the endophytic fungus *Aspergillus terreus* isolated from the roots of *Carthamus lanatus* (Asteraceae). displayed activity against *L. donovani* (Elkhayat et al., 2015). Three active compounds—(22*E*,2i4*R*)-stigmasta-5,7,22-trien-3-$\beta$-ol, Terrenolide S, and Stigmast-4-ene-3-one—showed $IC_{50}$ values of 11.24, 15.32, and 27.27 μM, respectively, and $IC_{90}$ values of 14.68, 40.56, and 167.03 μM, respectively. The fungal extract collected from *Drechslera rostrata* (di-2-ethylhexyl phthalate) and *Eurotiumtonpholium* (1,8-dihydroxy-3-methoxy-6-methyl-anthraquinone) showed a significant antileishmanial activity against *L. major*; $IC_{50}$ was of 3.2 and 10.38 μg/mL, respectively (Awaad et al., 2014). The oral administration of di-2-ethylhexyl phthalate and 1,8-dihydroxy-3-methoxy-6-methyl-anthraquinone (50 mg/kg) showed very good activity when compared with the antileishmanial drug Pentostam (125 mg/kg). Complete healing activity was displayed after 13 to 17 days of treatment, whereas complete healing activity of Pentostam was obtained after 28 days with no complications in the liver and kidney functions in tested animal models (Awaad et al., 2014).

The acetonitrile fraction extracted of *Fusarium* sp. TA54 and *Verticillium* sp. TH28 showed strong antileishmanial activity against promastigotes of *L. Mexicana* (Gamboa-Angulo et al., 2013). The inhibition ranged from $IC_{50}$ 6.36 and $IC_{100}$ 25 μg/ml for *Fusarium* sp. and $IC_{50}$ 10.44 and $IC_{100}$ 50 μg/ml for *Verticillium* sp. Twelve endophytic fungi isolated from plants grown in the Antarctica region exhibited inhibition of the proliferation of *L. amazonesis* at a low $IC_{50}$ of between 0.2 and 12.5 μg/ml. The fungus *Phaeosphaeria herpotrichoides* displayed strong leishmanicidal activity with an $IC_{50}$ of 0.2 μg/ml; the inhibitory value was equivalent to amphotericin B (Santiago et al., 2012). Six Basidiomycota fungi from the Atlantic rainforest in Brazil had higher activity against amastigote forms of *L. amazonensis*. Isolates of unidentified Basidiomycetes CCB369 (92%), *G.* cf. *areolatus* UFMGCB36 (91%), *I. lacteus* CCB196 (87%), *L.* cf. *strigosus* CCB162 (85%), *N. hygrophanus* CCB216 (80%), and *P. flabellatus* CCB210 (79%) could kill *L. amazonensis* (Rosa et al., 2009).

The endophytic fungi *Cochliobolus* sp., isolated from *Piptadenia adiantoides*, could kill 90% of the amastigote-like forms of *L. amazonensis*. The chromatographic fractionation resulted in the isolation of cochlioquinone A and isocochlioquinone A. Both compounds are active against *L. amazonensis*, with $EC_{50}$ values of 1.7 mM and 4.1 mM, respectively (Campos et al., 2008).

### 18.4.3 Cyanobacteria

Cyanobacteria are one of the richest sources of known and novel bioactive compounds including toxins with wide pharmaceutical applications (Singh et al., 2011). The secondary metabolites from cyanobacteria include a range of compounds showing toxicity on animals, and antibacterial, antifungal, anti-inflammatory, antiprotozoal, antiviral, and antitumor activities (Mayer et al., 2009).

Coibacins AD, isolated from Panamanian marine cyanobacterium of *Oscillatoria* sp., showed antileishmanial response against *L. donovani* with an $IC_{50}$ value of 2.4 μM (Balunas et al., 2012). In another study, antiprotozoal lipopeptides viridamine A and B isolated from marine *Oscillatoria nigro-viridis* showed antileishmanial activity against *L. mexicana* with an $IC_{50}$ value 1.5 μM. In an analysis of tropical diseases, it was proved that viridamine A is a potential component for inhibiting *Leishmania* (Simmons et al., 2008). The lipopeptides Dragonamide A and E and herbamide E extracted from marine cyanobacterium *Lyngbya majuscula* showed strong antileishmanial activity against *L. donovani* with $IC_{50}$ values of 6.5, 5.1, and 5.9 μM respectively (Balunas et al., 2010).

Symplocamide A, isolated from marine cyanobacterium *Symploca* sp., showed strong inhibitory activity against *L. donovani* with $IC_{50}$ > 9.5 μM (Linington et al., 2008). The organic extract of Almiramides A–C isolated from marine cyanobacterium *L. majuscula* showed strong in vitro activity against *L. donovani* without showing significant cytotoxicity to mammalian Vero cells. Biological evaluation of these three compounds showed that compounds 2 and 3 possessed strong in vitro antileishmanial activity with $IC_{50}$ = 2.4 and 1.9 μM, respectively, compared to sodium stibogluconate and miltefosine ($IC_{50}$ = 44.7 and 0.5 μM) (Sanchez et al., 2010). Alkaloid Nostocarboline (1) and its dimers 2–11 from the cyanobacterium *Nostoc* 78-12A showed antileishmanial activity at submicromolar values against *L. donovani*. The $IC_{50}$ values ranged from 0.2 μM to 34.7 μM (Barbaras et al., 2008).

### 18.4.4 Algae

Microalgae are found in almost all environments (both aquatic and terrestrial), and it has been suggested that several secondary metabolites identified in microalgae have high antimicrobial activities. *Nannochloris* sp. SBL1 isolated from the Red Sea showed inhibition of *L. infantum* promastigotes at a concentration of 250 μg/mL for 48 h, and cell viability determined by the MTT assay also showed significant reduction of promastigotes viability down to 62%, as compared with untreated cells. The *L. infantum* treated with amphotericin B as the standard drug exhibited a viability of 47% at a concentration of 0.23 μg/mL (Pereira et al., 2015). Organic extracts from *Laurencia microcladia*, *Dictyota caribaea*, *Turbinaria turbinate*, and *Lobophora variegata* showed promising inhibitory activity against L. *mexicana* promastigotes with $LC_{50}$ values ranging from 10.9 to 49.9 μg/ml. Further, these extracts had no cytotoxicity against *Artemia salina* (Freile-Pelegrin et al., 2008).

The lipophilic extract of *Laurencia dendroidea* from the southeastern Brazilian coast showed strong inhibitory response against *L. amazonensis* with an $IC_{50}$ value 17.9 ± 1.3 μg/mL (da Silva Machado et al., 2014). The production of sesquiterpenes by algae is the lead compound responsible for the inhibition of *Leishmania* (da Silva Machado et al., 2011). The presence of bioactive compounds elatol and obsutol in *L. dendroidea* crude extract were reported as major antileishmanial agents (da Silva Machado et al., 2011).

Antileishmanial compounds such as merediterpenoids, (3*R*)- and (3*S*)-tetra prenyltoluquinol and (3*R*)- and (3*S*)-tetraprenyltoluquinone, isolated from macroalgae *Cystoseira baccata*, inhibited the growth of the *L. infantum* promastigotes ($IC_{50}$ = 44.9 ± 4.3 and 94.4 ± 10.1 μM, respectively. Apart from growth inhibition, these two compounds also decreased the intracellular infection index 25% to 50% of the intracellular amastigotes at a concentration > 88.0 μM (Bruno de-Sousa et al., 2017).

The compounds Dolabelladienetriol and dolabellane diterpene isolated from *Dictyota pfaffii* are active against *L. amazonensis* during HIV infection (Soares et al., 2012). The authors found that Dolabelladienetriol suppressed the amastigote multiplication in a dose-dependent manner, inhibiting 56% and 61% of growth at concentrations of 50 mM and 100 mM of Dolabelladienetriol, respectively. The $IC_{50}$ of Dolabelladienetriol was recorded with 43.9 mM compared with control amphotericin B (1.1 mM), which inhibited 44.5% of the amastigote survival. A study reported that compounds 4α-acetoxy-9β,14α-dihydroxydolast-1 and 7-diene produced by brown algae *Canistrocarpus cervicornis* showed dose-dependent strong antileishmanial activity against *L. amazonensis* and different forms of *Leishmania* such as promastigote, axenic amastigote, and intracellular amastigote with an $IC_{50}$ value 2.0 μg/ml, 12.0 μg/ml, and 4.0 μg/ml, respectively (dos Santos et al., 2011).

The extract from the red alga *Osmundaria obtusiloba* and a mixture of isomers pachydictyol A and isopachydictyol A isolated from *Dictyota menstrualis* showed potential antileishmanial activity. The $EC_{50}$ activity ranged from 0.61 to 154.0 μg $mL^{-1}$. The extracts also showed a selectivity index better than pentamidine (Lira et al., 2015).

The ethanol extract of red algae *Asparagopsis taxiformis* showed powerful inhibitory effects against *L. infantum* (Vitale et al., 2015). Algal extract at 5 μg/ml showed 95% vitality, and percentage viability of *L. infantum* decreased ($LD_{50}$ values range from 9 μg/ml) as the extract concentration of *A. taxiformis* increased. Likewise, an increased concentration of algal reaching 100% of mortality of the *Leishmania* with an $LD_{50}$ value range from 25 μg/ml (Vitale et al., 2015). Natural polysaccharides play a significant role in medicinal therapy, pharmaceutical application, and drug delivery. The polysaccharide phycocolloids extract from marine macroalgae showed potential biocontrol activity against *L. infantum* (Minicante et al., 2016). Hexane and dichromomethane extract of Mediterranean red alga *A. taxiformis* and *A. armata* showed antileishmanial response against *L. donovani* with an $IC_{50}$ value of 17.00 μg/mL and 16.00 μg/mL, respectively. At 33.00 μg/mL and 32.00 μg/mL concentrations, respectively, both extracts showed 90% inhibitory effects (Genovese et al., 2009).

## 18.5 CONCLUSIONS

Recent years have witnessed extraordinary progress in diagnosing and treating *Leishmania* infection. Lack of efficiency, high cost, and growing resistance of current antileishmanials highlight the importance searching for new drugs against *Leishmania.* The challenge is to convert such studies in effective strategic programs aimed to control and eradicate the disease. Microorganisms are a very promising source of novel antileishmanial compounds, which can be further developed as drug prototypes for leishmaniasis treatment. This chapter shows a range of microbial extracts exhibiting interesting antileishmanial properties in vitro, seeming to validate their use in folk medicine. The potent leishmanicidal activities of certain molecules isolated from microbial origins represent an exciting advance in the search for novel antiprotozoal agents at a time when there is an urgent need for new innovative drug leads.

## REFERENCES

Alvar, J., Ivan, D. V., Caryn, B., Merce, H., Philippe, D., Cano J., Jannin, J., and Margriet, D. B. 2012. Leishmaniasis worldwide and global estimates of its incidence. *PLoS One* 7 (5):e35671.

Alvar, J., Yactayo, S., and Bern, C. 2006. Leishmaniasis and poverty. *Trends Parasitol.* 22:552–557.

Amóra, S. S. A., Bevilaqua, C. M. L., Feijó, F. M. C., Alves, N. D., and do V Maciel, M. 2009. Control of phlebotomine (Diptera: Psychodidae) leishmaniasis vectors. *Neotrop. Entomol.* 38:303–310.

Awaad, S. A., Al-Zaylaee, H. M., Alqasoumi, S. I., Zain, M. E., Aloyan, E. M., Alafeefy, A. M., Awad, E. S. H., and El-Meligy, R. M. 2014. Anti-leishmanial activities of extracts and isolated compounds from *Drechslera rostrata* and *Eurotium tonpholium*. *Phytother. Res.* 28:774–780.

Balunas, M. J., Grosso, M. F., Villa, F. A., Engene, N., McPhail, K. L., Tidgewell, K., Pineda, L. M., Gerwick, L., Spadafora, C., Kyle, D. E., and Gerwick, W. H. 2012. Coibacins A–D, antileishmanial marine cyanobacterial polyketides with intriguing biosynthetic origins. *Org. Lett.* 14:3878–3881.

Balunas, M. J., Linington, R. G., Tidgewell, K., Fenner, A. M., Ureña, L. D., Togna, G. D., Kyle, D. E., and Gerwick, W. H. 2010. Dragonamide E, a modified linear lipopeptide from *Lyngbya majuscula* with antileishmanial activity. *J. Nat. Prod.* 73:60–66.

Barbaras, D., Kaiser, M., Brun, R., and Gademann, K. 2008. Potent and selective antiplasmodial activity of the cyanobacterial alkaloid nostocarboline and its dimers. *Bioorg. Med. Chem. Lett.* 18:4413–4415.

Barreto, R. S., Marques, E. J., Gondim Jr., M. G. C., and Oliveira, J. V. 2004. Seleção de *Beauveria bassiana* (Bals.) Vuill. e *Metarhizium anisopliae* (Metsch.) Sorok. para controle do ácaro *Mononychellus tanajoa* (Bondar). *Sci. Agric.* 61:659–664.

Bruno-de-Sousa, C., Gangadhar, K. N., Morais, T. R., Conserva, G. A. A., Vizetto-Duarte, C., Pereira, H., Laurenti, M. D., Campino, L., Levy, D., Uemi, M., Barreira, L., Custódio, L., Passero, L. F. D., Lago, J. H. G., and Varela, J. 2017. Antileishmanial activity of meroditerpenoids from the macroalgae *Cystoseira baccata*. *Exp. Parasitol.* 174:1–9.

Campos, F. F., Rosa, L. H., Cota, B. B., Caligiorne, R. B., Teles Rabello, A. L., Almeida Alves, T. M., Rosa, C. M., and Zani, C. L. 2008. Leishmanicidal metabolites from *Cochliobolus* sp., an endophytic fungus isolated from *Piptadenia adiantoides* (Fabaceae). *PLoS Negl. Trop. Dis.* 2:e348.

Courtenay, O., Kovacic, V., Gomes, P. A., Garcez, L. M., and Quinnell, R. J. 2009. A long-lasting topical deltamethrin treatment to protect dogs against visceral leishmaniasis. *Med. Vet. Entomol.* 23:245–256.

da Silva Machado, F. L., Lima, P. L., Duarte, H., M. Rossi-Bergmann, B., Gestinari, L. M., Fujii, M. T., Kaiser, C. R., and Soares, A. R. 2014. Chemical diversity and antileishmanial activity of crude extracts of *Laurencia* complex (Ceramiales, Rhodophyta) from Brazil. *Rev. Bras. Farmacogn.* 24:635–643.

da Silva Machado, F. L., Pacienza-Lima, W., Rossi-Bergmann, B., de Souza Gestinari, L. M., Fujii, M. T., Campos de Paula, J., Costa, S. S., Lopes, N. P., Kaiser, C. R., and Soares, A. R. 2011. Antileishmanial Sesquiterpenes from the Brazilian Red Alga *Laurencia dendroidea*. *Planta Med.* 77:733–735.

Davies, C., Kaye, R. P., Croft, S. L., and Sundar, S. 2003. Leishmaniasis: New approaches to disease control. *BMJ* 326:377–382.

de Barjac, H., Larget, I., and Killick-Kendrick, R. 1981. Toxicity of *Bacillus thuringiensis var. israelensis*, serotype H14, to the larvae of *Phlebotomine* sandflies. *Bull. Soc. Pathol. Exot.* 74:485–489.

do Nascimento, A. M., Soares, G. M. A., da Silva Torchelsen, F. K. V., de Araujo, J. R. A., Lage, P. S., Duarte, M. C., Andrade, P. H. R., Ribeiro, T. G., Coelho, E. A. F., and do Nascimento, A. M. 2015. Antileishmanial activity of compounds produced by endophytic fungi derived from medicinal plant *Vernonia polyanthes* and their potential as source of bioactive substances. *World J. Microbiol. Biotechnol.* 31:1793–1800.

Dogan, N., Ozbel, Y., Toz, S. O., Dinleyici, E. C., and Bor, O. 2006. Sero-epidemological survey on canine visceral leishmaniasis and the distribution of sandfly vectors in northwestern Turkey: Prevention strategies for childhood visceral leishmaniasis. *J. Trop. Pediatr.* 52:212–217.

dos Santos, A. O., Britta, E. A., Bianco, E. M., Ueda-Nakamura, T., Filho, D. P. D., Pereira, R. C., and Nakamura, C. V. 2011. 4-Acetoxydolastane diterpene from the Brazilian brown alga *Canistrocarpus cervicornis* as antileishmanial agent. *Mar. Drugs* 9:2369–2383.

Elkhayat, E. S., Ibrahim, S. R., Mohamed, G. A., and Ross, S. A. 2016. Terrenolide, S, a new antileishmanial butenolide from the endophytic fungus *Aspergillus terreus. Nat. Prod. Res.* 30:814–820.

El-Sadawy, H. A., Abou El-Hag, H. A., Georgy, J. M., El Hossary, S. S., and Kassem, H. A. 2008. *In vitro* activity of *Bacillus thuringiensis* (H14) 43 kDa crystal protein against *Leishmania major. American-Eurasian J. Agric. Environ. Sci.* 3:583–589.

Freile-Pelegrin, Y., Robledo, D., Chan-Bacab, M. J., and Ortega-Morales, B. O. 2008. Antileishmanial properties of tropical marine algae extracts. *Fitoterapia* 79:374–377.

Frézard, F., Demicheli, C., and Ribeiro, R. R. 2009. Pentavalent antimonials: New perspectives for old drugs. *Molecules* 30:2317–2336.

Gamboa-Angulo, M., Molina-Salinas, G. M., Chan-Bacab, M., Peraza-Sánchez, S. R., Heredia, G., de la Rosa-García, S. C., and Reyes-Estebanez, M. 2013. Antimycobacterial and antileishmanial effects of microfungi isolated from tropical regions in México. *Parasitol Res.* 112:559–566.

Genovese, G., Tedone, L., Mark, T. M., and Morabito, M. 2009. The Mediterranean red alga *Asparagopsis*: A source of compounds against *Leishmania. Mar. Drugs* 7:361–366.

Gómez, R. J. V., Restrepo, G. E. C., Herrera, C. X. M., Ospina, V., Uribe, S., and Robledo, S. 2016. Antagonistic effect of bacteria isolated from the digestive tract of *Lutzomyia evansi* against promastigotes of *Leishmania infantum*, antimicrobial activities and susceptibility to antibiotics. *Adv. Microbiol.* 6:760–775.

González, U., Pinart, M., Sinclair, D., Firooz, A., Enk, C., Vélez, I. D., Esterhuizen, T. M., Tristan, M., and Alvar, J. 2015. Vector and reservoir control for preventing leishmaniasis. *Cochrane Database Syst. Rev.* 8:CD008736.

Guimaraes, D. O., Borges, W. S., Kawano, C. Y., Ribeiro, P. H., Goldman, G. H., Nomizo, A., Thiemann, O. H., Oliva, G., Lopes, N. P., and Pupo, M. T. 2008. Biological activities from extracts of endophytic fungi isolated from *Viguiera arenaria* and *Tithonia diversifolia. FEMS Immunol. Med. Microbiol.* 52:134–144.

Herwaldt, B. L., and Magill, A. J. 2010. *Leishmaniasis, cutaneous. Yellow Book. CDC Travelers' Health Chapter 5.*

Hurwitz, I., Hillesland, H., Fieck A., Das, P., and Durvasula, R. 2011. The paratransgenic sand fly: A platform for control of *Leishmania* transmission. *Parasite. Vector.* 4:82.

Hynes, R. K., and Boyetchko, S. M. 2006. Research initiatives in the art and science of biopesticide formulations. *Soil. Biol. Biochem.* 38:845–849.

Institute for Health Metrics and Evaluation 2010. The Global Burden of Disease*: Injuries, And Risk Factors Study* 2010 (GBO 2010). Seattle, WA: IHME.

Jhingran, A., Chawla, B., Saxena, S., Barrett, M. P., and Madhubala, R. 2009. Paromomycin: Uptake and resistance in *Leishmania donovani. Mol. Biochem. Parasitol.* 164:111–117.

Kishore, K., Kumar, V., Kesari, S., Dinesh, D. S., Kumar, A. J., Das, P., and Bhattacharya, S. K. 2006. Vector control in leishmaniasis. *Indian J. Med. Res.* 123:467–472.

Lainson, R., and Rangel, E. F. 2005. *Lutzomyia longipalpis* and the eco-epidemiology of American visceral leishmaniasis, with particular reference to Brazil—A review. *Mem. Inst. Oswaldo Cruz.* 100:811–827.

Lenta, B. N., Ngatchou, J., Frese, M., Yemeda, F. L., Voundi, S., Nardella, F., Michalek, C., Wibberg, D., Ngouela, S., Tsamo, E., Kaiser, M., Kalinowski, J., and Sewald, N. 2016. Purpureone, anantileishmanial ergo chrome from the endophytic fungus *Purpureocillium lilacinum. Z. Naturforsch.* 71:1095–1175.

Leon, L. L., Miranda, C. C., De Souza, A. O., and Durán, N. 2001. Antileishmanial activity of the violacein extracted from *Chromobacterium violaceum*. *J. Antimicrob. Chemother.* 48:449–450.

Linington, R. G., Edwards, D. J., Shuman, C. F., McPhail, K. L., Matainaho, T., and Gerwick, W. H. 2008. Symplocamide A, a potent cytotoxin and chymotrypsin inhibitor from the marine cyanobacterium *Symploca* sp. *J. Nat. Prod.* 71:22–27.

Lira, M. L. F., Lopes, R., Gomes, A. P., Barcellos, G., Verícimo, M., Osako, K., Ortiz-Ramirez, F. A., Ramos, C. J. B., Cavalcanti, D. N., Teixeira, V. L., and do Amaral, V. 2015. Anti-leishmanial activity of Brazilian green, brown, and red algae. *J. Appl. Phycol.* 28:591–598.

Martínez-Luis, S., Cherigo, L., Higginbotham, S., Arnold, E., Spadafora, C., Ibañez, A., Gerwick, W. H., and Cubilla-Rios, L. 2011. Screening and evaluation of antiparasitic and in vitro anticancer activities of Panamanian endophytic fungi. *Int. Microbiol.* 14:95–102.

Martínez-Luis, S., Della-Togna, G., Coley, P. D., Kursar, T. A., Gerwick, W. H., and Cubilla-Rios, L. 2008. Antileishmanial constituents of the Panamanian endophytic fungus *Edenia* sp. *J. Nat. Prod.* 71:2011–2014.

Mayer, A. M., Rodríguez, A. D., Berlinck, R. G., and Hamann, M. T. 2009. Marine pharmacology in 2005–6: Marine compounds with anthelmintic, antibacterial, anticoagulant, antifungal, anti-inflammatory, antimalarial, antiprotozoal, antituberculosis and antiviral activities; affecting the cardiovascular, immune and nervous systems and other miscellaneous mechanisms of action. *Biochim. Biophys. Acta* 1790:283–308.

Metwaly, A. M., Ghoneim, M. M., and Musa, A. 2015. Two new antileishmanial diketopiperazine alkaloids from the endophytic fungus *Trichosporum* sp. *Der. Pharma. Chemica* 7:322–327.

Minicante, S. A., Michelet, S., Bruno, F., Castelli, G., Vitale, F., Sfriso, A., Morabito, M., and Genovese, G. 2016. Bioactivity of phycocolloids against the Mediterranean protozoan *Leishmania infantum*: An inceptive study. *Sustainability* 8:1131.

Mishra, J., Saxena, A., and Singh, S. 2007. Chemotherapy of leishmaniasis past, present and future. *Curr. Med. Chem.* 14:1153–1169.

Moore, E. M., and Lockwood, D. N. 2010. Treatment of visceral leishmaniasis. *J. Glob. Infect. Dis.* 2:151–158.

Moraes, C. S., Seabra, S. H., Castro, D. P., Brazil, R. P., de Souza, W., Garcia, E. S., and Azambuja, P. 2008. *Leishmania* (Leishmania) *chagasi* interactions with *Serratia marcescens*: Ultrastructural studies, lysis and carbohydrate effects. *Exp. Parasitol.* 118: 561–568.

Neva, F. A., and Brown, H. W. 1994. Blood and tissue protozoa of human beings. In: Neva, F. A., and Brown, H. W. (eds.), *Basic clinical parasitology*. Appleton & Lange, Norwalk, CT, p. 57.

Noazin, S., Modabber, F., Khamesipour, A., Smith, P. G., Moulton, L. H., Nasseri, K., Sharifi, I., Khalil, E. A., Bernal, I. D., Antunes, C. M., Kieny, M. P., and Tanner, M. 2008. First generation leishmaniasis vaccines: A review of field efficacy trials. *Vaccines* 26: 6759–6767.

Pereira, H., Custódio, L., Rodrigues, M. J., de Sousa, C. B., Oliveira, M., Barreira, L., da Rosa Neng, N., Nogueira, J. M. F., Alrokayan, S. A., Mouffouk, F., Abu-Salah, K. M., Ben-Hamadou, R., and Varela, J. 2015. Biological activities and chemical composition of methanolic extracts of selected autochthonous microalgae strains from the Red Sea. *Mar. Drugs* 13:3531–3549.

Quinnell, R. J., and Courtenay, O. 2009. Transmission, reservoir hosts and control of zoonotic visceral leishmaniasis. *Parasitol.* 136:1915–1934.

Requena J. M., Iborra S., Carrion, J., and Alonso, M. S. 2004. Recent advances in vaccines for leishmaniasis. *Expert Opin. Biol. Ther.* 4:1505–1517.

Robert, L., Perich, M. J., Schlein, Y., Jacobson, J. L., Wirtz, R. A., Lawyer, P. G., and Githure, J. I. 1997. Phlebotomine sandfly control using bait-fed adults to carry the larvicide's *Bacillus sphaericus* to the larval habitat. *J. Am. Mosq. Cont. Assoc.* 13:140–144.

Roberts, M. T. M. 2006. Current understanding on the immunology of leishmaniasis and recent developments in prevention and treatment. *Br. Med. Ball.* 75:115–130.

Rosa, L. H., Machado, K. M. G., Rabello, A. L. T., Souza-Fagundes, E. M., Correa-Oliveira, R., Rosa, C. A., and Zani, C. L. 2009. Cytotoxic immunosuppressive, trypanocidal and antileishmanial activities of Basidiomycota fungi present in Atlantic Rainforest in Brazil. *Antonie* van *Leeuwenhoek* 95:227–237.

Sanchez, L. M., Lopez, D., Vesely, B. A., Togna, G. D., Gerwick, W. H., Kyle, D. E., and Linington, R. G. 2010. Almiramides A–C: Discovery and development of a new class of leishmaniasis lead compounds. *J. Med. Chem.* 53:4187–4197.

Santiago, I. F., Alves, T. M. A., Rabello, A., Sales Junior, P. A., Romanha, A. J., Zani, C. L., Rosa, C. A., and Rosa, L. H. 2012. Leishmanicidal and antitumoral activities of endophytic fungi associated with the Antarctic angiosperms *Deschampsia Antarctica* Desv. and *Colobanthus quitensis* (Kunth) Bartl. *Extremophiles* 16:95–103.

Simmons, T. L., Engene, N., Ureña, L. D., Romero, L. I., Ortega-Barría, E., Gerwick, L., and Gerwick, W. H. 2008. Viridamides A and B, lipodepsipeptides with anti-protozoal activity from the marine cyanobacterium *Oscillatoria nigro-viridis*. *J. Nat. Prod.* 71:1544–1550.

Singh, N., Kumar, M., and Singh, R. K. 2012. Leishmaniasis: Current status of available drugs and new potential drug targets. *Asian Pac. J. Trop. Dis.* 5:485–497.

Singh, R. K., Tiwari, S. P., Rai, A. K., and Mohapatra, T. M. 2011. Cyanobacteria: An emerging source for drug discovery. *J. Antibiot.* 64:401–412.

Soares, D. C., Calegari-Silva, T. C., Ulisses, G. L., Valeria, L. T., Izabel, C. N., de Palmer, P., Cirne-Santos, C., Bou-Habib, D. C. and Elvira, M. S. 2012. Dolabelladienetriol, a compound from *Dictyota pfaffii* algae, inhibits the infection by *Leishmania amazonensis*. *PLoS Negl. Trop. Dis.* 6:e1787.

Sundar, S. 2001. Drug resistance in Indian visceral leishmaniasis. *Trop. Med. Int. Health* 6:849–854.

Sundar, S., and Chatterjee, M. 2006. Visceral leishmaniasis—Current therapeutic modalities. *Indian J. Med. Res.* 123:345–352.

Vitale, F., Genovese, G., Bruno, F., Castelli, G., Piazza, M., Migliazzo, A., Simona, A. M., Antonio, M., and Marina, M. 2015. Effectiveness of red alga *Asparagopsis taxiformis* extracts against *Leishmania infantum*. *Open Life Sci.* 10:490–496.

Wahba, M. M. 2000. The influence of *Bacillus sphaericus* on the biology and histology of *Phlebotomus papatasi*. *J. Egypt Soc. Parasitol.* 30:315–323.

Wahba, M. M., Labib, I. M., and Hamshary, E. M. 1999. *Bacillus thuringiensis* var. *israelensis* as a microbial control agent against adult and immature stages of the sandfly, *Phlebotomus papatasi* under laboratory conditions. *J. Egypt Soc. Parasitol.* 29:587–597.

Warburg, A. 1991. Entomopathogens of phlebotomine sand flies: Laboratory experiments and natural infections. *J. Invertebr. Pathol.* 58:189–202.

# *Section III*

## *Microbial Control of Schistosomiasis Vector*

# 19 Biological Control of Aquatic Snail-Borne Diseases (Schistosomiasis)

*Mohamudha Parveen Rahamathulla*

**CONTENTS**

## 19.1 EPIDEMIOLOGY OF SCHISTOSOMIASIS

### 19.1.1 Introduction to Schistosomiasis

Schistosomiasis, also known as bilharziasis, is a disease caused by the infection of parasitic blood flukes (Ducklow et al. 1979). The occurrence of this disease was discovered by German surgeon Theodore Bilharz in Cairo. In 1851, he was the first individual to identify the etiological agent known as *Schistosoma haematobium*. The infectious form of parasites known as cercariae is responsible for contaminating water. These parasites reside in certain types of freshwater snails (Ducklow et al. 1979). The infection is transferred to human beings through direct contact of their skin with the contaminated water.

The waterborne transmission of schistosomal infection occurs when the larval cercariae penetrates into the skin, which is found in contaminated bodies of freshwater. The distribution of infective cercariae is determined on the basis of distributed schistosome eggs, which are responsible for causing infection among the snails. Approximately half of the eggs get excreted through urine and feces, while the rest get trapped in the body tissues causing significant damage to the human body (Ducklow et al. 1979). The eggs cause damage to different organs including the bladder and intestine. Figure 19.1 shows the microscopic examination of a schistosomal egg.

The reinfection from schistosomes occurs as a result of recontact with the infected environment, because they are not capable of dividing in the human body like in protozoa, fungi, viruses, and bacteria. The exposure to infested water due to recreational, domestic, and agricultural activities is the major risk factor for the development of schistosomiasis. However, the adverse effects of this disease can be minimized through snail control, drug treatment, health education, and improved sanitation.

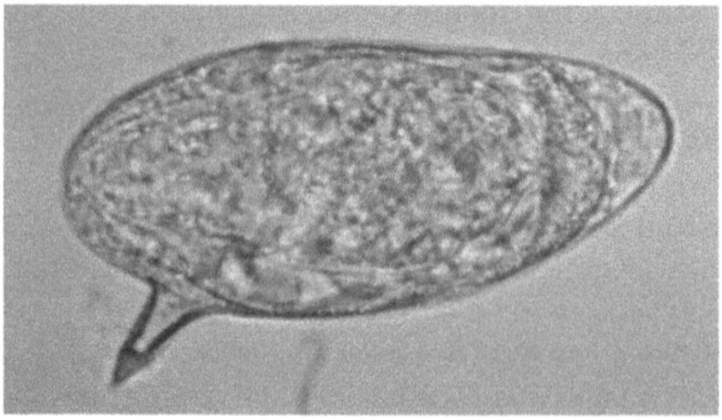

**FIGURE 19.1** A schistosomal egg. (From Centers for Disease Control and Prevention. DPDx: Laboratory identification of parasites of public health concern, accessed on April 2017, www.dpd.cdc.gov/DPDx/.)

Schistosomiasis is the third most devastating tropical disease after malaria and intestinal helminthiasis (worm infection) (Ducklow et al. 1979; WHO 2017). Schistosomiasis has been a major source of morbidity and mortality in developing countries in Asia, Africa, Middle East, Caribbean, and South America. Approximately, 200,000 deaths occur across the world due to schistosomiasis on the annual basis (Chistulo et al. 2004; Ducklow et al. 1979).

### 19.1.2 Etiology of Schistosomiasis

The larvae of parasites, released by freshwater snails, cause schistosomal infection through direct contact of skin with infested water. The larvae after penetrating in the bloodstream develops into an adult schistosome. The snails, responsible for transmittance of schistosomal infections, are amphibious and are mostly found in vegetation near water and wet grass. The snails infected with cercariae hatch into motile miracidia when they reach freshwater.

Multiple cercariae are developed from the miracidium that get released from the snails into water over a period of several weeks. In the human body, cercariae develop into schistosomula, which then migrate through the liver and lungs after maturation. Initially, schistosomula get settled in the lungs, and then migrate to the abdominal blood vessels as their final destination. The cycle begins after 2 to 3 months, when the fluke is matured and finds a mate. After maturation, the majority of the schistosomula is accumulated in venous plexus. The maturation process takes around 6 to 8 weeks, which causes ulceration of the host tissues and reaches lumen of bladder or intestines. However, the adult schistosomes live around 3 to 10 years in human body, but they are capable of surviving as long as 40 years (Colley et al. 2014; Chistulo et al. 2004).

### 19.1.3 Characteristics of Schistosomiasis

The immunologic reactions are caused due to schistosoma eggs that are trapped in the tissues. A granulomatous reaction, involving macrophages, eosinophils, and T-cells, is stimulated by the antigens of the egg, which causes progression of the disease. However, the location and number of eggs trapped in the tissues determine the signs and symptoms of the disease. The initial inflammatory reaction caused by the antigens is reversible. The later stages of the disease are characterized by fibrosis and deposition of collagen, which damage different organs due to irreversible inflammatory reactions.

The schistosomes have the ability to penetrate into the brain, skin, muscles, eyes, and adrenal glands. Moreover, if these eggs get penetrated into the urinary system of a female individual, they can easily reach female genital organs (ovaries, fallopian tube, and uterus) and form granulomas. The embolization of schistosomes to the central nervous system (brain and spinal cord) from portal mesenteric system occurs through the paravertebral venous plexus (Mohammed et al. 2007).

### 19.1.4 Pathophysiology of Schistosomiasis

Individuals infected with schistosomal infections develop a serum sickness-like disease after a few weeks of primary infection that is also known as acute schistosomiasis. Acute schistosomiasis is detected through the presence of circulating immune complexes and peripheral eosinophilia. Although, the symptoms of disease may resolve after few weeks, it can be fatal among a few individuals. Within a few hours of exposure to cercariae, the individual may develop maculopapular and mild skin lesions.

The inflammatory colonic polyposis, bloody diarrhea, and cramping are the symptoms associated with penetration of an egg and formation of granuloma in the digestive system. However, chronic schistosomiasis in intestines is presented through endemic bleeding, perforation, and appendicitis (Badmos et al. 2006; Mohammed et al. 2007). Hepatic diseases may also develop after the exposure to schistosomal infection due to heavy infestations. Fibrosis in hepatic cells is positively associated with development of portal hypertension due to ascites, splenomegaly, portosystemic collaterals, and esophageal variceal bleeding. The condition of pulmonary hypertension worsens due to fibrosis and pulmonary granulomatosis.

### 19.1.5 Epidemiology of Schistosomiasis

Schistosomiasis has been given much importance due to unhygienic water and various activities related to water resource development schemes (fishing, irrigation, and hydroelectricity) (Omonijo et al. 2016). The occurrence of schistosomiasis is highly prevalent in subtropical and tropical regions, where there is limited access to adequate sanitation and pure drinking water. The two major forms of schistosomiasis are urogenital schistosomiasis and intestinal schistosomiasis, which are caused due to infection of six main species of blood flukes. The schistosomal species and geographical distribution of schistosomiasis are presented in Table 19.1.

The intermediate hosts of schistosomes characterize the significant factors that influence the measure, controlling the ability of aquatic snails to survive seasonal drying of their natural habitats. The snails escape their natural predators during drying of the aquatic habitats by making way into the sheltered spots or bottom mud. It is believed that snails are able to survive up to 30 days after being exposed to wind and

**TABLE 19.1**
**Schistosomal Species and Geographic Distribution of Schistosomiasis**

| Form of Schistosomiasis | Schistosomal Species | Geographic Distribution |
|---|---|---|
| Intestinal schistosomiasis | *Schistosoma mansoni* | Africa, Middle East, Brazil, Venezuela, Suriname, and Caribbean |
| | *Schistosoma guineensis* and related *S. intercalatum* | Rain forest regions of central Africa |
| | *Schistosoma mekongi* | Lao People's Democratic Republic and several districts of Cambodia |
| | *Schistosoma japonicum* | Philippines, China, and Indonesia |
| Urogenital schistosomiasis | *Schistosoma haematobium* | Africa and Middle East |

*Source:* World Health Organization (WHO). Accelerating work to overcome the global impact of neglected tropical diseases: A roadmap for implementation. 2012.

sun with temperatures around 45°C. However, only a few of the snails could survive when the moisture content of soil surface dropped to 1.8% (Cridland 1967).

The introduction of new diseases is mitigated through migration of different populations to urban areas. The transmission of disease is facilitated by developmental schemes (corresponding need for water and power, increased population) and environmental modifications. Among women, urogenital schistosomiasis is a potential risk factor for the development of HIV infection.

The most prevalent form of schistosomiasis in endemic regions is chronic schistosomiasis, which develops as a result of continuous exposure to infectious cercariae. If schistosomal infection occurs at 2 years of age, the burden of infection subsequently increases within the time period of 10 years due to colonization of new worms. However, the intensity and prevalence of schistosomal infection is highest among young adolescents, which may decrease as the individual reaches maturation.

Individuals who are in frequent contact with water during daily activities have a higher prevalence rate. A majority of the long-term residents in endemic regions get infected with schistosomes at some point in their life. They affect approximately 20% to 40% of adults and 60% to 80% of school-going children in regions with typical transmission patterns. Therefore, the schistosomal infection is being recognized increasingly among infants and adolescents. However, this issue was previously overlooked because of low standard diagnosis and decreased output of parasite eggs at a young age. This pathological condition arises in the endemic pediatric population as a result of host immunomodulation at a young age and development of chronic antischistosome inflammation.

The development of schistosomiasis is common among regions that possess good snail habitats and increased density of human population. For instance, the transmission of this disease is intense during irrigation schemes. Moreover, natural habitats including lakes, ponds, and streams having direct contact with human beings may also result in schistosomal infection. Although, the intermediate hosts are not normally present in fast flowing waters, they are found in regions where there is a direct contact of humans with suitable microhabitats for the host snails.

The increase in distribution of schistosomiasis depends on the snails as they carry infected loci to new areas, which results in the contamination of uninfected areas through discharge of schistosomes into the water. Particularly, schistosomiasis affects the productive life and health status of rural families and impedes the development of a nation. Fibrotic and inflammatory responses define the severity of infection in the host cells that affect the urinary bladder, liver, intestines, and lungs.

The damage to the urinary tract is indicated through the presence of blood in urine as a result of urogenital schistosomiasis by *Schistosoma haematobium*. Schistosomal infection causes damage to the bladder ureters and kidneys, which makes the process of urination quite painful. The advanced case of this infection commonly develops into bladder cancer. As compared to urogenital schistosomiasis, the intestinal schistosomiasis starts to develop due to *Schistosoma mansoni*. The formation of fibrotic lesions around schistosomes causes progressive enlargement of the liver, spleen, and intestine. It may even cause hypertension, and conditions can be fatal if there is excessive bleeding from the vessels.

In endemic areas, all ages are at higher risk of developing schistosomal infection when exposed to freshwater through bathing, swimming, and wading in contaminated freshwater. However, schistosomal infection in humans cannot be acquired through salty water of seas and oceans.

### 19.1.6 Incidence of Schistosomal Infection in United Republic of Tanzania

The fecundity of snails is decreased due to presence of various parasites including schistosomes. The infection rate of schistosomes was raised by 50% even after preventing transmission of schistosoma by applying 600 million eggs to a small pound over a period of 7 months (Joe et al. 1974; Mohammed et al. 2007). After Nigeria, the United Republic of Tanzania is the country that is facing the highest burden of schistosomiasis. Across the world, around 207 million cases of schistosomiasis have been reported; out of which 93% occurred in United Republic of Tanzania (Steinmann et al. 2006).

The species of *Schistosoma mansoni* causing intestinal schistosomiasis and *Schistosoma haematobium* causing urogenital schistosomiasis are highly prevalent among the regions of Tanzania. For a long time the *Schistosoma haematobium* and *Schistosoma mansoni* have been found extensively in the regions of Tanzania. However, other species causing infection include *Schistosoma japonicum*, *Schistosoma mekongi*, and *Schistosoma intercalatum*. The species of *Schistosoma mansoni* are extensively found in the northwest regions surrounding Lake Victoria, and southern, central, northern, and southeast regions of the country. The hinterland regions of Nigeria are found to be highly endemic for presence of *Schistosoma haematobium*. Table 19.2 summarizes different areas where there is increased prevalence of schistosomal infection.

**TABLE 19.2**
**Areas with Human Schistosomal Infection**

| Species | Affected Areas |
|---|---|
| *Schistosoma mansoni* | • Africa, including southern and sub-Saharan regions<br>• Nile River valley in Egypt and Sudan.<br>• Caribbean, including Saint Lucia, Dominican Republic, Martinique, and Guadeloupe<br>• South America, including Venezuela, Suriname, and Brazil |
| *Schistosoma haematobium* | • Africa, including southern and sub-Saharan regions<br>• Nile River valley in Egypt and Sudan<br>• Middle Eastern areas |
| *Schistosoma intercalatum* | • Regions of Central and West Africa |
| *Schistosoma mekongi* | • Laos and Cambodia |
| *Schistosoma japonicum* | • Regions of Indonesia<br>• Some parts of Southeast Asia and China |

*Source:* World Health Organization (WHO). 2017. Schistosomiasis. Fact Sheet. Available from http://www.who.int/mediacentre/factsheets/fs115/en/, accessed on 11th April 2017.

### 19.1.7 Distribution of Schistosomiasis

The rainy season between the months of June and July is responsible for increased transmission of infection; whereas, in the dry season, the transmission rate of the disease is decreased. The administrative regions are endemic at varying transmission levels as urogenital and intestinal schistosomiases are considered major public health problems in Tanzania. The southeastern and eastern coasts, northwest zones, and islands of Pemba and Unguja are endemic for the intermediate host snail species to transmit *Schistosoma haematobium*. Due to absence of host snails and thermal exclusion, the species of *Schistosoma mansoni* are not found on the coastal area, but they are extensively found on the shores of Lake Victoria (Figure 19.2).

The distribution of schistosomiasis reflects the close association between individuals with the availability of surface water. The frequency of water contact is higher among the communal areas where there is no piped water as compared to the urban areas. The total population is at risk of developing schistosomal infection because of wide distribution of schistosomiasis. Moreover, the increase in population size subsequently increases the prevalence of schistosomal infection. The geographical distribution significantly supports the previous mapping of the prevalence of schistosomal in the country.

The distribution of schistosomiasis disease is of great importance and can be easily evaluated through the inadequate sanitation, infected humans, and competent snail vectors. The travel and immigration patterns of travelers reflect the cases of schistosomiasis on the basis of geographical distribution. However, travelers and tourists should be aware of the contamination of surface water sources, which are a major source of infection. The absence of schistosomal infection in freshwater should not be

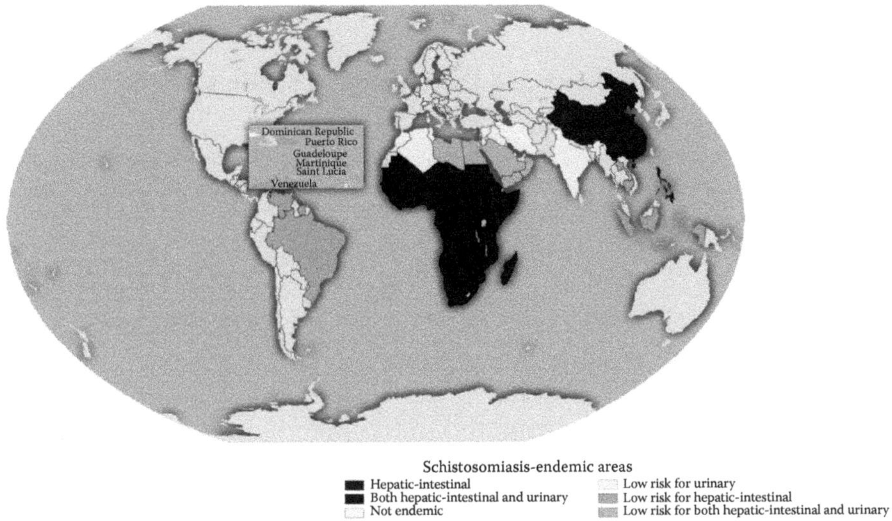


**FIGURE 19.2** Map indicating geographical distribution of schistosomiasis. (Montgomery, S., Infectious diseases related to travel, Centers for Disease Control and Prevention, 2015. https://wwwnc.cdc.gov/travel/yellowbook/2016/infectious-diseases-related-to-travel/schistosomiasis#4701.)

considered as a reliable statement. The infections of snails with human schistosomes need to be examined in laboratory for the identification of specific snail vectors.

### 19.1.8 Prevalence of Schistosomiasis Disease

The most critical factor responsible for the transmission of schistosomiasis is water contact. The cercaria develops into a worm after entering the human body. Therefore, the symptoms of schistosomal infection are severe due to increased augmented contact with the infection. As compared to adults, infants and adolescents are greatly exposed to schistosomal infection in the majority of the endemic foci. The intensity and prevalence of infection are common among children ages 10 to 14 years. The prevalence of *Schistosoma mansoni* is increased among adults, as compared with the prevalence of schistosome haematobium (Chandiwana et al. 1988; Steinmann et al. 2006). Therefore, young people are responsible for increased contamination of schistosome eggs in the environment, which facilitates transmission of the disease.

In nature, the number of infected snails is decreased significantly because few snails have the ability to provide enough cercariae to spread infection among a large population of people. It is due to maximum contact of people with infected water and an increased number of cercariae produced by the infected snails. However, the prevalence of schistosomiasis is affected by the age and sex of the individuals. The gender differences vary on the basis of cultural habits of the people. Moreover, the age-related pattern of infection characterizes the immunity development. The major factor influencing the prevalence and distribution of schistosomiasis is explained in Table 19.3.

**TABLE 19.3**
**Factors Influencing Prevalence and Distribution of Schistosomiasis**

| Factors | Affect/Influence |
|---|---|
| Climatic and environmental factors | Rainfall and temperature differ on the basis of different areas, and produce conditions that negatively affect the snail population and amount of human contact with water |
| Water availability and permanence | Significant for the production of large snail populations |
| Human–water contact; behavior | Defines the extent of contamination and incidence of schistosomal infection |

*Source:* Taylor, P., and O. Makura, Prevalence and distribution of schistosomiasis in Zimbabwe, *Annals Tropical Medicine and Parasitology* (1985) 79 (3):287–299.

The intensity of infection and prevalence of schistosomiasis is determined on the basis of different factors such as ecology, socioeconomic, biological factors, and human behavior. All these factors affect the association between animal host (snails), their life cycle stages, and human beings. The prevalence of *Schistosoma haematobium* and *Schistosoma mansoni* is the highest in the presence of high temperature and increased rainfall. The increased water conservation in the commercial farming sector (with more dams) and the historical allocation of better farming land (with more water) to farmers are responsible for the high prevalence of *Schistosoma haematobium* and *Schistosoma mansoni* in these areas. Moreover, the prevalence of schistosomal infection has increased due the historical allocation of efficient farming land for commercial farmers.

The prevalence of schistosomiasis depicts the different patterns of water contact and development of acquired immunity. Schistosomal infection in males is more severe as compared to females (Chandiwana et al. 1988). Control of schistosomal infection is most effective when there is a decrease in the infected snail population between the months of June and August. The decrease in temperature during June and August depicts decreased rate of recruitment and increased pre-patent period. A pre-patent period is defined as the time period between the infection in snails and transmission of that infection to others.

## 19.2 LIFE CYCLE OF SCHISTOSOME PARASITE AND ITS TRANSMISSION TO HUMAN BEINGS

### 19.2.1 Snails, the Intermediate Host

The development period of cercariae in the snail is influenced by different environmental factors. Temperature has been regarded as a significant factor, affecting the development of cercariae in the intermediate hosts. Cercariae are approximately 1 mm in length and possess a fork-shaped tail. Usually, the cercariae are released from an infected snail over a period of several weeks. The schistosome egg appears in urine and culminates in human water–contact activity. The cercariae have a life span of approximately 24 hours and are capable of covering a distance of 4 meters per hour. When an individual comes in contact with cercariae-infected water, the cercariae are attached to the skin and penetrate into the skin after water evaporation. In the human host, no multiplication occurs; for instance, a single cercaria releases a single fluke.

Generally, the snails require slow moving or stagnant water for the transmission of schistosomiasis. In the habitat of freshwater, they also possess an increased degree of tolerance for variation of temperature. Their optimal activity is between 22°C and 26°C, but they are able to withstand a high temperature for some time (Meyer-Willerer and Santos-Soto 2006). Usually, the snails are found near the shores of shallow water and are very rare in depths that exceed 1.5 to 2.0 meters. However, plants provide shelter and protection to the snails from mechanical effects of currents and high temperature due to intense sunlight. The approximate life span of an adult aquatic snail is 12 to 15 months. However, various factors, including temperature, level of water, and rainfall, determine the population and life cycle of the snail.

### 19.2.2 Ecology of Snails

Snails live in freshwater including small ponds, streams, rivers, and lakes. The distribution of snails is varied and their recognition needs proper examination of different sites. Snails usually live on mud and water plants that are rich in decayed organic matter. They might also live on stones/rocks that are covered with different types of debris and algae. The snails exist in freshwater, where aquatic plants are in abundance and the water is mildly polluted with traces of urine and feces.

### 19.2.3 Schistosomal Infection

Schistosomiasis is known as the "disease of poverty" because it produces disability, and decreases the educational and work performance of the individuals suffering from schistosomal infection. Schistosomal infection differs from other waterborne diseases in that it does not occur through ingestion of contaminated water. It rather occurs through direct contact of the skin with contaminated freshwater. Therefore, the purification of drinking water is not effective to prevent the transmission of schistosomal infection because certain activities and occupations may expose the individual toward infection. The procedure by which the schistosome parasite infects human beings is among nature's complex biological system (Evan Secor 2014).

### 19.2.4 The Life Cycle of Schistosome Parasite

A cercaria penetrates into the skin when skin is in direct contact with contaminated freshwater. The tail of a cercaria gets detached and the head penetrates into the body of the human being. After transformation, the head of the cercaria is transformed into the larval stage, which is known as schistosomula. The immature worm travels

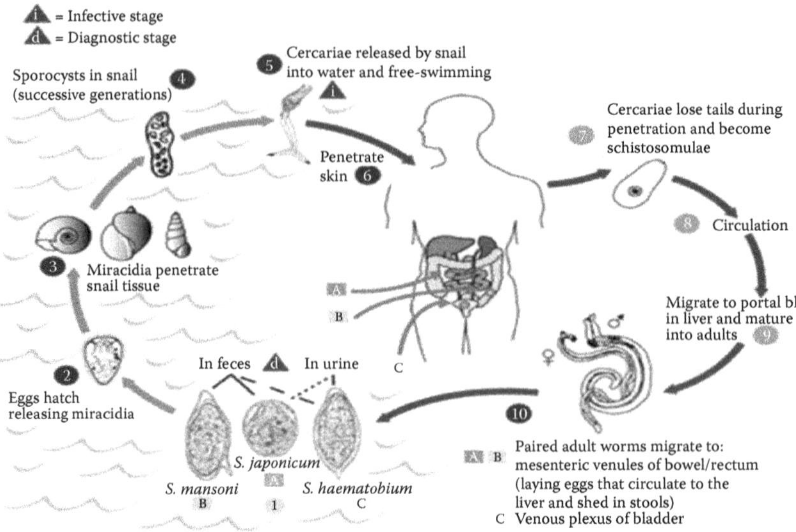


**FIGURE 19.3** Life cycle of schistosome egg. (From Centers for Disease Control and Prevention. DPDx: Laboratory identification of parasites of public health concern, accessed on 14th April 2017, www.dpd.cdc.gov/DPDx/.)

through different vital organs for about 4 to 6 weeks. These worms mature into adult worms while residing in the blood vessels close to the bladder and intestines. A mature adult worm is about 10 to 15 mm long. (See Figure 19.3.)

The female and male worms live together in copula, and female worms are capable of producing approximately 300 to 3000 eggs per day (Evan Secor 2014). The majority of the eggs produced by the female worm penetrate into the tissues and undergo granulomatous reactions, leading to the pathological condition of schistosomiasis.

Miracidium is defined as the infectious form of the intermediate snail host. If the miracidium does not enter the body of an aquatic snail, it would die within 6 to 24 hours (Evan Secor 2014). The existence of specific species of snail is necessary to transmit infection for each of the human schistosomes. The miracidium is transformed into mother sporocysts after infecting the snails. The mother sporocysts are responsible for the production of daughter sporocysts that are capable of increasing the population of parasites through asexual replication. After 4 to 6 weeks of infection with miracidia, the daughter sporocysts are responsible for releasing infectious cercariae from the snail (Evan Secor 2014). Sporocysts are observed as a thick-walled and rounded structure. Within the snail two generations of daughter sporocysts are produced, which considerably affects the snail by decreasing its life expectancy and fecundity. However, fork-tailed cercariae are developed in these sporocysts and then they leave the body of the snail.

The adult schistosome parasite lays eggs while residing in the blood vessels of final hosts. The eggs are capable of penetrating through the wall of the blood vessels into the bladder or intestine, depending upon the species of schistosome. The rupture of eggs in tissues causes infection, which is indicated through various symptoms including blood in feces, blood in urine, nausea, and headaches. Approximately 90% of the eggs are trapped in the body of the final host, causing damage to the tissues. The development of miracidia is arrested as they enter different species of snails.

### 19.2.5 Transmission of Schistosomiasis in Human Beings

The main species of schistosomes causing infection in human beings are

- *Schistosoma haematobium*
- *Schistosoma mansoni*
- *Schistosoma japonicum*

Each species of schistosome possesses a specific range of suitable snail hosts; therefore, the distribution of these species is defined by the habitat of host snails. The *Biomphalaria* and *Bulinus* species of freshwater snails are needed by *Schistosoma mansoni* and *Schistosoma haematobium.*

#### 19.2.5.1 Urogenital System

The *Schistosoma haematobium* has the ability to penetrate into the urinary bladder, seminal vesicles, and lower ureters. It affects the female genital system, vas deferens, and prostate gland less frequently. The adult worms migrate through portosystemic anastomosis at the third lumbar vertebra and live in the perivesicle venous plexus (Barsoum et al. 2013). The female schistosomes are carried by the oxygen gradient generated through the contents of the urinary system to the bladder. The granuloma may form tubercles, masses, and nodules that can cause ulceration. The inflammatory process significantly affects the muscle layers and submucosa, whereas the surrounding mucosa remains hyperemic (Barsoum et al. 2013). The clinical manifestation of the disease is diagnosed through dysuria and increased frequency of micturition. The inflammation is indicated through blood traces in the urine or dark-colored urine.

#### 19.2.5.2 Schistosomal Infection in Intestines

The *Schistosoma mansoni*, *Schistosoma mekongi*, *Schistosoma japonicum*, and *Schistosoma intercalatum* infect the colorectal region in human beings. The adult worms after maturation in hepatic sinusoids reside in the portal vein and its tributaries (particularly in inferior mesenteric vein). The schistosomal infection affects all parts of colon; however, the sigmoid, descending colon, rectum, and inferior mesenteric vein are the significant sites of infection and inflammation in 90% of cases (Barsoum et al. 2013). The formation of granuloma, edema, congestion, and polyp formation take place as a result of egg deposition. It may lead to severe pathological conditions, including dysentery, bloody diarrhea, and abdominal cramping.

#### 19.2.5.3 Asymptomatic Ectopic Formation of Granuloma

The asymptomatic ectopic formation of granuloma occurs in kidneys. In the upper urinary tract, visible lesions are observed in majority of the cases (62%)

(Barsoum et al. 2013). However, impaired kidney function is also preserved in the majority of cases. These alterations in the human body result in edema of ureterovesical junctions, near to the region of heaviest oviposition known as trigone. These reactions are mostly reversible as they resolve when the inflammation subsides.

#### 19.2.5.4 Schistosomal Infection in the Liver

Hepatic sinusoids are the main sites for the growth of schistosomula. As soon as the worms mature, they leave hepatic cells and migrate to spacious and well-oxygenated portal veins. The liver later gets involved in the schistosomal infection as the ova travels through the blood stream and gets settled in the portal tracts humans (Barsoum et al. 2013; Ross et al. 2007). The liver is mostly infected with *Schistosoma japonicum*, *Schistosoma mansoni*, *Schistosoma mekongi*, and *Schistosoma intercalatum.* However, it may get infected with *Schistosoma haematobium* if the ectopic ova escapes from the blood stream and gets settled in the portal venous system. As a result of schistosomal infection, the liver gets tender and enlarged. After a few weeks, the lymphoid hyperplasia causes enlargement of the spleen.

#### 19.2.5.5 Central Nervous System

The ectopic ova enters the central nervous system via anastomosis present between the lumbar veins. Lumbar veins are the branches of the internal vertebral venous plexus and inferior vena cava, where the eggs are deposited and formation of granuloma is proved adjacent to the spinal cord. These schistosomes may even travel cephalad and invade into the brain tissue due to straining or coughing. The migration of *Schistosome japonicum* is facilitated to the brain because of its smaller size, which allows it to penetrate into the brain tissue causing cerebral schistosomiasis. Schistosomal infection in the brain may affect the basal ganglia, cortex, internal capsule, and subcortical white matter in humans (Barsoum et al. 2013; Ross et al. 2007). Moreover, *Schistosoma haematobium* and *Schistosoma mansoni* significantly target the myelopathy of the lumbosacral region.

Neuroschistosomiasis is characterized as a severe condition corresponding to the schistosomal infection. It causes myelopathy, increased intracranial pressure, and radiculopathy. If the condition is not treated, it may result in lesions that leave irreversible glial scars. Encephalopathy, visual impairment, headache, ataxia, delirium, motor deficit, and seizures are a few of the complications associated with neuroschistosomiasis in humans (Barsoum et al. 2013; Ross et al. 2007).

#### 19.2.5.6 Genital Schistosomiasis

The *Schistosome haematobium* is responsible for the development of genital schistosomiasis, which most commonly occurs in endemic areas and tourists. The accumulation of schistosomes and lesions in fallopian tubes and ovaries lead to infertility. The genital organs in males, including the spermatic cord, testicles, prostate gland, and epididymis, may suffer schistosomal infection. Ulcerative and hypertrophic lesions in the vulva, cervix, and vagina are major symptoms of schistosomal infection in females.

### 19.2.6 Development of Fibrotic Lesions in Human Body

The formation of granulomas leads to the development of fibrosis and calcification. The development of fibrotic lesions interferes with the muscular layer of the bladder causing motor disabilities in bladder functioning and abnormalities in the internal vesicle. Moreover, the schistosomes form nodules and polyps leading to mucosal scarring, which may even lead to obstruction in the left colon or rectum.

Chronic hepatic schistosomiasis is much more serious condition as compared to chronic schistosomal colitis. Chronic hepatic schistosomiasis is characterized with fibrotic and thickened portal veins, which causes shrinkage of the liver associated with presinusoidal hypertension. The *Schistosoma haematobium* travels from the perivesical toward the right cardiac chamber through the inferior vena cava. However, *Schistosoma japonicum* and *Schistosoma mansoni* are capable of penetrating across the portosystemic shunts. The accumulation of these schistosomes in the shunts results in formation of granuloma, which provokes the immune-mediated endothelial proliferation in the capillaries.

## 19.3 TRANSMISSION OF SCHISTOSOMIASIS BY SNAILS AND NATURAL HABITAT OF SNAILS

### 19.3.1 Classification of Snails

Snails are in the class Gastropoda; snails and slugs make up to 80% of all the mollusks included in this class (Holmes 1999). The class of gastropods is found all over the world including Antarctic Ocean, Arctic Ocean, and equatorial regions. Snails are capable of surviving on land as well as in water (Figure 19.4). Therefore, snails have been known as the most widely distributed invertebrates across the world. Snails are able to survive in regions like the sub-Antarctic, where the temperature is extremely low.

**FIGURE 19.4** A snail. (From *Wikipedia*. Snail. https://en.wikipedia.org/wiki/Snail.)

### 19.3.2 Snail as a Gastropod

The class of gastropods belongs to the phylum Mollusca/Mollusks, which classifies the invertebrates on the basis of a soft unsegmented body covered with a shell or exoskeleton.

Snails possess eyes and olfactory organs, but have no ears; therefore, they are not able to hear. The olfactory organs are the most important sensory organs for snails because they use their sense of smell to search for food. Moreover, snails are hermaphrodites, because a single snail possesses both female and male reproductive organs. The approximate life span of snails is 3 to 7 years, but some of the larger species may live over 10 to 15 years in the wild (Holmes 1999). A distinct feature of aquatic snails is that a few of their species breathe air from the atmosphere that goes in the lungs for supply of oxygen.

### 19.3.3 Habitat of Snails

Snails have diverse habitats, as they have adapted to life on land, salt water, and freshwater. The adaptations allow the snails to respire, move, digest, and protect themselves from injuries. Also, snails can be carnivores, herbivores, or omnivores.

- Land snails—Land snails are capable of living in a wide range of temperatures (steam jungles to cool mountains). They eat vegetables, fruits, and leaves.
- Freshwater snails—Freshwater snail species are populated in ponds, rivers, lakes, and streams. Decaying matter and algae is the diet of these snails.
- Marine snails—The oceans possess the largest, most colorful and intricate-shaped marine snail species. Saltwater snails reside in warm tropical reefs and black ocean depths. The Australian trumpet, known as *Syrinx aruanus*, is the world's largest marine snail with a shell ranging from 4 to 27 inches long.

#### 19.3.3.1 Snail Adaptation to Habitats

Snails use their shells as protection against predators. They take coverage inside the shell by using the operculum, which is a doorlike part guarding the opening and closing of the shell. Snails possess two eyes placed at the top of each tentacle, which helps in searching of food, shelter, and spawning grounds. The rasp tongue of the snail is known as the radula, which is lined with rough and tiny teeth. It helps the snail grasp food particles, particularly vegetation.

### 19.3.4 Transmission of Schistosomiasis Disease by Snails

The intermediate snail hosts of schistosomes are amongst the essential components of freshwater aquatic environments. The diverse species of schistosomes are responsible for infecting a wide variety of mammals, including human beings. The species of intermediate snail hosts reside in a well-defined habitat; for instance, they may be found in still water environments and in running waters. It is necessary to discuss the snail biology before considering the transmission of schistosomiasis disease by snails.

#### 19.3.4.1 Snail Biology

The intermediate snail hosts carrying *Schistosoma mekongi*, *Schistosoma malayensis*, and *Schistosoma japonicum* belong to the family Pomatiopsidae and subclass Prosobranchiata. These snails have a maximum height of 10 mm and possess small subconical and conical shells with approximately 4 to 8 dextral whorls. The surface of the shell has fine or strong axial growth lines. The calcareous operculum (shell) of the snail contains a paucispiral or concentric ring. The gills help them to acquire dissolved oxygen directly from the water. They need to capulate for producing eggs because they are dioecious. However, snails belonging to the genus *Neotrichula* are prone to get infected with *Schistosoma mekongi*, the genus *Robertsiella* get infected by *Schistosoma malayensis*, and the genus *Oncomelania* are infected by *Schistosoma japonicum* (Gryseels et al. 2006).

#### 19.3.4.2 Snail Hosts for *Schistosoma haematobium* and *Schistosoma mansoni*

*Schistosoma haematobium* and *Schistosoma mansoni* are primarily infected with *Biomphalaria* and *Bulinus*. These snail hosts are pulmonary mollusks, which obtain oxygen directly from the atmosphere. However, a small amount of oxygen is obtained through diffusion across the epithelium of exposed tissues. Although, the schistosomes are hermaphrodites, their male reproductive system is not fully developed. They are capable of producing thousands of eggs in their life span through cross-fertilization and self-fertilization. The maximum production of eggs occurs in temperatures up to 30°C to 35°C; however, the production of eggs seizes in temperatures below 18°C. The mortality of eggs and snails increase above 35°C (Gryseels et al. 2006; Ross et al. 2007).

The eggs are laid in clusters with yellow, transparent, and gelatinous masses. The two genera of snails produce two different types of eggs:

- *Bulinus* eggs—Elongated and surround the curved surfaces of the stems of plants
- *Biomphalaria* eggs—Oval/circular and remain attached to the flat surfaces

The size and mass of egg corresponds to the size of the snail itself. The hatching and development of a snail to maturity significantly depends on the temperature. In appropriate environmental conditions, the egg is hatched in approximately 5 to 10 days and measures about 0.5 to 1.0 mm in length. However, it gets mature in about 4 to 12 weeks (Gryseels et al. 2006). A mature snail is about 5 mm in height and diameter.

#### 19.3.4.3 Transmission of Schistosomal Infection in Humans, Reservoir Hosts

The schistosomes including *Schistosoma mekongi*, *Schistosoma mansoni*, and *Schistosoma japonicum* affect humans and other species living in the wild and are exposed to farming practices such as rice paddies. The association between snail intermediate hosts of the parasite and predators of the snails are responsible for the transmission of schistosomiasis. Humans are targeted by schistosomal infection through direct skin contact with schistosome larvae (cercariae). Infected aquatic snails are responsible for releasing the larvae in freshwater. The cycle continues as the infected human being excretes urine and feces contaminated with schistosoma infection into the habitat of freshwater snails.

Unlike adult worms, schistosome eggs are responsible for inducing morbidity in humans as a result of schistosomal infection. The majority of the eggs are not excreted through urine and feces but get retained in liver, bladder, and intestines of human. The lymphocytes, macrophages, and eosinophils are responsible for inducing the granulomatous host immune response. Tissue necrosis is prevented by the proteolytic enzyme present in the granulomas. The clinical manifestations of schistosomiasis disease are indicated when the process of granuloma formation results in chronic inflammation (Colley et al. 2014). Acute schistosomiasis occurs among individuals who are exposed to the schistosome parasite for the first time. This condition occurs after a few weeks to months of infection. The process of development of acute schistosomal infection is illustrated in Figure 19.5.

The condition of acute schistosomal infection is also known as Katayama syndrome, which is clinically presented through fever, malaise, eosinophilia, myalgia, abdominal pain, headache, and fatigue (Ross et al. 2007). The progression of the infection to chronic intestinal schistosomal infection is characterized by the downregulated granulomatous response of schistosome eggs. It leads to various gastrointestinal complications like mucosal hyperplasia and pseudopolyposis. Intestinal schistosomiasis is poorly immune regulated to the antigens of parasitic eggs, which results in development of fibrosis and hepatosplenic disease associated with periportal fibrosis.

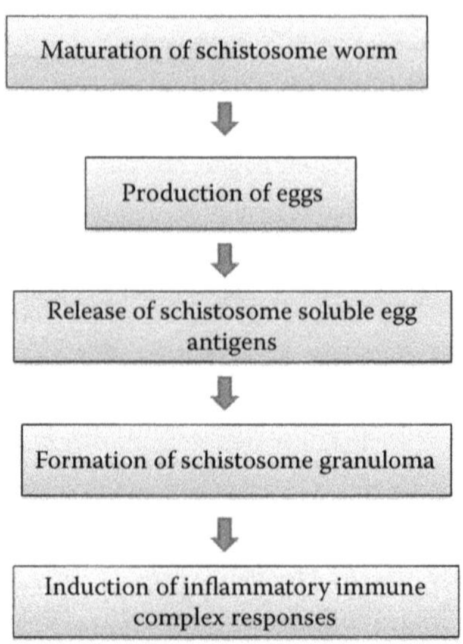


**FIGURE 19.5** Process of acute schistosomal infection.

Patients suffering from advanced hepatic fibrosis disease experience pulmonary hypertension due to presence of granuloma in the pulmonary arteries. The time period required for development of advance fibrosis from initial infection is approximately 5 to 15 years (Ross et al. 2007). *Schistosoma haematobium* significantly affects the reproductive health of women, as it is responsible for causing schistosomiasis in the female genital system. The schistosome eggs present in the vesicle plexus are transferred to the genital tract causing inflammatory lesions in the ovaries, fallopian tube, and vagina. The schistosomal infection in females is clinically manifested through pain, infertility, stress incontinence, and high risks of abortion. However, in males, schistosomal infection is manifested with orchitis, oligospermia, prostatitis, haematospermia, and dyspareunia (Leutscher et al. 2009). The species of schistosomes that rarely affect the genital tract include *Schistosoma japonicum* and *Schistosoma mansoni*.

### 19.3.5 Transmission of Schistosomal Infection and Its Environmental Influence

Different relationships have been identified between snails and different environments associated with densities of snail populations and the impact of seasonal changes. There are different environmental factors that affect the viability and density of snail populations:

- The density of a snail population is controlled by water levels, which may vary on the basis of different seasons. The optimal habitat of snail is defined as the narrow zone of elevation, which is specific above the mean low level of water. However, adult snails may drown as a result of flooding. It shows that any change in the water environment negatively influences snail populations.
- The speed of riparian environments determines the densities of snail populations. The large populations are relocated down the river when there are high water tides. The habitat of snail is devoid by flood-driven currents. The control of a snail population is difficult in lentic ecosystems because snails easily repopulate in the flooded areas.
- Temperature significantly affects the reproduction of snails. The reproduction of snails is inhibited in the spring season of a subtropic environment below 10°C. Temperatures greater than 30°C result in destruction of the schistosomes.
- The density of a snail population can be easily determined through the elevation, specifically in the regions with low water levels such as rivers and lakes. The optimal habitat for survival of snails needs to be flat, midlevel land, with thick grass. This type of environment is found in the ecotonal zones of rivers (freshwater).

## 19.4 BIOLOGICAL NAMES OF THE SNAILS AND THEIR LIFE CYCLE

### 19.4.1 OVERVIEW OF GASTROPODS

Gastropoda is considered as the largest family of mollusks. The life cycle of gastropods differ depending on their habitats (land, freshwater, and ocean). The family of gastropods includes snails, slugs, sea hares, and limpets. These animals are characterized by the presence of an asymmetrical body and a single shell. The major subclasses of gastropods are

- Prosobranchia—Ocean snails
- Pulmonata—Freshwater and land snails and slugs
- Opisthobranchia—Sea hares, sea butterflies, and sea slugs

It is impossible to describe the life cycle of each individual species belonging to the Gastropoda class. The reproductive cycle and biology of these species have become complex since the evolution of different species. There are approximately 1000 different species of snails present across the world in different regions (Cowie 1984). Snails are found in a variety of habitats, but the majority of snail populations is found in regions where there is a plenty of vegetation. The various characteristics of snails are listed in Table 19.4.

**TABLE 19.4**
**Snail Facts**

| | |
|---|---|
| Kingdom | Animalia |
| Phylum | Mollusca |
| Class | Gastropoda |
| Order | Achatinoidea |
| Common name | Snail |
| Scientific/biological name | Achatinoidea |
| Occurrence | Worldwide |
| Diet | Herbivore |
| Size | 0.5–80 cm (0.2–32 inches) |
| Weight | 0.01–18 kg (0.02–40 lb) |
| Number of species | 1000 |
| Average life span | 1–20 years |
| Type of skin | Calcareous shell |
| Habitat | Well vegetated areas |
| Main prey | Stems, fruits, and leaves |
| Predators | Birds, rodents, and frogs |
| Distinctive features | Thin long eye stems and armored shell |

*Source:* *Wikipedia.* Snail. https://en.wikipedia.org/wiki/Snail.

### 19.4.2 Life Cycle of the Snail

The life cycle of snails starts from hatching and progresses to the age of reproduction, which is characterized with mating of snails and laying eggs. The cycle of a snail involves the succession of one generation to the next generation through the process of reproduction. The initiation of a snail's life cycle takes place in winter, when the snail is hatched from egg an inch below the surface of the ground. The snail acquires vegetation during the summer season. The reproductive organs of snails mature in spring, whereas, the process of mating starts with arrival of autumn. Thereby, the life cycle of snail is characterized on the basis of the following steps:

- Birth of snail and its development
- Attaining sexual maturity
- Search for mating partner
- The process of mating
- Gestation period
- Release of the egg
- Hatching of the egg

#### 19.4.2.1 Process of Courting and Mating

Snails require about 2 years of maturing before the process of reproduction. Land snails engage in various processes to attract males. Before mating, snails spend much time courting. While courting, they tend to twist themselves around each other and surround themselves with frothy slime. The process of courting is associated with a lot of touching. The breeding capacity of snails range from 4 to 12 hours (Cowie 1984).

#### 19.4.2.2 Process of Breeding

The breeding of snails is associated with sexual reproduction. During the process of copulation, both the mates have the ability to transfer sperms. During intercourse, the sperm exits the body of the snail through spermatophores. This process is carried out by pressurizing the front part of the muscular foot and shooting the sperm-filled darts. The released sperm enters the body of other snail to carry on the process of fertilization with the ova.

#### 19.4.2.3 Process of Fertilization

The association of sperm and ova results in the development of an egg, which is known as the process of fertilization. The egg is laid in the surroundings after fertilization depending on the snail species. Snails have the ability to store sperm for a long time period. Snails produce around 100 to 400 eggs at a time (Cowie 1984).

#### 19.4.2.4 Process of Hatching

Snails often lay eggs under low-lying plants and shrubs that keep the soil moist. The eggs hatch in about 4 weeks. The eggs are produced in a large quantity, but only a few of them reach the stage of maturity. The eggs are highly resistant to heat and cold temperatures as a protective layer; however, a majority of eggs are either washed away by water/rain or eaten by predators.

#### 19.4.2.5 Process of Maturing

The development of shell starts in the embryo inside the egg. The eggs are hatched after about 2 weeks or 1 month under favorable conditions. Snails possess a soft shell as they emerge from the egg. Therefore, as soon as the snails are hatched, they need to get calcium. The immature snail possesses a soft shell and a transparent body. The color of its body changes as it grows and acquires strength. Initially, the transparent body of a snail changes to a bluish color, which later changes to brown. Young snails are more vulnerable to numerous predators because of their incomplete development. Moreover, the snail may develop a thickened lip around its shell aperture when it matures into full adult size. However, adverse environmental conditions delay the onset of sexual maturity in a few snail species.

#### 19.4.2.6 Hibernation and Estivation in Snails

Hibernation and estivation are defined as the period of lethargy, which is characterized by a decreased metabolic rate, decreased heart rate, and decreased rate of respiration than normal. The period of lethargy is known as hibernation. Snails secrete a layer of mucus known as epiphragm to maintain moisture and remain protected from predators while the snails are retracted into the shell. This mucus layer is responsible for opening and closing of the shell (Adamson et al. 2015). The body temperature of these creatures rises in extreme high temperatures; therefore, they need a moist and shady place to protect their body against the dysregulation of temperature.

## 19.5 SPREAD OF SCHISTOSOMIASIS

Schistosomiasis is a water-related disease that is capable of producing acute and chronic infections. It affects more than 250 million individuals across the world (WHO Expert Committee 2002). The spread of schistosomiasis is prevalent in subtropical and tropical regions, particularly in poor communities. It has been estimated that the spread of schistosomiasis is most common in Africa (WHO Expert Committee 2002).

### 19.5.1 Water as Means of Spreading Schistosomiasis

Water has been regarded as the major source of schistosomiasis transfer and spread. Schistosomal infection is transferred from human to environment when infected people contaminate freshwaters with excretions (urine and feces) containing the parasitic eggs. However, schistosomal infection is transferred from the environment to human as a result of exposure to infested water while carrying out daily activities. The spread of schistosomiasis disease is a complicated cycle in humans, which has been illustrated in Figure 19.6.

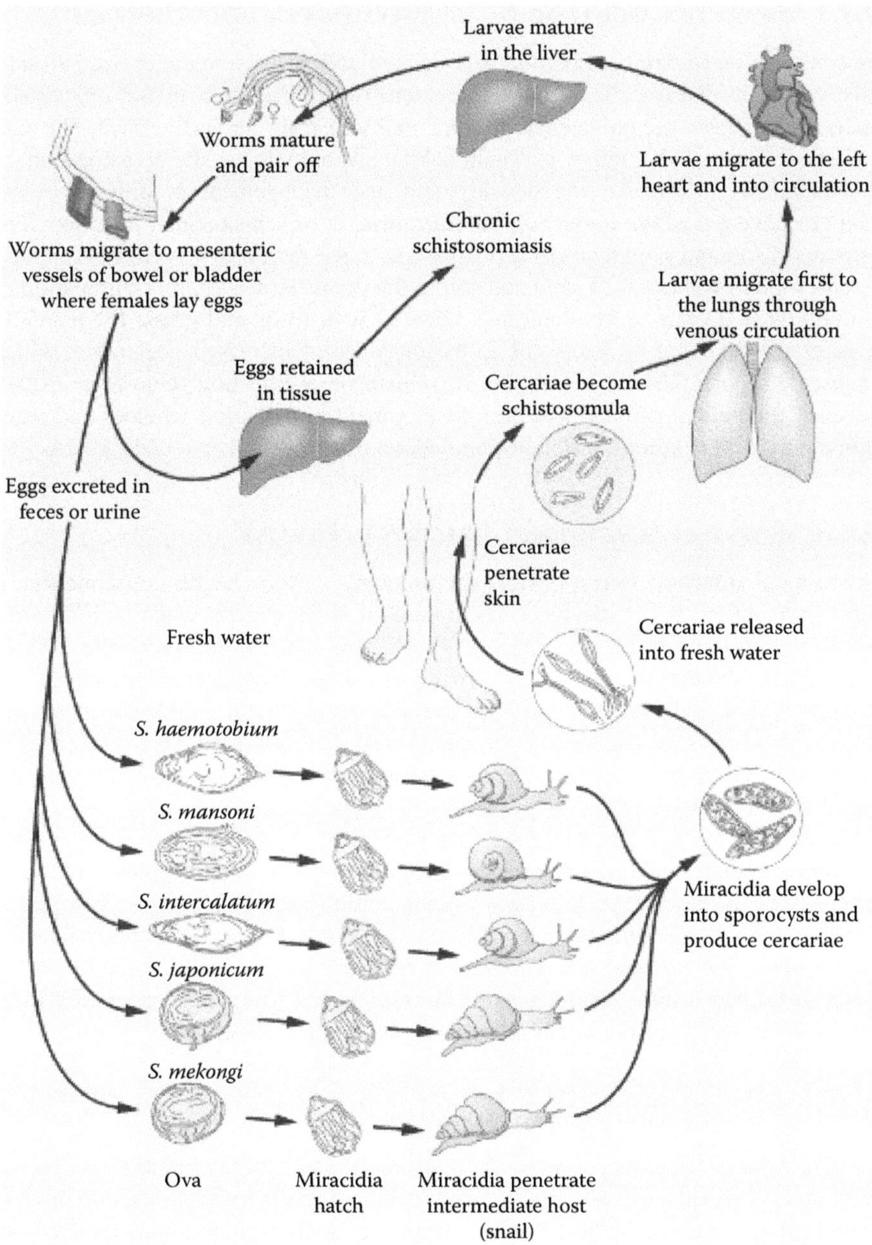


**FIGURE 19.6** The spread of schistosomal disease in humans. (From Gray, D. J., A. G. Ross, Y. S. Li, and D. P. McManus, Diagnosis and management of schistosomiasis, *British Medical Journal*, 2011, 342:d2651.)

### 19.5.2 Effect of Large Dams on the Prevalence of Schistosomiasis

The construction of dams is positively associated with a surge in the transmission of schistosomiasis disease. Dams built for extension of irrigation in the dry season noticeably increases the prevalence of urogenital schistosomiasis (Li 2007). The dam is likely to extend the range of snail habitats and increase the transmission of schistosomiasis. Moreover, the development and management of different water resources have a positive impact on the transmission of schistosomal infection. The upstream of soil and sand tends to deposit as silt in the reservoir, which subsequently decreases the downstream of sand and soil in the dams. However, if accumulation of silt deposits is decreased, the land gets covered with reeds and grass. It eventually increases the breeding of snails due to the presence of increased vegetation, which increases the potential of transmission of schistosomal infection (Guo et al. 2006). However, the pattern of infection after development of irrigation schemes and dams significantly differs among different populations due to different social factors.

### 19.5.3 Spread of Schistosomal Infection in Humans

Schistosomal infection is responsible for inducing various health consequences if adequate treatment is not received. The spread of disease starts when an individual enters a freshwater body containing snails that grow schistosoma sporocytes, which are capable of developing into freely moving cercariae. The cercariae get attached to the human skin and penetrate through the skin to reach the portal blood system through blood vessels and lung blood capillaries. The cercariae develop into adult female and male parasitic worms from a schistosomula as they travel through the portal blood vessels.

The immature eggs excreted through urine and feces mature in the freshwater body and complete their cycle by entering the body of a host snail and causing infection. The species of *Schistosoma japonicum* usually complicate the spread of schistosomal infection by infecting domestic and wild animals, which serve as a host body carrying schistosomes. *Schistosoma haematobium* is responsible for causing infection in the tissues of human urinary bladder, whereas other species of schistosomes are associated with infection of the bowel tissues (Colley et al. 2014).

### 19.5.4 Pathogenesis of Schistosomal Infection

The pathogenesis of schistosomiasis is unusual among other helminthic infections. Much of the pathogenesis is due to the eggs rather than larvae or adults, and most of the pathology is caused by host immune responses such as granulomatous reactions and delayed-type hypersensitivity.

The spread of schistosomal disease is carried out by the formation of polyps (granulomatous reaction) in the submucosal layers due to deposition of the schistosomal eggs. The fluke eggs loosen the connective tissues and dissociate the superficial bounding of these tissues with firm tissues. This results in granuloma formation and associated fibrotic changes. A barrier is formed due to fibrosis in the submucosa and muscularis mucosa, which allows the transit of ova from the mesenteric veins into the lumen of gut ((Elbaz and Esmat 2013; Olveda et al. 2014).

### 19.5.5 Spread of Schistosomal Infection in Intestinal Tissues

The major site for schistosomal infection in intestines is the distal rectosigmoid colon. The ova usually deposits in the minute vessels of submucosa. The ova release miracidia into the lumen of gut as a result of cytolytic enzymes and mechanical pressure to be removed from the body in the form of urine and feces. Fibrosis in the portal tract occurs if the ova is retained in the host body (human) and migrate to the liver via portal vein. Moreover, the retaining of ova in intestinal tissues causes disintegration of miracidia with fibrosis in the tissues (Olveda et al. 2014).

### 19.5.6 Spread of Schistosomal Infection in Liver Tissues

The hyperplasia and formation of granulomas in reticuloendothelial tissues are extensively observed among the organs including spleen and liver. The embryo, enclosed in the ova, presents in the intrahepatic portal radicles that are capable of surviving for around 2 to 3 weeks. The embryo is responsible for releasing a soluble antigen, which initiates the granulomatous reaction. It may eventually cause blockage of the intrahepatic vessels, which may lead to a condition known as thrombosis.

### 19.5.7 Spread of Schistosomal Infection in Cardiac Tissues

The ova are directed toward the right side of the heart through the collateral venous channels. The ova are further directed to the pulmonary arterial tree; therefore, it is believed that cardiovascular schistosomiasis develops among patients suffering hepatosplenic schistosomiasis. The granuloma formation starts to develop around the ova that spreads across the lumen of pulmonary vessels. Initiation of the inflammatory process destroys the wall of arteries. Moreover, the veins are dilated as a result of newly formed arteries in the para-arterial and intra-arterial granulomas, which may even lead to pulmonary obstructive arteriolitis.

## 19.6 HISTORICAL EVIDENCE OF MAN-MADE WATER SYSTEMS IN DEVELOPING SCHISTOSOMIASIS

The ecological disturbances and environmental changes have potentially exerted a significant influence toward the emergence and proliferation of various parasitic diseases. Various activities of humans including deforestation, road construction, and water control projects are responsible for increasing the prevalence of parasitic disease (such as schistosomiasis disease). The ecological balance is altered due to different natural phenomena, environmental changes, and human interventions. These changes may facilitate the breeding of parasites, leading to development and transmission of the disease.

### 19.6.1 Deforestation Contributes to Schistosomiasis Disease

Deforestation has been regarded among the disruptive alterations that influence the populations of parasitic vectors. The clearance of forests may result in erosion that can wash away the thin layer of nutrients. However, the movement of human populations and expansion of different human settlements needing an adequate supply of food, leads to change in the growing of vegetation. It causes change in different ecological conditions, which proliferates the existing parasites (Patz et al. 2000).

### 19.6.2 Crop Farming and Ranching Support Parasitic Growth

A supportive habitat for parasites and host vectors is created by replacing the forests with farming and ranching. Moreover, the introduction of different animal species contributes to an increase in the transmittance of schistosomal infection.

### 19.6.3 Water Bodies in Disrupted Areas Cause Schistosomal Infection

The type of water bodies created in the disrupted areas is determined by the type of soil and its elevation. The floor of forests tends to be littered and shaded with a thickened layer of organic matter, which is capable of absorbing water. It leads to the accumulation of water, resulting in formation of puddles. Such puddles are accumulated with clear water with pH less than 5.5 (Patz et al. 2000). However, the growth and fecundity of freshwater snails is increased due to reduced salinity of water and conversion to alkaline condition from acidic environment. It leads to the enhancement of transmission related with schistosome parasite (flukes) (Southgate 1997).

### 19.6.4 Movement of Humans Increases Prevalence of Schistosome Parasites

Human settlements change with the replacement of deforested areas with agriculture and irrigation development. The indigenous population of forest dwellers is immune to forest-dwelling parasites. However, the introduction of a new human population is particularly vulnerable to parasitic diseases, which are endemic to the particular area. The humans are exposed to new factors in the new environment by fragmentation of landscape and urbanization. The interactions between parasitic vectors are increased as a result of settlement of a human population from a sparsely settled area to a densely populated area. It occurs due to the presence of dense human reservoirs, facilitating exchange and transmission of parasitic diseases.

### 19.6.5 Vector Competence and Spread of Schistosomal Infection

The level of vector competence among species varies extensively. The change in ecology of any area also alters the presence of living species in that particular area. Each population of species is different in behavior on the basis of different geographical areas and frequency of chromosomal inversion as a response to man-made water systems.

### 19.6.6 Schistosomiasis and Different Water Control Projects

The spread of parasitic diseases, including schistosomiasis, is positively associated with reservoirs, dams, and irrigation canals. Different species of snails vary in their larval stage and adult stage based on their habitat requirements. The majority of snail species need rich vegetation to fulfill their nutrient demand. The diversion of riverbed flow out results in formation of riverine pools, which facilitate breeding of snails. Such a wide range of environmental conditions ensures the transmittance of parasitic diseases, which flourish particularly in the tropical areas. Prevalence of schistosomiasis is increased with water resources development projects (Brinkmann et al. 1988).

### 19.6.7 Water Resources Development and Schistosomiasis

It has been estimated that approximately 779 million people are at risk of developing schistosomiasis, out of which around 106 million (13.6%) of the affected population reside close to irrigation schemes and large dam reservoirs (Steinmann et al. 2006). The irrigation schemes increase the output of crops, reduce the dependency on fossil fuels, and facilitate the generation of export earnings. Moreover, water scarcity is maintained by increasing storage capacity through reservoirs. However, these developments in subtropical and tropical climatic zones results in transmission of schistosomiasis in previously nonendemic areas (Keiser and Utzinger 2008). Schistosomiasis is regarded as a potential indicator to monitor the ecological transformations because it is distributed widely and the rate of spread of infection promptly changes.

The habitats of intermediate host snails are expanded as a result of development of dams and implementation of irrigation systems (Figure 19.7). The transmission cycle of schistosomiasis can be stopped by improving the water supply and sanitation. The population growth near irrigated areas and near dam reservoirs certainly

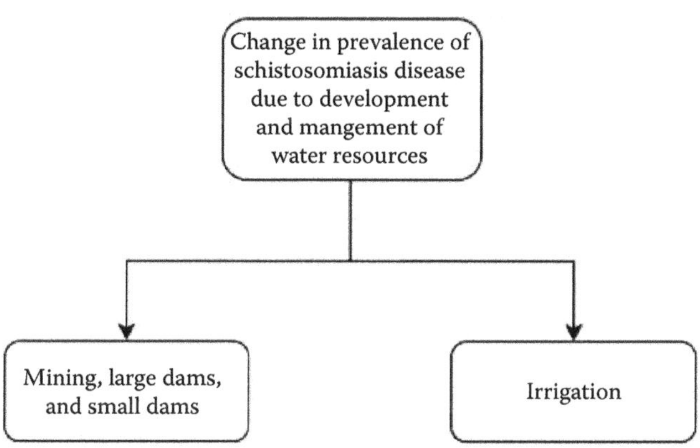


**FIGURE 19.7** Interface of schistosomiasis and development of water resources.

underestimates the results. Moreover, there is an increased risk of acquiring infection in rural settings as compared to the national average. Particularly, the level of irrigation and dam construction is not proportional to the fraction of the population living in the respective endemic areas.

## 19.7 PROBLEMS IN CONTROLLING SNAIL POPULATION

Parasite transmission is blocked from the effective interruption of the *Schistosoma* life cycle. Therefore, prevention of individual schistosomiasis is appropriately determined over the long-term process. Environmental contamination is reduced to restrict the transmission of existing mass treatment campaigns for controlling the population of aquatic snails and parasite eggs. By contrast, the focused and highly effectual process of transmission has provided the core essence of reducing environmental contamination in those areas where the observation of mass drug treatment is not much highly apparent. Thereby, the consideration and control of the aquatic snail population is adhered on the alternative means for interruption and to entail the approaches that can be used to mitigate and reduce the intermediate snail hosts from local habitats. Alternative approaches are consequent in preventing water contamination by using sanitation.

### 19.7.1 Regional Schistosomiasis Control Programs

An important component for many regional schistosomiasis control programs is to monitor snail control and prevention of *Schistosoma* transmission. Control of aquatic snails is emphasized with the transmission-interruption method that reviews the perceived weaknesses and strengths of the programs. The intention of these programs is to control the *Biomphalaria* and *Bulinus* host snail species. The most common approach of controlling the snail population is with the application of chemical molluscicides, which is eventually covered by blanket and focal treatments (Evan Secor 2014).

Niclosamide became the most frequently used compound for aquatic snail control, even though sodium pentachlorophenate and copper sulfate were available. In this regard, managers have to become proficient and familiar with the local snail genera to obtain effective snail control programs. The desired effect of the control program is accomplished using the appropriate application of molluscicide. The centralized administration for control of vector-borne diseases can be considered as one of the classical approaches within the project phase. The control of schistosomiasis is monitored and regulated by a supervisor and five laborers throughout the control phase of the project.

### 19.7.2 Control of Snail Population

The optimal population of snails is targeted by the developed operating procedures, which eventually locate the habitats throughout critical times. The rapid control of snails is possible with the use of molluscicides, which are consequently beneficial in the transmission of schistosomiasis.

The active cooperation of the population through mass chemotherapy and health education are required for the prevention of schistosomiasis transmission. In addition, the use of molluscicides is essential in controlling the vectors of trematode infections, as they are important aspects for the economy of domestic animals. Therefore, these trematode infections, particularly fascioliasis, contribute massively to the control and protection of animal health (Evan Secor 2014). On the contrary, it has been stated, that because of the problem of reinfestation, the application of molluscicides, except in completely isolated places, has to be continually repeated and is, therefore, costly.

The largest expense in this scenario is ultimately the labor costs, which generates a warning to the long-term use of molluscicides. It is observed that the budget for local mollusciciding accounted for three-fourths in the labor costs in southern Rhodesia and Tanzania (King and Bertsch 2015).

The influence of consistent multilayer niclosamide molluscicide is apparently essential to reducing and controlling the population of aquatic snails. However, the prevalence of mollusciciding is apparently higher in the high- and low-risk untreated areas. Numerous programs have been found noteworthy in many countries, where they provide effective results in controlling the population of snails and reducing the impact of schistosomiasis.

The potential effect on the reduction of mollusciciding is observed from scheduled mass drug administration and informed timing of delivery, associated to peak transmission season. Ultimately, the impact of these approaches is further observed in maximizing their effects on transmission. Chemicals, predators, and plant-derived saponin molluscicide are experimentally used to control the population of snails and to prevent schistosomiasis transmission.

In East Africa, a sustainable local product is experimentally used to control the population of snails and to prevent schistosomiasis transmission. Retrospectively, studies and their outcomes have indicated that the reduction of schistosoma transmission can be effectively reduced by informed snail control. It is likely that informed snail control can be favorably compared to those of modern-day mass drug delivery programs, if they are appropriately implemented to prevent schistosomiasis transmission (King and Bertsch 2015).

The population of snails can be controlled if attractive implementation of alternatives to chemicals are developed by extracting indigenous plants. Schistosomiasis transmission can be effectively controlled with implementation of the soapberry plant; however, it has not received great attention for control of aquatic snails. The success and failure of this approach relies on the participation rates, as there is a fundamental requirement of community involvement in the health interventions.

Alteration of snail population is further controlled from numerous other methods. These methods consequently include lining canals with cement, draining water bodies, and removing vegetation. The successful implementation of vegetation removal has been seen in recent years. On the contrary, there is a negative or appalling effect shown to augmenting risk of infection among employees, who may not have appropriate tools and clothing.

In China, financial incentives have been employed to alter the population of snails as well as to control the impact of schistosomiasis infection. However, adverse effects of these employments have been observed on workers trying to move snails. Thereby, the effectiveness of this system is further indistinguishable. The extent of infected snails is frequently low, which indicates that the maintenance of the snail life cycle is sustained from few positive snails in areas with high schistosomiasis transmission. Therefore, the predictable extent of removing and altering intermediate snail hosts is undetermined.

The expensive and impractical applications of environmental modifications for resource-constrained areas are not successful in controlling the snail population (King and Bertsch 2015). Thus, there is a need of appropriate programs and chemicals that can be highly effectual in controlling the population of snails for resource-constrained areas.

## 19.8 TREATMENT MEASURES FOR SCHISTOSOMIASIS

Patients suffering schistosomal infection need to have antischistosomal drugs and corticosteroids. The corticosteroid drugs are responsible for reducing the inflammation and helpful in the suppression of changes that result from the killing of parasitic worms (flukes). A secondary course of treatment is necessary because the matured schistosomes have decreased susceptibility toward the therapy as compared to the adult worms. Usually, the second course of treatment starts a few weeks after the first course of therapy.

All individuals are susceptible to schistosomal infection; however, children are at higher risk of getting reinfected after treatment, as compared to the adult population. Although, the process of immunization is of great research interest, the probability of its success is not definite. The three purposes served by treating schistosomiasis are

- Reversal of acute or early chronic schistosomiasis disease
- Prevention of complications associated with chronic infection
- Prevention of neuroschistosomiasis

Treatment against schistosomal infection aims to reduce the rate of morbidity and mortality by reducing the production of eggs through reduction in the accumulation of parasitic worm (Richter 2003).

### 19.8.1 Drugs Used for Treatment of Schistosomiasis

The drugs used for the treatment of schistosomiasis possess a history of successful usage at the clinical level. The few drugs that can be taken orally for treating schistosomal disease include the following:

- Praziquantel—This drug is effective with a single dose against all types of schistosomiasis. The irreversible damage caused by the schistosomal infection can be treated with this drug. The cure rate for schistosomal infection after treatment with praziquantel is approximately 65% to 90% (Doenhoff et al. 2008). The drug targets the membrane permeability of the parasite resulting in vacuolation of the tegument. The drug imposes its effects by paralyzing the worm and exposing it to be attacked by the host (human) immune system. Moreover, the use of this drug is highest compared to other drugs used for treating schistosomiasis because of its decreased toxicity, increased efficacy, and oral administration. However, this drug is not effective for targeting underdeveloped schistosomula. It can be consumed by pregnant as well as lactating mothers. The side effects of praziquantel are mild and usually last for approximately 24 hours.

- Oxamniquine—This drug is extensively used for the treatment of intestinal schistosomiasis. It is an alternative drug used for treating *Schistosoma mansoni.* The drug is effective, as the parasitic worms that become resistant to praziquantel respond to oxamniquine. However, oxamniquine is more expensive and possesses more side effects as compared to praziquantel.
- Metrifonate—This drug was initially developed as an insecticide, but now it is used as a safe and effective treatment of urinary schistosomiasis. This drug is only effective for the treatment of urinary schistosomiasis, which is caused by *Schistosoma haematobium*. The cure rate of metrifonate is lower compared to praziquantel because its treatment schedule has a decreased rate of compliance. This drug needs to be administered thrice in about 2 weeks.

The rapid diagnosis and treatment of infected individuals decrease the environmental contamination with schistosome eggs. The decrease in number of infected cases is maintained for approximately a year without intervention. But in regions where there is continuing transmission of schistosomal infection, individuals may get reinfected in 4 to 6 months (Doenhoff et al. 2008).

Schistosomal infection affecting the central nervous system can be treated with praziquantel and glucocorticoids. Corticosteroid drugs are responsible for the reduction of inflammation and edema caused due to accumulation of schistosome eggs. The decrease in egg excretion is evaluated to check the response to treatment. However, the amount of schistosome eggs accumulated in the human body do not decrease in the initial 2 weeks after the treatment because the eggs laid before initiation of the treatment need approximately 2 weeks to get excreted from human body. Moreover, the eggs that are viable get excreted from the body in about 6 to 8 weeks after treatment.

The treatment of schistosomiasis reverses the process of hydronephrosis, resulting in depressed periportal fibrosis and portal hypertension. However, the treatment of schistosomal infection does not benefit late-stage reversible fibrosis development in the urogenital and hepatic system. Patients need to be monitored while undergoing the therapy for occurrence of any seizers or neurological consequences as a result of dying cercariae (Doenhoff et al. 2008). As compared to adult parasitic worms, the susceptibility of developing schistosomes toward chemotherapy is decreased.

### 19.8.2 Surgical Care for Schistosomiasis

A surgical procedure is required in chronic or severe cases of schistosomiasis. The removal of tumors, portacaval shunt surgeries, and ligation of esophageal varices are included in the procedures of surgical care. The different surgical procedures that are applied in severe cases of schistosomal infection include:

- Placement of distal splenorenal shunt to maintain portal hypertension
- Resection of bladder
- Resection of colonic polyps
- Partial colectomy for polyposis
- Partial colectomy for fibrosis
- Resection of cerebral cortical granulomas
- Removal of obstructive uropathy

The schistosomal disease remains an important cause of liver damage in the areas that are presented as nonendemic regions.

### 19.8.3 Clinical Management of Intestinal Schistosomiasis

Schistosomiasis possesses a broad spectrum of clinical presentations and approximately 10% of the patients experience severe hepatosplenic complications. The determination of morbidity levels through biological markers and diagnostic tests characterizes the improvement in staging and follow up of intestinal schistosomiasis. In some cases of intestinal schistosomiasis, there is a persistence process of inflammation for a long period of time (few years).

Antischistosoma drugs and drugs used as inhibitors of fibrosis are prescribed for the treatment of intestinal schistosomiasis. However, new challenges are imposed due to the presence of comorbidities that include the use of immune-suppressive drugs, organ transplantation, and viral hepatitis. The basic aim for management of schistosomiasis depends on the detection of active infection and stage of liver damage.

## 19.9 CHEMICAL AND BIOLOGICAL CONTROL MEASURES OF SCHISTOSOMIASIS

Recently, the use of control measures has been effective in controlling snail populations and preventing schistosomiasis infection. To date, biological and chemical measures have been effective. Copper sulfate has acquired noteworthy significance in controlling snails in ponds. The applicability of copper sulfate has shown restricted expansion to the National Councils for Research (NCR), as warm temperature is mostly dependent on the effectiveness of biological measures, and, therefore, such implications are much positively evaluated in controlling ramshorn snails. As snails retreat into bottom substrate, the effectiveness of copper sulfate has never been comprehensive to alter the snail population. In small populations, the reduction of snails is effectively employed through controlled release copper glass.

Numerous biological factors have been studied that have impact on the alteration of snail population, including water movement, pond management, and wind. These factors further affect the stability and application rate of biological measures in small populations. The extensive accumulation of copper assists in decreasing the natural food of larval fish. In addition, the toxicity of these biological controls is associated with the concerns of these chemicals in the form of food resources and cultured fish resources.

Smaller populations are directly affected by the application of copper sulfate for algae control. This application suggested that copper sulfate concentrations are much higher in the control of snail population. Lime or lime slurry treatments have also been studied as snail controls. However, it is noticed that snails avoid the effects of lime by migrating to the bottom of the pond as compared to copper treatment. Water alkalinity is considered to be the major drawback in controlling the number of snails in smaller populations. In addition, the drawback can significantly influence the ecology of the pond.

Another biological measure that is used to control the snail population is Bayluscide, found in ornamental fish production ponds. However, regulatory agencies have not provided approval to use Bayluscide in food fish for the control of specific populations. Vulgarone was further observed as a molluscicide and has been effectively used to control the snail population. Osmotic stress reduction can be significantly effective in increasing salinity. However, the potential osmotic and snail control benefits prohibit the maintenance of ponds.

## 19.10 BIOLOGICAL METHODS FOR CONTROL OF SCHISTOSOMIASIS

The control strategy for schistosomiasis and other parasitic diseases in tropical regions have greatly varied. Schistosomiasis has been subjected to different infrastructural and economic conditions in endemic regions. The control of schistosomiasis primarily depends on the control of its intermediate hosts or vectors (snail). The highest health authorities have advocated the necessity of eradication and control of certain species.

Control of the intermediate snail hosts population utilizing existing molluscicides has been recognized since the 17th century (Frandsen 1987). The aim of control strategies have been altered on the basis of availability of new effective drugs, which shifted the control of disease toward the direction of morbidity control. The vaccine strategy has been regarded as a possible way of controlling tropical parasitic diseases. The field of microbiology is concerned with the possibility of development of a vaccine against schistosomiasis.

The immunological developments associated with schistosomiasis have a variety of positive attributes. Horticultural and agricultural pest control is attributed to biological methods for the control of parasitic diseases. It is expected that the results would have an absolutely encouraging effect on the control of schistosomiasis. However, biological control methods are comprised in the primary health care system and can assist in development of vaccines against schistosoma. There are different biological methods used for controlling schistosomiasis:

- Application of bacterial pathogens—The snail host may get attacked by different microbial pathogenic organisms. Different investigations are carried out in endemic regions by utilizing the recently collected data together with the microbiologist.

- Plant-derived saponin molluscicides—The fecundity of snails is decreased by different parasites including schistosomes. Various trials for the prevention of schistosomiasis have failed (Frandsen 1987). The density of a snail host population in a specific kind of habitat can be reduced or eliminated by termolluscan competition.
- Predators—The population of snail hosts can be suppressed by some vertebrae predators such as fish in a specialized habitat (such as fish ponds). However, the evidence of efficacy of the fish is severely lacking. The major constraint is their inability to respond to rapid changes in the density of prey (snail). Apart from fish, turtles and birds are considered as possible control agents for schistosomiasis. Moreover, sciomyzid flies are predators of the mollusks that are utilized against the host snail species (Frandsen 1987).
- Competitor snails—The most promising group of snail host antagonists are identified as the competitors. They directly affect the number of aquatic snails. Moreover, the presence of competitors affects the success rate of miracidial penetration through the decoy effect.
- Genetic manipulation of snails—This approach refers to direct engineering changes in the genome of an organism that is effectually used through biotechnology applications. Genetic makeup of cells are modified using these technologies as well after the transfer of genes takes place to generate enhanced and novice organisms.

## 19.11 BIOLOGICAL MEASURES OF CONTROLLING AQUATIC SNAILS

The genetic control technique is used for controlling the snail population and, thereby, the core focus of this biological measure is determinedly attributed toward small population areas. The genetic regulation of this control measure is focused on isolating the infectivity types of the schistosome population from normally transmitted areas. The genetic control technique encompasses mass-rearing characteristics to alternate their genetic production in freshwaters and small inherited areas. The specific control measure drives to alter the descendants resulting in fewer snails and forcing larval schistosomes to segregate from the minute population. Thereby, the genetic control measure is in the position to eradicate the extent of new infections in the definitive host population.

The genetic control of schistosomes in the resistant strain of snails comes from three approaches. At the disease transmission site, the first approach is to observe and then select the incompatible features throughout the population of snails in inherited areas. The second approach can be beneficial in the migration of snails from one area to another, based on their local strain of parasites. The chances of crossbreeding between compatible snails and conspecific snails can be applied in the third approach; however, hybrid resistance is used for incompatible snails to explore the local schistosomes.

The selection of unaffected snails might be a major process in identifying the genetic or other markers associated with compatible states. The genetic control measure can be appropriately used to isolate field-collected snails, based on individual or pairwise screenings. The screening is to identify the cercarial output that lasts for at least 8 weeks. Unparasitized snail eggs are segregated to reduce the influence on the metacercarial infection and echinostome transmission. Unparasitized snails eggs are then utilized potentially to identify the first laboratory generation. The investigation of first-generation snails are tested based on cohorts to resist the derived parasitized samples of the parental snail population. The offspring of resistant adult snails aid in creating a second laboratory generation.

A contaminated freshwater source spreads schistosomiasis infection among the human population, but ecological processes have been identified to reduce the effects of schistosomiasis infection. The distribution of schistosome eggs determines the distribution of ineffective cercariae beneath the surface of the water that infects the production of snails. Biological and physical factors are important in reducing and altering the movement and infections of these water snails. Predator–prey is an important control measure within biological controls that determine the association between the predator of these snails and the snail intermediate hosts of the parasite. The effectiveness of these measures are adhered to mitigate the frequency of infected snails and to control schistosomiasis transmission in dense areas.

The alteration of schistosomiasis infection is dynamically changed by predator–prey measures. For instance, the modifications in schistosomiasis incidence among individuals are highly related to the modifications in the excessive number of specialized predators of snail prey. The abundance of molluscivorous fish is a dominant predator in releasing the predation of *Bulinus Nyassanus* snails in Lake Mallawi. Molluscivorous fish augment the schistosomiasis hemoglobin transmission among individuals and increase the abundance of intermediate host. The predator–prey paradigm has been proven, as Louisiana crayfish contributed significantly in reducing the extent of compatible snails. *Proserpinus clarkia* is another dynamic biological controller that reduces the abundance of host snails and eradicated the frequency of schistosomiasis among schoolchildren in Kenya. Predators have positively contributed to schistosomiasis transmission patterns, and, therefore, the influence of the specific control measure is focused on the snail predation and snail antipredator behavior. Biological controls have enhanced the efficacy of schistosomiasis transmission in lowering the risks of infected snails among the human population.

The employment of biological controls came to be when chemical pesticides brought concerns about effectiveness, costs, and adverse impacts on the environment. Biological controls have provided excessive and wide-ranging efforts to mitigate the effects of schistosomiasis fever in schoolchildren. Ecological and extrinsic parameters are the dependent factors on which the success and failure of a biological process is based. At both individual and population levels of snails, biological controls are interacted with intrinsic attributes of infected snails to determine the carrying capacity of the habitat. Effective biological controls are sought for their attributes that appeal to the utilization of these controls in the abundance of trematode parasites and *Schistosoma* spp.

Parasites, pathogens, and predators are effective species that are used as alternatives to control snail intermediate hosts. However, the efficacy and effectiveness of these species have not apparently been investigated outside laboratory model systems. Among predator species, fish and insects have been investigated explicitly to determine their impact on altering the abundance of snail intermediate hosts. *Tilapia Melanopleura*, *Clarias* sp., and *Astatoreochromis alluaudi* are malacophagous species that are predictively examined in the alteration of schistosomiasis infection. Other predator fish are used as an alternative source of scarce protein in fish ponds, where generation of snail intermediate hosts is excessively found. The effectiveness of fish species, such as voraciously vegetarian and malacophagous, are explored to investigate the impact on biocontrol of snail host populations. The biocontrol of the snail intermediate host population is comprehensively based upon management regimes and ecological factors that assist in governing the biota dynamics of snail populations in freshwater systems.

Sciomyzid flies have also received much emphasis among snail-eating insects that are sometimes regarded as obligatory feeders on mollusk larvae and gastropods. However, there are numerous snail-eating insects that can be widely used to control the snail population. Sciomyzids are further investigated to determine their impact on the aquatic Hemiptera in Egypt. From the investigation, the contribution of obligatory and nonspecific feeders is dominant in considering release of any exotic organism beyond its normal range.

### 19.11.1 Competitor Snails (Apple Snail)

*Pila globosa*, or the apple snail, is one of the most dominant competitor snails that contributes to maintaining the balance of aquatic ecosystem. In Nepal and Southwestern Asia, their abundant numbers exist to control the environment and aquatic ecosystem. The adaptation of apple snails is practiced more than other competitors, as they have the potential to exhibit from summer periods to excessive rainfall periods. Temporary and permanent types of *Pila globosa* are represented in ditches, ponds, and canals. However, the lower habitat and fish poisoning caused reductions to their numbers in Nepal and Western Asia, even though the efficacy of these snails is widespread in controlling schistosomiasis infection.

*Pila globosa* has been used as a protein supplement in some parts of India and Bangladesh. The poisoning and infections of the aquatic weed *salvinia molesta* were further investigated using *Pila globosa* in Bangladesh and India. The role of apple snails in population regulation is dominantly established, even though the natural pathogens of these snails are not reported as symbionts and commensals in these regions. Occurrence of disease in *Pila globosa* may pose a threat to some groups of people residing within its range.

### 19.11.2 *Biomphalaria glabrata*

Bacterial pathogens are excessively determined and utilized to control and isolate the frequency of snail populations in specific regions. *Biomphalaria glabrata* is an important constituent of bacterial pathogens that controls and isolates the occurrence of new microbial pathogens of the freshwater snail. The classification of *Biomphalaria glabrata* can be related to the attributes of the *Paenicacillus* genus, which is effective in invading and proliferating the genetic occurrence of snails. The discovery and usage of *Biomphalaria glabrata* is an important aspect in contributing its role as an intermediate host and in transmitting schistosomiasis infection. The objective of the World Health Organization consequently encourages the development of approaches to eradicate the frequency of schistosomiasis infection.

The miracidium transforms the initial sporocysts that effectually increase the asexual production of secondary sporocysts in producing cercariae. The production of cercariae is effective in leaving the genus of snails and invading the vertebrate definitive lost. The focus of investigation is apparently determined on the immunology of infections, transmitted from *Biomphalaria* sp., which approach to recognize the interacted snail resistance to the parasite among key genes. Lethal effects have been exhibited with bacteria at inoculation at higher densities, and, therefore, the development of immune response is shown among host snails. Different insect pests including freshwater snails are investigated through *Bacillus thuringiensis*, which is a spore-forming and gram-positive bacterium used to secrete toxins. The widespread attributes and usage of *Biomphalaria glabrata* are further effective in secreting and invading vector snails such as *Biomphalaria alexandrina*. The ovotestis of *Australorbis glabratus* is used as a potential biological control agent, which alters the infectious influence of *Bacillus pinotti* in the human population. The exertion of molluscicidal and egg hatching prevention is negatively influenced using *Biomphalaria thuringiensis*.

### 19.11.3 *Sargochromis codringtonii*

Various actions have been taken to reduce the effects of schistosomiasis in causing illness to humans. Different medications have been developed against the disease, but the effectiveness of the medication campaign is lessened due to the high cost of repeated medication. Therefore, biological controls and application of molluscicides are often used to control schistosomiasis. Biological control of snails is frequently monitored through fish in irrigation canals, rice–fish culture, and aquaculture ponds. The experimental evidence of snail control using fish reflects the development of effective approaches against schistosomiasis, despite the fact that snail populations can be directly and indirectly affected by fish. The success of this approach is evident from the assorted degrees of black carp and the influence of African cichlids and shell-cracker sunfish.

*Astatoreochromis alluaudi* is the best studied snail eater in Africa, and, therefore, is regarded as a potential candidate when it comes to control of the snail population. The performance of *Astatoreochromis alluaudi* is much emphasized from laboratory findings and outer-field evidences. In Eastern Africa, this fish is commonly found in lakes and rivers and specifically in Lake Victoria, where the extent of compatible snails are found excessively in feeds and Littoral zones. The thick-shelled *Melanoides tuberculata* crushes the shells with its pharyngeal jaws to control the excessive amount of snails.

The instigation of *Astatoreochromis alluaudi* was effective in mitigating the snail population in the Nyanza province of Kenya. The effects of *Astatoreochromis alluaudi* were further observed in Kenya at dam sites to explore whether these species still exist to control the numbers of specific types of snail. *Astatoreochromis alluaudi* were migrated from Uganda to the South Cameroon to conduct a field experiment for controlling snail populations. The evidence indicated that snails were effectively controlled and biological attributes of *Astatoreochromis alluaudi* successfully adapted in the culture of Nile tilapia.

Another beneficial predator, used for altering schistosome-transmitting snails, is molluscivorous fish. Effective and encouraging results have been found from studies conducted in Zimbabwe. The quality of this predator is the selection criteria, which differentiates its features from other predators. In addition, smaller snails of 3 to 8 mm are controlled and eliminated using the widely distributed fish *Geophagus brasilensis*. The role of natural predators of snails is determined from the evidence as well as the reduction and control of snail populations evident using the molluscivorous fish. In the case of biological control of snails using competitor snails, species with higher fecundity rates, voraciously feeding snails, snails resistant to parasites, species with longer life spans, and snails that are harmless to other animals and surrounding crops are preferable as competitors.

## 19.12 ADVANTAGES AND DISADVANTAGES OF BIOLOGICAL METHODS

### 19.12.1 Advantages and Disadvantages of Bacterial Pathogens

Bacterial pathogens are an important biological control used to regulate the natural population. The significance of bacterial pathogens is widely examined to control the frequency of snail in denser areas as well as in eradicating the extent of schistosomiasis infection. Bacterial pathogens are beneficial due to their specific attributes used in the control process of pests. Bacterial pathogens are effectively compatible with pesticides and can be culturally adapted genetically. Pesticide technology is essential for bacterial pathogens in reducing snail populations. The development of bacterial pathogens in commercial areas is more easily covered. However, these bacterial pathogens are often negatively affected by weather, which restricts the process from quickly completing. The expansion of bacterial pathogens is optimally spread among high population areas. The advantages and disadvantages of different biological methods of snail control are given in Table 19.5.

### 19.12.2 Advantages and Disadvantages of Plants

Plants, specifically molluscicides, are also emphasized in the reduction process of snail populations. The control operations of molluscicides are effective in interrupting schistosomiasis infection. The cost-effective and plausibly efficient method provides a comprehensive approach to reduce the population of snails. Control of vectors can be supervised using simple equipment even though specialized operational schemes are present to control the snail population. However, several drawbacks restrict the

**TABLE 19.5**
**Advantages and Disadvantages of Biological Methods**

| Biological Methods | Cost Effective | Adaptability | Natural Process | Affected by Weather | Risk for Human Population | Time Conservation |
|---|---|---|---|---|---|---|
| Bacterial pathogens | ✓ | ✓ | ✓ | X | X | X |
| Plant-derived saponin molluscicides | ✓ | X | ✓ | ✓ | X | ✓ |
| Predators | ✓ | X | ✓ | ✓ | X | X |
| Competitor snails | ✓ | X | ✓ | ✓ | ✓ | ✓ |
| Genetic manipulation of snails | X | ✓ | X | X | ✓ | ✓ |

practicality of plants and create complexity for snail eradication. Plant assessment and deployment in field areas are time-consuming and, therefore, effective results are not obtained due to this limitation. The variation in transmission sites requires appropriate application procedures for the technical capacity.

### 19.12.3 Advantages and Disadvantages of Predators

Different paradigms are examined to explore the effectiveness of snail controlling. Predators are widely used in the biological control process, specifically *Sciomyzidae*, *Astatoreochromis alluaudi*, and *Sargochromis codringtonii* are the common predators that are utilized to control the snail populations. These control process is effectual as compared to other biological control processes. These controls allow pests to be culturally controlled from prey and are cost-effective. The drawback of these biological control processes is that they exhibit the polyphagy of indigenous species. The long-term effect of biological control is slow.

### 19.12.4 Advantages and Disadvantages of Competitor Snails

Competitor snails are usually used as a biological agent for controlling an excessive amount of snails. The number of *Biomphalaria* snails is reduced by using the biological control process, which is determinedly revealed as intermediate hosts to schistosomiasis infection. Larvae or *schistosoma haematobium* is developed inside the *Biomphalaria* snails to represent miracidium. The benefit of transforming into *Biomphalaria* is to reduce the probability of schistosomiasis being spread. Potential ecological impacts of these species are examined in various regions of world particularly in North America. Furthermore, invasive aquatic plant genes are controlled using competitor snails as biological agents.

### 19.12.5 Genetic Manipulation of Snails

Genetic engineering is comprehensively used in the protection of biodiversity and has recently been studied to control the population of snails. The extent of reducing infectious diseases is possible using genetic engineering as a biological agent. The hybrid process is an effectual approach for controlling snail populations. However, the risks associated with these approaches possess uncertain effects, such as creation of entities.

## 19.13 THE BEST BIOLOGICAL METHOD

Control of aquatic snails plays a major role in reducing the risk of various diseases and public health concerns. Differential biological methods have been discussed to identify certain methods that can be implemented to control aquatic snails. Some of the common biological methods discussed earlier are bacterial pathogens, plant-derived saponin molluscicides, predators, competitor snails, and genetic manipulation of snails. All of these methods have their own sort of advantages and disadvantages. Most of the biological methods are associated with effective outcomes; however, some are associated with higher costs. Referring to the advantages and disadvantages of biological methods, the use of bacterial pathogens is identified as the most effective biological method for controlling and reducing the prevalence of aquatic snails to limit the emergency of schistosomiasis.

### 19.13.1 Bacterial Pathogens

Duval et al. (2015) described the utilization of bacterial pathogen *Biomphalaria glabrata*, which is identified as a potential weapon for schistosomiasis control. The World Health Organization put out a call to professionals to help eradicate schistosomiasis. The professionals encouraged an integrated approach that included hygiene and health education, chemotherapeutic control, and snail control (WHO 2012). Recently, praziquantel has been used across the globe as the most effective approach for controlling the prevalence of schistosomiasis along with the relative comorbidities (Chai 2013; Cioli et al. 2014). On the contrary, a major concern has emerged regarding the differential parasitic resistance toward praziquantel. Unfortunately, no other vaccination therapy is available to control such aspects (Doenhoff et al. 2002; Melman et al 2009). At the same time, different chemical compounds have been utilized to control the prevalence of the snail population. Considering all such methods, the control of the snail population remains problematic. Therefore, a significant number of alternatives have been developed.

The family of Paenibacillaceae is widely distributed within the territorial and aquatic environment (McSpadden Gardener 2004; Lal and Tabacchioni 2009). In the current era, the origin of a new microbial pathogen known as *Candidatus Paenibacillus glabratella* is unknown. The appropriate screening of healthy aquatic snails using molecular approaches has mainly excluded the chances of snail microbiota origin. At the same time, bacterial communities have also been cultivated and characterized from *B. glabrata* that were gram-negatives (Silva et al. 2013).

*Paenibacillus larvae* and *Paenibacillus popilliae* have been identified as invertebrate pathogens among bacterial species belonging to the Paenibacillaceae family (Harrison et al. 2000). *P. larvae* is a spore-forming bacterium that is a larval pathogen of honeybees. These species are identified as ethological agents (Fries et al. 2006). This bacterium has the ability to grow in the midgut of insects after being ingested by honeybee larvae in the form of a spore. Afterward, the vegetative form secretes proteases, which usually facilitates tissue invasion and also results in the liquefaction of the host (Garcia-Gonzalez et al. 2014). It has been further evaluated that the most significant histopathological feature of *Paenibacillus* after the interference with *B. glabrata* is the development within the hepatopancreas tissues. This feature is mainly helpful for developing the ability of the snail to nourish itself.

At the same time, the existence of *Paenibacillus* within the circulatory system of an organism indicates that the bacteria can follow the path of hemolymph for reaching systemic organs. In accordance with this perspective, *Paenibacillus* has been observed within the ovotestis. The presence of bacterium in ovotestis would lead toward the suppression of gametogenesis along with the partial destruction of ovotestis acini. Therefore, the reproductive capacity of a snail can be easily affected by compromising the eggs laid during the advanced stage of infection. *Paenibacillus* has also been observed among secondary reproductive organs, which mostly include muciparous and albumin glands. This is a major sign of vertically transmitted infection. It has been evaluated that this pathogenic agent has been found in eggs of snails to affect development and hatching (Duval et al. 2015; Garcia-Gonzalez et al. 2014). Therefore, such histological observations have suggested that the major pathogenic impact of *Candidatus Paenibacillus glabratella* is related with the strong compression of tissues that can easily cause significant damage to the soft tissue organs, including liver and ovotestis.

### 19.13.2 Reasons behind Effectiveness of Bacterial Pathogens

Bacterial pathogens have been identified as a major approach helpful in controlling aquatic snails and therefore schistosomiasis. The use of bacterial pathogens is also cost-effective. Other biological measures are directly associated with the higher costs, which has a major impact on economic perspectives. Therefore, the application of bacterial pathogens will be cost-effective in comparison with other biological measures. Similarly, the control of aquatic snails along with control of schistosomiasis can be achieved through natural processes.

The use of bacterial pathogens will affect the population of aquatic snails through natural processes. As suggested earlier, the bacterial pathogens would affect the reproductive capacity of snails, which refers to the compromised eggs laid during infection. Similarly, *Paenibacillus* would also affect the secondary reproductive organs, which include muciparous and albumin glands. At the same time, the bacterial pathogens will also affect the eggs of snails during the development and hatching. Such natural phenomenon will be helpful to control the population of aquatic snails accordingly. Another rationale behind the effectiveness of bacterial pathogens is associated with adaptability. It has been evaluated that the use of bacterial pathogens for the control of aquatic snails can be easily adapted by scientists and natural processes.

Scientists and microbiological professionals can also understand the entire process with absolute security. The process is entirely safe, which is considered another major benefit of biological measures. Safe procedures or handling has direct relevance to better execution of control measures. Therefore, it can be said that among the different biological measures, the utilization of bacterial pathogens is significant in controlling the emergence of schistosomiasis and population of aquatic snails. It will also be beneficial for the human population to get rid of the symptoms and complications associated with schistosomiasis. At the same time, the cost-effective benefit will allow the government and organization to incorporate other aspects for the control of aquatic snails. Thus, the best method to control the population of aquatic snails, which would ultimately result in a decreased prevalence rate of schistosomiasis, is the use of bacterial pathogens.

## 19.14 CONCLUSION

Schistosomiasis is a serious infection that occurs through parasitic blood flukes. The infectious form of parasites is responsible for contaminating water, which is also known as cercariae. Such parasites reside in freshwater snails. The larvae of parasite usually causes schistosomal infection by direct contact of skin with infested water. A granulomatous reaction, which involves macrophages, eosinophils, and T-cells, is stimulated through the antigens of the egg causing progression of disease. The infected individuals develop a serum sickness-like disease after a few weeks, which is also known as acute schistosomiasis. The symptoms of inflammatory colonic polyposis, bloody diarrhea, and cramping are directly associated with penetration of eggs along with the formation of granuloma in the digestive system. The emergence rate of schistosomiasis infection was raised by 50%, even after preventing transmission of schistosoma.

The treatment of schistosomiasis can reverse the process of hydronephrosis, which can result in depressed periportal fibrosis along with portal hypertension. On the contrary, the treatment of schistosomal infection cannot be helpful in stage reversible fibrosis development within the urogenital and hepatic system. Recently, the utilization of differential control measures has been done for controlling the snail population and preventing schistosomiasis. A significant number of biological factors have been studied, which had an effective impact on the alteration of the snail population. Such factors have also affected the stability and application rate of biological measures. Controlling the emergence of schistosomiasis is completely dependent on the control of its intermediate hosts or vectors (snails). Therefore, horticultural and agricultural pest control is attributed to biological methods for the control of parasitic diseases.

Some of the significant biological measures for controlling aquatic snails include bacterial pathogens, plant-derived saponin molluscicides, predators, competitor snails, and genetic manipulation of snails. The selection of unaffected snails is a major process for identifying genetic or other markers that are associated with compatible states. All of the identified methods have many advantages and disadvantages. Referring to the advantages and disadvantages of different biological methods, the utilization of bacterial pathogens has been identified as the best biological method for controlling and reducing the prevalence of aquatic snails. *Paenibacillus larvae* and *Paenibacillus popilliae* are invertebrate pathogens belonging to the Paenibacillaceae family. This bacterium can grow in the midgut of insects after being ingested by honeybee larvae in the form of a spore. Histological observations have suggested that the major pathogenic impact of *Candidatus Paenibacillus glabratella* is directly associated with strong compression of tissues that can easily cause significant damage to the soft tissue organs, including the liver and ovotestis. Therefore, it can be concluded that the utilization of bacterial pathogens is relatively helpful for controlling the emergence of schistosomiasis and population of aquatic snails among differential biological measures.

## 19.15 FUTURE ASPECTS

From the future perspective, it can be said that professionals and welfare organizations must focus on the biological features to control the growth of aquatic snails. It is a fact that aquatic snails are a major reason for the spread of schistosomiasis among the human population. For this purpose, scientists and relative professionals should discover certain beneficial alternatives, like bacterial pathogens, to reduce the emergence and prevalence of the disease. At the same time, disease control authorities should also identify environmental factors that can contribute to the control of the aquatic snail population. Different possible features should be analyzed comprehensively to develop a cost-effective and healthy alternative for reducing the risk of schistosomiasis among the human population. As identified earlier, the prevalence rate of schistosomiasis is increasing at a constant rate in African regions; therefore, future policies, standards, and alternatives need to ensure a better quality of life for humans.

## REFERENCES

Adamson, K. J., T. Wang, B. Rotgans, T. Kruangkum, A. V. Kuballa, K. B. Storey, and S. F. Cummins. Genes and associated peptides involved with aestivation in a land snail. *General and Comparative Endocrinology* (2015).

Badmos, K. B., A. O. Komolafe, and O. Rotimi. Schistosomiasis presenting as acute appendicitis. *East African Medical Journal* (2006) 83:528. doi: 10.4314/eamj.v83i10.9464.

Barsoum, R. S., G. Esmat, and T. El-Baz. Human schistosomiasis: Clinical perspective: Review. *Journal of Advanced Research* (2013) 4:433–444. doi: 10.1016/j.jare.2013.01.005.

Brinkmann, U. K., R. Korte, and B. Schmidt-Ehry. The distribution and spread of schistosomiasis in relation to water resources development in Mali. *Tropical Medicine and Parasitology: Official Organ of Deutsche Tropenmedizinische Gesellschaft and of Deutsche Gesellschaft fur Technische Zusammenarbeit (GTZ)* (1988) 39(2):182–185.

Centers for Disease Control and Prevention. DPDx: Laboratory identification of parasites of public health concern. (Accessed on 3rd April 2017) www.dpd.cdc.gov/DPDx/.

Chai, J.-Y. Praziquantel treatment in trematode and cestode infections: An update. *Infection & Chemotherapy* (2013) 45, no. 1:32–43. doi: 10.3947/ic.2013.45.1.32.

Chandiwana, S. K., and N. O. Christensen. Analysis of the dynamics of transmission of human schistosomiasis in the highveld region of Zimbabwe: A review. *Tropical Medicine and Parasitology: Official Organ of Deutsche Tropenmedizinische Gesellschaft and of Deutsche Gesellschaft fur Technische Zusammenarbeit (GTZ)* (1988) 39:187–193.

Chistulo, L., P. Loverde, and D. Engels. Disease watch: Schistosomiasis. *TDR Nature Reviews Microbiology* (2004) 2:12. doi: 10.1038/nrmicro801.

Cioli, D., L. Pica-Mattoccia, A. Basso, and A. Guidi. Schistosomiasis control: Praziquantel forever? *Molecular and Biochemical Parasitology* (2014) 195, no. 1:23–29. doi: 10 .1016/j.molbiopara.2014.06.002.

Colley, D. G., A. L. Bustinduy, W. Evan Secor, and C. H. King. Human schistosomiasis. *The Lancet* (2014) 383:2253–2264. doi: 10.1016/s0140-6736(13)61949-2.

Cowie, R. H. The life-cycle and productivity of the land snail Theba pisana (Mollusca: Helicidae). *Journal of Animal Ecology* (1984): 311–325.

Cridland, C. C. Resistance of Bulinus (Physopsis) globosus, Bulinus (Ph.) africanus, Biomphalaria pfeifferi and Lymnaea natalensis to experimental desiccation. *Bulletin of the World Health Organization* (1967) 36, no. 3:507.

Doenhoff, M. J., D. Cioli, and J. Utzinger. Praziquantel: Mechanisms of action, resistance and new derivatives for schistosomiasis. *Current Opinion in Infectious Diseases* (2008) 21, no. 6:659–667.

Doenhoff, M. J., J. R. Kusel, G. C. Coles, and D. Cioli. Resistance of Schistosoma mansoni to praziquantel: Is there a problem? *Transactions of the Royal Society of Tropical Medicine and Hygiene* (2002) 96, no. 5:465–469.

Ducklow, H. W., P. J. Boyle, P. W. Maugel, C. Strong, and R. Mitchell. Bacterial flora of the schistosome vector snail Biomphalaria glabrata. *Applied and Environmental Microbiology* (1979) 38, no. 4: 667–672.

Duval, D., R. Galinier, G. Mouahid, E. Toulza, J. F. Allienne, J. Portela, C. Calvayrac, A. Rognon, N. Arancibia, G. Mitta, A. Theron, and B. Gourbal. A novel bacterial pathogen of Biomphalaria glabrata: A potential weapon for schistosomiasis control? *PLoS Neglected Tropical Diseases* (2015) 9, no. 2:e0003489. doi: https://doi.org/10.1371/journal.pntd.0003815.

Elbaz, T., and G. Esmat. Hepatic and intestinal schistosomiasis: Review. *Journal of Advanced Research* (2013) 4, no. 5:445–452.

Evan Secor, W. Water-based interventions for schistosomiasis control. *Pathogens and Global Health* (2014) 108:246–254. doi: 10.1179/2047773214y.0000000149.

Frandsen, F. Control of schistosomiasis by use of biological control of snail hosts with special reference to competition. *Memorias do Instituto Oswaldo Cruz* (1987) 82:129–133.

Fries, I., A. Lindström, and S. Korpela. Vertical transmission of American foulbrood (Paenibacillus larvae) in honey bees (Apis mellifera). *Veterinary Microbiology* (2006) 114, no. 3:269–274.

Garcia-Gonzalez, E., L. Poppinga, A. Fünfhaus, G. Hertlein, K. Hedtke, A. Jakubowska, and E. Genersch. Paenibacillus larvae chitin-degrading protein Pl CBP49 is a key virulence factor in American Foulbrood of honey bees. *PLoS Pathogens* (2014) 10, no. 7:e1004284.

Gray, D. J., A. G. Ross, Y. S. Li, and D. P. McManus. Diagnosis and management of schistosomiasis. *British Medical Journal* (2011) 342:d2651.

Gryseels, B., K. Polman, J. Clerinx, and L. Kestens. Human schistosomiasis. *The Lancet* (2006) 368:1106–1118. doi: 10.1016/s0140-6736(06)69440-3.

Guo, J., Y. Li, D. Gray, A. Ning, G. Hu, H. Chen, and G. M. Williams. A drug-based intervention study on the importance of buffaloes for human Schistosoma japonicum infection around Poyang Lake, People's Republic of China. *American Journal of Tropical Medicine and Hygiene* (2006) 74, no. 2:335–341.

Harrison, H., R. Patel, and A. A. Yousten. Paenibacillus associated with milky disease in Central and South American scarabs. *Journal of Invertebrate Pathology* (2000) 76, no. 3:169–175.

Holmes, K. J. *Snails*. Capstone. 1999.

Joe, L. K., C. R. Schneider, S. Sornmani, G. R. Lanza, and P. Impand. Biological control by trematode antagonism. II. Failure to control Schistosoma spindale in a field trial in northeast Thailand. *Southeast Asian Journal of Tropical Medicine and Public Health* (1974) 5:60–64.

Keiser, J., and J. Utzinger. Efficacy of current drugs against soil-transmitted helminth infections: Systematic review and meta-analysis. *JAMA* (2008) 299(16):1937–1948.

King, C. H., and D. Bertsch. Historical perspective: Snail control to prevent schistosomiasis. *PLoS Neglected Tropical Diseases* (2015) 9, no. 4:e0003657.

Lal, S., and S. Tabacchioni. Ecology and biotechnological potential of Paenibacillus polymyxa: A minireview. *Indian Journal of Microbiology* (2009) 49, no. 1:2–10. doi: 10.1007/s12088-009-0008-y.

Leutscher, P. D. C., E. Høst, and C. M. Reimert. Semen quality in Schistosoma haematobium infected men in Madagascar. *Acta Tropica* (2009) 109:41–44. doi: 10.1016/j.actatropica.2008.09.010.

Li, Y., G. Raso, Z.-Y. Zhao, Y.-K. He, M. K. Ellis, and D. P. McManus. Large water management projects and schistosomiasis control, Dongting Lake Region, China. *Emerging Infectious Diseases* (2007) 13, no. 7:373–379.

McSpadden Gardener, B. B. Ecology of Bacillus and Paenibacillus spp. in agricultural systems. *Phytopathology* (2004) 94, no. 11:1252–1258. doi: 10.1094/PHYTO.2004.94.11.1252.

Melman, S. D., M. L. Steinauer, C. Cunningham, L. S. Kubatko, I. N. Mwangi, N. Barker Wynn, M. W. Mutuku, D. M. S. Karanja, D. G. Colley, C. L. Black, W. Evan Secor, G. M. Mkoji, and E. S. Loker. Reduced susceptibility to praziquantel among naturally occurring Kenyan isolates of Schistosoma mansoni. *PLoS Neglected Tropical Diseases* (2009) 3, no. 8:e504. doi: 10.1371/journal.pntd.0000504.

Meyer-Willerer, A. O., and A. Santos-Soto. Temperature and light intensity affecting egg production and growth performance of the Apple Snail Pomacea patula (Baker, 1922). *Avances en Investigación Agropecuaria* (2006) 10:41–58.

Mohammed, A. Z., S. T. Edino, and A. A. Samaila. Surgical pathology of schistosomiasis. *Journal of the National Medical Association* (2007) 99:570.

Montgomery, S. *Infectious diseases related to travel. Centers for Disease Control and Prevention.* 2015. https://wwwnc.cdc.gov/travel/yellowbook/2016/infectious-diseases-related-to-travel/schistosomiasis#4701.

Olveda, D. U., R. M. Olveda, D. P. McManus, P. Cai, T. N. P. Chau, A. K. Lam, Y. Li, D. A. Harn, M. L. Vinluan, and A. G. P. Ross. The chronic enteropathogenic disease schistosomiasis. *International Journal of Infectious Diseases* (2014) 28:193–203.

Omonijo, A. O., Samuel Oreoluwa Asaolu, and Ifeanyi Emmanuel Ofoezie. Ecology of schistosoma snail vectors in ado-ekiti local government area, Ekiti State, Nigeria. *International Journal of Pure and Applied Zoology* (2016) 4:77–84.

Patz, J. A., T. K. Graczyk, N. Geller, and A. Y. Vittor. Effects of environmental change on emerging parasitic diseases. *International Journal for Parasitology* (2000) 30(12): 1395–1405.

Richter, J. The impact of chemotherapy on morbidity due to schistosomiasis. *Acta Tropica*, (2003) 86, no. 2:161–183.

Ross, A. G., D. Vickers, G. R. Olds, S. M. Shah, and D. P. McManus. Katayama syndrome. *The Lancet Infectious Diseases* (2007) 7:218–224. doi: 10.1016/s1473-3099(07)70053-1.

Silva, T. M., E. S. Melo, A. C. S. Lopes, D. L. Veras, C. R. Duarte, L. C. Alves, and F. A. Brayner. Characterization of the bacterial microbiota of Biomphalaria glabrata (Say, 1818) (Mollusca: Gastropoda) from Brazil. *Letters in Applied Microbiology* (2013) 57, no. 1:19–25. doi: 10.1111/lam.12068.

Southgate, V. R. Schistosomiasis in the Senegal River Basin: Before and after the construction of the dams at Diama, Senegal and Manantali, Mali and future prospects. *Journal of Helminthology* (1997) 71, no. 2:125–132.

Steinmann, P., J. Keiser, R. Bos, M. Tanner, and J. Utzinger. Schistosomiasis and water resources development: Systematic review, meta-analysis, and estimates of people at risk. *The Lancet Infectious Diseases* (2006) 6, no. 7:411–425.

Taylor, P., and O. Makura. Prevalence and distribution of schistosomiasis in Zimbabwe. *Annals Tropical Medicine and Parasitology* (1985) 79(3):287–299.

*Wikipedia*. Snail. https://en.wikipedia.org/wiki/Snail.
World Health Organization (WHO). Accelerating work to overcome the global impact of neglected tropical diseases: A roadmap for implementation. 2012.
World Health Organization (WHO). 2017. Schistosomiasis. Fact Sheet. Available from http://www.who.int/mediacentre/factsheets/fs115/en/ accessed on 11th April 2017.
World Health Organization (WHO) Expert Committee. Prevention and control of schistosomiasis and soil-transmitted helminthiasis. World Health Organization Technical Report Series, 912, i. 2002.

# *Section IV*

## *Microbial Control of Black Fly Vector*

# 20 Microbial Approaches for Black Fly-Targeted Onchocerciasis Control

*Sankarappan Anbalagan*

## CONTENTS



## 20.1 INTRODUCTION

According to a World Health Organization report (2016), vector-borne diseases account for one-fourth of all infectious diseases present in the world and cause one million deaths annually. If any living organism transmits an infectious disease to human and animals, it is known as a vector. Many of these vectors are blood-feeding insects, which carry disease-producing microorganisms from infected host to new host. In blood-feeding insects, mosquitoes are well known to everybody, which cause several human diseases. Next to mosquitoes, black flies (Simuliidae) receive major focus due to their wide distribution and potential as vectors in transmitting agents of lethal diseases (Anbalagan et al., 2014). Many species of female black flies are hematophagous and are central to the transmission of the parasitic nematode *Onchocerca volvulus*, the causative organism of human onchocerciasis (river blindness), in tropical Africa, Central America, South America, and Yemen where 30 million people are infected. Larval black flies constitute a vital part of aquatic habitats, which are employed as bioindicators for the quality of these habitats because of their high sensitivity to environmental degradation (Harwood and James, 1979).

Black flies belong to the family of Simuliidae and order Diptera, including 2204 living and 15 fossil species under 32 genera (26 genera from living species and 6 genera from fossil species) distributed throughout the world (Adler and Crosskey, 2016). The genus *Simulium* accounts for 80% of species. Most black fly species are blood feeders and others feed on nectar. Davies and Crosskey (1992) grouped the blood-feeding nature of black flies under two categories: *Simulium damnosum* and *Simulium neavei.* The vector species of *S. damnosum* complex is under four categories: category 1, 2, 3, and 4, studied in West Africa, Sierra Leone/Guinea, Zaire basin, and Eastern Africa, respectively. The four categories consist of 40 vector species. The *S. neavei* group has all simuliids, consisting of eight described species and some forms of uncertain taxonomic status. Among these, three species are considered as vectors. The *S. neavei* group occurs widely but sporadically in Africa from Ethiopia southward to Malawi and westward through the Zaire River basin to Cameroon and Liberia.

The greatest public health importance of black flies lies in their roles as vectors of the filarial nematode *Onchocerca volvulus* Leuckart, the causative organism of human onchocerciasis (river blindness) and several clinical symptoms. Due to climate change, urbanization, industrialization, and anthropogenic impact, it is possible to spread this disease throughout the world. Hence, control measures for the black fly as well as onchocerciasis are essential. Many researchers have applied chemical and biological techniques to control larvae and adult black flies. These strategies face challenges to control the vector in the field. In this chapter, the microbial approaches for black fly-targeted onchocerciasis control are discussed.

## 20.2 BLACK FLY LIFE CYCLE

The black fly life cycle has four distinct developmental stages: egg, larva, pupa, and adult. The adult female is ready to mate after emergence and oviposit usually after egg maturation, for which a blood meal or nectar is required. After mating, females lay their eggs in a variety of submerged substrates (e.g., boulders, coarse particulate organic matter, pebbles, grasses) in water of streams and rivers. A single female fly can lay 200 to 500 eggs. The hatching time varies with the species and ranges from 3 to 5 days or longer if eggs pass through diapause. Larvae are found behind the surface of flowing water. Larvae usually pass through seven molts, but 4 to 9 molts in some species. Water temperature and quality and quantity of food influence the growth rate of larvae. The larval period differs from less than 1 to 7 months in over wintering species. The last stage larva (pharate pupa) spins a cocoon, which serves to anchor and protect the developing pupae. These cocoons attach on slightly submerged substrates. The pupal period lasts from 4 to 7 days or from 3 to 5 weeks, depending on species or water temperature. Adults emerge from the pupal case, rise to the water surface in a bubble of air, expand their wings, and fly away. The adult flies immediately to a nearby support to rest and allow its cuticle to harden. Mating generally occurs shortly after the initial flight. The complete life cycle, from egg to adult, varies from 1 to 4 months and the number of generations per year ranges from one to eight, depending on climatic conditions and species (Figure 20.1).

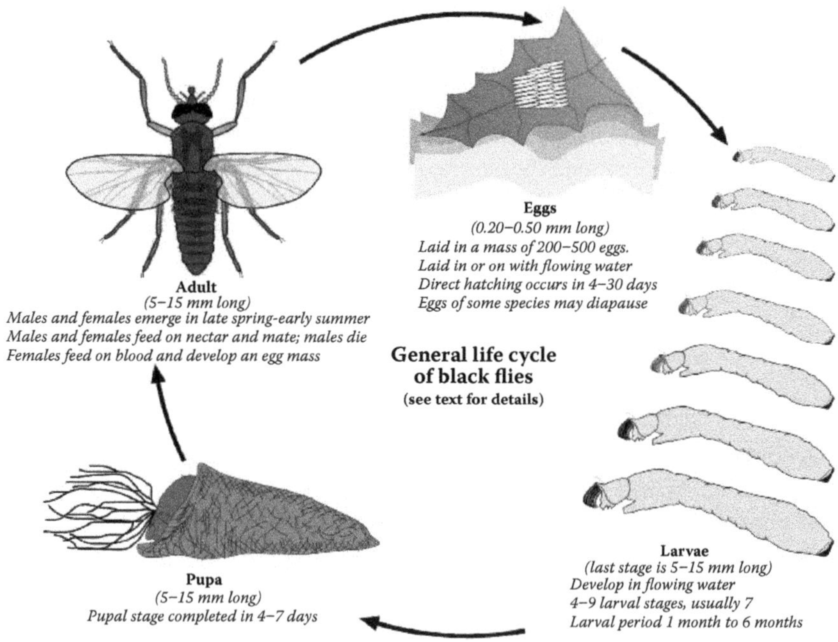


**FIGURE 20.1** Life cycle pattern of black fly. (Illustration by: Scott Charlesworth, Purdue University, based in part on Peterson, B.V., in: *Manual of Nearctic Diptera*, Vol. 1.)

## 20.3 PARASITE TRANSMISSION

Black flies are of medical and veterinary importance and they transmit various pathogenic organisms to humans, mice, rabbits, cattle, and birds. The diseases caused by vector black fly species include human and bovine onchocerciasis, and leucocytozoonosis in birds. Black flies are vital to the transmission of the parasite *Onchocerca volvulus*, which causes onchocerciasis, or river blindness, to humans. It acts as the larval host for the nematode and acts as the vector by which the disease is spread. Transmission of the parasite occurs through the bite of a black fly when feeding on human blood. Bovine onchocerciasis is caused by two main nematodes (*O. gutturosa* and *O. lienalis*). Microfilariae are found in the dermis and on rare occasions circulating in peripheral blood. The microfilariae lack a sheath and are 200 to 250 μm long with a short, sharply pointed tail, which transmit by vector black fly species from infective host to new host. Leucocytozoonosis of birds is caused by *Leucocytozoon* sp. Acute outbreaks of leucocytozoonosis have been reported in chickens (Asia and Africa), turkeys (North America), waterfowl (North America and Europe), and a number of free-living and captive avian species throughout the world. Species in domestic birds include *L. simondi* in waterfowl; *L. smithi* in turkeys; and *L. caulleryi*, *L. sabrazesi*, *L. andrewsi*, and *L. schoutedeni* in chickens. *L. caulleryi* can be highly pathogenic, causing a lethal hemorrhagic disease of chickens in Southeast

Asia. Numerous Leucocytozoon sp. infect nondomestic birds (e.g., blood smears from raptors often contain gametocytes). Wild birds are reservoirs in some areas and are responsible for initiating infection in young birds each year. Acute disease is seen more often in the young when they have high parasitemia, and when black flies or biting midges are most abundant. Black fly attacks on people, cattle, horses, and pigs tend to be concentrated around the ears and head. In addition to the blood loss, effects of the insect saliva can cause a variety of problems, with swelling and intense skin irritation most common (Cranshaw et al., 2005). Pruritic dermatitis with marked edema and considerable discomfort appeared mainly on the lower legs.

Allergenic asthma, nausea, and more systemic effects can also occur, a condition known as "black fly fever."

## 20.4 ONCHOCERCIASIS

The blood-feeding female black flies are serious biting pests and obligate vectors of three nematodes (*Dirofilaria*, *Mansonella*, and *Onchocerca*) to humans. *Dirofilaria* is a genus of nematodes in the family Onchocercidae, which causes dirofilariasis. Three species—*D. ursi*, *D. subdermata*, and *D. striata*—are commonly transmitted to humans. *D. ursi* is transmitted by black flies and the other two transmitted by mosquitoes. Humans are not common hosts and may become so after being fed upon by infected black flies. *Mansonella* is a genus of nematodes in the family Onchocercidae and it causes serous cavity filariasis or mansonelliasis in humans. *M. perstans*, *M. streptocerca*, and *M. ozzardi* are commonly transmitted to humans by black flies and mosquitoes. *Onchocerca* is a genus of nematodes in the family Onchocercidae. The species *Onchocerca volulus* causes the disease of onchocerciasis or river blindness in humans. The following six parasites are commonly transmitted to various animals: *Onchocerca cervicalis* (infecting horses), *O. gutturosa* and *O. lienalis* (cattle), *O. eberbardi* (sika deer: *Cervus nippon*), *O. skrjabini* (sika deer and serows: *Capricornis crispus*), and *O. suzukii* (serows) (Takaoka et al. 2004). *O. dewittei japonica*, a parasite of wild boar, is reported as zoonotic onchocerciasis (Ishizawa et al., 2015).

### 20.4.1 Human Onchocerciasis

Human onchocerciasis is the second leading cause of blindness due to infection in the world. Black flies are the important vector of *Onchocerca volvulus*. This disease is commonly found in Africa, and Central and South America. In Africa and South America, *Simulium damnosum* complex is the major vectors of *O. volvulus*, whereas the *S. naevi* group is in East Africa and *S. ochruceum* in Southern Mexico and Guatemala are responsive vectors of *O. volvulus*. The large black fly populations and strong bite reactions can be life threatening and have been reported to kill domestic animals as well as humans, and the injection of venom into the skin causes intense itching, local swelling and soreness. Severe complications (swelling) from female black fly bites are possible in allergic individuals and rare cases of death from toxemia or anaphylactic shock. The bites of female black fly are seriously painful and due to the release of pharmacologically active substances such as histamine, leukotrienes, prostaglandins, platelet activating factor, and eosiniphilic chemotactic factor (ECF) from IgE-sensitized basophils and mast cells after contact with antigens in the flies' saliva acute, urticaria, which is one of the disorders due to hypersensitivity, may occur. The condition is characterized by local wheals and erythema in the dermis. In urticaria, pruritis is the first symptom followed by the appearance of wheals. Acute urticaria is a self-limited condition and treatment is palliative. In hypersensitive persons, bites can be fatal.

#### 20.4.1.1 Life Cycle of *Onchocerca volvulus*

A parasitic worm of *Onchocerca volvulus* in the family Onchocercidae lives in the human body for up to 14 years (Figure 20.2). Males of *O. volvulus* range from 2 to 3 cm long and the female is about 60 cm long. Adults occur in the subcutaneous tissues and in nodules. The size of microfilariae is 300 × 8 mm, and 1000 to 3000 are produced per day per adult female worm. Adult worms have longevity of 10 to 15 years. *O. volvulus* has no common animal reservoir. It develops only in man, although it has been transmitted experimentally to the chimpanzee. The filaria passes through several stages during a development cycle that takes place partly in man and partly in black fly. Figure 20.3 represents the life cycle of *O. volvulus* in man. During a blood meal, an infected black fly (genus *Simulium*) releases third-stage filarial larvae into the skin of the human host (1). The larvae (2) develop into adult filariae in subcutaneous tissues. Adult worms are then surrounded by connective tissue and can be palpated as nodules (3). Some of these nodules contain numerous male and female worms. Adults can reach an age of up to 15 years. Females measure 33 to 50 cm in length and 270 to 400 m in diameter, while males only reach a length of 19 to 42 mm and a diameter of 130 to 210 m. Female worms are capable of producing their offspring, the so-called microfilariae, for approximately 9 years. These microfilariae have a length of 220 to 360 m, a diameter of 5 to 9 m and can survive for up to 2 years. Microfilariae are typically found in the skin and in the lymphatics of connective tissues (4). Occasionally, they may be seen in the peripheral blood, urine, or sputum. During a blood meal, black flies ingest some of these microfilariae (5). After ingestion, the microfilariae migrate from the black fly's midgut through the hemocoel and

**FIGURE 20.2** Morphology of *Onchocerca volvulus*.

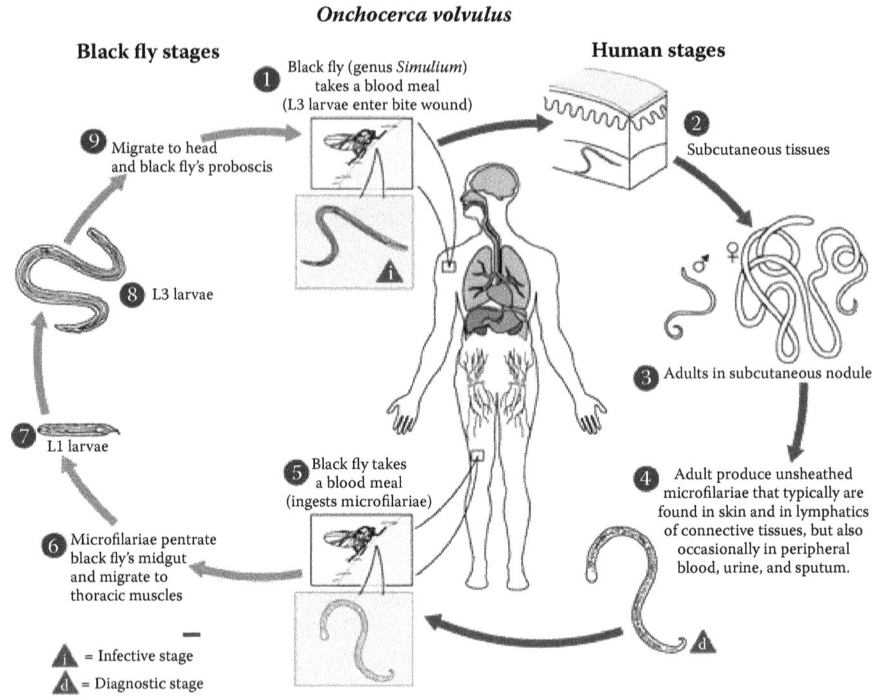


**FIGURE 20.3** Life cycle of *Onchocerca volvolus* (from DPDx: CDC's web site for laboratory identification of parasites).

invade the thoracic muscles (6). There they live as intracellular parasites and first develop into the so-called sausage stage (7), subsequently into second- and third-stage infective larvae (8). The third-stage larva is the infective larva. It migrates to the black fly's proboscis (9) and is able to infect another human when released during a fly's blood meal (1).

### 20.4.1.2 Symptoms

Onchocerciasis can have a variety of clinical presentations, which normally appear 9 to 24 months after the bites of the infected flies. Inflammatory responses to dead or dying microfilaria are responsible for the majority of the clinical manifestations, which can range from the dermatological to the ocular. Approximately 10% of those infected with the disease are completely asymptomatic. The first signs that a person has been infected with the *O. volvulus* parasite include fever, neuralgic pain in joints, and temporary hives on the trunk and face. Followed by the two main symptoms that occur in skin: onchocercal dermatitis and onchocercomas. Also ocular symptoms occur. Initially, dead microfilariae in the eye cause inflammation and keratitis. If the infection persists over a number of years, sclerosing keratitis, iridocyclitis, and uveitis occur, resulting in permanent visual impairment, glaucoma, or blindness.

### 20.4.2 Zoonotic Onchocerciasis

An infectious disease that is transmitted from animals to humans is called a zoonotic disease. Several nematodes are transmitted from animals to humans by black flies (e.g., *O. lupi, O. gutturosa, O. cervicalis, O. dewittei japonica*, and *O. jakutensis*). *O. dewittei japonica* from wild boar in Japan is commonly transmitted to humans by black fly bites. The black fly *Simulium bidentatum*, anthropophilic and zoophilic, is a natural vector of *O. dewittei japonica* (Fukuda et al., 2008, 2010a, 2010b). Human zoonotic onchocerciasis is rare and few cases have so far been reported in the world and most of the cases are from Japan (Uni et al., 2010).

## 20.5 TREATMENT AND CONTROL: MICROBIAL PERSPECTIVES

Black flies are a constant threat to health of humans and cause human onchocerciasis. This disease is controlled by two major methods:

1. Parasite control—The killing of microfilariae and reduction of longevity of adult females of *O. volvulus* in man using the drug ivermectin.
2. Vector control—The destruction of the vector simuliid with larvicidal and microbial methods.

The control the larval black flies by chemical methods creates serious ecological problems and pesticide-resistant black flies populations have evolved. Also beneficial organisms present in the same ecological habitat are eliminated, and consequently environmental pollution is increased. Therefore, scientists have been vigorously recommending a new and environmentally safe technology to control black flies as well as the diseases by the application of microbes (Table 20.1). The role of microbes in earth is the most imperative and extensively useful to man. Previously, microbes in animal systems were underappreciated in the aspect of the animal life cycle. Recently, experimentally it has been proved that microbes have wide-ranging association with animal behavior, for example, host–microbe interactions, ecological perspective, medical perspective, and nutrient supplementation. Medically, biological control of several diseases is recommended to cure and prevent the diseases. Microbes act as an adjuvant, which is prescribed with chemotherapy.

### 20.5.1 Parasite Control

The parasite control is managed by physical, chemical, and biological methods. Nodulectomy is the effective physical method to remove popular nodules in the human body. It helps to decrease microfilaria production from the worm of *O. volvulus*, because female adult worms can produce up to 1900 microfilaria a day. Chemically, ivermectin is an effective drug for killing of microfilariae and longevity of adult females (Basanez et al., 2008) and repeated dosing at intervals from 3 to 12 months would be recommended for at least 10 to 12 years. In addition, the drugs of doxycycline, rifampin, and azithromycin have been tested (Bah et al., 2013). Further, moxidectin, which is closely related to ivermectin, is an antiparasitic drug that was

**TABLE 20.1**
**Microbial Applications to Control Human Onchocerciasis**

| Microbe Used | Control Organism | Reference |
|---|---|---|
| Primary powder R 153-78 made of *Bacillus thuringiensis* var. *israelensis* | Larvae of *Simulium damnosum* | Guillet and de Barjac (1979) |
| *Bacillus thuringiensis* (H-14) | Larvae of *Simulium damnosum* | Lacey et al. (1982) |
| *Bacillus thuringiensis* (ONR-60A) | Larvae of *Simulium verecundam, Cnephia ornithophilia, C. mutate, S. vittatum*, and *Prosimulium mixtum* | Undeen and Nagel (1978) |
| *Bacillus thuringiensis* var. *israelensis* | Larvae of *Simulium ochraceum* | Undeen et al. (1981) |
| *Bacillus thuringiensis* | Larvae of *S. vittatum, S. argus, S. piperi, S. tescorum, S. aureum, and S. virgatum* | Lacey et al. (1978) |
| *Bacillus thuringiensis* var. *israelensis* | *S. vittatum* | Frommer et al. (1981) |
| Thirteen strains of *Bacillus thuringiensis* | *S. vittatum* larvae | Lacey and Mulla (1977) |
| *Bacillus thuringiensis* | Larvae of *S. maculatum, S. reptans*, and *S. erythrocephalum* | Bartninkaitė et al. (2006) |
| *Bacillus thuringiensis israelensis* | Larvae *of S. jenningsi* group | Jackson et al. (2002) |
| *Bacillus thuringiensis* var. *israelensis* | *S. vittatum* larvae | Iburg et al. (2011) |
| *Wolbachia* + Mectizan | *Onchocera volvulus* | Higazi et al. (2005), Tamarozzi et al. (2011) |
| *Wolbachia* + Doxycycline | *Onchocera volvulus* | Pfarr and Hoerauf (2006), Hoerauf et al. (2008), Turner et al. (2010) |

studied by the World Health Organization (WHO) (Siva, 2009). In the biological perspective, the endosymbiotic *Wolbachia* bacterial species targeting control of onchocerciasis was very effective. The targeting of *Wolbachia* with antibiotics therapy provides superior efficacy for individual treatment and has achieved the "holy grail" of a safe macrofilaricidal therapy that has so far eluded the field of onchocerciasis chemotherapy (Tamarozzi et al., 2011).

### 20.5.2 Vector Control

In order to control onchocerciasis, vector control is significant. Initially, DDT (dichlorodiphenyltrichloroethane) was used (Brown, 1962), however, beneficial organisms seriously suffered and resulted in environmental pollution. This method was withdrawn due to forbidding of DDT by government order. An individual water-flow and integrated-water flow manipulation have been suggested (Car, 1983; De Moor, 1982), and these methods are unsuccessful due to long distance of water-flow and impracticality throughout the river length. At this time, a microbial approach by using *Bacillus thuringiensis* (Bti) was employed to control black fly larvae. Laboratory and field trials by Car (1984) also showed that Bti was effective in controlling black flies, but that its toxicity was considerably reduced in polluted rivers with a high sewage level and high chloride concentration. Subsequent trials in the field during 1983 confirmed the efficacy of Bti against black flies and its low toxicity to nontarget organisms (Car and De Moor, 1984). Current discoveries of new bacterial isolates precise to new hosts and recent genetic improvements in these isolates have formed the potential for a wide range use of these biological control agents. The proficient microbial control of black flies is achieved, but an appropriate knowledge of factors relating to the safe and successful use of these biological control agents is obligatory (de Barjac and Sutherland, 2012).

Further, Trichomycetes are another alternative control for black fly larvae. These fungi are commonly live in the digestive tract of arthropods including larval black flies. Knowledge of Trichomycetes and their relationship with black flies might be a valuable tool in biological control of vector populations (Rios-Velasquez and Hamada, 2002). Arthropods are infected by Trichomycetes after ingestion of trichospores or zygospores (in the order Harpellales). These spores attach themselves by a holdfast ("foot") to the peritrophic matrix in the midgut and to the chitinous intima in the hindgut without penetrating the walls of the intestine; in this way, no physical injury is caused to the wall of the intestine. Black fly larvae have been reported as hosts of a large number of Trichomycetes species in the genera *Paramoebidium, Harpella, Genistellospora, Graminelloides, Pennella, Simuliomyces, Smittium*, and *Stipella.* In general, the Trichomycetes are commensalistic, but the type of relationship with the host can vary according developmental characteristics of the host and environmental parameters of the host's habitat. Species in the order Harpellales have a mutualistic relationship with their host, but some species have been reported as parasites of black fly ovaries or as pathogenic to the larval host.

## 20.6 CONCLUSION

As per a WHO report, out of the estimated 18 million infected people with onchocerciasis worldwide more than 80% live in Africa. The aims to control onchocerciasis entail eliminating the vector and hence the disease. Treatment of the disease in endemic areas is successfully administrated by ivermectin with endosymbiotic *Wolbachia* and this method provides valuable outputs. Likely, vector control is most effective by using *Bacillus thuringiensis* and Trichomycetes. Although, effective ongoing black fly control with microbial applications has achieved the target level, practically it has failed several times in the field due to a variety of breeding habitats, migratory behavior, seasonal fluctuations, and environmental factors. Hence, the incessant research to adjust and develop control programs for the black fly is fundamental.

## ACKNOWLEDGMENTS

This work was supported by the Science and Engineering Research Board (SERB), Government of India (Ref. No. ECR/2016/000191).

## REFERENCES

Adler, P. H. and Crosskey, R. W. 2016. World blackflies (Diptera: Simuliidae): A comprehensive revision of the taxonomic and geographical inventory. http://blackflies.info/content/world-inventory-simuliidae.

Anbalagan, S., Arunprasanna, V., Dinakaran, S. and Krishnan, M. 2014. Two new species of *Simulium* (*Gomphostilbia*) (Diptera: Simuliidae) from Peninsular India with keys to Peninsular Indian members of the genus *Simulium*. *Zootaxa* 3861(5): 451–465.

Bah, G. S., Ward, E. L., Srivastava, A., Trees, A. J., Tanya, V. N. and Makepeace, B. L. 2013. Efficacy of three-week oxytetracycline or rifampicin monotherapy compared with a combination regimen against the filarial nematode *Onchocerca ochengi*. *Antimicrobial Agents and Chemotherapy* 58(2): 801–810.

Bartninkaitė, I., Bernotienė, R., Pakalniškis, S. and Žygutienė, M. 2006. Control of blood-sucking black fly (Simuliidae) populations in Lithuania. *Ekologija* 4: 70–75.

Basanez, M. G., Pion, S. D., Boakes, E., Filipe, J. A., Churcher, T. S. and Boussinesq, M. 2008. Effect of single-dose ivermectin on *Onchocerca volvulus*: A systematic review and meta-analysis. *Lancet Infectious Diseases* 8(5): 310–222.

Brown, A. W. A. 1962. A new insecticide for the control of black flies (Diptera: Simuliidae). *Bulletin of the World Health Organization* 27: 51–522.

Car, M. 1983. The influence of water-level fluctuation on the drift of *Simulium chutteri* Lewis 1965 (Diptera: Nematocera) in the Orange River, South Africa. *Onderstepoort Journal of Veterinary Research* 50: 173–177.

Car, M. 1984. Laboratory and field trials with two *Bacillus thuringiensis* var. *israelensis* products for *Simulium* (Diptera: Nematocera) control in a small polluted river in South Africa. *Onderstepoort Journal of Veterinary Research* 51: 141–144.

Car, M. and De Moor, F. C. 1984. The response of Vaal River drift and benthos to Simulium (Diptera: Nematocera) control using *Bacillus thuringiensis* var. *israelensis* (H-14). *Onderstepoort Journal of Veterinary Research* 51: 155–160.

Cranshaw, W. S., Peairs, F. B. and Kondratieff, B. 2005. Biting flies. *Colorado State University Journal of Cooperative Extension-Horticulture* 5: 582–591.

Davies, J. B. and Crosskey, R. W. 1992. *Simulium—Vectors of onchocerciasis*. Geneva: World Health Organization.

de Barjac, H. and Sutherland, D. J. 2012. *Bacterial and control of mosquitoes and black flies*. Springer.

De Moor, F. C. 1982. *A community of Simulium species in the Vaal River near Warrenton*. Ph.D. thesis, University of the Witwatersrand, Johannesburg, South Africa.

Frommer, R. L., Hembree, S. C., Nelson, J. H., Remington, M. P. and Gibbs, P. H. 1981. The evaluation of *Bacillus thuringiensis* var. *israelensis* in reducing *Simulium vittatum* (Diptera: Simuliidae) larvae in their natural habitat with no extensive aquatic vegetative growth. *Mosquito News* 41(2): 339–347.

Fukuda, M., Otsuka, Y., Uni, S., Bain, O. and Takaoka, H. 2010a. Genetic evidence for the presence of two species of *Onchocerca* from the wild boar in Japan. *Parasite* 17: 33–37.

Fukuda, M., Otsuka, Y., Uni, S., Bain, O. and Takaoka, H. 2010b. Molecular identification of infective larvae of three species of *Onchocerca* found in wild-caught females of *Simulium bidentatum* in Japan. *Parasite* 17: 39–45.

Fukuda, M., Takaoka, H., Uni, S. and Bain, O. 2008. Infective larvae of five *Onchocerca* species from experimentally infected *Simulium* species in an area of zoonotic onchocerciasis in Japan. *Parasite* 15: 111–119.

Guillet, P. and de Barjac, H. 1979. Toxicity of *Bacillus thuringiensis* var. *israelensis* for *Simulium* larvae, vectors of onchocerciasis. *Comptes Rendus de l'Académie des sciences, Serie D, Sciences Naturelles* 289(6): 549–552.

Harwood, R. F. and James, M. T. 1979. *Entomology in human and animal health.* 7th edition. New York: Macmillian.

Higazi, T. B., Filiano, A. F., Katholi, C. R., Dadzie, Y., Remme, J. H. and Unnasch, T. R. 2005. *Wolbachia* endosymbiont levels in severe and mild strains of *Onchocerca volvulus. Molecular and Biochemical Parasitology* 141(1): 109–112.

Hoerauf, A., Specht, S., Büttner, M., Pfarr, K., Mand, S., Fimers, R., Marfo-Debrekyei, Y., Konadu, P., Debrah,.Y., Bandi, C., Albers, A., Larbi, J., Batsa, L., Adjei, O. and Buttner, D. 2008. *Wolbachia* endobacteria depletion by doxycycline as antifilarial therapy has macrofilaricidal activity in onchocerciasis: A randomized placebo-controlled study. *Medical Microbiology and Immunology* 197: 295–311.

Iburg, J. P., Gray, E. W., Wyatt, R. D., Cox, J. E. Fusco, R. A. and Noblet, R. 2011. The effect of seston on mortality of *Simulium vittatum* (Diptera: Simuliidae) from insecticidal proteins produced by *Bacillus thuringiensis* subsp. *Israelensis. Environmental Entomology* 40(6): 1417–1426.

Ishizawa, H., Kato, S., Nishimura, H., Tanaka, R., Fukuda, M., Takaoka, H. and Tsutsum, Y. 2015. Zoonotic onchocerciasis. *Pathology International* 65: 271–273.

Jackson, J. K., Horwitz, R. J. and Sweeney, B. W. 2002. Effects of *Bacillus thuringiensis israelensis* on black flies and non-target macroinvertebrates and fish in a large river. *Transaction of the American Fisheries Society* 131(5): 910–930.

Lacey, L. A., Escaffre, H., Philippon, B., Seketeli, A. and Guillet, P. 1982. Large river treatment with Bacillus thuringiensis (H-14) for the control of Simulium damnosum s.l. in the Onchocerciasis Control Programme. *Tropenmedizin Parasitologie* 33(2): 97–101.

Lacey, L. A. and Mulla, M. S. 1977. Evaluation of Bacillus thuringiensis as a biocide of blackfly larvae (Diptera: Simuliidae). *Journal of Invertebrate Pathology* 30(1): 46–49.

Lacey, L. A., Mulla, M. S., and Dulmage, H. T. 1978. Some affecting the pathogenicity of *Bacillus thuringiensis* Berliner against blackflies. *Environmental Entomology* 7: 583–588.

Pfarr, K. M. and Hoerauf, A. M. 2006. Antibiotics which target the *Wolbachia* endosymbionts of filarial parasites: A new strategy for control of filariasis and amelioration of pathology. *Mini-Reviews in Medicinal Chemistry* 6: 203–210.

Rios-Velasquez, C. and Hamada, N. 2002. Trichomycete fungi (Zygomycota) associated with the digestive tract of *Simulium goeldii* Cerqueira & Nunes de Mello and *Simulium ulyssesi* (Py-Daniel & Coscarón) (Diptera: Simuliidae) Larvae, in Central Amazônia, Brazil. *Memoria do Instituto Oswaldo Cruz, Rio de Janeiro* 97(3): 423–426.

Siva, N. 2009. WHO researchers start trial on a new drug for river blindness. *BMJ* 339: b2755.

Takaoka, H., Bain, O., Uni, S. Korenaga, M., Kozek, W. J., Shirasaka, C., Aoki, C., Otsuka, Y., Fukuda, M., Eshita, Y. and Daa, T. 2004. Zoonotic onchocerciasis caused by a parasite from wild boar in Oita, Japan. A comprehensive analysis of morphological characteristics of the worms for its diagnosis. *Parasite* 11(3): 285–292.

Tamarozzi, F., Halliday, A., Gentil, K., Hoerauf, A., Pearlman, E. and Taylor, M., J. 2011. Onchocerciasis: The role of *Wolbachia* bacterial endosymbionts in parasite biology, disease pathogenesis, and treatment. *Clinical Microbiology Reviews* 24(3): 459–468.

Turner, J. D., Tendongfor, N., Esum, M., Johnston, K. L., Langley, R. S., Ford, L., Faragher, B., Specht, S., Mand, S., Hoerauf, A., Enyong, P., Wanji, S. and Taylor, M. J. 2010. Macrofilaricidal activity after doxycycline only treatment of *Onchocerca volvulus* in an area of *Loa loa* co-endemicity: A randomized controlled trial. *PLOS Neglected Tropical Diseases* 4(4): e660.

Undeen, A. H. and Nagel, W. L. 1978. The effect of *Bacillus thuringiensis* ONR-60A strain (Goldberg) on *Simulium* larvae in the laboratory. *Mosquito News* 524–527.

Undeen, A. H., Takaoka, H. and Hansen, K. 1981. A test of *Bacillus thuringiensis* var. *israelensis* De Barjac as a larvicide for *Simulium ochraceum*, the central American vector of onchocerciasis. *Mosquito News* 41(1): 37–40.

Uni, S., Boda, T., Daisaku, K., Ikura, Y., Maruyama, H., Hasegawa, H., Fukuda, M., Takaoka, H. and Bain, O. 2010. Zoonotic filariasis caused by *Onchocerca dewittei japonica* in a resident of Hiroshima Prefecture, Honshu, Japan. *Parasitology International* 59: 477–480.

World Health Organization (WHO). 2016. *Vector-borne diseases*. http://www.who.int/mediacentre/factsheets/fs387/en/.

# *Section V*

## *Microbial Control of Flea Vector*

# 21 Effective Microbial Compounds for Controlling Flea-Borne Rickettsial Diseases

*Arunachalam Chinnathambi**
*and Ravikumar Mathrubutham*

## CONTENTS



## 21.1 INTRODUCTION

Rickettsiae are the group of microorganisms that phylogenetically fills the gap between bacteria and viruses. Rickettsiae are vector-borne zoonotic infections caused by obligate intracellular bacteria of the genus *Rickettsia* belonging to alpha group of the phylum Proteobacteria. They are divided into four groups based on their genotypic characteristics, namely, spotted fever group (SFG) (*R. rickettsii, R. conorii, R. parkeri*, and several others), typhus group (TG) (*R. prowazekii* and *R. typhi*), transitional group (*R. felis, R. akari*, and *R. australis*), and the nonpathogenic ancestral group (*R. bellii* and *R. canadensis*) (Mansueto et al., 2012). *R. typhi* (known in the old literature as *R. mooseri*) and *R. felis* are the etiological agents of murine typhus and flea-borne spotted fever. Both bacteria produce a flu-like febrile syndrome, frequently accompanied by rash.

Rickettsia is the member of the TG of rickettsiae and closely resembles the other member of that group *R. prowazekii* in its genetical, biological, and immunological

* Specified for Corresponding author : e mail:dr.arunmicro@gmail.com

characteristics. Normally, commensal rats belonging to the genus *Rattus* are the primary zoonotic reservoir of *R. typhi*, followed by the oriental rat flea, *Xenopsylla cheopsi*, which is the classic vector that transmits murine typhus to humans.

Humans are the prime reservoir for *R. prowazekii* where the pathogen persists despite humoral and cellular immune responses. While clinically seeking, epidemic typhus is fatal in 10% to 50% of untreated patients and is more likely to be fatal in human communities with concurrent malnourishment (Eremeeva and Dasch, 2015). *Ctenocephalides felis* yet more than 20 other species of fleas, ticks, and mites species have been reported to harbor *R. felis*, which may pose an adverse impact on human health with changes to the ecology and distribution of *R. felis* (Abdad et al., 2011).

### 21.1.1 Natural History of Flea-Borne Rickettsial Diseases

Flea-borne bacterial diseases include cat-scratch disease, plague, murine typhus, and flea-borne spotted fever (FBSF) due to *R. felis*. Rickettsia in cat flea *Ctenocephalides felis* was first described and published in 1918, and it was authentically named *R. ctenocephali* followed by a nomenclature pattern and rediscovered in 1990. Further, in the course of discovery about rickettsial types, with the advent of optical instrumentation practices and with the advance of microscopy techniques, a new rickettsial species was observed in the ticks of the carrier of these species through electron microscopy (Adams et al., 1990). Later in 1994, the same rickettsial bacterium was detected at a larger extent by polymerase chain reaction (PCR) following genomic sequence comparison based on the 17-kDa protein gene sequencing, and was considered to be a new SFG *Rickettsia* named *R. felis* (Higgins et al., 1996). *R. felis* was thoroughly characterized and validated as a unique SFG *Rickettsia* in 2001 when culture conditions using *Xenopislaevis* tissue cells and mosquito cells at 28°C were established (La Scola et al., 2002).

Murine typhus is a flea-borne typhus caused by the bacterium *Rickettsia typhi* and it is also known as endemic typhus or flea-borne typhus. It has been known since the early 20th century and is described as an illness of ports and urban areas where synanthropic rodents, particularly *Rattus rattus* and *R. norvegicus*, are classically involved as primary reservoirs and amplifying hosts of both *Rickettsia* and the flea vector *Xenopsylla cheopis*. However, wild and domestic animals and their respective ectoparasites have also been implicated in the disease cycle (Raoult and Parola, 2009). The disease is prevalent in many areas of the world, including Southeast Asia, Southern Europe, the Mediterranean, Northern Africa, South America, and the United States (Labruna, 2009; Renvoise and Raoult, 2009).

Flea-borne spotted fever (cat flea typhus) was described recently as human infection (Table 21.1). The clinical characteristics are very similar to murine typhus, except that an inoculation eschar is occasionally found; no fatalities have been reported (Raoult and Parola, 2009; Reif and Macaluso, 2009). The cat flea (*Ctenocephalides felis*) appears to be the main vector and reservoir of *R. felis* with a prevalence of infection ranges between 4% and 90% levels. The cosmopolitan condition of *C. felis* and its mammalian hosts (cats and dogs) probably explains the multiple reports arising from many countries like Egypt, France, Germany, South Korea, Spain, Taiwan, Thailand, the United States, Mexico, and Brazil, linked clearly an emerging rickettsioses of medical importance (Perez-Osorio et al., 2008; Reif and Macaluso, 2009).

**TABLE 21.1**
**Spotted Fever Groups**

| Disease | Rickettsial Agents | Geographic Location | Reference |
|---|---|---|---|
| Cat flea rickettsiosis, flea-borne typhus | *Rickettsia felis* | Worldwide | Perez Osorio et al., 2008 |
| Mild spotted fever | *Rickettsia parkeri* | North and South America | Nava et al., 2008 |
| Rocky Mountain spotted fever | *Rickettsia rickettsii* | North, Central, and South America | Galvao et al., 2003 |
| Flinders Island spotted fever, Thai tick typhus | *Rickettsia honei* | Australia and Southeastern Asia | Graves and Stenos, 2003 |
| Far eastern spotted fever | *Rickettsia heilongjiangensis* | Far East | Mediannikov et al., 2004 |
| North Asian tick typhus, Siberian tick typhus | *Rickettsia sibirica* | Asia, Europe, and Africa | Fournier et al., 2005 |
| Tick-borne lymphadenopathy, necrosis, erythema | *Rickettsia slovaca* | Europe | Selmi et al., 2008 |
| African tick-bite fever | *Rickettsia africae* | Sub-Saharan Africa, Caribbean | Mediannikov et al., 2010 |
| Mediterranean spotted fever, boutonneuse fever, Israeli spotted fevers, Astrakhan fever, Indian tick typhus | *Rickettsia conorii* | Southern Europe, Southern and Western Asia, and Africa | Rovery et al., 2008 |
| Japanese spotted fever or Oriental spotted fever | *Rickettsia japonica* | Japan and Eastern Asia | Chung et al., 2006 |
| Rickettsialpox | *Rickettsia akari* | Worldwide | Saah, 2000 |
| Queensland tick typhus | *Rickettsia australis* | Australia | Sexton et al., 1991 |

**TABLE 21.2**
**Species of Flea Connected by *Rickettsia felis***

| Species of Flea | Geographic Location | Reference |
|---|---|---|
| *Ctenocephalides canis* | Africa, South America | Horta et al., 2006a; Venzal et al., 2006 |
| *Chlamydophila felis* | Asia, Australia, Africa, United States, Europe, South America, | Rolain et al., 2003; Marie et al., 2006; Schloderer et al., 2006; Hawley et al., 2007 |
| *Archaeopsylla erinacei* | Africa, Europe | Bitam et al., 2006; De Sousa et al., 2006 |
| *Anomiopsyllus nudata* | United States | Stevenson et al., 2005 |
| *Ctenophtalmus* sp. | Europe | De Sousa et al., 2006 |
| *Pulex irritans* | Europe | Brouqui et al., 2007 |
| *Xenopsylla cheopis* | Asia | Jiang et al., 2006 |

However, it is likely to be underdiagnosed, unlike murine typhus and other rickettsioses, because of the absence of commercially available diagnostic methods that are applicable during the acute stage. Further it is also found that serological diagnosis is complicated by cross-reactivity of the human humoral immune response against different rickettsiae (Raoult and Parola, 2009; Lim et al., 2012).

Moreover, *R. felis* is the only known species of SFG that is transmitted by fleas. Today several species of fleas have been associated with *R. felis* and the cat flea *C. felis* is one of the most frequent external parasites of companion animals worldwide (Table 21.2). *C. felis* is generally regarded as the predominant species to find as an ectoparasite on dogs, cats, and opossums. *R. felis* DNA has been sequenced at molecular levels in the United States with an infection rate of 3.8% and 7.6%, in Israel and up to 12% in the United Kingdom (Bauer et al., 2006; Znazen and Raoult, 2007). Antibodies to *R. felis* have been detected in cats, but the current infection rate using polymerase chain reaction (PCR) assay or cell-culture techniques has not been fully described (Case et al., 2006; Hawley et al., 2007). Although data suggests that rickettsial isolation through *R. felis* may be prevalent worldwide, their transmission through fleas and mammals still seems unclear (Parola et al., 2005).

### 21.1.2 Indian Scenarios

Rickettsial diseases in India have been documented from Jammu and Kashmir, Himachal Pradesh, Uttaranchal, Rajasthan, Assam, West Bengal, Maharashtra, Kerala, and Tamil Nadu (Mathai et al., 2001; Sundhindra et al., 2004; Mahajan et al., 2006; Batra, 2007) and also reported a high magnitude of scrub typhus, spotted fever, and Indian tick typhus caused by *R. conorii* (Batra, 2007). An extensive study on tick-borne rickettsiosis in Maharashtra revealed that Indian tick typhus exist as a zoonosis (Padbidri et al., 1984).

## 21.2 LABORATORY FINDINGS

No single laboratory finding is specific for early diagnosis. Various laboratory abnormalities found in rickettsial disease are frequently associated with (or) revealed in hematology, biochemistry, serology (Weil-Felix test and immunofluorescence assay [IFA]), polymerase chain reaction assay, and immunohistochemistry and cell culture techniques (laboratory animals), which are equally restricted to research laboratories (Rathi and Rathi, 2010).

The Weil-Felix test is still not clearly providing correct results and has to be interpreted or described through correct clinical diagnosis (Mahajan et al., 2006). It has been found to be insignificant for initiating therapy in suspected cases.

### 21.2.1 Differential Diagnosis

Serological tests are the more precise frequently used and widely available methods for the diagnosis of rickettsial patients. The Weil-Felix test was developed 90 years ago but still continuously used by many clinical laboratories throughout the world. Microimmunofluorescence (MIF) is widely accepted as the reference method and is reliable but does not allow differentiation of infection among SFG rickettsiae (Hechemy et al., 1983, 1989).

Western blot assays and cross-absorption studies can also be used when differences in titers values between several antigens are lower than two dilutions and in cases of cross-reactivity. However, these techniques are time- and antigen-consuming reactive steps (Znazen and Raoult, 2007). Rickettsia can also be detected and identified by PCR and gene sequencing methods targeting the enzyme citrate synthase (*gltA*), the 16S rRNA, the rickettsia genus-specific 17-kDa antigen, and SFG-specific outer membrane protein A (*ompA*) or protein B (*ompB*) genes (Pornwiroon et al., 2006; Znazen and Raoult, 2007). Recently, an advanced version of quantitative real-time polymerase chain reaction (qPCR) assay detecting outer membrane protein B genes (*ompB*) was proposed for the detection of the rickettsia (Henry et al., 2007).

Rickettsial diseases can be easily confused with a variety of viral (measles, enteroviral exanthems, dengue, infectious mononucleosis), protozoal (malaria), bacterial (meningococcemia, typhoid, leptospirosis, toxic shock syndrome, scarlet fever), and collagen vascular (Kawasaki disease, other vaculities) diseases, and adverse drug reactions. It is evident that invasive meningococcal disease may not be reliably distinguished from rickettsial disease clinically (Chapman et al., 2006).

## 21.3 TREATMENT OF FLEA-BORNE RICKETTSIAL DISEASES

Many rickettsial diseases cannot be easily diagnosed with commercially available diagnostic test kits during the acute stage of the disease. Therefore, early empiric antibiotic therapy should be prescribed in any suspected flea-transmitted rickettsiosis before confirmation of the diagnosis. Doxycycline (200 mg per day) remains the treatment of choice for flea- and tick-transmitted SFG rickettsioses, including in children (Raoult and Roux, 1997; Purvis and Edwards, 2000; Holman et al., 2001; Masters et al., 2003). In cases of allergy to tetracyclines, chloramphenicol or josamycin (a macrolide not available in the United States) may be administered. Fluoroquinolones and newer macrolides may also be used (Rolain et al., 1998; Cascio et al., 2002; Segura and Anton, 2002). To treat pregnant women, josamycin or newer macrolides can be used effectively. In patients with severe disease, doxycycline should be administered intravenously up to 24 h after apyrexia. The exact treatment duration is not fully determined. Usually, therapy should be prescribed for up to 2 or 3 days after the patient's fever has abated. Kodama et al. (2002) highlighted the controversies over the use of corticosteroids in severe forms at the acute stage. Definitive treatment should be instituted on the basis of clinical and epidemiological clues as early as possible to avoid severe disease and fatal outcome (Kirkland et al., 1995).

Various antibiotics useful for treating rickettsial diseases are tetracyclines, preferably doxycycline, chloramphenicol, macrolides specially azithromycin, clarithromuycin, roxythromycin, telithromycin, rifampin and fluroquinolones, ciprofloxacin, ofloxacin, pefloxacin, and levofloxacin, which show promising effects in controlling the degree of intensity of the infection cycle (Rolain et al., 2002; Siberry and Dumler, 2007).

Doxycycline is the drug of choice in Louse-borne (epidemic) typhus. Oral treatment is used unless the patient is vomiting or obtunded. The dose is 5 mg/kg/day in two divided doses for children under 45 kg, and 200 mg/day in two divided doses for children over 45 kg. Duration of therapy should be at least 3 days after defervescence or a minimum of 5 to 7 days. The conventional antibiotic regimen for SFG rickettsioses is a 7- to 14-day oral course of doxycycline (200 mg/day) (Table 21.3). All reported cases of FBSF rickettsioses have been treated with doxycycline (Chung et al., 2000; Znazen and Raoult, 2007).

Tetracycline derived from *Streptomyces aureofaciens* can also be used in a dosage of 500 mg 6 hours orally or intravenously for 7 days. Infected pregnant patients must

**TABLE 21.3**
**Suggested Antibiotics for Spotted Fever Group of Rickettsiosis**

| Suggested Antibiotic | Duration | Recommended Dose | Condition | Reference |
|---|---|---|---|---|
| Doxycycline | 1 day | 200 mg in single or two doses p.o. | Adult and child above 8 years | Bella-Cueto et al., 1987; Bella et al., 1990 |
| Doxycycline | 7–14 days | 100 mg b.i.d. p.o. (i.v. in case of severe disease) | Adult and in child above 8 years | Meloni and Meloni, 1996; Ruiz Beltran et al., 1992 |
| Doxycycline | 1 day | 4 mg/Kg/day in two doses p.o. (no more than 200 mg) | Child of less than 8 years old | Bella-Cueto et al., 1987; Bella et al., 1990 |
| Fluoroquinolone | 10–14 days | Ofloxacin 200 mg b.i.d. p.o.; ciprofloxacin 500 mg b.i.d. p.o. | Adult and child above 8 years | Gudiol et al., 1989; Jabarit Aldighieri et al., 1992 |
| Macrolide | 7 days | Josamycin 50 mg/Kg/day b.i.d. p.o. | Adult and child above 8 years | Bella et al., 1990 |
| Chloramphenicol | 7–14 days | 500 mg every 6 hours | Child of less than 8 years old; adult and child above 8 years | Holman et al., 2001 |

be evaluated individually and either chloramphenicol (early trimester) or doxycycline (late trimester) may be used (Kim et al., 2007; Mahajan, 2008).

Chloramphenicol produced from *Streptomyces venezuelae* has more side effects. However, it is not widely used in the because of its potential bone marrow toxicity. It may not be as effective against rickettsiae as doxycycline (Mahajan, 2008).

Treatment with macrolide compounds clarithromycin and azithromycin can be considered a valid alternate therapy to tetracycline and chlormaphenicol in the treatment for children aged below 8 years and pregnant females (Alvarez-Eicoro and Enzler, 1999; Mahajan, 2008).

Azithromycin seems to be an effective agent against scrub typhus because it efficiently penetrates polymorphonuclear leukocytes and macrophages, which are target cells for *Orientia tsutsugamushi*. In addition, a long tissue half-life and the long lasting postantibiotic effects of azithroymcin may explain no relapse despite the use of a single dose (Chung et al., 2000; Van Bambeke and Tulkens, 2001; Mahajan et al., 2008).

Recently telithromycin 800 mg was administrated twice on the first day, then once daily from the next day and was continued for 2 to 3 days after the fever disappeared. It could be considered as a promising new antibacterial agent for treatment of scrub typhus (Kim et al., 2007).

Rickettsiae are also sensitive to rifampicin and quinolones. Occasional cases with resistance to doxycycline are treated with macrolides or rifampin. However, quinolones have been associated with clinical failures despite good *in vitro* activity (Rathore and Maraqa, 2009).

Sulfonamides are contraindicated in rickettsial diseases, as they increase morbidity and mortality either by delaying institution of appropriate antibiotics or directly stimulating the growth of organisms (Cascio et al., 2001; Claudid et al., 2006). No effective vaccine has been developed for scrub typhus, Q fever, and Rocky Mountain spotted fever.

### 21.3.1 Preventive Measures

Avoidance of tick bite by wearing protective clothing and use of tick repellants is more feasible than eliminating exposure by not entering flea-containing environments. Studies have demonstrated the effectiveness of permethrin-treated clothing in both controlled and field conditions (Miller et al., 2011; Vaughn et al., 2014). Prevention efforts are aimed at reducing the conditions associated in contact with the fleas of cats, dogs, rodents, and hedgehogs. To date, no vaccine or vaccination is available.

### 21.3.2 Control and Surveillance of Flea-Borne Rickettsial Diseases

Despite the increasing availability of a variety of treatments for domestic animals with long-term action against fleas, flea-borne zoonoses continue to pose a potential risk for human illness. Prevention of flea-borne zoonoses requires an understanding of the role of both wild and domestic animals in maintenance or transmission of disease, and effective measures to ensure flea control. To minimize the risk of disease transmission, pet owners and veterinarians need to remain vigilant in treating pets for fleas (Blagbum and Dryden, 2009). Consideration of these risks may also impact pet selection. For example, *Bortonella henselae* bacteremia seems more common among cats from shelters, therefore veterinarians should consider the cat's source when recommending pets or pet care to immunocompromised clients (Boulouis, 2005; Brunt, 2006). Finally, because veterinarians can be at risk for occupational exposure to flea-borne pathogens, veterinary clinics should employ infection-control practices to minimize exposures, including flea control, use of gloves, and other types of personal protective measures, and, for plague, measures to prevent aerosol and small droplet transmission when examining cats presenting with suspect illness (Brunt, 2006; Elchos, 2008; Wright, 2008).

## 21.4 CONCLUDING REMARKS

Flea-borne microorganisms that are widely distributed throughout the world fall under zoonotic infections type, where the fleas' life cycle partly merges with Rickettsial types. However, flea-borne diseases could reemerge in epidemic form because of changes in vector–host ecology due to environmental and human behavior modifications. While local environmental changes are frequent, global climate change may influence parameters of flea development, distribution, and disease transmission on a much larger scale. For many fleas, temperature and humidity are crucial for development and survival. The warmer temperatures predicted through most climate change scenarios could lead to an increased expansion of flea vectors into the Northern Hemisphere (Gage et al., 2008).

The factors influencing the emergence and recognition of flea-borne rickettsial diseases in recent years are to be seriously viewed. First, the major role of primary physicians should be emphasized, including their careful case history recording as well as their complete physical and laboratory examinations, since it was essential for the description of some emerging SFG rickettsioses. Second, the recent availability of molecular biology techniques have been shown to greatly facilitate the description of emerging human rickettsioses and the investigation of their epidemiology all over the world. Third, improved culture systems have also allowed for incrimination of a new rickettsial species in human diseases. Finally, people are more involved in outdoor activities and there is an increased in the development of tourism in rural and remote areas resulting in increased contact with flea and flea-borne pathogens.

Rickettsia are dangerous bacterial pathogens and, unfortunately, specific serological tests are only available in selected specialized laboratories, and further, it is imperative to have a high index of suspicion for rickettsial diseases and make a clinical diagnosis based on prudent history taking and appropriate physical findings. In order to control the intensity of such zoonotic diseases, preventive antimicrobial therapy has been strongly suggested because tetracycline and chloramphenicol are highly considered to be rickettsiostatic and nonrickettsiacidal. A therapeutic trial in these rickettsial patients is justified because a delay in initiating treatment may end up fatal. A rapid and favorable response is suggested for correct diagnosis.

Finally, numerous *Rickettsia* spp., have been identified in fleas throughout the world. Although their pathogenicity for people has yet to be demonstrated, they are good candidates to be involved in human diseases since many of them have been detected in fleas that readily bite people. Therefore, effective studies could lead to the description of emerging human flea-borne rickettsial diseases in the future.

## REFERENCES


Abdad M. Y., Stenos, J., and Graves, S. 2011. *Rickettsia felis*, an emerging flea transmitted human pathogen. *Emerg Health Threats J* 4: 7168–7175.

Adams, J. R., Schmidtmann, E. T., and Azad, A. F. 1990. Infection of colonized cat fleas, *Ctenocephalidesfelis* (Bouché), with a rickettsia-like microorganism. *Am J Trop Med Hyg* 43: 400–409.

Alvarez-Eicoro, S., and Enzler, M. J. 1999. The macrolide: Erythromycin, clarithromycin and azithromycin. *Mayo Clin Proc* 74: 613–634.

Batra, H. V. 2007. Spotted fever and typhus fever in Tamil Nadu—Commentary. *Indian J Med Res* 126: 101–103.

Bauer, O., Baneth, G., Eshkol, T., Shaw, S. E., and Harrus, S. 2006. Polygenic detection of *Rickettsia felis* in cat fleas (*Ctenocephalidesfelis*) from Israel. *Am J Trop Med Hyg* 74: 444–448.

Bella, F., Font, B., Uriz, S., Munoz, T., Espejo, E., Traveria, J., Serrano, J. A., and Segura, F. 1990. Randomized trial of doxycycline versus josamycin for Mediterranean spotted fever. *Antimicrob Agents Chemother* 34: 937–938.

Bella-Cueto, F., Font-Creus, B., Segura-Porta, F., Espejo-Arenas, E., Lopez-Pares, P., and Munoz Espin, T. 1987. Comparative, randomized trial of one-day doxycycline versus 10-day tetracycline therapy for Mediterranean spotted fever. *J Infect Dis* 155: 1056–1058.

Bitam, I., Parola, P., and De La Cruz, K. D. 2006. First molecular detection of *Rickettsia felis* in fleas from Algeria. *Am J Trop Med Hyg* 74: 532–535.

Blagbum, B. L., and Dryden, M. W. 2009. Biology, treatment and control of flea and tick infestations. *Vet Clin N Am Small Anim Pract* 39: 1173–1200.

Boulouis, H. J. 2005. Factors associated with the rapid emergence of zoonotic Bartonell infections. *Vet Res* 36: 383–410.

Brouqui, P., Parola, P., Fournier, P. E., and Raoult, D. 2007. Spotted fever rickettsioses in southern and eastern Europe. *FEMS Immunol Med Microbiol* 49: 2–12.

Brunt, J. 2006. American Association of Feline Practitioners 2006 panel report on diagnosis, treatment, and prevention of *Bartonella* spp. infections. *J Feline Med Surg* 8: 213–226.

Cascio, A., Colomba, C., Di Rosa, D., Salsa, L., Di Martino, L., and Titone L. 2001. Efficacy and safety of clarithromycin in the treatment of Mediterranean spotted fever in children: A randomized controlled trial. *Clin Infect Dis* 33: 409–411.

Cascio A., Colomba C., Antinori S., Paterson D. L., and Titone L. 2002. Clarithromycin versus azithromycin in the treatment of Mediterranean spotted fever in children: A randomized controlled trial. *Clin Infect Dis* 34: 154–158.

Case, J. B., Chomel, B., Nicholson, W., and Foley, J. E. 2006. Serological survey of vector-borne zoonotic pathogens in pet cats and cats from animal shelters and feral colonies. *J Feline Med Surg* 8: 111–117.

Chapman, A. S., Bakken, J. S., and Folk, S. M. 2006. Diagnosis and management of tickborne rickettsial diseases: Rocky Mountain spotted fever, ehrlichioses, and anaplasmosis—United States: A practical guide for physicians and other health-care workers and public health professionals. *MMWR Recomm Rep* 55: 1–27.

Chung, M. H., Han, H. W., and Choi, M. G. 2000. Comparison of a 3-day course of azithromycin with doxycycline for treatment of scrub typhus. *Korean J Infect Dis* 23: 433–438.

Chung, M. H., Lee, S. H., Kim, M. J., Lee, J. H., Kim, E. S., Kim, M. K., Park, M. Y., and Kang, J. S. 2006. Japanese spotted fever, South Korea. *Emerg Infect Dis* 12(7): 1122–1124.

Claudid, C., Laura, S., Valentino, F. P., Raffaella, R., and Lucio, T. 2006. Mediterranean spotted fever: Clinical and laboratory characteristics of 415 Sicilian children. *BMC Infect Dis* 6: 60.

De Sousa, R., Edouard-Fournier, P., Santos-Silva, M., Amaro, F., and Bacellar, F. 2006. Molecular detection of *Rickettsia felis*, *Rickettsia typhi* and two genotypes closely related to *Bartonellaelizabethae*. *Am J Trop Med Hyg* 75: 727–731.

Elchos, B. S. 2008. Compendium of veterinary standard precautions for zoonotic disease prevention in veterinary personnel. *J Am Vet Med Assoc* 233: 415–432.

Eremeeva, M. E., and Dasch, G. A. 2015. Challenges posed by tick-borne rickettsiae: Eco-epidemiology and public health implications. *Front Public Health* 3(55): 1–17.

Fournier, P. E. F., Gouriet, P., Brouqui, F., Lucht, and Raoult, D. 2005. Lymphangitis associated rickettsiosis, a new rickettsiosis caused by *Rickettsia sibirica mongolotimonae*: Seven new cases and review of the literature. *Clin Infect Dis* 40(10): 1435–1444.

Gage, K. L., Burkot, T. R., Eisen, R. J., and Hayes, E. B. 2008. Climate and vector borne diseases. *Am J Prev Med* 35: 436–450.

Galvao, M. A., Mafra, C. L., Moron, C., Anaya, E., and Walker D. H. 2003. Rickettsiosis of the genus Rickettsia in South America. *Ann NY Acad Sci* 990: 57–61.

Graves, S. and Stenos, J. 2003. *Rickettsia honei*: A spotted fever Rickettsia on three continents. *Ann NY Acad Sci* 990: 62–66.

Gudiol, F., Pallares, R., Carratala, J., Bolao, F., Ariza, J., Rufi, G., and Viladrich, R. F. 1989. Randomized double-blind evaluation of ciprofloxacin and doxycycline for Mediterranean spotted fever. *Antimicrob Agents Chemother* 33: 987.

Hawley, J. R., Shaw, S. E., and Lappin, M. R. 2007. Prevalence of *Rickettsia felis* DNA in the blood of cats and their fleas in the United States. *J Feline Med Surg* 9: 258–262.

Hechemy, K. E., Anacker, R. L., Carlo, N. L., Fox, J. A., and Gaafar, H. A. 1983. Absorption of *Rickettsia rickettsia* antibodies by *Rickettsia rickettsia* antigens in four diagnostic tests. *J Clin Microbiol* 17: 445–449.

Hechemy, K. E., Raoult, D., Fox, J., Han, Y., Elliott, L. B., and Rawlings, J. 1989. Cross-reaction of immune sera from patients with rickettsial diseases. *J Med Microbiol* 29: 199–202.

Henry, K. M., Jiang, J., Rozmajzl, P. J., Azad, A. F., Macaluso, K. R., and Richards, A. L. 2007. Development of quantitative real-time PCR assays to detect *Rickettsia typhi* and *Rickettsia felis*, the causative agents of murine typhus and flea-borne spotted fever. *Mol Cell Probes* 21: 17–23.

Higgins, J. A., Radulovic, S., Schriefer, M. E., and Azad, A. F. 1996. *Rickettsia felis*: A new species of pathogenic rickettsia isolated from cat fleas. *J Clin Microbiol* 34: 671–674.

Holman, R. C., Paddock, C. D., Curns, A. T., Krebs, J. W., McQuiston, J. H., and Childs J. E. 2001. Analysis of risk factors for fatal Rocky Mountain spotted fever: Evidence for superiority of tetracyclines for therapy. *J Infect Dis* 184: 1437–1444.

Horta, M. C., Chiebao, D. P., and De Souza, D. B. 2006a. Prevalence of *Rickettsia felis* in the fleas *Ctenocephalidesfelis* and *Ctenocephalidescanis* from two Indian villages in Sao Paulo Municipality, Brazil. *Ann NY Acad Sci* 1078: 361–363.

Jabarit Aldighieri, N., Torres, H., and Raoult, D. 1992. Susceptibility of *R. conorii*, *R. rickettsii* and *C. burnetii* to CI- 960 (PD 127,391), PD 131,628, pefloxacin, ofloxacin and ciprofloxacin. *Antimicrob Agents Chemother* 36: 2529–2532.

Jiang, J., Soeatmadji, D. W., Henry, K. M., Ratiwayanto, S., and Bangs, M. J. 2006. *Rickettsia felis* in *Xenopsyllacheopis*, Java, Indonesia. *Emerg Infect Dis* 12: 1281–1283.
Kim, D. M., Yu, K. D., and Lee, J. H. 2007. Controlled trial of a 5-day course of telithromycin versus doxycycline for treatment of mild to moderate scrub typhus. *Antimicro Agents and Chemother* 51: 2011–2015.
Kirkland, K. B., Wilkinson, W. E., and Sexton, D. J. 1995. Therapeutic delay and mortalityin cases of Rocky Mountain spotted fever. *Clin Infect Dis* 20: 1118–1121.
Kodama, K., Senba, T., Yamauchi, H., Chikahira, Y., Katayamam, T., Furuya, Y., Fujita, H., and Yamamoto, S. 2002. Fulminant Japanese spotted fever definitively diagnosed by the polymerase chain reaction method. *J Infect Chemother* 8: 266–268.
La Scola, B., Meconi, S., Fenollar, F., Rolain, J. M., and Roux, V. 2002. Emended description of *Rickettsia felis*, a temperature-dependent cultured bacterium. *Int J Syst Evol Microbiol* 52: 2035–2041.
Labruna, M. 2009. Ecology of Rickettsia in South America. *Ann NY Acad Sci* 166: 156–166.
Lim, M., Brady, H., Hambling, T., Sexton, K. 2012. *Rickettsia felis* infections, New Zealand. *Emerg Infect Dis* 18: 167–169.
Mahajan, S. K., Kashyap, R., Kanga, A., Sharma, V., Prasher, B. S., and Pal, L. S. 2006. Relevance of Weil-Felix test in diagnosis of scrub typhus in India. *J Assoc Phys India* 54: 619–621.
Mahajan, S. K., Rolain, J. M., and Raoult, D. 2008. Fatal scrub typhus despite azithromycin during pregnancy in Indian Himalayas. Proceedings 5th International conference of Rickettsiae and Rickettsial Diseases. Marseilles, France, May 18–20.
Mansueto, P., Vitale, G., Cascio, A., and Seidita, A. 2012. New insight into immunity and immunopathology of rickettsial diseases. *Clin Dev Immunol* 2012: article 967822.
Marie, J. L., Fournier, P. E., Rolain, J. M., Briolant, S., and Davoust, B. 2006. Molecular detection of *Bartonellaquintana*, *B. Elizabethae*, *B. Koehlerae*, *B. Doshiae*, *B. Taylorii*, and *Rickettsia felis* in rodent fleas collected in Kabul, Afghanistan. *Am J Trop Med Hyg* 74: 436–439.
Masters, E. J., Olson, G. S., Weiner, S. J., and Paddock, C. D. 2003. Rocky Mountain spotted fever: A clinician's dilemma. *Arch Intern Med* 163: 769–774.
Mathai, E., Lloyd, G., Cherian, T., Abraham, O. C., and Cherian, A. M. 2001. Serological evidence of continued presence of human rickettsiosis in southern India. *Ann Trop Med Parasitol* 95: 395–398.
Mediannikov, O. Y., Sidelnikov, Y., Ivanov, L., Mokretsova, E., Fournier, P. E., Tarasevich, I., and Raoult, D. 2004. Acute tick-borne rickettsiosis caused by *Rickettsia heilongjiangensis* in Russian Far East. *Emerg Infect Dis* 10(5): 810–817.
Mediannikov, O., Trape, J. F., Diatta, G., Parola, P., Fournier, P. E., and Raoult, D. 2010. *Rickettsia africae*, Western Africa. *Emerg Infect Dis* 16(3): 571–573.
Meloni, G., and Meloni, T. 1996. Azithromycin vs. doxycycline for Mediterranean spotted fever. *Pediat Inf Dis J* 15: 1042–1044.
Miller, N. J., Rainone, E. E., Dyer, M. C., Gonzalez, M. L., and Mather, T. N. 2011. Tick bite protection with permethrin-treated summer-weight clothing. *J Med Entomol* 48: 327–333.
Nava, S., Elshenawy, Y., Eremeeva, M. E., Sumner, J. W., Mastropaolo, M., and Paddock, C. D. 2008. *Rickettsia parkeri* in Argentina. *Emerg Infect Dis* 14(12): 1894–1897.
Padbidri, V. S., Rodrigues, J. J., and Shetty, P. S. 1984. Tick-borne rickettsiosis in Pune district, Maharastra, India. *Int J Zoonoses* 11: 45–52.
Parola, P., Davoust, B., and Raoult, D. 2005. Tick- and flea-borne rickettsial emerging zoonoses. *Vet Res* 36: 469–492.
Perez-Osorio, C., Zavala-Velazquez, J., Arias, J., and Zavala-Castro, J. 2008. *Rickettsia felis* as emergent global threat for humans. *Emerg Infect Dis* 14: 1019–1023.

Pornwiroon, W., Pourciau, S. S., Foil, L. D., and Macaluso, K. R. 2006. Rickettsia felis from cat fleas: Isolation and culture in a tick-derived cell line. *Appl Environ Microbiol* 72(8): 5589–5595.

Purvis, J. J., and Edwards, M. S. 2000. Doxycycline use for rickettsial disease in pediatric patients. *Pediatr Infect Dis J* 19: 871–874.

Raoult, D., and Parola, P. 2009. Informa Healthcare; *Rickettsial Diseases*. New York, pp. 37–46.

Raoult, D., and Roux V. 1997. Rickettsioses as paradigms of new or emerging infectious diseases, *Clin Microbiol Rev* 10: 694–719.

Rathi, N., and Rathi, A. 2010. Rickettsial infections: Indian perspective. *Indian Pediatr* 47: 157–164.

Rathore, M. H., and Maraqa, N. F. 2009. Rickettsial Infection. Medscape Reference. [updated May 11, 2009; cited July 20, 2011]. http://emedicine.medscape.com/article /968385.

Reif, K., and Macaluso, K. 2009. Ecology of *Rickettsia felis*: A review. *J Med Entomol* 46: 723–736.

Renvoise, A., and Raoult, D. 2009. An update on rickettsiosis. *Med Mal Infect* 39: 71–81.

Rolain, J. M., Franc, M., Davoust, B., and Raoult, D. 2003. Molecular detection of *Bartonellaquintana, B. koehlerae, B. henselae, B. clarridgeiae, Rickettsia felis and Wolbachiapipientis*in cat fleas, France. *Emerg Infect Dis* 9: 338–342.

Rolain, J. M., Maurin, M., Vestris G., and Raoult, D. 1998. In vitro susceptibilities of 27 rickettsiae to 13 antimicrobials. *Antimicrob Agents Chemother* 42: 1537–1541.

Rolain, J. M., Stuhl, L., Maurin, M., and Raoult, D. 2002. Evaluation of antibiotic susceptibilities of three rickettsial species including *Rickettsia felis* by a quantitative PCR DNA assay. *Antimicrob Agents Chemother* 46: 2747–2751.

Rovery, C., Brouqui, P. and Raoult, D. (2008). Questions on Mediterranean spotted fever century after its discovery. *Emerg Infect Dis* 14(9): 1360–1367.

Ruiz Beltran, R., and Herrero-Herrero, J. I. 1992. Evaluation of ciprofloxacin and doxycycline in the treatment of Mediterranean spotted fever. *Eur J Clin Microbiol Infect D* 11: 427–431.

Saah, A. 2000. Rickettsia akari (Rickettsialpox). In Mandell, G. L., Bennet, J. E., Doalin, R., eds., *Principles and Practice of Infectious Diseases*. Philadelphia: Churchill Livingstone, pp. 2042–2043.

Schloderer, D., Owen, H., Clark, P., Stenos, J., and Fenwick, S. G. 2006. *Rickettsia felis* in fleas, Western Australia. *Emerg Infect Dis* 12: 841–843.

Segura, F., and Anton, E. 2002. Clarithromycin for the treatment of Mediterranean spotted fever. *Clin Infect Dis* 34: 560.

Selmi, M., Bertolotti, L., Tomassone, L., and Mannelli, A. 2008. Rickettsia slovaca in *Dermacentor marginatus* and tick-borne lymphadenopathy, Tuscany, Italy. *Emerg Infect Dis* 14(5): 817–820.

Sexton, D. J., Dwyer, B., Kemp, R., and Graves, S. 1991. Spotted fever group rickettsia infections in Australia. *Rev Infect Dis* 13(5): 876–886.

Siberry, G. K., and Dumler, J. S. 2007. Rickettsial infections. In: Kliegman, R. M., Behrman, R. E., Jenson, H. B., Stanton, B. F., eds., *Nelson Textbook of Pediatrics*, 18th ed. Pennsylvania: Saunders, pp. 1289–1301.

Stevenson, H. L., Labruna, M. B., Montenieri, J. A., Kosoy, M. Y., and Gage, K. L. 2005. Detection of *Rickettsia felis* in a New World flea species, *Anomiopsyllus nudata* (Siphonaptera: Ctenophthalmidae). *J Med Entomol* 42: 163–167.

Sundhindra, B. K., Vijaykumar, S., and Kutti, A. K. 2004. Rickettsial spotted fevers in Kerala. *Natl Med J India* 17: 51–52.

Van Bambeke, F., and Tulkens, P. K. 2001. Macrolides: Pharmacokinetics and pharmacodynamics. *Int J Antimicrob Agents* 18: S17–S23.

Vaughn, M. F., Funkhouser, S. W., Lin, F., Fine, J., Juliano, J. J., Apperson, C. S., and Meshnick, S. R. 2014. Long-lasting permethrin impregnated uniforms: A randomized-controlled trial for tick bite prevention. *Am J Prev Med* 46: 473–480.

Venzal, J. M., Perez-Martinez, L., Felix, M. L., Portillo, A., and Blanco, J. R. 2006. Prevalence of *Rickettsia felis* in *Ctenocephalides felis* and *Ctenocephalides canis* from Uruguay. *Ann NY Acad Sci* 1078: 305–308.

Wright, J. G. 2008. Infection control practices and zoonotic disease risks among veterinarians in the United States. *J Am Vet Med Assoc* 232: 1863–1872.

Znazen, A., and Raoult, D. 2007. Flea-borne spotted fever. In Raoult, D., and Parala, P., eds., *Rickettsial Diseases*. New York: Informa Health Care, pp. 87–96.

# Index

## D



## E

## H



## I



## J



## K



## L

## M

## N

## Q



## R

## S

## T

Lightning Source UK Ltd.
Milton Keynes UK
UKHW021437231019
352120UK00014B/357/P

9 781138 055810